Smart Energy Grid Engineering

Smart Energy Grid Engineering

Edited by

Hossam A. Gabbar

University of Ontario Institute of Technology (UOIT), Oshawa, ON, Canada

AMSTERDAM • BOSTON • HEIDELBERG • LONDON
NEW YORK • OXFORD • PARIS • SAN DIEGO
SAN FRANCISCO • SINGAPORE • SYDNEY • TOKYO
Academic Press is an imprint of Elsevier

Academic Press is an imprint of Elsevier
125 London Wall, London EC2Y 5AS, United Kingdom
525 B Street, Suite 1800, San Diego, CA 92101-4495, United States
50 Hampshire Street, 5th Floor, Cambridge, MA 02139, United States
The Boulevard, Langford Lane, Kidlington, Oxford OX5 1GB, United Kingdom

Notices

Knowledge and best practice in this field are constantly changing. As new research and experience broaden our understanding, changes in research methods, professional practices, or medical treatment may become necessary.

Practitioners and researchers must always rely on their own experience and knowledge in evaluating and using any information, methods, compounds, or experiments described herein. In using such information or methods they should be mindful of their own safety and the safety of others, including parties for whom they have a professional responsibility.

To the fullest extent of the law, neither the Publisher nor the authors, contributors, or editors, assume any liability for any injury and/or damage to persons or property as a matter of products liability, negligence or otherwise, or from any use or operation of any methods, products, instructions, or ideas contained in the material herein.

Library of Congress Cataloging-in-Publication Data
A catalog record for this book is available from the Library of Congress

British Library Cataloguing-in-Publication Data
A catalogue record for this book is available from the British Library

ISBN: 978-0-12-805343-0

For information on all Academic Press publications
visit our website at https://www.elsevier.com/

ELSEVIER
Book Aid
International
Working together
to grow libraries in
developing countries
www.elsevier.com • www.bookaid.org

Publisher: Joe Hayton
Acquisition Editor: Lisa Reading
Editorial Project Manager: Maria Convey
Production Project Manager: Julie-Ann Stansfield
Cover Designer: Greg Harris

Typeset by SPi Global, India

I dedicate this book to my family for their great support, inspiring and motivating me to complete the book.

Contents

Contributors

N. Ayoub
The University of Missouri, Columbia, MO, United States; Helwan University, Cairo, Egypt

A.W. Bizuayehu
University of Beira Interior, Covilhã, Portugal

J.P.S. Catalão
University of Beira Interior, Covilhã; University of Porto, Porto; University of Lisbon, Lisbon, Portugal

M.H. Cintuglu
Florida International University, Miami, FL, United States

A. Eldessouky
University of Ontario Institute of Technology (UOIT), Oshawa, ON, Canada

A. Elsayed
Florida International University, Miami, FL, United States

D.Z. Fitiwi
University of Beira Interior, Covilhã, Portugal

H.A. Gabbar
University of Ontario Institute of Technology (UOIT), Oshawa, ON, Canada

Y. Koraz
University of Ontario Institute of Technology (UOIT), Oshawa, ON, Canada

W.J. Miller
MaCT USA, Washington, DC, United States

O. Mohammed
Florida International University, Miami, FL, United States

A.M. Othman
University of Ontario Institute of Technology (UOIT), Oshawa, ON, Canada; Zagazig University, Zagazig, Egypt

K.C. Ruland
University of Siegen, Siegen, Germany

J. Runge
University of Ontario Institute of Technology (UOIT), Oshawa, ON, Canada

S.F. Santos
University of Beira Interior, Covilhã, Portugal

J. Sassmannshausen
University of Siegen, Siegen, Germany

M. Shafie-khah
University of Beira Interior, Covilhã, Portugal; University of Salerno, Salerno, Italy

K. Sayed
University of Ontario Institute of Technology (UOIT), Oshawa, ON, Canada; Sohag University, Sohag, Egypt

M. Xiaoli
China Electric Power Research Institute, Beijing, China

T. Youssef
Florida International University, Miami, FL, United States

A. Zidan
University of Ontario Institute of Technology (UOIT), Oshawa, ON, Canada; Assiut University, Assiut, Egypt

About the Authors

HOSSAM A. GABBAR

Dr Hossam A. Gabbar is a professor in the Faculty of Energy Systems and Nuclear Science, and the Faculty of Engineering and Applied Science (cross-appointed) at the University of Ontario Institute of Technology (UOIT), as well as the director of the Energy Safety & Control Lab at UOIT. He is a world-leading scholar in the fields of smart energy grid engineering and process systems engineering, with a focus on plasma engineering, micro energy grids, and energy safety and control. He is a certified TUV functional safety engineer, a fellow of the Reliability, Availability, Maintainability, and Safety Professionals (RAMSP), and a senior member of IEEE. He is the founding chair of the Toronto Chapter of the IEEE Nuclear & Plasma Science Society, and the founding chair of the Technical Committee on Intelligent Green Production Systems at IEEE, Systems, Man, & Cybernetics Society. Dr Hossam A. Gabbar is the chair and cofounder of the Symposium on Plasma and Nuclear Systems, and the founder of the IEEE International Conference on Smart Energy Grid Engineering. He serves as the editor-in-chief of the International Journal of Process Systems Engineering. He also founded the RAMSP Society, and currently serves as its vice president of safety. Dr Hossam A. Gabbar has successfully managed the completion of 57 theses, has more than 212 academic journal publications to his name, holds several inventions/patents, has published several books, and is regularly invited as a speaker at international symposiums and conferences. His previous successfully completed projects include Modeling & Simulation of Green Hybrid Energy Production/Supply Chains With Grid Integration, Automated Control Recipe Design for Flexible Chemical Batch Production Plants, Biomass Production Chain Planning, and Plastic Production Chain With Recycling.

TAREK A. YOUSSEF

Florida International University, USA

Tarek A. Youssef was born in Cairo, Egypt. He received his BS and MS degrees in electrical engineering from Helwan University, Cairo, Egypt. Mr Youssef has more than 10 years of experience in the communication and security field. He worked as a researcher at the Université Libre de Bruxelles, in Belgium. In 2012, He moved to The Energy Systems Research Laboratory, at Florida International University. He is currently a PhD candidate. His PhD topic is on the Co-Design of Security Aware Power System Distribution Architecture as a Cyber Physical System.

His research interests include cyber physical systems, communication, wide-area measurement, smart grid security, and real-time monitoring of power systems. Furthermore, he also has interests in artificial intelligence and signal-processing applications in the power and energy systems area.

NASSER AYOUB

Missouri University, USA; and Helwan University, Egypt

Nasser Ayoub is a tenured associate professor on the Production Technology Department Faculty of Industrial Education, at Helwan University, Egypt. Currently he is a research scientist within the Bioengineering Department of The University of Missouri, Columbia, United States. He holds a PhD in environmental engineering from the Tokyo Institute of Technology, Japan, an MSc in engineering materials from the Eindhoven University of Technology in the Netherlands, and a BSc in industrial engineering from the Production and Industrial Engineering Department of Zagazig University in Egypt. He has worked as a research associate from 2007 to 2010 at the Tokyo Institute of Technology. Dr Ayoub is specialized in systems development, smart grids infrastructure, and risk-based life cycle assessment. He has worked in several international projects in Japan, the United States, and Qatar in the area of biomass supply chains and networks, life cycle assessment, and environmental policymaking.

KARL CHRISTOPH RULAND

University of Siegen, Germany

Karl Christoph Ruland studied mathematics, physics, and computer science at the University of Bonn, Germany, receiving a diploma and PhD in mathematics. He worked in the industry for six years, before becoming a professor of data communications at the University of Applied Sciences in Aachen, Germany, in 1982. He founded the company, KryptoKom, in 1988. In 1992, he became a full professor at the University of Siegen, for data communications systems. His main research area is the integration of security and cryptographs in communications systems, mainly for real-time and industrial-oriented applications.

SÉRGIO F. SANTOS

University of Beira Interior, Portugal

Sergio F. Santos received his BSc and MSc degrees from the University of Beira Interior (UBI), Covilhã, Portugal, in 2012 and 2014, respectively. He was a researcher between 2013 and 2015 on the EU FP7 Project, SiNGULAR, at the Sustainable Energy Systems Lab of the same university. He is currently pursuing his PhD degree under the UBI/Santander Totta doctoral incentive grant in the Engineering Faculty of UBI. His research interests are in planning and operation optimization of distribution systems and smart grid technologies integration.

MEHMET HAZAR CINTUGLU

Florida International University, USA

Mehmet Hazar Cintuglu received his BS and MS degrees in electrical engineering from the Gazi University, in Ankara, Turkey, in 2008 and 2011, respectively. He is currently pursuing his PhD degree from the Energy Systems Research Laboratory in the Electrical and Computer Engineering Department of the College of Engineering and Computing at Florida International University, Miami, FL, United States. From 2009 to 2011, he was a power systems project engineer at an electric utility company in Turkey. His current research interests include agent systems, distributed control, cyber-physical systems for active distribution networks, and microgrids.

WILLIAM J. MILLER

MaCT, USA

William is the President of Maximum Control Technologies (MaCT), a Miller WJ & Associates Company, which provides international consulting and systems integration services, with offices in the United States and Canada. He is a graduate of Pennsylvania State University with a degree in Electrical Engineering and Telecommunications. He has designed, managed, and installed numerous distributed process control systems for the power, paper, chemical, and cement industries with over 35 years of experience. He was the chairman for the IEEE P2030 TF3 SG2 for the Interoperability of the Smart Grid, and is now the chairman of the joint ISO/IEC/IEEE P21451.1.4 (Sensei/IoT*) Smart Transducer Interface for Sensors, Actuators, and Devices use of eXtensible Messaging and Presence Protocol (XMPP) for Network Device Communications and the 1st Semantic Web 3.0 standard for M2M and the Internet of things (IoT).

YAHYA KORAZ

University of Ontario Institute of Technology, Canada

Yahya Koraz obtained his BSc and MSc degrees in electrical engineering at the Engineering Faculty at the University of Mosul. He completed his PhD at the Engineering Faculty at the University of Baghdad in 2005. From 2001 to 2008 he was a lecturer at the Computer Engineering Department at the University of Mosul. He is currently a PhD candidate in electrical engineering on the Faculty of Engineering and Applied Science at the University of Ontario Institute of Technology. His current research interests are in the area of safety design and self-healing for interconnected resilient micro energy grids. He has been awarded the ICCD Student Mobility Award 2016 from Italy.

AHMED M. OTHMAN

University of Ontario Institute of Technology, Canada; and Zagazig University, Egypt

Ahmed M. Othman is an Assistant Professor in the Electrical Power and Machine Department, Faculty of Engineering, Zagazig University, in Zagazig, Egypt. He was awarded a PhD in 2011 from Aalto (Helsinki University of Technology), Finland. Dr Ahmed has published many articles in refereed international journals and conferences. He was awarded distinctions in international publishing from the Zagazig University in 2013 and 2015. The research activities of Dr Othman are concerned with the application of intelligent techniques to power systems and microgrids.

XIAOLI MENG

China Electric Power Research Institute, China

Xiaoli Meng is working in the Power Distribution Department at the China Electric Power Research Institute in Beijing, China. From 2012 to 2013, she was a visiting scholar at the Energy Safety and Control Lab (ESCL), at the University of Ontario Institute of Technology, Canada.

OSAMA MOHAMED

Florida International University, USA

Dr Osama Mohammed is a Professor of Electrical Engineering, and is the Director of the Energy Systems Research Laboratory at the Florida International University, in Miami, Florida. He received his Masters and Doctoral degrees in Electrical Engineering from Virginia Tech in 1981 and 1983, respectively. He has performed research on various topics in power and energy systems in addition to design optimization and physics-passed modeling in electric drive systems and other low-frequency environments. Professor Mohammed is a world-renowned leader in electrical energy systems. He has performed research in the area of electromagnetic signatures, wideband gap devices, power electronics, and ship power systems modeling and analysis. He has current active research projects with several Federal agencies dealing with power system analysis and operation, smart grid distributed control and interoperability, cyber physical systems, and co-design of cyber and physical components for future energy systems applications.

Dr Mohammed has published more than 450 articles in refereed journals and other IEEE refereed International conference records. He also authored a book and several book chapters. Dr Mohammed is an elected fellow of IEEE and is an elected fellow of the Applied Computational Electromagnetic Society. Professor Mohammed is the recipient of the prestigious IEEE Power and Energy Society Cyril

Veinott Electromechanical Energy Conversion Award, and the 2012 outstanding research award from Florida International University.

Dr Mohammed has lectured extensively with invited and plenary talks at major research and industrial organizations worldwide. He serves as editor of several IEEE Transactions including the IEEE Transactions on Energy Conversion, the IEEE Transactions on Smart Grid, IEEE Transactions on Magnetics, and the IEEE Transactions on Industry Application. Dr Mohammed served as the International Steering Committee chair for the IEEE International Electric Machines and Drives Conference (IEMDC), and the IEEE Biannual Conference on Electromagnetic Field Computation (CEFC). Professor Mohammed was the general chair of six major international conferences in his areas of expertise in addition to being the general chair for two future IEEE major conferences.

JOÃO P. S. CATALÃO

University of Beira Interior, Portugal; Faculty of Engineering of the University of Porto, Portugal; and INESC-ID, Inst. Super. Tecn., University of Lisbon, Portugal

João P. S. Catalão (M'04-SM'12) received the MSc degree from the Instituto Superior Técnico (IST), Lisbon, Portugal, in 2003, and his PhD and Habilitation for Full Professor (Agregação) from the University of Beira Interior (UBI), Covilha, Portugal, in 2007 and 2013, respectively.

Currently, he is a Professor in the Faculty of Engineering at the University of Porto (FEUP) in Porto, Portugal, and a researcher at INESC TEC, INESC-ID/IST-UL, and C-MAST/UBI. He was the primary coordinator of the EU-funded FP7 project SiNGULAR ("Smart and Sustainable Insular Electricity Grids Under Large-Scale Renewable Integration"), a 5.2-million euro project involving 11 industry partners. He has authored or coauthored more than 450 publications, including 147 journal papers, 271 conference proceedings papers, 23 book chapters, and 11 technical reports, with an *h*-index of 27 (according to Google Scholar), having supervised more than 40 post-docs, PhD, and MSc students. He is the editor of the books titled Electric Power Systems: Advanced Forecasting Techniques and Optimal Generation Scheduling and Smart and Sustainable Power Systems: Operations, Planning and Economics of Insular Electricity Grids (Boca Raton, FL, United States: CRC Press, 2012 and 2015, respectively). His research interests include power system operations and planning, hydro and thermal scheduling, wind and price forecasting, distributed renewable generation, demand response, and smart grids.

Dr Catalão is an editor of the IEEE Transactions on Smart Grid and the IEEE Transactions on Sustainable Energy, and an associate editor of the IET Renewable Power Generation. He was the guest editor-in-chief for a special section on "Real-Time Demand Response" of the IEEE Transactions on Smart Grid, published in December 2012, and the guest editor-in-chief for a special section on "Reserve and Flexibility for Handling Variability and Uncertainty of Renewable Generation" of the IEEE Transactions on Sustainable Energy, published in April 2016. He was the recipient of the 2011 Scientific Merit Award UBI-FE/Santander Universities, and the 2012 Scientific Award UTL/Santander Totta. Also, he has won four Best Paper Awards from the IEEE Conferences.

JASON RUNGE

University of Ontario Institute of Technology, Canada

Jason Runge was born in Toronto, Canada in 1987. He received his BEng (with honors) in energy systems engineering from the University of Ontario Institute of Technology in 2014 and his MASc degree in electrical engineering from the University of Ontario Institute of Technology in 2016.

AHMED S. ELDESSOUKY

Canadian International College, Egypt; and University of Ontario Institute of Technology, Canada

Ahmed S. Eldessouky is an assistant professor at the Canadian International College, in Egypt. From 2014 to 2015, he visited The Energy Safety and Control Lab (ESCL), at the University of Ontario Institute of Technology, Canada, where he worked on a research project concerning building energy conservation and management.

JOCHEN SASSMANNSHAUSEN

University of Siegen, Germany

Jochen Sassmannshausen was born in 1991 and started studying computer science at the University of Siegen in 2011. After receiving his bachelor's degree in 2015, he is now working on his master's thesis at the institute for data communication systems. The main focus of his work is communication and security in smart grids and the Internet of things.

KHAIRY SAYED

University of Ontario Institute of Technology, Canada; and Sohag University, Egypt

Khairy Sayed received his BS in electrical power and machines in 1997 from Assiut University, in Assiut, Egypt. He got his Masters degree at the Electrical Energy Saving Research Center in the Graduate School of Electrical Engineering at Kyungnam University in Masan, Korea, 2007. He got his PhD from Assiut University in 2013. He is working as an assistant professor in the Department of Electrical Engineering at Sohag University in Egypt. His research interests include:

Renewable energy, solar PV, wind energy, fuel cells, utility interactive power conditioners for renewable energy sources, soft switching DC/DC power converter topologies, smart energy grids, and the protection and control of smart microgrids. He has more than nine years of experience in SCADA/DMS, obtained during his work at the Middle Egypt Electricity Distribution Company as a system integrator for the control center project. Currently, Dr Khairy is a visiting scholar at the University of Ontario Institute of Technology UOIT, 2016.

MIADREZA SHAFIE-KHAH

University of Beira Interior, Portugal and University of Salerno, Italy

Miadreza Shafie-khah (S'08, M'13) received his MSc and PhD degrees in electrical engineering from Tarbiat Modares University, Tehran, Iran, in 2008 and 2012, respectively. He received his first postdoc from the University of Beira Interior (UBI), Covilha, Portugal in 2015, while working on the EU-funded FP7 project SiNGULAR. He is currently pursuing his second postdoctoral studies at the University of Salerno in Salerno, Italy. His research interests include power market simulations, market power monitoring, power system optimization, operation of electricity markets, price forecasting, and smart grids.

AHMED T. ELSAYED

Florida International University, USA

Ahmed T. Elsayed received his BSc and MSc degrees from the Faculty of Engineering, Benha University, Egypt in 2006 and 2010, respectively. From 2006 to 2012, he was a research/teaching assistant in the Faculty of Engineering at Benha University. Currently, he is a PhD candidate and a research assistant in the Electrical and Computer Engineering Department in the College of Engineering and Computing at Florida International University in Miami, Florida, United States. His current research interests include DC distribution architectures, multiobjective optimization, and flywheel energy storage.

ABEBE W. BIZUAYEHU

University of Beira Interior, Portugal

Abebe W. Bizuayehu (M'15) received his MSc degree in mechanical engineering, in the specialty of industrial mechanic from José Antonio Echevarría Higher Polytechnic Institute (I.S.P.J.A.E.), Havana, Cuba, in 1989, and his PhD in Renewable Energy and Energetic Efficiency from the University of Zaragoza, Spain, in 2012. He received his postdoc from the University of Beira Interior (UBI) Covilha, Portugal in 2015, while working on the EU-funded FP7 project SiNGULAR.

DESTA Z. FITIWI

University of Beira Interior, Portugal

Desta Z. Fitiwi received his BSc degree in electrical and computer engineering from Addis Ababa University in Ethiopia in 2005, and his MSc degree in electrical and electronics engineering from Universiti Teknologi Petronas in Tronoh, Malaysia, in 2009. He worked as a supervisory control and data acquisition (SCADA) and adaptation engineer on the Ethiopian Load Dispatch Center Project from 2005 to 2007. He is pursuing an Erasmus Mundus Joint Doctorate in sustainable energy technologies and strategies (SETS), jointly offered by the Comillas Pontifical University, the KTH Royal Institute of Technology of Stockholm, and the Delft University of Technology. He actively participated in the realization of four EU-funded projects. He is currently working as a researcher at the University of Beira Interior in Covilhã, Portugal. His research interests include regulation and economics of the power industry, transmission expansion planning, sustainable energy modeling and strategic planning, artificial-intelligence applications, and mathematical optimizations for power systems.

ABOELSOOD ZIDAN

University of Ontario Institute of Technology, Canada; and Assiut University, Egypt

Aboelsood A. Zidan was born in Sohag, Egypt, in 1982. He received his BSc and MSc degrees in electrical engineering from Assiut University, Egypt, in 2004 and 2007, respectively, and his PhD in electrical engineering from the University of Waterloo, in Waterloo, Ontario, Canada in 2013. He is currently an assistant professor at the Assiut University in Egypt. His research interests include distribution automation, renewable DG, distribution system planning, and smart grids.

Foreword

The editor, Dr Hossam A. Gabbar, is the founding general chair of the annual IEEE Smart Energy Grid Engineering Conference. He launched this conference to meet the increasing demands to help the public and industry to understand the foundations of smart energy grids and micro energy grids; and how it can help overcome the limitations and challenges of energy supply and infrastructures, such as high cost, reliability, environmental stresses, and operability factors. There are increasing needs to discuss the foundations of smart energy grid engineering, including design methods, control strategies, and operation and maintenance mechanisms that facilitate the migration from current infrastructures to future smart energy grids. The contributors in this book are from a number of countries, universities, and research groups that includes Germany, Italy, Portugal, China, Canada, and the United States. This book addresses the critical issues of smart energy grids, such as adaptive infrastructures, micro energy grid configuration, planning, scheduling, distributed energy resource placement, control, protection, optimization, security, monitoring, and automation, which are organized into different chapters. This book also reflects real implementations of smart energy grids and micro energy grids, including a real demonstration originated by Dr Hossam A. Gabbar and his research group at the Energy Safety and Control Lab (ESCL) at the University of Ontario Institute of Technology, Canada, which was the first of its kind worldwide to present a micro energy grid of integrated thermal, gas, and electricity to meet different loads with intelligent control systems.

This book will benefit national and international energy communities that wish to understand the critical issues, and develop effective deployment plans of smart energy grids and micro energy grids for different scales and load profiles, meeting national and international standards.

Hossam A. Gabbar
Editor

Preface

Smart energy grids are energy networks that offer a high-performance and bidirectional energy supply with the use of energy storage to match generation with loads. It includes distributed energy resources, such as renewable energy, energy storage, and other emerging energy technologies, to offer cheap and efficient energy infrastructures to support residential, commercial, and industrial facilities and transportation infrastructures. This requires intelligent monitoring and control of different components within the distribution and transmission lines, as well as other systems from utilities such as natural gas, thermal energy, electricity, transportation, and water on the one side, to the end user on the other side, while maintaining energy quality, security, reliability, and safety with minimum environmental impacts. Micro energy grids are essential components within smart energy grids that provide islanded energy grids that can support local loads. Governments around the world are investing heavily in smart energy grids to ensure optimum energy use and supply, to enable better planning for outage responses and recovery, and to facilitate the integration of heterogeneous technologies, such as renewable energy systems, electrical vehicle networks, and smart homes, around the grid. Smart energy grids present enormous engineering challenges in the design, operation, and maintenance of integrated energy grids, as well as challenges for communication networks and information technologies. In addition, it provides higher-levels of security and privacy for different components within smart energy grids that ensure reliable integration of smart metering, web, and mobile applications.

This book is the first edition to provide researchers, students, and professionals from the industry the foundations and technical details of the design, control, and operational aspects of smart energy grids, which will pave the way for wider and more successful implementations worldwide of smart energy grids and their associated technological infrastructures.

Hossam A. Gabbar
Editor

Acknowledgments

The editor would like to thank all contributors to this book, and the research team at ESCL-UOIT for their dedication, research outcomes, and their support for the demonstration of micro energy grids at ESCL-UOIT, Canada. Also, the editor would like to thank UOIT for their continuous support for this research work.

CHAPTER

Introduction

1

H.A. Gabbar

University of Ontario Institute of Technology (UOIT), Oshawa, ON, Canada

1.1 INTRODUCTION

Recently, the world has been witnessing an increased demand in electricity and energy. It is clear that energy use has a direct negative impact on climate changes, which provoked governments and industries to invest more in renewable energy (RE) technologies as alternative energy sources and power supply systems. There are increasing tendencies to utilize renewable green energy sources (solar, photovoltaic (PV), wave, tidal, fuel cell, biogas, and hydrogen), which are supported by economic and environmental factors. The increasing reliance on fossil fuels with an increasing rate of resource depletion is causing a basic shift to green RE alternatives, clean fuel replacement, and energy displacement of conventional fossil energy sources to new green renewable, environmentally safe and friendly counterparts [1,2]. PV and distributed energy resources (DERs) generation schemes are considered the most viable and economic choices for microgrid (MG) electrical energy generation [3]. The focus is on small isolated standalone integrated AC-DC grid power systems that utilize a combination of PV, fuel cell, and a diesel generator with a local grid backup with a capacity usually ranging in size from 15 to 1500 kW [4,5]. Typical applications include electricity supply to remote isolated islands/villages with a limited utility grid, heating, water pumping, and ventilation and air conditioning systems. Diesel generators are the backup source of electricity in most remote cases.

As part of the future implementation of smart energy grid (SEG), it is important to demonstrate regional SEGs and micro energy grids (MEGs) with RE as distributed generation. This decision involves capital cost for changing infrastructure to support the target dynamic and scalable SEG. In order to achieve such a target, it is essential to provide a modeling and simulation environment to understand existing grid structures and support the design and evaluation of the target SEG. There is no current modeling and simulation technology that can support the dynamic MGs with distributed generation structures as mapped to a geographic information system (GIS), which created a bottleneck to move forward in the design and implementation of true SEGs. In addition, there is no technology to support the evaluation of different grid

Smart Energy Grid Engineering.

applications in different scales within the grid, such as hybrid-electric vehicle. Moreover, most of the design and operation decisions should be based on safety and reliability of grid physical structures; however, current modeling and simulation technologies don't support asset integrity models as linked with power/energy modeling, which created limitations to evaluate different design and operational scenarios in terms of safety, reliability, and availability. In addition, most of the current modeling tools don't provide data analysis features that are needed to tune SEG physical structure models with dynamic simulation capabilities. This will lead to limitations on performing accurate optimization of grid performance.

The planning of resilient SEGs can be achieved via automated and intelligent monitoring, planning, and distributed modeling and simulation to support the design and operation of SEG. The solution will support real-time modeling and simulation of the superstructure of SEG in two levels: top-level grid model and low-level component level. A distributed data management module will be developed to construct and tune the following SEG models: grid superstructure physical and asset models; power/energy models; grid asset integrity and reliability models; safety and protection models; and operation and control models.

The proposed technology will support the following business functions: multi-sensor intelligent metering design support, distributed control and operation support, dynamic peak analysis and optimization, distributed energy generation (DEG) planning and optimization, asset degradation and upgrade scenario synthesis and verification, and design and implement security components in the component levels.

Technological features will be proposed to automatic and dynamic identification of practical distribution and partitioning strategies of MGs with the ability to define operational and alternative scenarios based on performance indicators and protection strategies. The target application will be a web-based interface with a simulation engine and distributed collaborative supply chain simulation models to allow local and global optimization for different SEG performance indicator parameters for design and operation scenarios. User interface will be accessible via cell phones to enable mobile access from remote locations to identify optimal ways to support energy supply and use within SEGs. Deterministic and predictive SEG models will be developed and associated with static and dynamic parameters of physical SEG superstructure and represented as energy semantic network (ESNs), which will be used to synthesize practical and optimal scenarios for MG generation, protection, and application integration. The proposed modeling and simulation will be mapped to GIS for geographical and environmental data analysis, and to monitor and analyze real-time risk index calculation with different energy supply chain scenarios.

There are several scenarios that can be adopted to analyze energy efficiency related to the DERs within MGs with different thermal-gas-electricity conversion and utilization strategies to cover regional energy needs. These strategies include (a) MG design and control; (b) thermal storage systems; (c) natural gas (NG) applications for transportation and facilities; (d) gas-electricity conversion technologies; and (e) energy conservation within MGs. Researchers, technology providers, as well as industries have been studying effective technologies in each of these tracks.

Investors have also been pushing to deploy several technologies for hydrogen production and conversion so that it can be used as clean and cheap fuel along with NG to compensate the challenges of electric vehicle deployment for regional clean transportation.

The analysis of energy efficiency of the integration of MGs with existing energy infrastructures is most likely to impact operation, in terms of control, protection, energy quality, and sustainability. Fig. 1 shows an integrated view of SEGs where DERs are integrated in each stage in the interconnected processes from sources, transportation, treatment, conversion, generation, storage, and utilization. There are many challenges to facilitate the proper and smooth implementation of MGs and their integration with grid topologies, as islanded and grid-connected modes. This requires detailed investigation of the different integration scenarios of thermal-gas-electricity networks and technologies to ensure optimal energy savings and efficiency, along with minimizing lifecycle costs (LCCs). There are several scenarios to implement fully integrated thermal-gas-electricity grids, and demonstrate local distributed energy systems within MGs, which requires proper analysis of the different energy supply and production (ie, generation) scenarios based on lifecycle assessment (LCA) and cost analysis [6], as well as energy policies and regulations of thermal-gas-electricity networks and their operation [7]. This will also require conducting detailed analysis of the design and operational challenges of building interconnected MGs to ensure their optimal performance with clear determination of potential and efficient protection and control systems in view of existing grid infrastructures, energy resources, and the current forecasted demand in the local region [8,9].

The huge investment of upgrading transmission lines or distribution lines motivated governments to invest in monitoring energy efficiency and power losses in all transmission and distribution lines, as well as MG infrastructures. The proposed DER will support SEG infrastructures that include thermal-gas-electricity networks, and can provide the most efficient energy supply to meet local or regional demands.

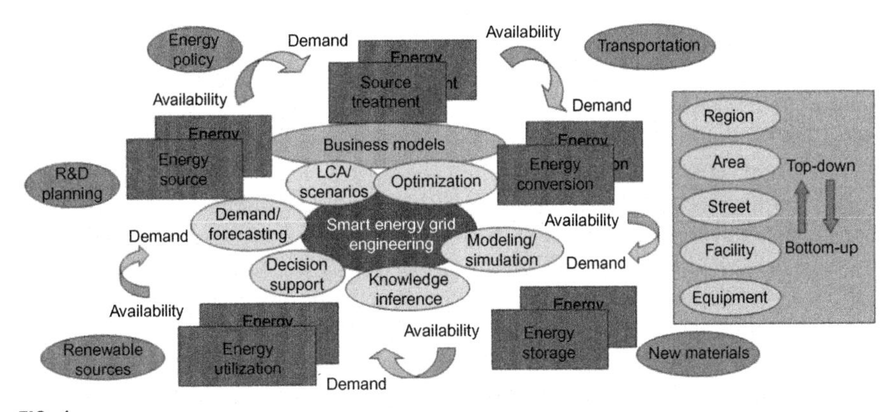

FIG. 1

Energy (micro) grid engineering.

In addition, an energy data center will support the evaluation of design, control, and operation scenarios of these integrated thermal-gas-electricity networks. It will also support the optimization of smart grids' and MGs' process parameters to achieve the most economical and potential solution to cover regional energy demands.

Energy data centers will include low- or medium-energy supply components, for example, thermal or voltage, as part of the energy distribution networks. Energy-efficiency parameters will be defined and analyzed for DERs connected at the distribution level, which will reflect energy supply system stability, reliability, energy quality, sustainability factors, LCC, and environmental factors. Preliminary studies at the Energy Safety and Control Laboratory, University of Ontario Institute of Technology completed the identification and modeling of energy efficiency and key performance indicators (KPIs) for all layers of MGs. In addition, initial infrastructure of energy data center is designed and well described and ready for implementations.

Energy efficiency will cover smart grids and MGs, such as: (a) an MG design based on available energy resources, existing grid topology, and load and energy demand forecasting; (b) dynamic protection with adequate risk-based fault propagation analysis in view of the dynamic nature of MGs in islanded and grid-connected modes; (c) energy supply quality and performance maximization in view of different operation parameters; (d) cost-benefit analysis for different energy technologies with realistic profit planning; and (e) considerations of international standards and national regulations for energy-efficiency limits and targets with multiview analysis based on detailed business models. In order to provide logical justification of building interconnected MGs, modeling and simulation tools are required, as explained in the following section.

1.2 SEGs INFRASTRUCTURES

Typically, an electricity grid includes transmission and distribution lines, where transformers are used to integrate them with loads in commercial, residential, and transportation levels.

SEGs can be modeled hierarchically in different regions and zones, which can include interconnected MEGs. A regional zone is a combination of a few subregional zones, and the subregional zones are the summation of cells. MEGs have been considered inside cells as shown in Fig. 2.

In order to achieve adaptive SEG infrastructures, interconnected MEGs are proposed to dynamically transfer energy between MEGs and transmission and distribution lines in view of loads, demand, and other performance indicators, such as price. Fig. 3 shows the proposed superstructure of SEGs with interconnected MEGs. To upgrade the traditional electric power system to an SEG, it is essential to make several enhancements at various levels of operation and control infrastructure, which includes the integration of intelligent electronic devices, synchrophasor-based devices, advanced communication infrastructure, and efficient monitoring and control algorithms that would make optimal use of these devices; however, before

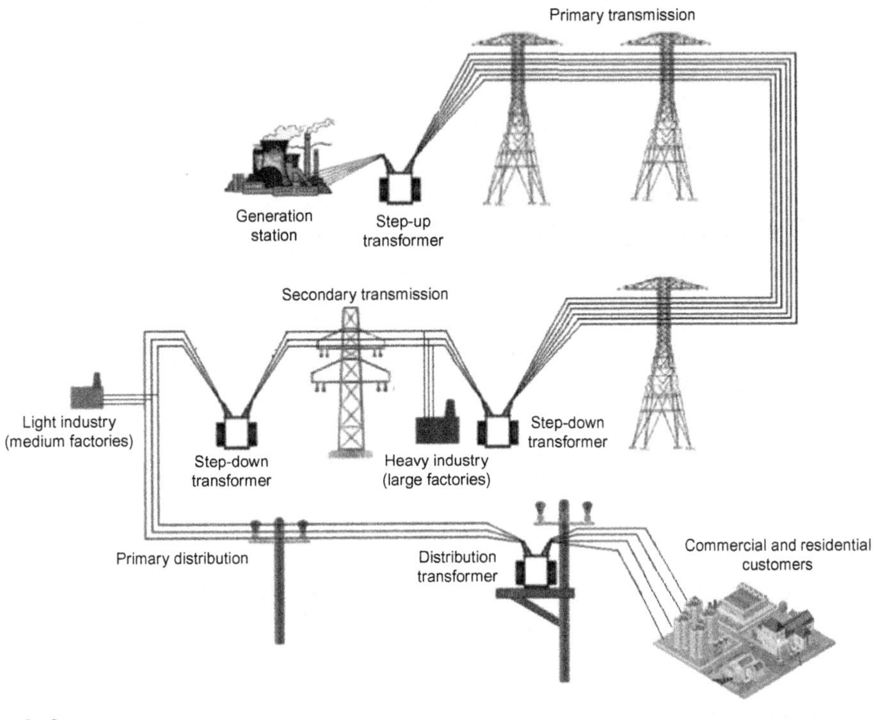

FIG. 2

Electricity grids.

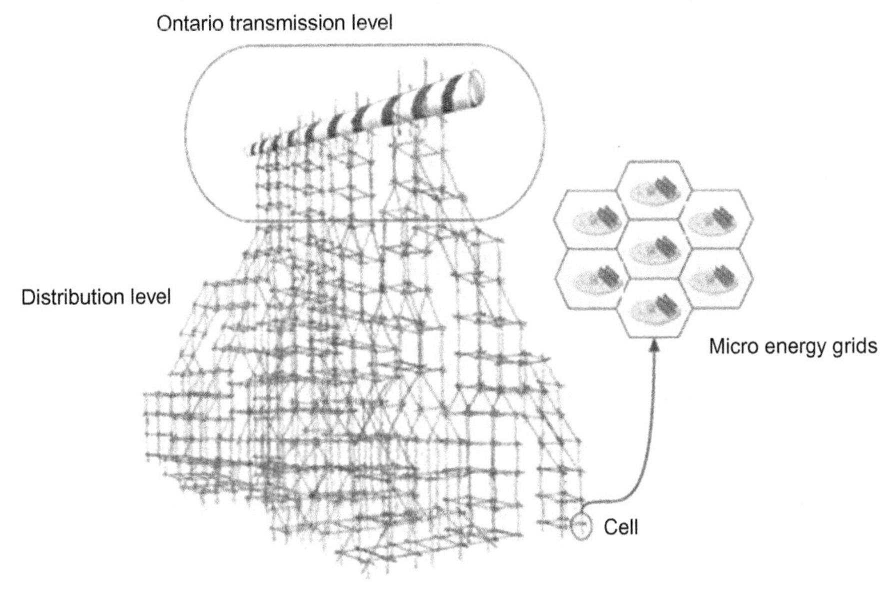

FIG. 3

SEG with transmission, distribution, cell, and interconnected MEGs.

deploying smart devices and intelligent systems in the actual power grid, it is important to test and validate their capabilities and functionality, as well as their accuracy. This will ensure the reliability and accuracy of these SEG technologies under different operating scenarios.

Several testbeds of SEG demonstrations have been reported, such as in Ref. [4], where state-of-the-art research facilities are used for testing and validating SEG deployment capabilities. This includes SEG devices, such as phasor measurement units (PMUs), which are tested under different operating conditions. In addition, it is important to test the interoperability of different hardware devices for energy/power system automation. The proposed SEG platform will be used to demonstrate interconnected MEGs and distribution network with distributed control systems.

1.3 MICRO ENERGY GRID

There has been a global movement in the direction of adoption and deployment of distributed and renewable resources. RE sources provide clean energy supply alternatives that differ from conventional fossil fuel sources. RE has dynamic behavior that cannot be scheduled. On the other hand, RE can provide relatively smaller energy supply capacities, which is smaller than conventional energy generation from typical power stations. RE is often connected to the electricity distribution system rather than the transmission system. The integration of such time-variable distributed or embedded sources into electricity network requires special consideration. Due to the ever-increasing demand for high-quality and reliable electric power, the concept of DERs has attracted widespread attention in recent years [1]. DERs consist of relatively small-scale generation and energy storage devices that interface with low- or medium-voltage distribution networks and can offset the local power consumption, or even export power to the upstream network if their generation surpasses the local consumption. One philosophy of operation, which is expected to enhance the utilization of DERs, is known as the MG concept [2,3]. MGs should widely utilize renewable energy resources (RERs) such as wind, solar, and hydrogen, to play a significant role in the electric power systems of the future, for cleaner air, reduced transmission and distribution costs, and incorporation of energy-efficiency enhancement initiatives. In addition, using energy storage devices such as batteries, energy capacitors, flywheels, and controllable loads. MGs including DGs will provide flexible and economic management of energy to meet local demands [4,5]. From a customer's point of view, MGs similar to traditional local voltage distribution networks not only provide their thermal and electricity needs, but in addition, they enhance local reliability, reduce emissions, improve power quality by supporting voltage and reducing voltage dips, and lead to lower costs of energy supply [10]. Much research has been performed to optimize the operation, load dispatch, and management of the energy storage system of the MGs. Particle swarm optimization method accordingly is employed in Ref. [11] to minimize the cost of MGs with controllable loads and battery storage. This is done by selling the stored energy at high prices to shave peak loads of the larger system. Linear programming algorithm is used in Refs. [12–14] to optimize MG operation cost and battery charge states. Maximizing benefits owing to the energy pricing differences between on-peak and off-peak periods are obtained by electrical and thermal storage charge scheduling in Refs. [15,16].

The important drawback of the above study is that it does not consider all uncertainties of the problem. Although employing RERs obviates environmental concerns and fossil fuel consumption, they introduce uncertain and fluctuated power because of the stochastic wind and solar variation [17]. In addition, this open-access power market and diverse commercial, residential and industrial consumer types, daily random nature of load demand [18]. Moreover, in an open-access power market, the degree of uncertainty of the load forecast error and market price can be even more perceptible. Engineers require computational methods that could provide solutions less sensitive to the environmental effects. So the techniques should take the uncertainty into account to control and minimize the risk associated with design and operation [19]. In order to consider uncertainty in optimal energy management planning of MGs effectively, the optimization problem should be solved for a suitable range of each uncertain input variables instead of just one estimated point. Using a deterministic optimization problem, a large computational burden is required to consider every possible and probable combination of uncertain input variables. Hence, from the system planning point of view, it turns out to be convenient to approach the problem of energy management planning of the MGs as a probabilistic problem. This leads to the problem known as energy management planning under uncertainty, where the output variable of an MEG objective function obtained as a random variable, and thus, it becomes easy to identify the possible ranges of the total operating cost.

There are several techniques to deal with problems under uncertainty. The three main approaches are analytical, simulation (Monte Carlo simulated), and approximate methods [20]. The vast majority of techniques have been analytically based, and simulation techniques have taken a minor role in specialized applications. All the proposed solutions for energy management of MEGs dealt with the MEG as a micropower grid comprised of electric supplies and loads and not as an MEG, in spite of existence of thermal and fuel supplies and loads.

This project is aiming at the development of a support tool for the design and operation of MEGs with the main objective of improving energy efficiency with conservation strategies in communities, commercial and residential buildings, industrial facilities, and transportation. This will support the deployment of high-performance energy supply grids with more penetration of NG applications such as boilers, NG vehicle, and NG-fuel cells. In addition, it will enable the penetration of high-performance solar systems and their applications in water and air heating, and cogeneration. The proposed project will include sensor network infrastructure to monitor and control MEGs and loads in efficient ways.

The team successfully developed energy modeling and a simulation environment with a knowledge base for energy conservation strategy evaluation as integrated within SEGs. In the new proposal, we are focussing on supporting energy and engineering consulting companies to evaluate the design and support the operation of

MEGs in buildings, facilities, communities, and transportation infrastructures. In addition, we are presenting potential strategies to increase the deployment of NG applications in infrastructures worldwide through NG-fuel cell, NG-boilers, NG-vehicles.

The main objective of the research is to design an optimization tool for MEG that is able to analyze, configure, simulate, and evaluate an optimal combination of available energy resources. The design tool allows different disciplinary expertise (eg, geographical, weather, and energy) to collaborate and share knowledge. MEG designer selects, modifies, and/or adds a cost function that matches the desired objective of the MEG. The design tool uses both the collected knowledge and the energy demand requirements to develop an optimal energy configuration that minimizes predefined cost function. In addition, the tool provides reported analyses and evaluation for an optimal MEG. The objective can be achieved by the following:

1. Business modeling and requirement analysis of implementation of MEG using IDEF0 and gap analysis, based on local demand dynamic profiles, DER dynamic profiles, business, technical, and regulation constraints, and target performance criteria; support the lifecycle engineering of MEG design and operation.
2. Automatic synthesizing of ESN of the target MEG with appropriate model representation techniques (utilizing feature extraction) of different energy resources and loads. The archive allows a clear view of the performance of different energy resources and loads for matching purposes of increasing MEG efficiency and reducing its dependency on energy storage components.
3. Performing a comprehensive study of energy storage technologies available in the market. While it is a fact that energy storage technologies offer wide varieties of storage types, the questions of what, how, where, and when storage can be implemented are highly dependent on the MEG design structure, load profile, and cost, and the dynamic behavior of the storage component itself.
4. Generating element modeling of MEG components based on the structure and type of each element.
5. Designing a knowledge base ESN system for MEG.
   - **(a)** Designing a database mechanism to accept geographical data, weather data, and other data concerning possible supply of energy resources, such as NG and fossil fuel.
   - **(b)** Designing a dynamic inference engine that is able to add and modify energy resources rules.
   - **(c)** Developing a fuzzy inference mechanism to allow reasoning and uncertainty about both analytic and black-box models.
   - **(d)** Integrating artificial neural network (ANN) to allow dynamic nonlinear mapping between ESN nodes. The ANN allows a nonlinear relation between class nodes of ESN (recovering limited alternatives logical relations). In addition, ANN reinforces the related relations while weakening undesired relations between class nodes. Hence, a reduced

effective energy scenario can be generated, allowing reasonable and efficient computational burden.

(e) ESN generating of all possible with most effective design parameters for the desired MEG with their dynamic range forming the possible scenarios of the energy resources, storage elements, dynamic connection between sources and loads, and load-source scheduling. In addition, the developed ESN will generate the most affected KPIs by the design parameter to reduce the computational burden.

6. Developing multiobjective functions that describe the desired KPIs of MEG for optimization process.
7. Achieving the most efficient combination and configuration of different MEG resources that satisfy the minimization of the predefined cost function.
8. Finding the optimal application of the energy resources within the MEG among different scenarios.
9. Applying power electronic technology to control and protect power flow within MEG in an optimal dynamic fashion as an attempt to improve KPIs by matching the source's dynamics to the load dynamics.
10. Comparing the performance of the optimal scenarios of the energy resources and its interconnection to the load with the commonly used MG structure, based on efficiency, reliability, gas emission, and other possible factors.
11. Applying and validating protection scenarios to achieve resilient MEGs that can provide the required energy supply capacity in different normal and abnormal conditions.

1.4 ENERGY SEMANTIC NETWORK

Energy is typically modeled in view of foundations of thermodynamic laws, where energy can be represented in the form of thermal, electricity, and fuel, as well as work. And energy profiles could be modeled for actual consumption, simulated data, demand profiles, and historical data, as shown in Fig. 4.

In order to map energy loads to supply technologies and nodes, ESNs are proposed to synthesize different mapping scenarios from supply to loads, with the utilization of conversion and storage, as shown in Fig. 5. The proposed model is implemented in a computer-aided design tool, called ESN, in Visio; the corresponding ESN knowledge base is implemented in Microsoft in a database based on process object-oriented modeling (POOM) and in view of ISA S95. Each building block of the proposed energy network is modeled using POOM, where process variables (inputs, internal, and output) are represented in a systematic way so that KPIs can be represented in each detailed building block, such as solar system, wind mill, thermal storage unit, and fuel cell unit and battery banks, as shown in Fig. 6.

One of the critical components of our MG design is power storage, and one type of storage device is a lithium ion battery. Modeling these components is part of our revised requirements. These models will allow for better design decisions, as well as playing a role in the simulations. The equation models of the chosen battery model are derived as shown below. Similar models are developed for different energy generation and storage of thermal, gas, and electricity, and linked with performance indicators.

The proposed representation will enable linking process models with control models for dynamic energy supply utilization based on dynamic energy generation and utilization schemes, and in view of business rules, policies, and regulations. Fig. 7 shows the proposed ESN quantitative modeling and simulation approach

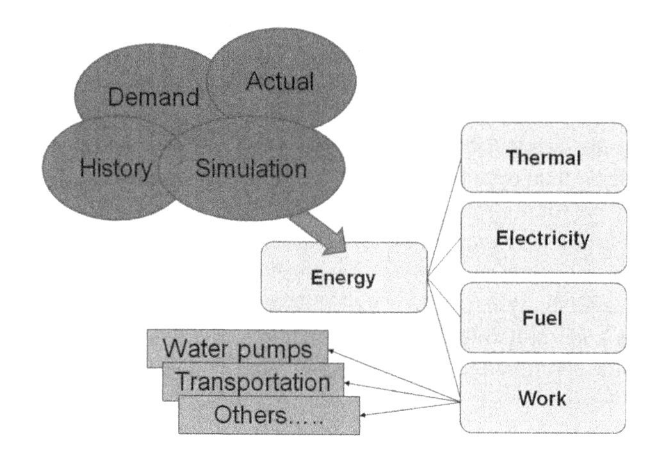

FIG. 4

Energy modeling in infrastructures.

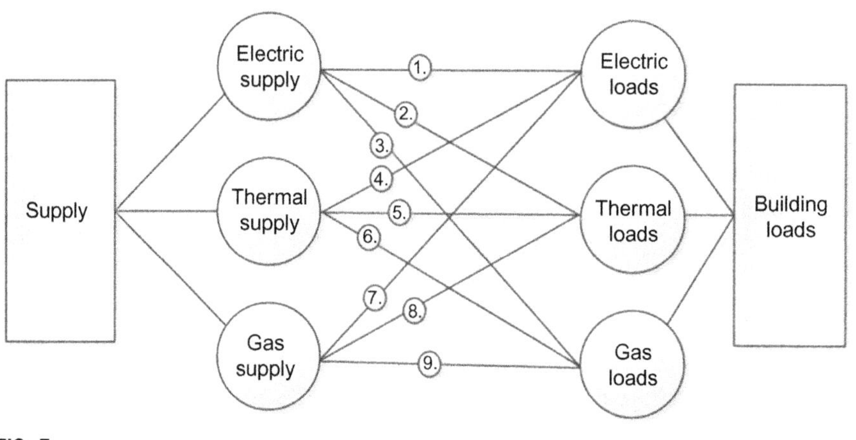

FIG. 5

ESN to link energy supply with loads, conversion, and storage.

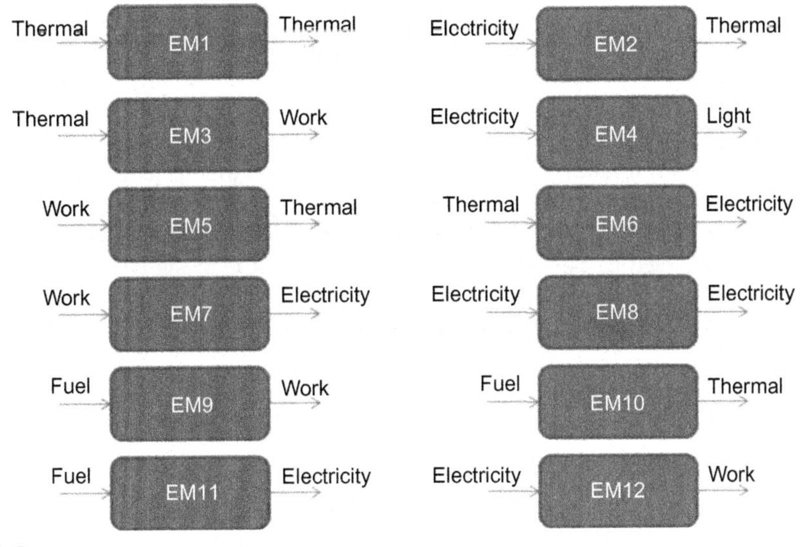

FIG. 6

Modeling of energy node within ESN using POOM.

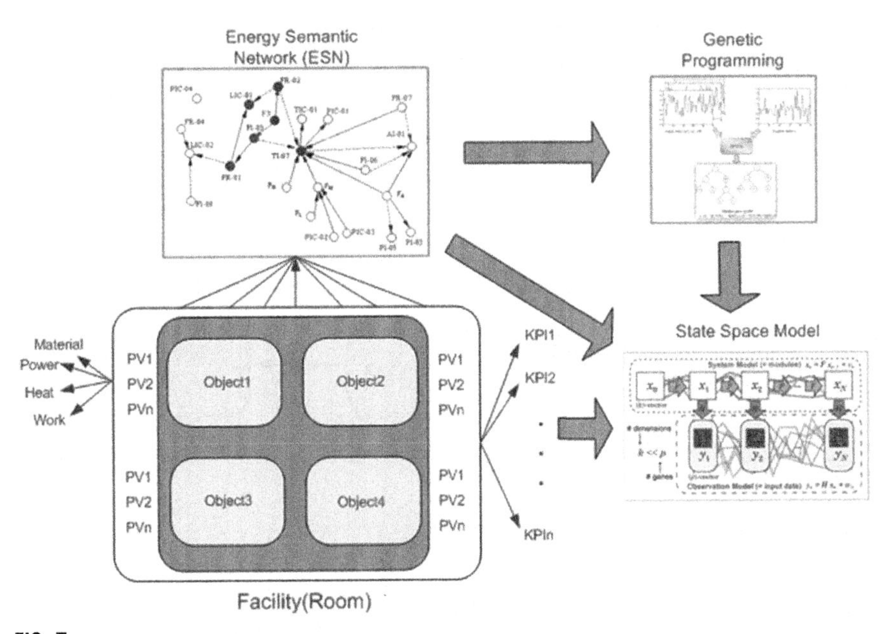

FIG. 7

ESN data modeling.

where process variables are represented for energy nodes, and real-time simulation model parameters and errors are fine-tuned using intelligent algorithm based on genetic programming. This will allow developing accurate energy models using real-time data, which provides continual learning and tuning of ESN while system is running.

This proposal introduces a novel integrated simulation tool to support the design of MEG based on the optimization of power, thermal and fuel supplies and loads with the considerations of intermittent natural of DERs and the profiles of the different types of loads (electric, thermal, and fuel loads) using LCC of the DERs, as shown in Fig. 8.

MEG, as shown in Fig. 9, is an emerging concept in intellectual networks that integrates different energy resources like electricity, heat, hydrogen, and NG. MEG needs to be more reliable, secure, economic, ecofriendly, and safer. To introduce an integrated simulation tool for efficient cost and optimization of MEG, the following steps should be followed:

1. Establishing an image archive using the feature extraction for thermal, gas, and electrical energy loads and sources (sources and loads signature or figure print). The archive allows a clear view of the performance of different energy resources and loads for matching purposes to increase MEG reliability and reduce its dependency on energy storage components.
2. Performing a comprehensive study of energy thermal, gas, and electrical storage technologies available in the market. While it is a fact that energy storage technologies offer wide varieties of storage types, the questions of what, where, and when storage can be implemented are highly dependent on

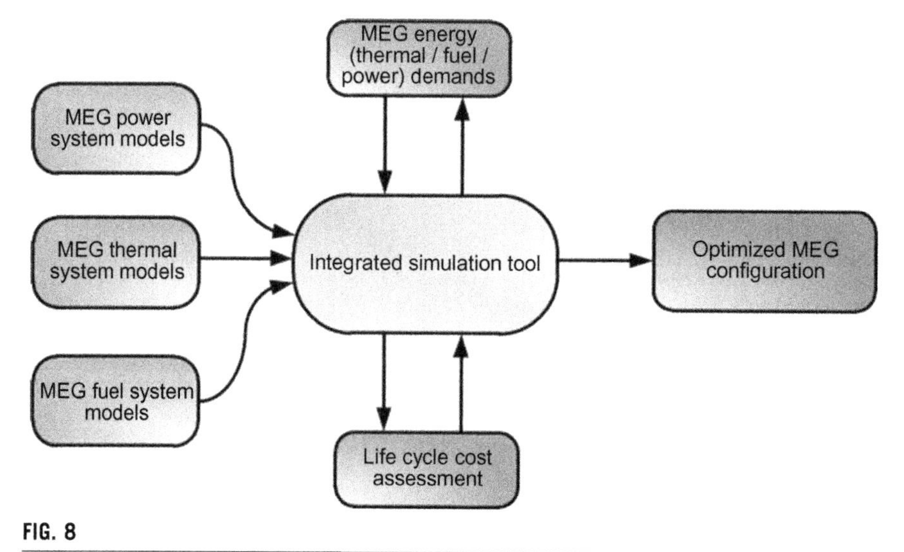

FIG. 8

Integrated simulation tool for MEG design and evaluation.

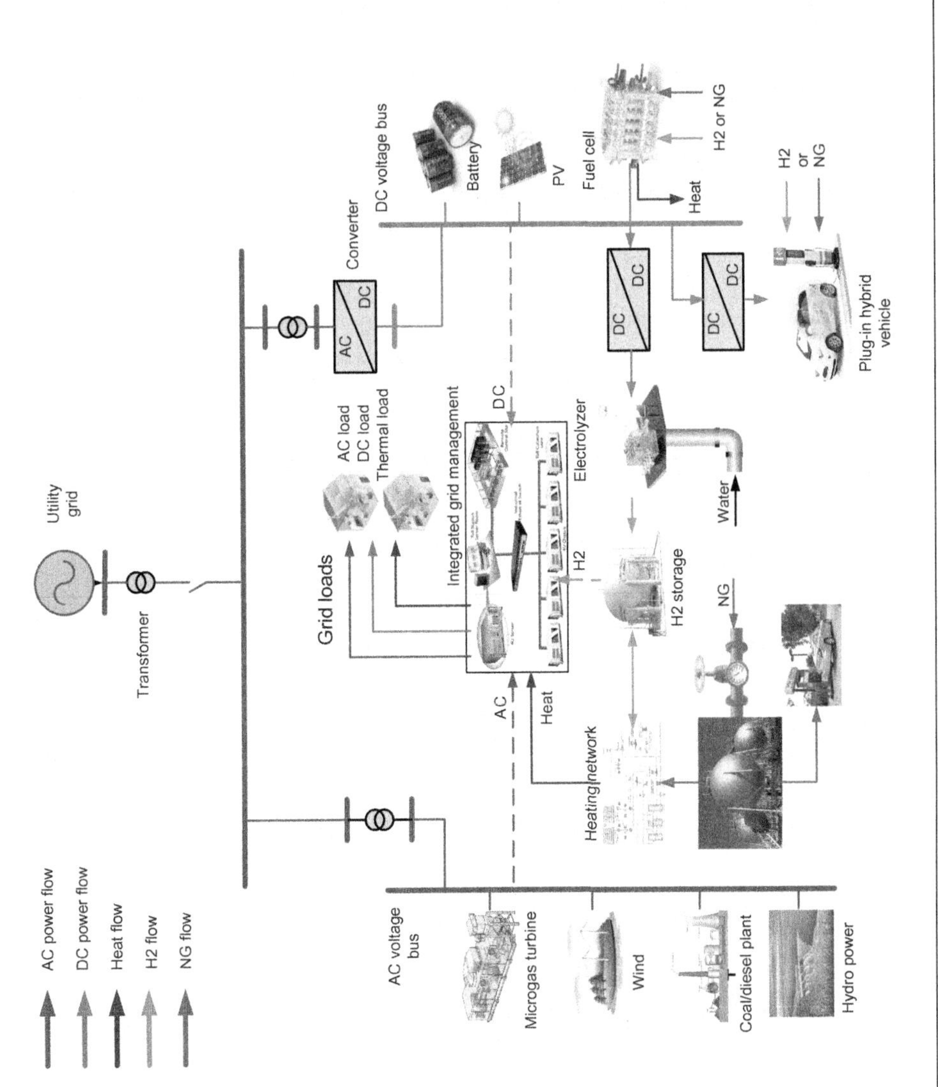

FIG. 9

Typical MEG structure (electricity, heat, hydrogen, and NG).

MEG structure, the load profile, and the cost and dynamic behavior of the storage components themselves.

3. Implementing an MEG structure that grants the insertion and removal of energy resources and storage, hence the evaluation of multiple MEG scenarios and its dynamic modification to compensate for possible risks and examination of its resiliency.
4. Designing dynamic power flow within MEG, applying power electronic technologies. The dynamic power flow within MEG would improve KPIs by matching sources' dynamics to load dynamics. In addition to the optimal structure of MEG that can be achieved by points 1–3, the dynamic power flow between source and load components, to match the load-source profile, minimizes the use of a storage system and the gap between load demand and available power generated by different sources during the on-pick and off-pick periods.
5. Developing a dynamic self-learning knowledge base ESN. The structure of the knowledge base ESN includes the following:
   (a) Physics-based (analytic model), data-driven (black-box model) in a combined mode for thermal, gas, and electrical energy resources and loads. A GUI (graphical user interface) is used to receive information about the physical structure, and dynamics of the MEG site allow delivery of the analytic model to ESN. An interface with sensors is used to allow the knowledge base ESN to form data-driven (black-box model).
   (b) Integrating GIS to supply ENS with the geographical information and physical possible distribution of energy resources, affecting both the possible generated energy and KPIs of the building, community, or site of the MEG.
   (c) Fuzzy inference mechanisms, allowing reasoning and uncertainty about both analytic and black-box models.
   (d) ANN allowing dynamic nonlinear mapping between ESN nodes. The ANN allows a nonlinear relation between class nodes of ESN (recovering limited alternatives logical relations). In addition, ANN reinforces the related relations, while weakening undesired relations between class nodes. Hence, a reduced effective energy scenario can be generated, allowing a reasonable and efficient computational burden.
6. Forming an efficient cost function that describes the MEG objective well. The most effective component of the optimization process is the cost function. In a multidimensional optimization space, the selection of the cost function is complicated and vague. The development of KPIs is an effort to form efficient cost function for MEGs. Merging the KPIs of the MEG into the dynamic self-learning knowledge base, ESN allows the selection of most effective KPIs and ignoring the least effective ones, hence the optimization computational burden will be reduce and it performance will be enhanced.
7. Finding the optimal application of the energy resources within the MEG among different scenarios using optimization algorithms (Particle Swarm or Genetic algorithms).

8. Applying power electronic technology to control the power flow among the MEG in an optimal dynamic fashion as an attempt to improve KPIs by matching the source's dynamics to the load dynamics.
9. Comparing the performance of the optimal scenarios of the energy resources and its interconnection to the load with the commonly used MG structure, based on efficiency, reliability, gas emission, and other possible factors.
10. Applying and validating protection scenarios to achieve resilient MEGs that can provide the required energy supply capacity in different normal and abnormal conditions.

The different MEG energy models are integrated in a single simulation tool to decide the best MEG configuration based on the combinations of the different models, supplies, and load, taking into account the intermittent nature of DERs and variable demand profiles. The LCC is used to assess the best MEG configuration. The proposed MEG simulation framework starts with MEG model library development for all potential DERs that can be utilized within MEG. This is followed by collecting demand requirements of thermal, fuel, and power within MEG. Based on the demand requirements, possible MEG design and configuration alternatives will be synthesized for thermal, fuel, and power systems. For each design and configuration scenario, LCC will be estimated for each component, and MEG LCC will be evaluated and optimized using an MEG simulation engine developed in Matlab, as shown in Fig. 10.

1.4.1 LCC ASSESSMENTS OF MEG

How do the locations of manufacture and deployment of RE technologies affect the lifecycle energy requirements, greenhouse gas emissions, and economic costs?

- To conduct LCAs of the following energy technologies:
  - Solar PV
  - Small wind
  - Fuel cells
  - Energy storages
  - Extension of the electricity grid
- To review published LCAs of the above technologies that were conducted under standard test conditions (STCs)
- To compare the LCAs for the above technologies under STC and local conditions and determine which has the lowest lifecycle impact
- To optimize LCC of MEG design and operation alternatives and provide the most economic MEG configuration to support stakeholders' decision

LCAs for RE technologies are widespread and well established. Many of these LCA studies have been conducted using STC. On one hand, STC provides a standard against which different technologies and innovations can be compared. On the other hand, STC may not reflect actual performance in the field, especially when field

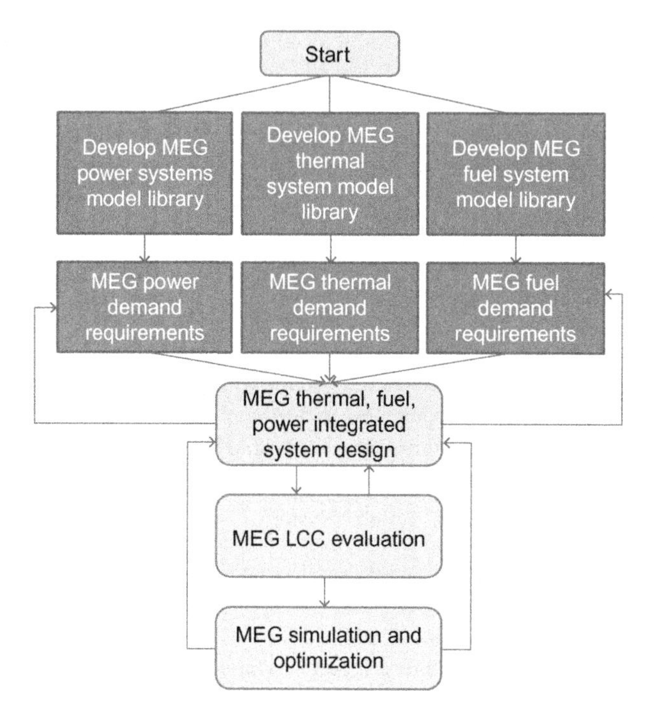

FIG. 10

The proposed MEG integrated simulation.

conditions differ substantially from STC. The proposed work would be the first LCA of solar PV and wind turbine systems.

The ISO 14040 standard provides a general framework for conducting LCAs, but it does not specify a particular procedure. The ISO framework identifies four phases of an LCA. In the first step, the LCA goals, scope, and system boundaries are defined. The second phase is inventory analysis, in which material and energy inputs and outputs are analyzed. This is the main data-collection phase. The third phase, impact assessment, describes the environmental significance of the inventory analysis. The LCC assessment is conducted on different MEG configuration scenarios to provide decision support to the deployment alternatives of RE within MEGs.

1.4.2 MEG INTERCONNECTIONS DESIGN AND OPERATION METHOD

The MEG of building, facility, community, and transportation, has two levels of interconnection: local and regional interconnection. Local interconnection is interconnection between different components of MEG, including energy resources, energy transformers, and loads. The local interconnection controller is responsible for monitoring the load demand and dynamically connecting the load to energy resources to satisfy the optimal value of MEG KPIs.

Regional interconnection is interconnection with other nearby MEGs or a national utility grid. In the next few years, the integration of many MGs inside the same distribution network will be the main target of improving MEG performance. The interconnection of multiple MEGs with each other by a private line based on their dynamic profile of both load and sources will highly improve the overall MEGs' performance; however, this requires supervisory control level to monitor and control the synchronization of the local and regional sources in an operation called (synchrophasors). The synchrophasors allow synchronized real-time measurements of multiple remote measurement points on MEG. A PMU is one of the most important measuring devices in the future of power systems that can be a dedicated device for this purpose, or its function can be incorporated into a protective relay or other device. With the development of cloud computing and storage and Internet access capabilities, the remote online sensing and controlling of multiple MEG sites becomes feasible, as shown in Fig. 11.

1.4.3 FAULT DIAGNOSIS, ISOLATION, AND SELF-HEALING OF MEG

One of the most important design consideration of MEG is fault diagnosis and isolation. The protection system should isolate the smallest part of the MG when it clears the fault. In addition, the protection system should isolate the MG from the main grid and all other connected MEGs as rapidly as necessary to protect the MG loads. Modern MGs consist of many different DERs like solar PV, wind

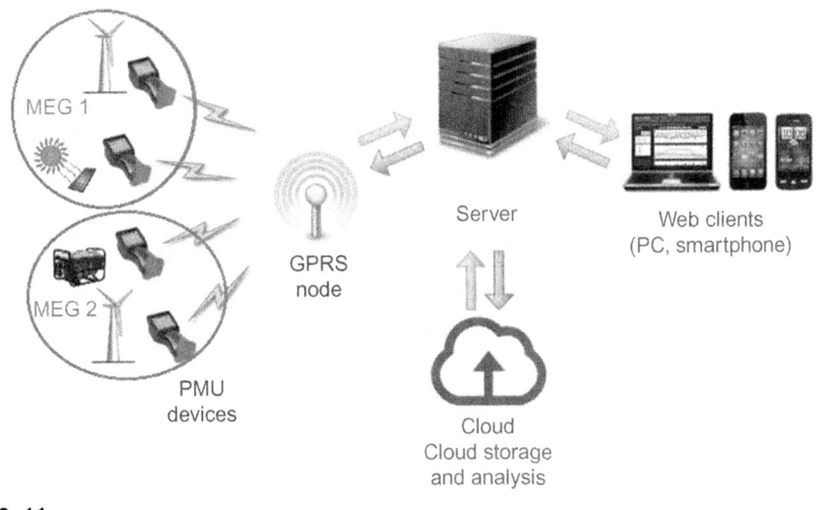

FIG. 11

Design and operation domain knowledge of ESN.

turbines, fuel cells, small-scale hydro, tidal, and wave generators, microturbines, combined heat power systems, and energy storage. When a different type of distributed resource is connected to an MEG, nearby MEGs and utility grids, the DR contributes fault current to the system, and the contribution level depends on distribution resource type. To ensure the safe operation of the MG, the protection equipment should be updated accordingly.

So the dynamic structure of MGs and their various operating conditions require the development of adaptive protection strategies. Here, the central control unit communicates with all relays and distributed generators in the MG to record their status as ON/OFF, their rated current, and their fault current contribution. Communication with the relay is required to update the operating current and to detect the direction of the fault currents, thus mitigating the fault properly. The control unit also records the status of the utility grid as connected, or the MG is islanded for adaptive protection.

1.5 TECHNOLOGICAL INFRASTRUCTURE OF SEGs

Fig. 12 shows a proposed technological information infrastructure of the target SEGs, where five layers are proposed to manage SEGs: application layer, GIS layer, data management layer, communication/control layer, and grid physical layer. These

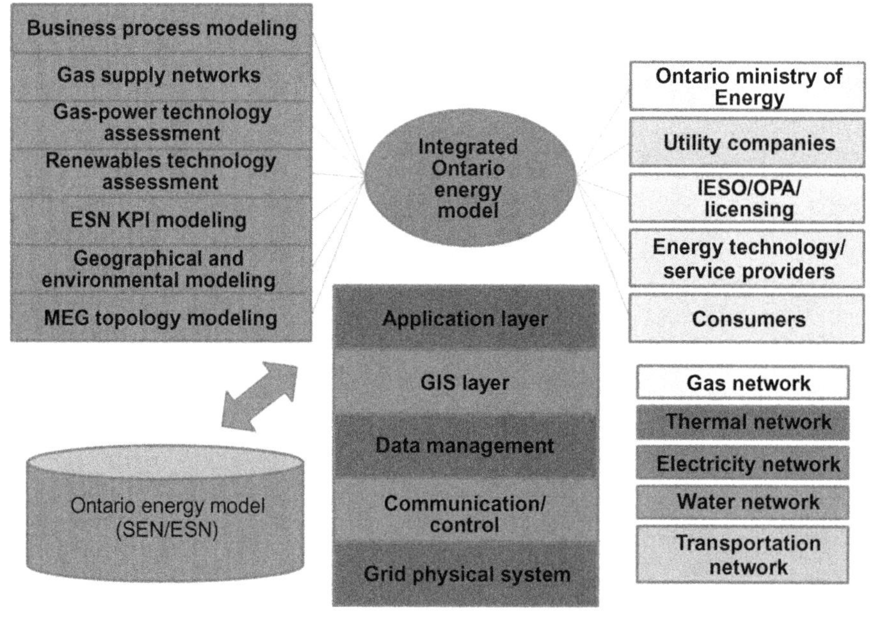

FIG. 12

Technological information infrastructure for SEG implementation.

layers will support energy networks (gas, electricity, thermal, water, and transportation). The integrated system will provide views to utilities, regulators, technology providers, and consumers/end users to interact and negotiate engineering activities around SEGs. The proposed integrated system will provide functionality to support and manage gas networks, electricity networks, gas-power grids, modeling KPIs for all components, and assessment of MEG technologies, including the environmental impacts.

Fig. 13 shows the system architecture for smart energy grid engineering. It shows all necessary components that should be implemented in the application layer to manage electricity, gas, and thermal grids, as integrated with water and transportation networks. Integrated SEG simulation is presented to link power simulation with thermal and gas simulation in a steady state and dynamic/transient modes. One important component is outage management, where fault propagation and diagnosis knowledge will be accumulated around the physical layers, and safety protection layers are mapped in each propagation path to ensure safety and resiliency of SEGs.

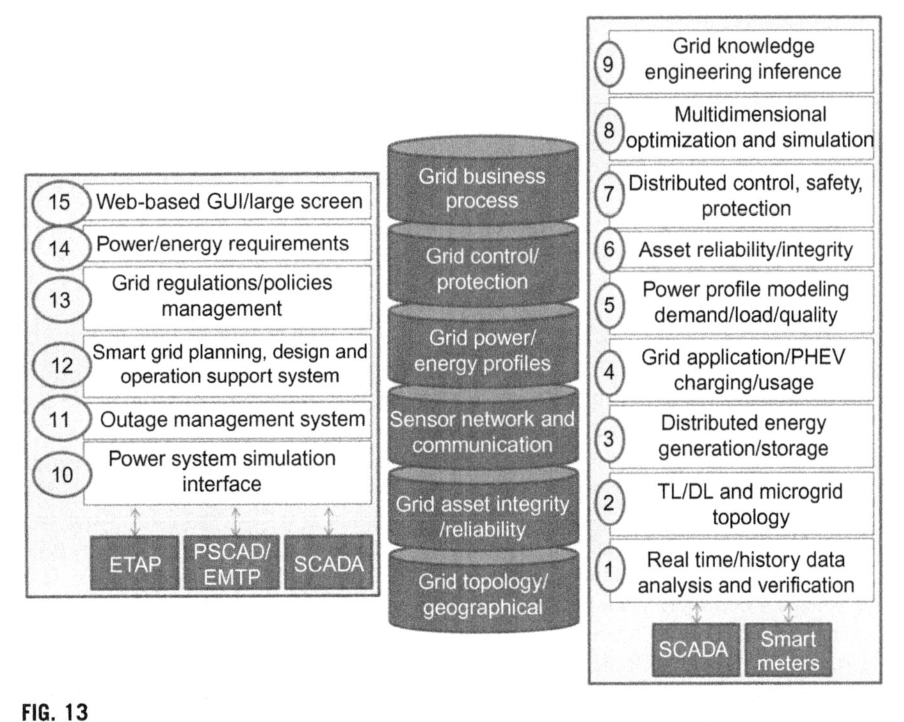

FIG. 13

Smart energy grid engineering—system architecture.

1.6 CONCLUSION

This book is developed to enable professionals and scientists to understand and review foundations of SEGs, and the associated technologies. Important engineering activities will be studied and discussed using several case studies of different components of SEGs.

An overview of challenges and opportunities of SEGs is presented, along with implementations using adaptive modeling of energy grids on the basis of ESNs. An engineering design approach of SEGs is presented and discussed with different implementation strategies. The proposed modeling technique is able to model interconnected MEGs. Control design methods are presented and applied on different systems and components within SEGs. In order to achieve resilient SEGs and MEGs, fault-tolerant control methods are presented in view of protection layers and fault semantic network representation. Information technology is adopted to offer the smooth exchange of smart grid information.

Finally, foundations and technologies of the Internet of Things are discussed to offer smooth integration of smart meters with different layers within SEGs.

REFERENCES

[1] Gabbar HA, Sayed HE, Yamashita Y, Gruver W, Kamel M. Engineering design aspects of hybrid energy supply unit (HENSU), In: 2nd international workshop on computational intelligence & applications, Okayama, Japan; 2006. pp. P12:1–P12:12.

[2] Kurohane K, Uehara A, Senjyu T, Yona A, Urasaki N, Funabashi T, et al. Control strategy for a distributed DC power system with renewable energy. Renew Energy 2011;36 (1):42–9.

[3] Wang C, Jia Q-Q, Li X-B, Dou C-X. Fault location using synchronized sequence measurements. Int J Electr Power Energy Syst 2008;30(2):134–9.

[4] Firouzjah KG, Sheikholeslami A. A current independent method based on synchronized voltage measurement for fault location on transmission lines. Simul Modell Pract Theory 2009;17(4):692–707.

[5] Santiangeli A, Orecchini F. Beyond smart energy grids—the need of intelligent energy networks for a higher global efficiency through energy vectors integration. Int J Hydrogen Energy 2011;36(8):26–33.

[6] Marszal AJ, Heiselberg P, Bourrelle JS, Musall E, Voss K, Sartori I, et al. Zero energy infrastructure—a review of definitions and calculation methodologies. Energy Infrastruct 2011;43:971–9.

[7] Mishra A, Irwin D, Shenoy P, Kurose J, Zhu T. GreenCharge: managing renewable energy in smart infrastructures. IEEE J Sel Areas Commun 2013;31(7):1281–93.

[8] Runge J, Gabbar H. Solar windows control system for an apartment infrastructure in Toronto with battery storage, In: International conference on power engineering and renewable energy (ICPERE), 2014, Bali; 2014.

[9] Gabbar HA. Engineering design of green hybrid energy production and supply chains. Environ Model Software 2009;24:423–35.

[10] Hammons TJ. Integrating renewable energy sources into European grids. Electr Power Energy Syst 2008;30:462–75.
[11] Battaglini A, Lillistam J, Hass A, Patt A. Development of SuperSmart energy grids for a more efficient utilization electricity from renewable sources. J Clean Prod 2009;17:911–8.
[12] Pouresmaeil E, Gomis-Bellmunt O, Montesinos-Miracle D, Bergas-Jane J. Multilevel converters control for renewable energy integration to the power grid. Energy 2011;36:950–63.
[13] Clayton JM, Young FS. Estimating lightning performance of transmission lines. IEEE Trans Power App Syst 1964;83:1102–10.
[14] Bouquegneau C, Dubois M, Trekat J. Probabilistic analysis of lightning performance of high-voltage transmission lines. Electr Power Syst Res 1986;102(1–2):5–18.
[15] Wade NS, Taylor PC, Lang PD, Jones PR. Evaluating the benefits of an electrical energy storage system in a future smart energy grid. Energy Pol 2010;38:7180–8.
[16] Divya KC, Ostergaard J. Battery storage technology for power systems—an overview. Electr Power Syst Res 2009;79:511–20.
[17] Transition to sustainable buildings: strategies and opportunities to 2050. International Energy Agency; 2013. https://www.iea.org/media/training/presentations/etw2014/publications/Sustainable_Buildings_2013.pdf.
[18] Schultz CC, PE CC, Scott G. Energy conservation in existing infrastructures. Eng Syst 2011;28(10):34–40.
[19] Difsa K, Bennstamb M, Trygga L, Nordenstamb L. Energy conservation measures in infrastructures heated by district heating—a local energy system perspective. Energy 2010;35(8):3194–203.
[20] Smeds J, Wall M. Enhanced energy conservation in houses through high performance design. Energy Infrastruct 2007;39:273–8.

CHAPTER

Smart energy grid infrastructures and interconnected micro energy grids

2

H.A. Gabbar

University of Ontario Institute of Technology (UOIT), Oshawa, ON, Canada

2.1 BACKGROUND

2.1.1 HISTORICAL DEVELOPMENT OF THE ENERGY GRID

In 1886, the first alternating current (AC) electricity system was installed in Great Barrington, Massachusetts. The grid was a centralized unidirectional system with transmission, distribution, and demand-driven control.

In the 20th century, local electricity grids have been increased over time and interconnected with each other for economic and reliability reasons. By the 1960s, electricity grids of developed countries became large, mature, and highly interconnected with huge numbers of central generation power stations (ie, coal, gas, and oil-fired power stations). These grids were delivering power to major load centers through high-capacity power lines, which were then branched to provide power to smaller industrial, commercial, and residential customers over the entire supply area.

Generation power stations were located close to fuel sources or close to rail, road, or port supply lines. Locations of hydroelectric dams in mountainous areas influenced the structure of the electricity grid. Nuclear power plants were located based on the availability of cooling water. As fossil-fuel-fired power stations were initially highly polluting, they were located as far as possible from populated areas. By the late 1960s, the electricity grid reached the majority of the developed countries' populations. Due to the continuous demand growth, higher numbers of power stations were installed. In some places, the supply of electricity, especially at peak demand time, could not cover this demand, resulting in poor quality of service. Increasingly, societies have full dependence on electricity for industry, heating, communication, lighting, and entertainment. Also, consumers demanded high-levels of reliability.

Smart Energy Grid Engineering.

2.1.2 MODERNIZATION OPPORTUNITIES

Recently, great improvements in electric, electronic, and communication technologies have been achieved to resolve the limitations and costs of the electrical grid. For instance, technological limitations on metering of electricity consumption no longer force peak power prices to be averaged out and passed on to all consumers equally. On the other side, growing concerns due to the environmental impact from fossil-fuel-fired power stations have led to a desire to use large amounts of clean renewable energy. As wind and solar power are highly variable, there is a need for more sophisticated control systems to facilitate their connection to the main electrical grid. Their rapidly falling costs indicate a major change from the centralized grid topology to a distributed one, with power being both generated and consumed in the same place. Furthermore, the growing concern about reliable service has led to calls for a more robust energy grid that is less dependent on centralized power stations.

2.1.3 DEFINITION OF SMART ENERGY GRID

Energy supply systems such as electricity, heat, natural gas, and hydrogen are usually considered individual subsystems; however, as shown in Fig. 1, electric power, heat, natural gas, and hydrogen networks are linked at different points through coupling

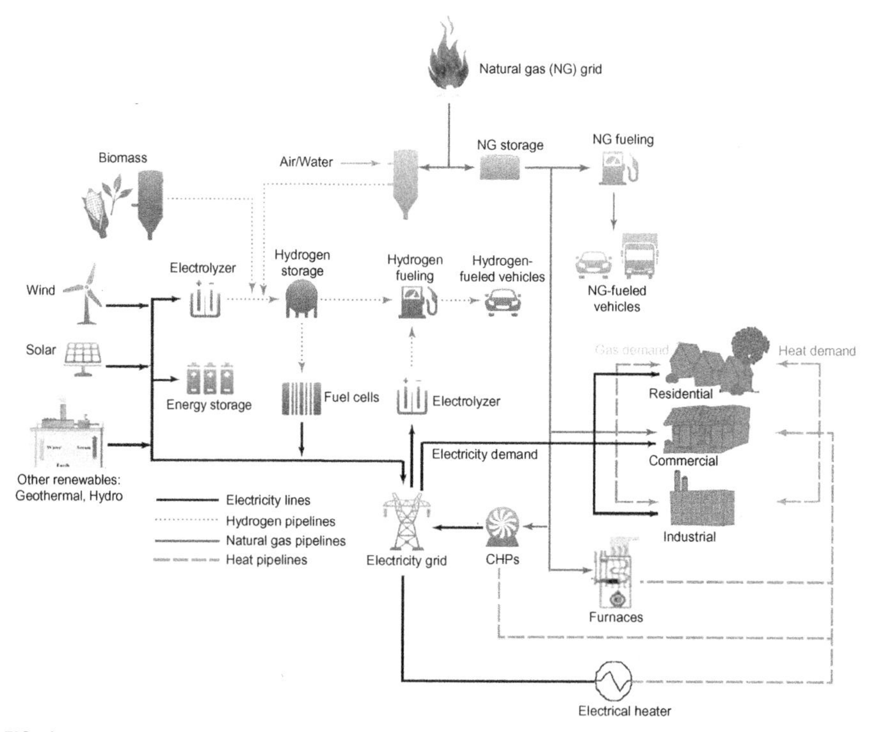

FIG. 1

Integrated multienergy systems.

components (eg, CHP units, furnaces, electric heaters, and circulation pumps). CHP units generate electricity and heat simultaneously. Furnaces and electric heaters convert natural gas and electricity to heat, respectively. Similarly, circulation pumps consume electricity to circulate water in the district heating network. In the present smart grid vision, considering multienergy systems as an integrated whole with energy storage could:

- contribute to more efficient utilization of distributed energy resources (DERs)
- increase the flexibility for equalizing the fluctuations from renewable energy resources
- control the amount of energy to be supplied from alternative sources; thus, security of energy supply could be improved
- determine the most efficient operating procedure for the integrated energy systems; thus, energy losses, costs, and emissions could be minimized

Smart energy grids (SEGs) are energy networks that promise to enhance the operational efficiency of energy supply via distributed generation with bidirectional energy flow. This objective is achieved by allowing intelligent monitoring and control of different components within the multienergy systems (ie, electricity, natural gas, thermal energy, and water), while maintaining the quality, security, reliability, and safety with minimal environmental impacts.

Governments around the world are investing heavily in SEGs to ensure optimal energy use and supply, enable better planning for outage responses and recovery, and facilitate the integration of heterogeneous technologies such as renewable energy systems, electrical vehicle networks, and smart homes around the grid. For example, electric utilities are making three classes of transformations: improvement of infrastructure; addition of the digital layer, which is the essence of the SEG; and business process transformation, necessary to capitalize on the investments in smart technology.

2.1.4 TECHNOLOGICAL INNOVATIONS

The concept of a smart grid is to add monitoring, analysis, control, and communication capabilities to multienergy systems in order to maximize efficiency and reduce consumption. SEG innovation will enable consumers to better manage energy consumption and allow energy-dependent business sectors to be more competitive. New technologies contribute to better demand response and load control, allowing utilities to lower costs by shifting loads to less expensive generation during peak demand. These technologies also assist ratepayers by adjusting consumption and usage patterns to save costs. Technologies such as embedded sensors and automated controls help utilities better anticipate, detect, and respond to system problems. Thus, these technologies are self-healing and can automatically avoid or at least mitigate power outages, power quality problems, and service disruptions, thereby providing ratepayers with a more reliable energy system. Renewable energy resources have an intermittent nature. SEG technologies enable power systems to better predict availability of these sources and to utilize them in an optimal and cost-efficient manner.

As an example, automatic meter reading was used for monitoring loads and evolved into the advanced metering infrastructure, whose meters could store how electricity was used at different times of the day. Smart meters add continuous communications to facilitate monitoring in real time. Thus, they can be used as a gateway to demand response-aware devices. Furthermore, devices such as air conditioners, refrigerators, and heaters can adjust their duty cycle to avoid running during grid peaking conditions.

2.2 SMART ENERGY GRID STRUCTURE

2.2.1 ENERGY SUPPLY AND PRODUCTION CHAINS

Currently, energy production chains are mainly based on oil and coal, which have negative impacts on the environment in terms of greenhouse gases. In addition, there is a limited availability of fossil fuels compared with the continuous energy demand growth of the world. Problems of increased oil prices, global warming, and environmental pollution highlighted the need for practical and flexible green energy production chains based on clean and renewable sources. There is a wide range of renewable energy sources, such as biomass, solar, nuclear, wind, geothermal, and hydropower, where they have less, or zero, impacts on the environment. Biomass is widely used as a strategic renewable energy source. Biomass sources include wood, which is produced from recycled wood products, garbage, or residual waste. Biomass has been categorized as a green energy source where the total CO_2 emitted during energy generation is almost equal to CO_2 absorbed by plants during the photosynthesis process. Other examples of green energy production strategies include the use of windmills; however, cost is still a major factor to decide whether to migrate to the new energy technologies. In addition, there are other energy production strategies such as from-waste-to-energy, where recycle scenarios are studied to generate energy from waste.

In order to optimize the energy supply locally in each region and as interconnected region, process object-oriented modeling (POOM) methodology was introduced and used to model energy generation, supply topology, and link to usage nodes. For better representation of energy grid networks, energy semantic networks (ESNs) are used to represent energy grids as physical models, as well as associated constraints and business rules. Fig. 2 shows the proposed modeling of regional energy supply chain, which was originally explained in Ref. [1]. The corresponding energy model is developed to analyze energy production from thermal, electricity, and fuel, which are mapped to energy usage. Table 1 shows the details of the model.

2.2.2 MODELING OF ENERGY GENERATION AND CONVERSION

In order to support process control functions, it is essential to provide systematic modeling methodology that can abstract all objects for the underlying process in terms of static views, which include facility; dynamic views, which include state transitions and messages; and operation views, such as functions, methods, etc.

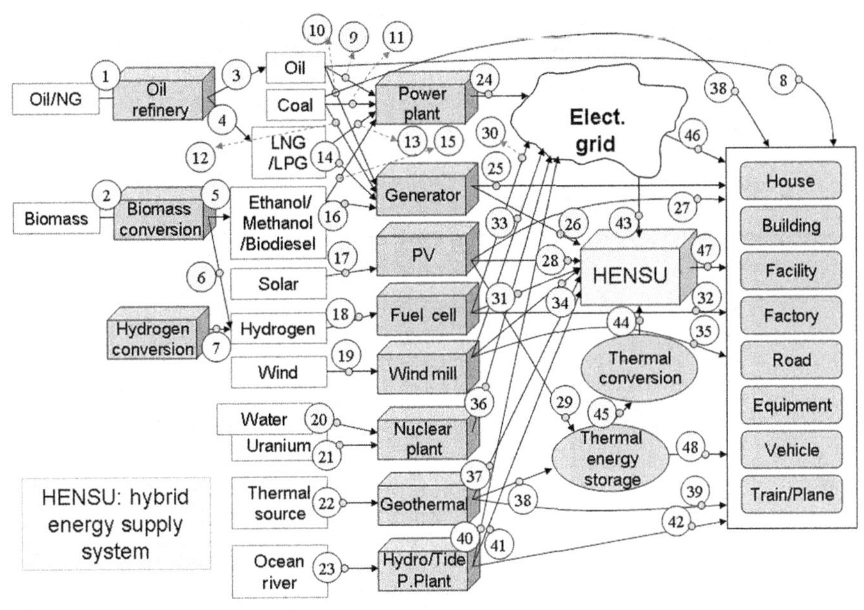

FIG. 2

Energy production and supply chain model.

Table 1 Energy Model Details

Model Element	Definition	Links
Oil and natural gas	Oil and natural gas supply to oil refinery production process	It supplies oil and coal
Biomass	Biomass supply from different domestic sources into biomass conversion to produce ethanol, methanol, biodiesel	It supplies biofuel to power plants and generators
Hydrogen	Hydrogen generation from water or other sources	Supplies hydrogen to fuel cell; also can supply to transportation (vehicles, rail, etc.)
Water/uranium	Supplies to nuclear facilities to produce power and thermal	It supplies power to the energy grid as well as hybrid energy supply unit (HENSU) at different levels
Thermal	It supplies thermal energy with technologies such as geothermal	It supplies thermal energy to thermal loads and thermal storage
Ocean/rivers	It supplies power via hydro power plants and other tide technologies	It supplies power to energy grid and HENSU
PV	It provides electricity and thermal from solar	It supplies electricity and thermal to energy grid, HENSU, and other thermal loads
All	It supplies energy to different loads	Facilities, houses, transportation, infrastructures, equipment, etc.

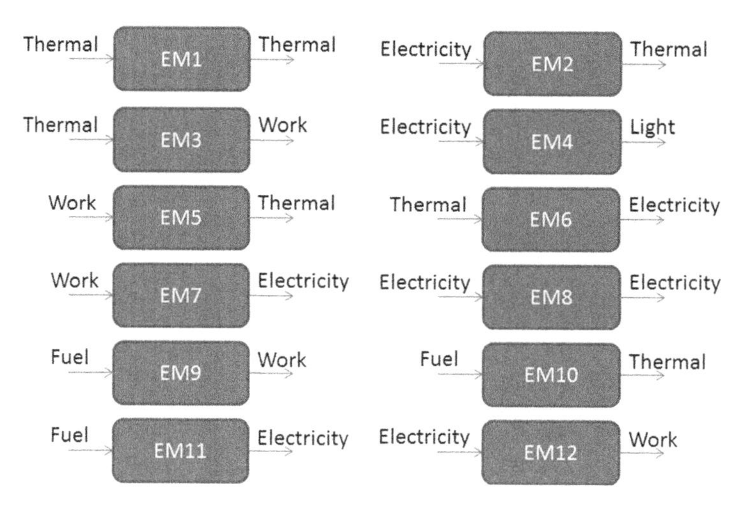

FIG. 3

Modeling of energy node within ESN using POOM.

Robust modeling methodology called POOM [2]. In POOM, performance indicators are defined as part of the behavior view to evaluate the current model at each iteration during design and operation stages. The proposed model formalization methodology, POOM, is based on key features from the commonly known object-oriented modeling methodology used in software engineering. POOM can support process design, operation design, process safety, and design for environment.

The proposed model is implemented in a computer-aided design tool in Visio, and the corresponding ESN knowledge base is implemented in Microsoft database based on POOM and in view of ISA S95. Each building block of the proposed energy network is modeled using POOM, where process variables (inputs, internal, and output) are represented in a systematic way so that key performance indicators (KPIs) can be represented in each detailed building block, such as solar system, wind mill, thermal storage unit, and fuel cell unit and battery banks, as shown in Fig. 3.

2.2.3 ENERGY SEMANTIC NETWORK

It is essential to provide domain knowledge structure with rule-based reasoning to be able to support decisions related to design and operation of the underlying energy production chain. For example, for given equipment, what are the energy sources used with respect to operation and/or behavior. This is essential to link process models with system models. There are several knowledge tools that are available and can support knowledge structuring and management.

A dynamic knowledge structure is established, which is called ESN [3]. ESN is an essential methodology to design an efficient SEG. ESN encodes the knowledge of a wide spectrum of expertise to be available for energy designer. For a complex building with complicated energy demands, the energy conservation measures and sources and loads design parameters are a challenge to identify.

ESN should be open, dynamic, and minimal. Open means its standard structure allows the incorporation of new components. Dynamic means its ability to include new properties of existent components. Minimal is to reduce the computational burden during the simulation and evaluation process. ESN aggregates metadata (structural and descriptive) and data (profile, measured, and statistical) regarding both thermal technologies of building envelope materials and the energy sources technologies, and arranges them in knowledge layers. Moreover, inference mechanism is used to extract decisions and conclusions by navigating through knowledge layers based on the supplied desired requirement.

ESN is the mapping tool between the problem domain and the design domain as shown in Fig. 4. The inputs to ESN are the characters of the site to implement the SEG (type, size, capabilities, activities, governmental constrains, environmental constrains, and customer and stockholder constrains). The output of ESN is the possible scenarios of energy design and energy conservation measures, in addition to a list of KPIs with high potential to be affected by the design structure. The structure of ESN is a collection of nodes representing the classes of both the problem domain instants, objects to the design domain instants, and objects with relation links encode the relation between the classes and objects. To improve the presentation of ESN, a fuzzy inference mechanism will be added for better presentation of uncertainty about both analytic and black-box models. An example of ESN for buildings is shown in Fig. 5.

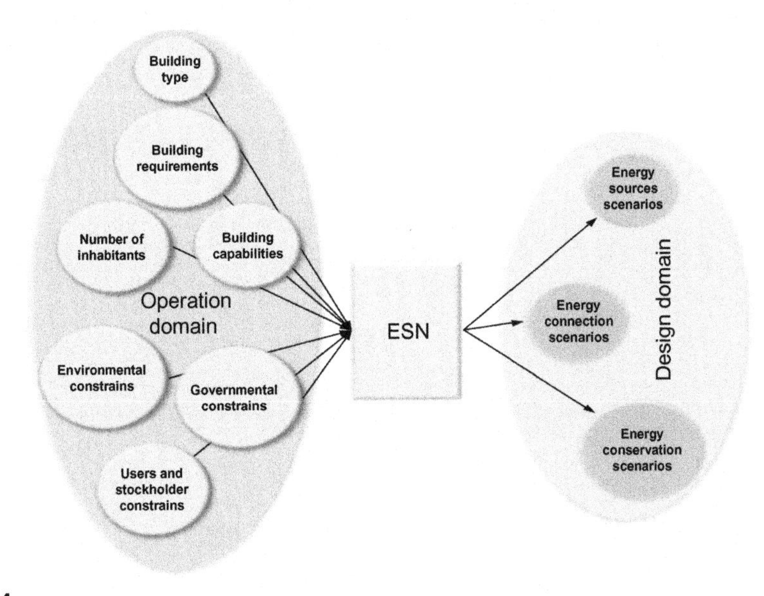

FIG. 4

Design and operation domain knowledge of ESN.

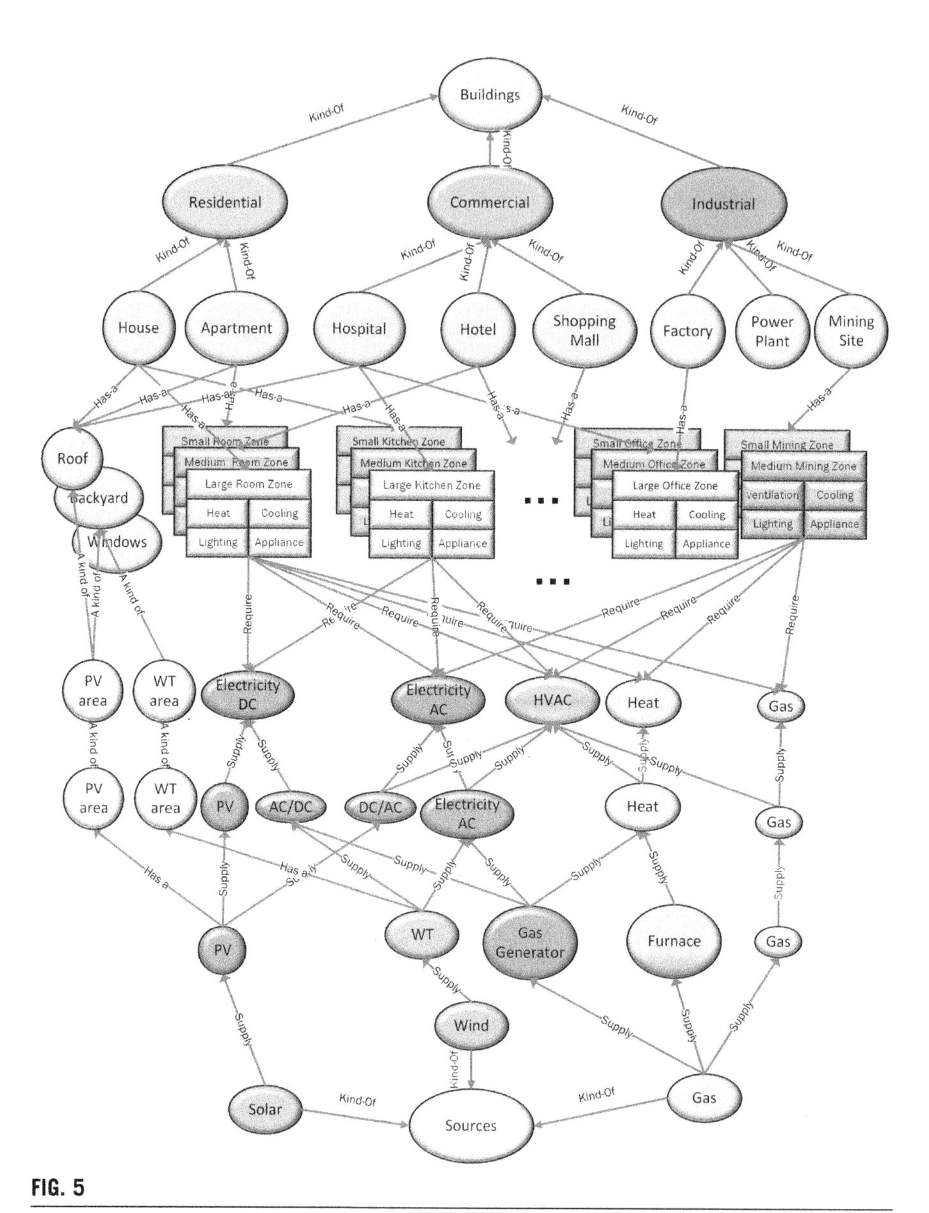

FIG. 5

Energy semantic network (ESN) structure.

2.2.4 MICRO ENERGY GRID ENGINEERING DESIGN

Until now, energy systems have been based on centralized energy generation with transmission and distribution systems; however, with the rapid increase in fuel prices, the capital cost of central generation plants, and electricity/head demand growth, an alternate generating system with higher efficiency of energy use is needed. Micro energy grid (MEG) is a relatively small-scale localized energy network that includes loads, a control system, and a set of energy resources, such as generators and energy storage devices [4]. MEG can operate in a grid-connected mode where energy resources interact with the main electrical grid, or in an islanding mode where an MEG feeds its local loads without the use of the main electrical grid. As an alternative to centralized energy systems, MEGs provide energy locally by utilizing DGs (distributed generation) with minimal energy transmission from/to remote regions. Fig. 6 shows an integrated engineering design of MEG, which shows electricity, thermal, and gas networks with supply, generation, storage, and load components.

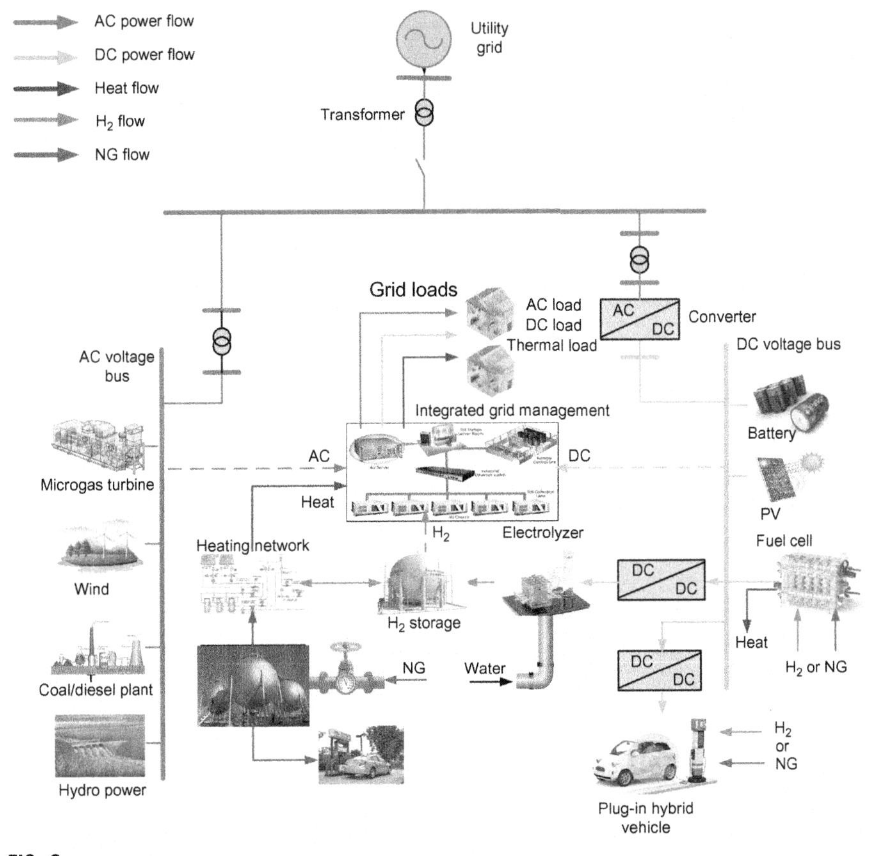

FIG. 6

Typical MEG structure (electricity, heat, hydrogen, and natural gas).

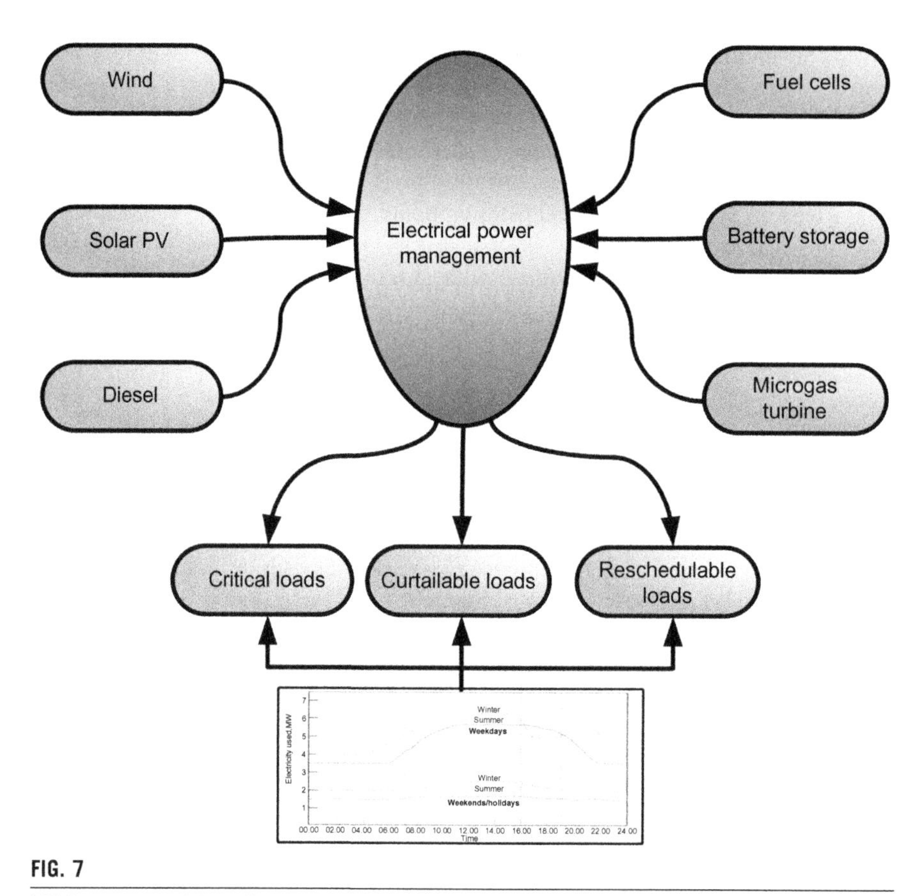

FIG. 7

MEG power model configuration.

MEG power model

In this power model, a simulation library of all energy resources such as wind, PV, microgas turbine, and fuel cells is introduced, taking into account the intermittent nature of these resources, as shown in Fig. 7. The modeling of all variable profiles of electric loads based on a constant power model can be achieved. The modeling is performed using the physics-based (analytic model), data-driven (black-box model) in a combined mode.

MEG thermal model

This module is based on building the ESN of MEG as shown in Fig. 8, for options of thermal network modeling. The ESN model takes into account all aspects of the thermal network modeling, which includes a variety of 73 different working fluids available, including the thermofluidic components such as boiler, pumps, thermal tanks, heat exchangers, condensers, and expansion valves.

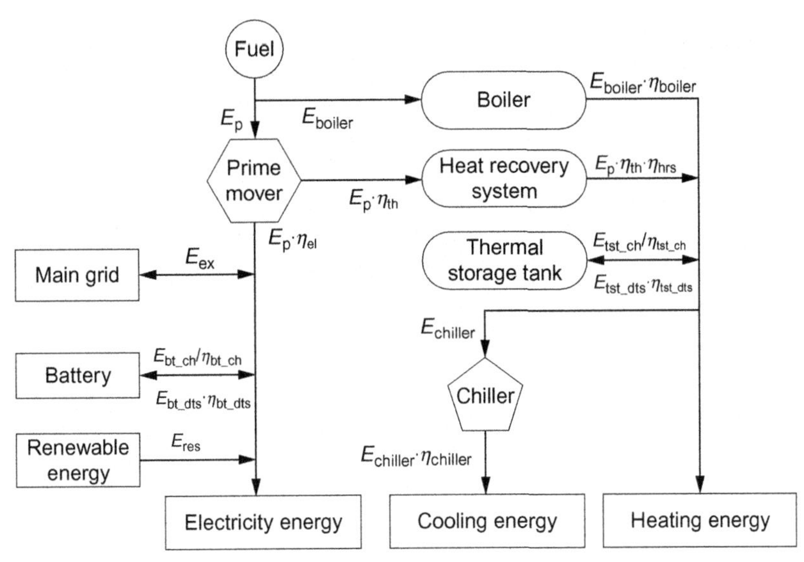

FIG. 8

MEG thermal model configuration.

MEG fuel model

Natural gas and hydrogen are the most clean fuel sources in the new communities. Fig. 9 shows the most generic and high-level representation for the use of fuel within MEGs, where blue lines indicate the electrical uses for MEG, and red lines indicate the thermal uses within MEG.

2.3 FEATURES OF THE SEG

SEG generally refers to modern technology used to bring energy delivery systems into the 21st century using computer-based control and automation. These systems are made possible by two-way communication and computer processing. For a century, utilities sent workers out to gather the needed data to provide energy. Workers read meters, looked for broken equipment, and measured voltage, for example. Most of the devices used have yet to be automated and computerized. Now, many products are being made available to the electricity and energy systems to be modernized. Benefits include enhanced cyber-security, handling sources of electricity like wind and solar power, and even integrating electric vehicles onto the grid.

2.3.1 RELIABILITY

Reliability is a fundamental requirement for energy systems. Reliability of an energy system is the capacity to perform its assigned function adequately and its resistance to failure. For instance, an energy source is considered reliable if it can be used to

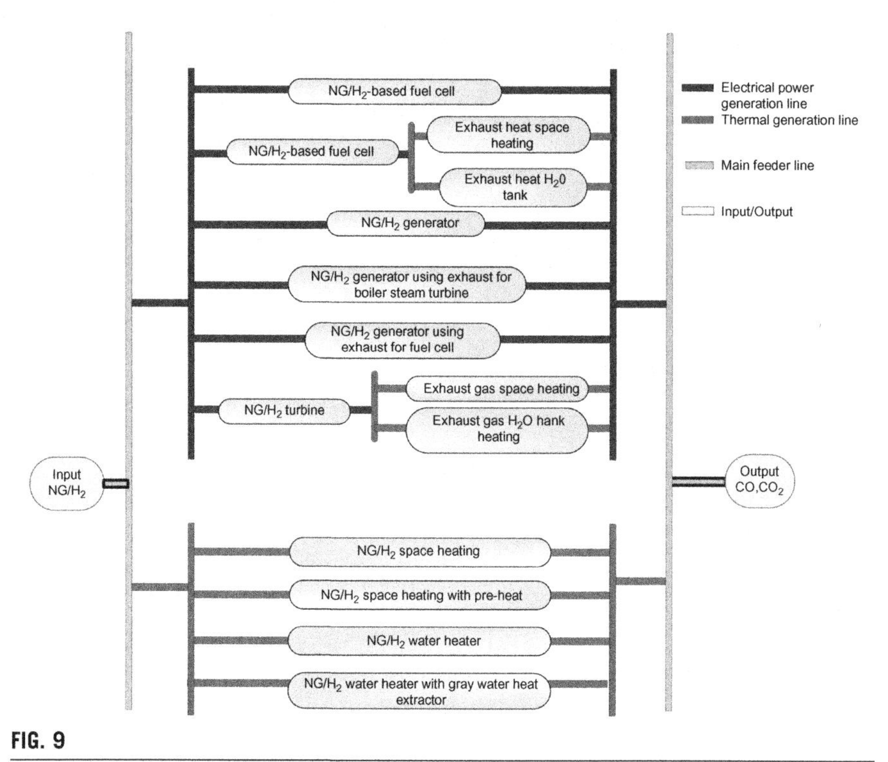

FIG. 9

MEG fuel model configuration.

generate a consistent output and is available to meet predicted peaks in demand. The SEG will make use of technologies, such as stating estimation that improves fault detection and facilitating self-healing of the network without the intervention of human operators [5]. The SEG instantly responds to system problems in order to avoid or mitigate power outages and power quality problems. This will be achieved by performing continuous self-assessment to detect, analyze, respond to, and, as needed, restore the grid components and network sections. It will act as an immune system in order to maintain grid reliability, security, affordability, power quality, and efficiency [5], thus providing a more reliable supply of electricity and energy, and reduced vulnerability to natural disasters or attack.

2.3.2 FLEXIBILITY IN NETWORK TOPOLOGY

Until now, electrical distribution systems have radial structure with unidirectional power flows and a simple protection scheme [6]. Due to the continuous demand growth and the environmental constraints to reduce the gas emissions, however, more DG units are required to be connected at the electrical distribution level, specially renewable ones (ie, solar and wind-based DG units). As the DG penetration level increases, the direction of power flow will be changed from unidirectional to bidirectional, which will raise safety and reliability issues [7]. Under the SEG vision, generation, transmission, and distribution infrastructure will be better able to handle possible bidirectional energy flows. Hence, an SEG aims to manage all these possible issues and allow for flexible network topology.

2.3.3 EFFICIENCY

Deploying SEG technology improves the efficiency of energy infrastructure. For example, some devices such as air conditioners based on a demand-side management program can be turned off during short-term spikes in electricity price, the outage management can be improved using data from advanced metering infrastructure systems, and we can better utilize generators, leading to lower power prices.

Load adjustment and balancing

As a natural characteristic, load demand can vary significantly over time due to the changed activities for customers [8]. Load balancing refers to the use of various techniques by energy system to provide generation that covers the required demand, such as storing excess energy during low-demand periods for release as demand rises. Smart grid technology can provide solutions to the load-balancing problem. For example, many consumers can communicate with the utility using digital means and could be switched on and off by the utility to run at off-peak hours [9]. Demand control can significantly lessen the need to run expensive peaking capacity power generation when there is a high demand for power.

Peak curtailment, leveling, and time-of-use pricing

Time-of-use (TOU) pricing reflects the costs of producing electricity at different times of the day [10]. For example, most utilities have three TOU periods: off-peak, when energy demand is low and less expensive sources of electricity are used; mid-peak, when the cost of energy and demand are moderate; and on-peak, when demand is highest and more expensive forms of electricity production are used. Under SEG vision, homes and businesses can be fitted with smart meters, which wirelessly transmit electricity use to utilities. TOU pricing coupled with smart meters allow each user to adjust his/her electricity use habits. Hence, each user has more control over the monthly electricity bill. This will reduce the amount of spinning reserve that utilities have to keep on stand-by and hence become more energy efficient.

2.3.4 SUSTAINABILITY

A key objective of the SEG is to permit greater penetration of renewable energy sources. Currently, the electrical network infrastructure is not suitable to allow for many distributed feed-in points. Furthermore, fluctuations in distributed generation due to weather conditions present significant challenges to power engineers who need to ensure stable and sustainable power levels. Therefore, smart grid technology is a necessary condition for large amounts of renewable electricity on the grid for the following reasons [11]: (1) varying the output of the controllable, such as gas turbines and hydroelectric generators; (2) using information and communication technology to intelligently integrate actions of the users; and (3) reducing peak loading and using grid capacity efficiently.

2.3.5 MARKET-ENABLING

Smart grid technologies allow for communication among suppliers (ie, energy price) and consumers. Consumers are expected to play an active role. This active participation in the electricity market will benefit the utility, customers, and environment and reduce the cost of electricity. Customers will have the ability to decide on their consumption according to the electricity price. Hence, this will shift the peak demand and help utilities to reduce or clip their peak demands, which in consequence will minimize the capital and operating costs. DG owners will be able to sell energy for maximum profit.

Demand response support

Demand response refers to intentional changes to electricity consumption patterns of end-use customers that alter the timing of energy usage, the level of instantaneous demand at critical times, and consumption patterns in response to market prices, or to incentive payment prices, or when system reliability is jeopardized [12]. Demand response support allows generation utilities and demand customers to interact automatically in real time. Demand response can reduce peak demand, hence it reduces the cost of adding reserve generators, wear and tear, extending the life of generators, and allowing users to lower their energy bills.

Platform for advanced services

SEG will use several technologies such as robust two-way communications, advanced sensors, and distributed computing technology. Therefore, it will improve the efficiency, reliability, and safety of power delivery. Furthermore, new services or at least improvements on existing ones can be included, such as fire monitoring and alarms, phone calls for emergency services, etc.

2.4 TECHNOLOGY

To build an SEG, several technologies need to be developed and implemented, including advanced sensing, communication, routing, protection, and control technologies. These technologies encompass generation, transmission, and distribution systems, as well as consumer appliances and equipment. Most of the SEG technologies are already used in other applications, such as manufacturing and telecommunications and are being adapted for use in SEG operations.

2.4.1 INTEGRATED COMMUNICATION

Two-way communication allows utilities to significantly improve their ability to monitor and control lines and equipment and to enable the participation of the customers [13]. Integrated communications will allow for real-time control, information and data exchange to optimize system reliability, asset utilization, and security, thus achieving more network automation and fewer blackouts. Communication is the key technology in the implementation of the SEG concept. The appropriate design for the physical, data, and network communication is still a topic of debate. Numerous types of communication technologies can be used, such as fiber optics, wireless, and wire line. The harsh and complex power system environment causes significant challenges to the reliability of wireless communications for SEG applications [14].

2.4.2 SENSING AND MEASUREMENT

The sensing and measurement technologies will monitor equipment, support control strategies, evaluate congestion and grid stability, and prevent energy theft. Technologies can include smart microprocessor-based meters, meter-reading equipment, wide-area monitoring systems, dynamic line rating, TOU, and real-time pricing tools, advanced switches, conductors, and cables, and digital protective relays.

Smart meters

SEG will replace analog mechanical meters with digital meters that record usage in real time. These digital meters, which are installed at all customer service locations, will have two-way communications to get the data back to the utility (ie, wires, fiber, Wi-Fi, cellular, or power line carrier). This technology is referred to as advanced metering infrastructure (AMI), which allows utilities to detect outages, accurately perform load forecasting, manage customer loads in real time, enhance the distribution system optimization, and eliminate the need to access customer property [15].

Phasor measurement units

Phasor measurement units (PMUs) are high-speed sensors that measure voltage and current synchrophasors of the power system with the accuracy in the order of one microsecond, which is much faster than the speed of existing SCADA technologies [16]. With large numbers of PMUs and the ability to compare voltage phase angles at key points on the grid, automated systems may be able to revolutionize the management of power systems by responding to system conditions in a rapid and dynamic fashion [17].

2.4.3 OTHER ADVANCED COMPONENTS

Innovated components can positively change fundamental abilities and characteristics of grids. Technologies within these broad R&D categories include: advanced power components, such as advanced power electronic devices, super-conductor-based equipment, distributed energy generation and storage devices, plug-in hybrid vehicles, smart houses, web services, grid computing, fault tolerance, and accurate weather forecasting technique.

Distributed power/gas flow control

Power flow control devices installed on power lines can control the flow of power within. These devices support greater use of renewable energy by providing more consistent, real-time control over how that energy is routed within the SEG [18]. Furthermore, these devices enable the SEG to store intermittent energy from renewables for later use.

Smart energy generation using advanced components

Smart energy generation is designed to match energy (ie, electricity and heat) production with demand using multiple generators that can start, stop, and operate efficiently based on the required demand. The SEG will accommodate a wide variety of generation options, especially the renewable energy-based DG units. Increasing DG penetration level will be beneficial in reducing the capital investment in generation and transmission. Also, integrating more renewable DG units will benefit the environment by reducing the impact of the fossil-based electricity generation.

2.4.4 ADVANCED CONTROL

Power system automation enables rapid diagnosis of and precise solutions to specific grid disruptions or outages. Advanced control methods include: (1) multiagent control systems to provide the decentralized/distributed and online control of the grid components instead of the current central fashion; (2) analytical tools such as software algorithms and high-speed computers; and (3) operational applications such as SCADA, substation automation, and demand response.

2.4.5 IMPROVED INTERFACES AND DECISION SUPPORT

This technology includes information systems that can reduce complexity and computational times to effectively and efficiently operate a grid with an increasing number of variables. Information technologies include visualization techniques to reduce large quantities of data and represent them into easily understood visual formats, software systems that provide multiple options when systems operator actions are required, and simulators for operational training and what-if analysis.

2.5 KPI MODELING OF SEGs AND MEGs

The SEG and/or MEG simulation framework starts with model library development for all potential DERs that can be utilized within SEG/MEG. This is followed by collecting demand requirements of thermal, fuel, and power within SEG/MEG. Based on the demand requirements, possible SEG/MEG design and configuration alternatives will be synthesized for thermal, fuel, and power systems. For each design and configuration scenario, KPIs will be estimated for each component and SEG/MEG level and will be evaluated and optimized using an SEG/MEG simulation engine that is developed in MATLAB.

It is challenging to find a unique DER technology that satisfies multiple considerations such as economic, technical, and environmental attributes. Thus, as shown in Fig. 10, a multiattribute decision approach can be used to assess the alternatives for DER technologies with respect to their attributes. Furthermore, a regional primary energy attribute can be included to reflect the potential of various types of energy sources in different regions. For example, Ontario relies on nuclear power generation to meet base load. Evaluation attributes (KPIs) are metrics that can be used to evaluate generation scenarios. They represent the cornerstone of evaluation algorithms and allow for quantitative and qualitative comparisons among different mixtures of gas and power technologies.

2.6 MODELING AND SIMULATION

Synchronization between real-time grid data and a simulation engine is important in two ways: (a) start simulation scenarios with initial conditions using real-time data from the grid and (b) fine tune simulation models using real-time data. This is achieved by building an intelligent interface program that captures real-time data from the grid and map to the state variables in the simulation engine, which will tune simulation models using genetic programming with limited iteration to reduce the error between simulation results and real time. Also, it is possible to predict future status of the grid loads/demands by using a smaller time step of the simulation starting from current condition of the grid as the initial condition. This will provide an accurate estimation of loads/demands and identify/plan best ways of recovery operation.

For supporting decision makers, intelligent algorithms can be developed to convert simulation data into qualitative models with symbolic representation that will support decision making. In terms of the simulation engine, and for interoperability purposes to support the wider range of commercialization, the simulation interface can be developed to interface with major simulation engines available in the market such as DIgSILENT Power Factory, CYMDIST, ETAP, Paladin Design, EMTP/PSCAD, Sim Power Systems, and PSS1.

2.6.1 SEG INFRASTRUCTURE MODELING FRAMEWORK

The modeling and simulation environment can allow hierarchical and network levels of abstraction and specifications of SEG components (ie, electricity grid as connected to other networks like thermal and gas networks). Each component can be linked with a knowledge base for different model components, such as business process, regulations, power, energy, asset integrity, physical topology, and

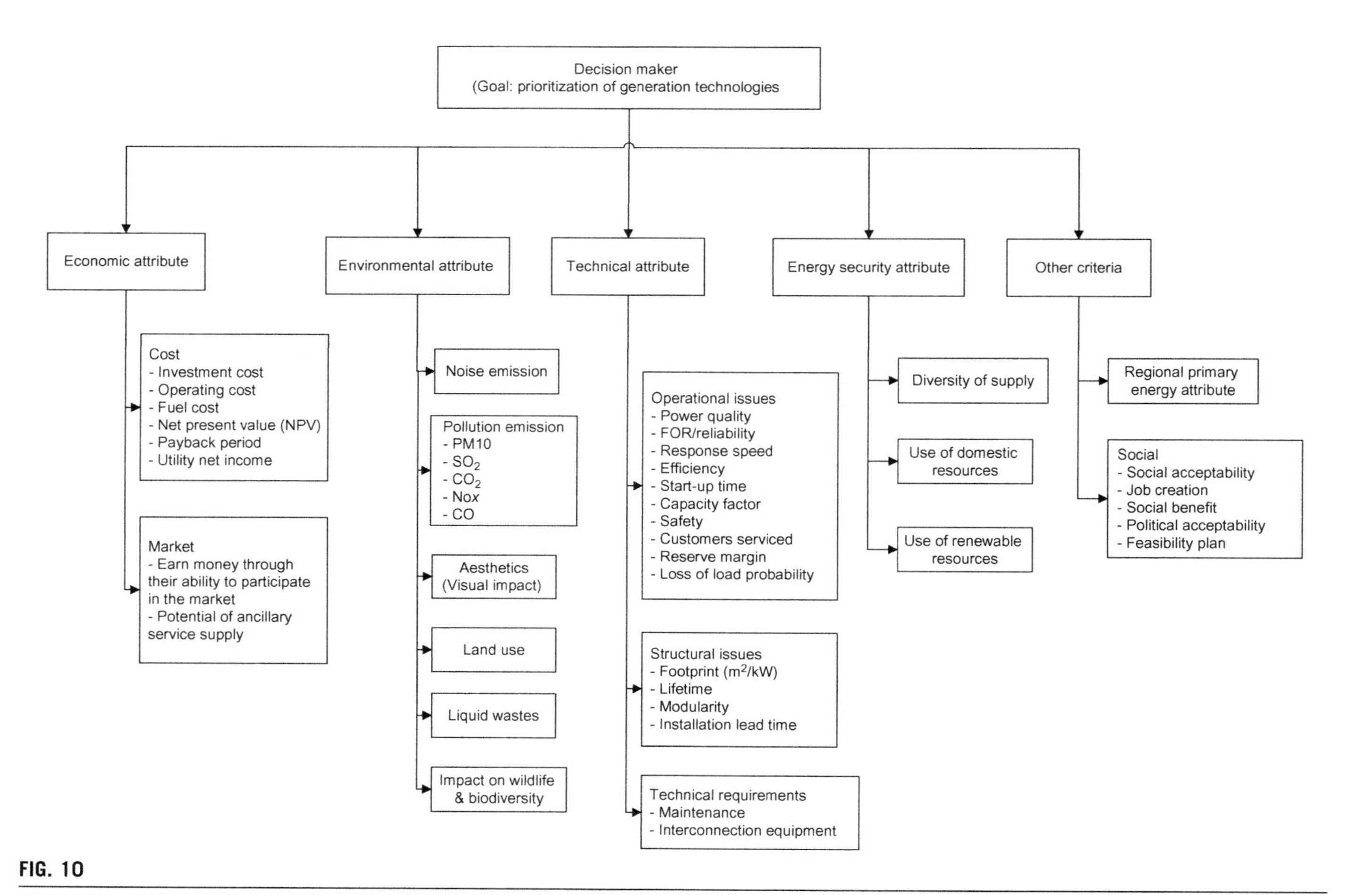

FIG. 10

Hierarchical KPI decision-making structure to prioritize generation technologies in SEG/MEGs.

geographical/environmental information. Utilities can build the detailed grid model and dynamically tune with different design and operational alternatives.

Fig. 11 and Table 2 show the typical infrastructure of SEGs with distributed generation/storage, and information technology infrastructure to integrate PMUs, which are implemented in regional power grids.

2.6.2 INTERCONNECTED MEGs MODELING AND SIMULATION

MEG of building, facility, community, and transportation, has two levels of interconnection: local and regional interconnection. Local interconnection is interconnection between different components of MEG, including energy sources, energy transformers, and loads. The local interconnection controller is responsible for monitoring the load demand and dynamically connecting the load-to-energy sources to satisfy the optimal value of MEG KPIs.

Regional interconnection is interconnection with other nearby MEGs and/or the national utility grid. In the next few years, integration of many MEGs inside the same distribution network will be the main target of improving the system performance. The interconnection of multiple MEGs with each other by a private line based on their dynamic profile of both demand and generation will highly improve the overall MEGs system performance; however, this requires supervisory control to monitor and control the synchronization of local and regional sources. With the development of cloud computing, storage, and Internet access capabilities, the remote online sensing and control of multiple MEG sites become feasible, as shown in Fig. 12.

2.7 SAFETY AND PROTECTION OF SEG

Safety analysis or safety design concept is important in SEG protection and their fault analysis. A proper safety model provides the appropriate level of confidence in protection system. The intelligent control and monitoring unit needs to meet the safety requirements, which provide a basis of safety design criteria. Central control and protection systems should be designed to ensure required safety.

2.7.1 FAULT SEMANTIC NETWORK

The concept of semantic network was first proposed as a network structure that represents relations between concepts. The concept of semantic network was further developed where semantic network was developed in a tree structure consisting of nodes and arcs. The nodes represent concepts, and the connections show relations between nodes. The fault semantic network (FSN) was originally realized as a means of representing fault knowledge based on relationships among fault objects as mapped to physical systems. In FSN, the nodes correspond to different faults/causes/consequences, and the links between them describe the dependencies. Initially, FSN is constructed based on ontology structure of fault models on the basis

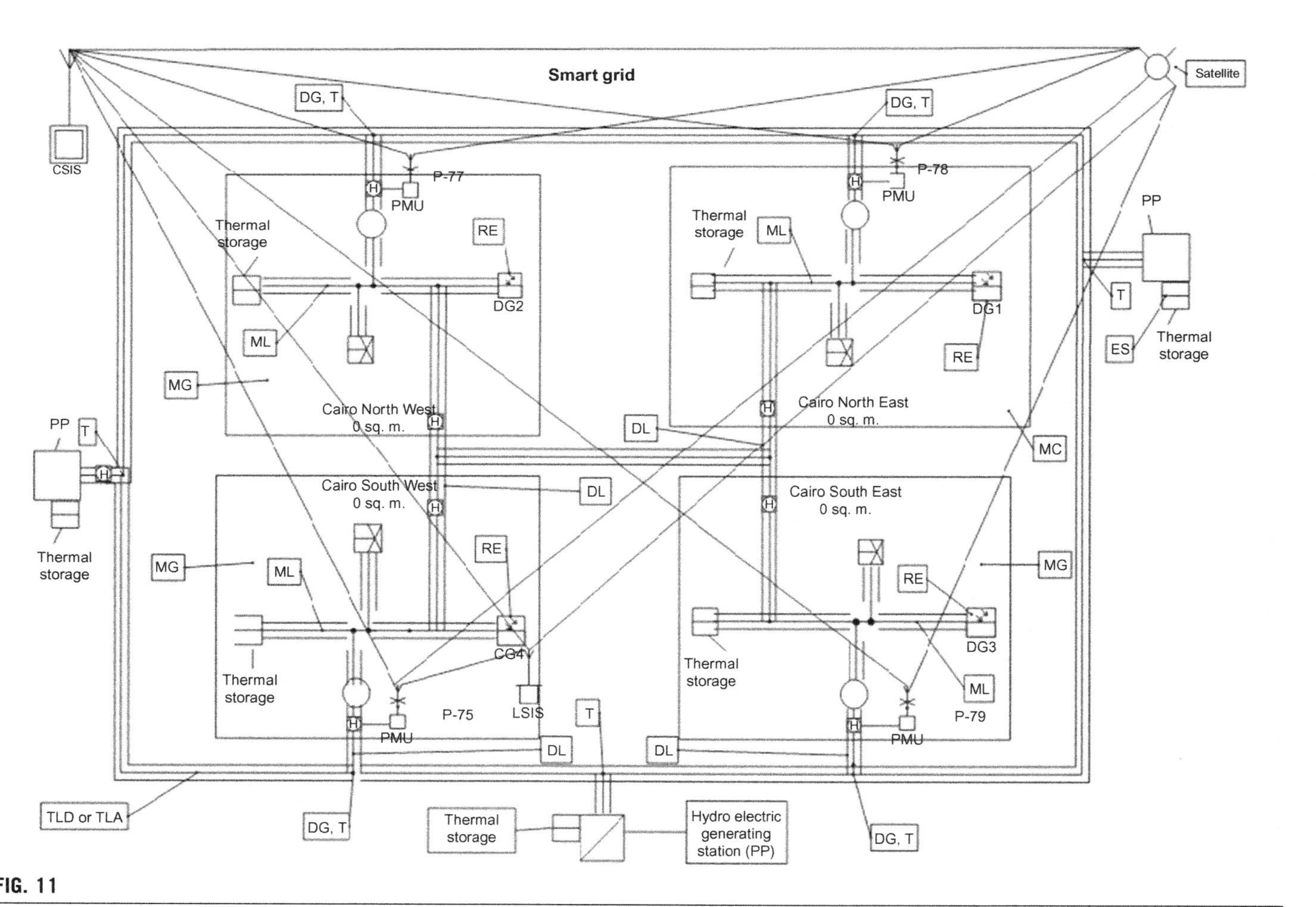

FIG. 11

Smart energy grid infrastructure editor.

Table 2 Smart Energy Grid Infrastructure

TLD	High-voltage direct-current line (HVDC)	T	Transformer
TLA	High-voltage alternating current line (HVAC)	RE	Renewable sources such as Solar PV, wind farms, etc.
DL	Low-voltage distribution line	ES	Energy storage
DG	Distribution grid/substation	CU	Fueling/charging unit (gas/PHEV)
MG	Microgrid	DLM	Medium-voltage distribution line
ML	Microgrid lines	DLL	Low-voltage distribution line
PP	Power plant such as nuclear, hydro, coal, etc.		

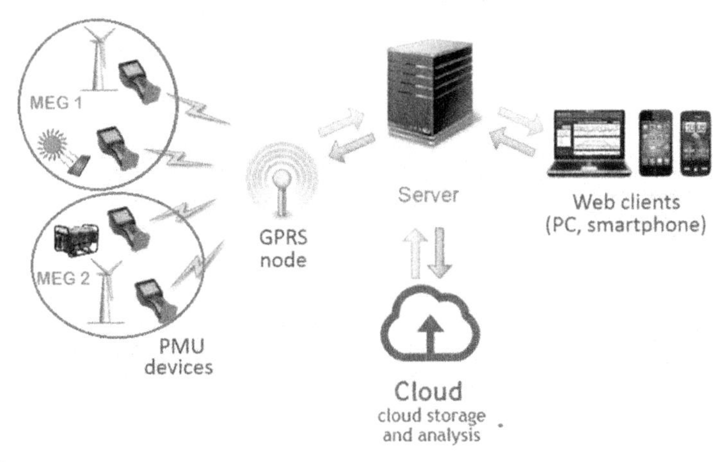

FIG. 12

Operation of interconnected MEGs.

of POOM where failure mode is described using symptoms, enablers, variables, causes, consequences, and repair actions. Quantitative (probabilistic) or qualitative rules are associated with each transition of the causation model within FSN.

2.7.2 FRAMEWORK FOR SAFETY AND PROTECTION OF SEG

The safety and protection framework is based on identifying all possible fault propagation scenarios using FSN, and estimating dynamic risks associated with each fault propagation scenario while mapping independent protection layers within each fault propagation scenario. FSN will be dynamically tuned with real-time data using computational intelligence algorithms, as shown in Fig. 13.

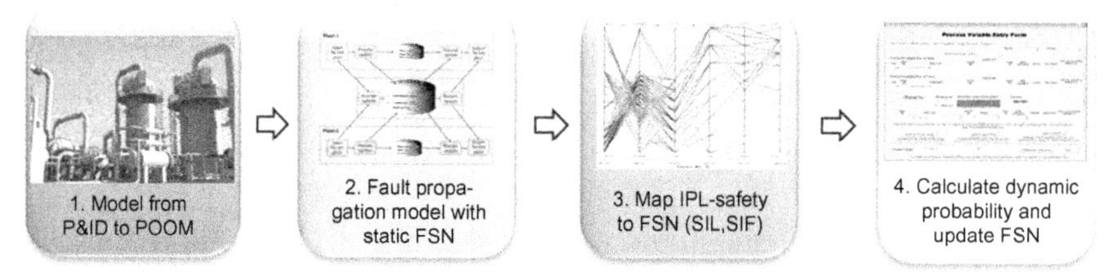

FIG. 13

Framework for FSN-based safety and protection design of smart energy grids.

2.8 CHALLENGES IN SEG IMPLEMENTATION

The implementation of the SEG faces many challenges. These can be categorized as procedural and technical challenges [19]. The procedural challenges range from the complexity of the SEG to unified SEG standards. Furthermore, due to the complexity and the large scale of real-power systems, the implementation of the SEG has to be gradual not to affect the operation of the existing system. The technical challenges present the need for new intelligent devices or modifying the currently used devices with smart capabilities.

Furthermore, SEG is different from traditional electric power systems, which only supply electrical loads as both electrical and heat loads may exist due to the implementation of CHP plants. A two-dimensional model of electricity and heat flows should be developed for SEGs with CHP plants. Moreover, the differences in the storage and charging/discharging characteristics of electricity and heat buffers should be investigated in SEGs modeling and optimization.

REFERENCES

[1] Gabbar HA. Engineering design for hybrid green energy production and supply chains. J Environ Modell Softw 2009;24:423–35.

[2] Gabbar HA, Chung PWH, Suzuki K, Shimada Y. Utilization of unified modeling language (UML) to represent the artefacts of the plant design model. In: International symposium on design, operation and control of next generation chemical plants (PSE Asia 2000), PS54, Japan; 2000. p. 387–92.

[3] Gabbar HA, Eldessouky AS. Energy semantic network for building energy management. Intell Indust Syst 2015;1(3):213–31.

[4] Zidan A, Gabbar HA, Eldessouky A. Optimal planning of combined heat and power systems within microgrids. Energy 2015;93:235–44. Part 1.

[5] Zidan A, El-Saadany EF. A cooperative multiagent framework for self-healing mechanisms in distribution systems. IEEE Trans Smart Grid 2012;3(3):1525–39.

[6] Zidan A, El-Saadany EF. Effect of network configuration on maximum loadability and maximum allowable DG penetration in distribution systems. In: IEEE electrical power & energy conference (EPEC); 2013. p. 1–6.

[7] Coster EJ, Myrzik JMA, Kruimer B, Kling WL. Integration issues of distributed generation in distribution grids. Proc IEEE 2011;99(1):28–39.

[8] Zidan A, El-Saadany EF. Incorporating customers' reliability requirements and interruption characteristics in service restoration plans for distribution systems. Energy 2015;87:192–200.

[9] Sinitsyn NA, Kundu S, Backhaus S. Safe protocols for generating power pulses with heterogeneous populations of thermostatically controlled loads. Energ Convers Manage 2013;67:297–308.

[10] Yang P, Tang G, Nehorai A. A game-theoretic approach for optimal time-of-use electricity pricing. IEEE Trans Power Syst 2013;28(2):884–92.

[11] Veldman E, Gibescu M, Slootweg HJG, Kling WL. Scenario-based modelling of future residential electricity demands and assessing their impact on distribution grids. Energy Policy 2013;56:233–47.

[12] Li S, Zhang D, Roget AB, O'Neill Z. Integrating home energy simulation and dynamic electricity price for demand response study. IEEE Trans Smart Grid 2014;5(2):779–88.

[13] Lu X, Wang W, Ma J. An empirical study of communication infrastructures towards the smart grid: design, implementation, and evaluation. IEEE Trans Smart Grid 2013;4(1):170–83.

[14] Gungor V, Lu B, Hancke G. Opportunities and challenges of wireless sensor networks in smart grid. IEEE Trans Indust Electron 2010;57(10):3557–64.

[15] Zhou J, Hu RQ, Qian Y. Scalable distributed communication architectures to support advanced metering infrastructure in smart grid. IEEE T Parall Distr Syst 2012;23(9):1632–42.

[16] Azizi S, Gharehpetian GB, Dobakhshari AS. Optimal integration of phasor measurement units in power systems considering conventional measurements. IEEE Trans Smart Grid 2013;4(2):1113–21.

[17] Vide PSC, Barbosa FPM, Barbosa FPM, Ferreira IM. State estimation model including synchronized phasor measurements. In: Proceedings of 46th International Universities' power engineering conference; 2011. p. 1–6.

[18] Available from [Online]: http://tdworld.com/test-monitor-control/distributed-power-flow-control-devices-integrated-georgia-power-s-ems. (accessed January 27, 2016).

[19] EPRI, Estimating the costs and benefits of the smart grid: a preliminary estimate of the investment requirements and the resultant benefits of a fully functioning smart grid, Technical report, March 2011.

CHAPTER

3 Optimal sizing and placement of smart-grid-enabling technologies for maximizing renewable integration

S.F. Santos*, D.Z. Fitiwi*, M. Shafie-khah*,†, A.W. Bizuayehu*, J.P.S. Catalão*,‡,§

University of Beira Interior, Covilhã, Portugal University of Salerno, Salerno, Italy† University of Porto, Porto, Portugal‡ University of Lisbon, Lisbon, Portugal§*

3.1 INTRODUCTION

Nowadays, the issue of integrating distributed generations (DGs; and renewable energy sources (RESs), in particular) is globally gaining momentum because of several techno-economic and environmental factors. In recent years, the size of DGs integrated into distribution systems has been increasing. And this trend is more likely to continue in the years to come, because it is now widely accepted that DGs bring wide-range benefits to the system in general; however, given the current setup of distribution networks (which are generally passive), large-scale DG integration is not technically possible, because this brings about tremendous challenges to the system operation, especially in undermining the power system quality and stability. Such challenges/limitations are expected to be alleviated when distribution networks undergo the anticipated evolutionary process from passive to active networks or smart-grids. This transition is expected to result in a system that is adequately equipped with appropriate technologies, state-of-the-art solutions, and a new operational philosophy totally different from the current "fit and forget" approach. And this is expected to offer sufficient flexibility and control mechanism to the system. Nevertheless, the process is not straightforward, as it demands exceptionally huge investments in smart-grid technologies and concepts to fully automate the system, and this should be accompanied by a new operational philosophy. Therefore, the whole transformation process, that is, the transformation of current distribution systems to full-scale smart-grids, might be very slow, if not impossible, and its realization might take several decades.

Given the techno-economic factors and global concerns about environmental issues, however, the integration of RESs cannot be postponed. It is likely that the

Smart Energy Grid Engineering.

integration of DGs in distribution systems will go ahead, along with smart-grid-enabling technologies that have the capability to alleviate the negative consequences of large-scale integration of DGs. In other words, in order to facilitate (speed up) the much-needed transformation of conventional (passive) distribution network systems (DNSs) and support large-scale RES integration, different smart-grid-enabling technologies, such as reactive power compensators, advanced switching, and storage devices are expected to be massively deployed in the near term. To this end, developing strategies, methods, and tools to maximize the penetration level of DGs (particularly, RESs) have become crucial to guide such a complex decision-making process. In this respect, this work focuses on the development of multistage mathematical models to determine the optimal sizing, time, and placement of energy storage systems (ESSs) and compensators, as well as that of RESs in distribution networks. The ultimate goal of this optimization work is to maximize the DG power absorbed by the system at a minimal cost while maintaining the power quality and stability at the required/standard levels.

3.2 STATE-OF-THE-ART LITERATURE REVIEW

Reducing fossil fuel dependence and mitigating climate change has led to increased pressure to change the current generation paradigm. It is expected that CO_2 emissions will increase from approximately 31 billion metric tons to 36 billion metric tons in 2020, reaching 45 billion metric tons by mid-2040 [1], an increase of 46%. Other associated concerns are an increase in global average temperature from 1°C to 5°C by the year 2100, increasing the average level of the sea water [2]. Global population will also increase, expected to be 9.6 billion in 2050, along with an increase in energy consumption by 56% between 2010 and 2040 [1]. The compounded effect of all these problems and challenges is triggering a policy shift all over the world, especially when it comes to energy production. Integration of DG, particularly RESs, in electric DNSs is gaining momentum. It is highly expected that large-scale DG integration will be one of the solutions capable of mitigating the aforementioned problems and overcoming the challenges. Because of this, governments of various nations have introduced targets to achieve large-scale integration of DGs. In particular, in the European Union, which strongly advocates the importance of integrating DGs (especially, renewables), that is expected to grow by 20% until 2020 and 50% of energy consumption by 2050.

Power distribution networks form a critical base to system reliability, power quality, and also energy cost [3,4]. One way of making the distribution system "less critical" is through the integration of DG systems that are small power sources connected near the end users. DG offers a more environmentally friendly option through great opportunities with renewable-enabling technologies such as wind, photovoltaic (PV), biomass, etc. RESs are abundant in nature, which leads to the attraction of the large-scale power generation sector. Nevertheless, there is no rule or partial rule on the DG unit's connection; typically, these are connected at the end of radial feeder systems or nodes with greater load on the distribution system. The size of DGs can vary from a few kilowatts to several megawatts depending on the voltage system level to which they are connected. The optimal planning of the DG unit's placement and sizing will become extremely important for energy producers, consumers, and network operators in technical and economic terms in the near future. There are many studies in the literature on this topic, yet most of them only consider the optimal location of a single DG unit or do not consider simultaneously positioning and sizing units, mainly due to the high dispatch unpredictability of these. The increase in DGs' penetration increases the uncertainty and the fluctuations of the system production. If the placement and proper sizing are not taken into account, the benefits of DG integration can be lost in efficiency losses, increasing the electricity cost and leading to energy losses.

Another major concern with the wide DG penetration is system reliability. In this paradigm, the use of ESSs has been seen as one of the viable options to mitigate the aforementioned concerns. The penetration of distributed systems can result in the degradation of power quality, particularly in cases of slightly meshed networks [5] or microgrids. Electricity production fluctuations can create voltage oscillations in a frequency range between 1 and 10 Hz. One possible approach for reducing voltage fluctuations in microgrids or slightly meshed networks is through a specific frequency damping; yet the use of ESS is required for a range of frequencies. Several ESS technologies are emerging, especially for demanding cases of charge and quick dispatch cycles; however, the smooth integration of ESSs in the grid requires power electronics-based interfaces. Normally, the ESS is connected at the RES coupling point. To perform energy smoothing, a comparison between the attenuation of fast power variation and regulation of the state of charge should be made. The latter is necessary to maximize the ability to deliver energy [6].

The DG allocation and sizing subject have received special interest from researchers in recent years, as shown in Ref. [7], a review on the subject until 2013. In Refs. [7,8], an analysis of several innovative techniques used on the DG impact investigation in the electrical system is presented. Most of these techniques analyze the distribution system to determine rules that can be used for DG integration [9–13]. Important issues related to the connection of DG units are the network topology, DG capacity, and suitable location, because each bus in the system has an optimal level of DG integration. And if the value surpasses this level, system losses can increase [14,15]. There are many ways proposed to formulate and analyze the optimal allocation of multiple DG units in a radial or meshed network.

Recently, several methods have been proposed for planning and operation or in some cases for both location and sizing of DGs in the distribution system. In general, these methods can be classified as heuristic-based [16–31], numerical-based [32–39], and analytical-based [40,41] methods.

Heuristic-based methods apply advanced artificial intelligence algorithms, such as genetic algorithms (GAs) [16–19], particle swarm optimization (PSO) [20–24], harmony search (HS) [25,26], and big bang crunch (BBC) [27–29]. In Ref. [16], GA is used to solve the expansion planning problem considering DG, uncertainties, and reliability in normal operating conditions. Another approach widely used with GA is the nondominated sorting genetic algorithm (NSGA-II) [17]. Another work that also uses the NSGA-II is Ref. [18], where a multiobjective integration approach is used for DG and ESS in distribution systems. A comparison between mixed-integer programming and GA methods for DG planning is presented in Ref. [19]. PSO is another method used, whether in its original version or an improved version, such as multiobjective particle swarm optimization (MOPSO) [20] or hybrid multi-objective particle swarm optimization [21]. In Ref. [22], a PSO algorithm was used to solve a distribution system expansion planning problem, considering ESSs, and DG systems. The work presented in Ref. [23] investigated the impact of ESSs in the distribution system multistage expansion planning problem, being formulated as an optimization problem and solved using PSO. This work indicates a positive impact of the ESS on the performance and costs associated with the network. A MOPSO approach is proposed in Ref. [21], to minimize the power system cost and improve the system voltage profiles by searching siting and sizing of storage units under consideration of uncertainties in wind power production. Authors in Ref. [24] presented a new approach to optimize the allocation and sizing of several DG units based on the maximization of loading systems using hybrid particle swarm optimization. HS is another heuristic-based methodology widely used, as in Ref. [25], which minimizes energy losses by distribution systems' reconfiguration in the presence of DG. Another work that uses HS is Ref. [26] where an improved multiobjective harmony search is used in order to obtain the optimal location of DGs in the distribution system. One method that has been commonly used is the BBC, which is a method based on the evolution of the universe that has been applied to solving the problems of DG placement and sizing in the distribution system. In particular [27], where the hybrid big bang crunch (HBBC), is used for reconfiguration and optimal allocation of DG in the distribution system. The work in Ref. [28] proposes an algorithm for modeling stochastically renewable-based DGs with the purpose of planning an unbalanced distribution network; the BBC algorithm is used to perform optimal DG placement. Authors in Ref. [29] use a modified BBC method to deal with the optimization problems incorporating multiple distributed generators for the sake of power, as well as energy loss minimization in balanced/unbalanced distribution systems. Two other algorithms that also have been used are water drop (WD) and fireworks (FW) algorithm. The intelligent algorithm WD [30] is used for the DG allocation and sizing, with the goal of minimizing energy losses and improving the voltage profile. In Ref. [31], authors present DG optimal allocation and distribution system reconfiguration, in order to minimize energy losses and voltage stability using FW.

Numerical methods are algorithms that seek numerical results for different problems, in particular to the problem in question. Some of the most recent works use mixed-integer nonlinear programming (MINLP) [32–34], mixed integer linear programming (MILP) [35,36], quadratic programming [37], and optimal power flow (OPF) [38]. Ref. [32] shows the allocation of DG using a decision made by the system approach planner (DMSP) based on the utilities and customer aspects under a deregulated environment; the problem uses OPFs' formulation and is solved with MINLP. In Ref. [33], one MINLP algorithm is presented to solve the optimal placement and sizing of DG with the goal of improving the voltage stability margin in the distribution system. A planning MINLP algorithm on a statistical basis is proposed in Ref. [34] to determine the optimal generation mix of different renewable DG unit types in an annual base to minimize the energy losses in the distribution system. One other approach within the numerical methods is through MILP algorithms. The authors of Ref. [35] focus on the problem in the optimal configuration/design of distributed resources, produced outside of buildings and sent to these through the distribution networks. The model provides the simultaneous optimal locations (ie, the place of production) as well as synthesizes (type, capacity, and number of equipment) and operational strategies for the entire system through an MILP model. An MILP algorithm is also used in Ref. [36], in a two-stage stochastic model multiperiod. The work in Ref. [37] presents a simultaneous optimization of ESS and DG in microgrids, and is solved by a nonsequential quadratic programming algorithm. The optimal installation of DG technologies to minimize energy losses in the distribution system is presented in Ref. [38], using an efficient analytical (EA) algorithm integrated with OPF algorithm, a new method EA/OPF. The DG unit planning in the distribution system is presented in Ref. [39] using a hierarchical agglomerative clustering algorithm.

The exhaustive search methods are based on the search for the optimal DG location for a given DG size under different load models. Therefore, these methods fail to represent accurately the DG optimization problem behavior involving two continuous variables, both for optimal DG size and location. In Ref. [40], authors present one technique with a probabilistic basis for determining the capacity and optimal placement of wind DG units to minimize energy loss in the distribution system. A sensitivity algorithm is presented in Ref. [41] for DG placement and sizing in the network.

Despite the many studies in the literature on areas related to DG placement and sizing problem, most of them only consider the optimal location of a single DG unit, mostly of conventional DGs. The simultaneous consideration of the placement, timing, and sizing of DG units (especially RESs), along with the placement, timing, and sizing of smart-grid-enabling technologies, seems to be far from being addressed in the literature. The increase in RES-based DG penetration escalates the uncertainty and fluctuations of the system production. If the placement and proper sizing are not taken into account, the benefits of DG integration may not be exploited; instead, this may result in the degradation of system efficiency, increased cost of electricity, and energy losses. Another major concern with the wide-range DG penetration is system reliability; however, the simultaneous investment planning of DGs, ESSs, and compensators is expected to significantly alleviate these challenges and increase the penetration level of RES-based DGs.

3.3 OBJECTIVES

The main objectives of this work are

- to develop a new joint multistage mathematical optimization model considering smart-grid-enabling technologies, such as ESS, compensators, and network switching and/or expansion to support DG integration.

- to determine the optimal sizing, time, and placement of ESSs and compensators as well as that of RESs in distribution networks. The ultimate goal of this optimization work is to maximize the DG power absorbed by the system at a minimal cost while maintaining the power quality and stability at the required/standard levels.
- to carry out case studies and test the developed model.
- to analyze simulation results and disseminate research outcomes.

3.4 MATHEMATICAL FORMULATION OF THE PROBLEM

3.4.1 INTRODUCTION

As mentioned earlier, the work here develops an integrated optimization model that simultaneously finds the optimal locations and sizes of installed DG power (particularly focusing on wind and solar), ESSs, and capacitor banks. The optimal deployment of the aforementioned enabling technologies should inherently meet the goal of maximizing the renewable power integrated/absorbed into the system. The entire model is formulated as a stochastic mixed-integer linear programming optimization. In addition, instead of the customary DC network models, a linearized alternating current (AC) model is used in the formulation to better capture the inherent characteristics of the network system.

3.4.2 OBJECTIVE FUNCTION

As mentioned earlier, the objective of this work is to maximize RES integration in DNS from the system perspective (or, from the distribution system operators' (DSOs') point of view) by optimally deploying different smart-grid-enabling technologies at a minimal cost. Here, it is assumed that the DSO owns some generation sources and ESSs.

The resulting problem is formulated as a multiobjective stochastic MILP with two objectives: maximization of integrated RES energy, as in Eq. (1a) and overall cost minimization in Eq. (1b). The problem can be considered as a minimax optimization; however, the first objective can be considered to be redundant if the cost of RES energy (tariff) is very small or the RESs are prioritized when carrying out the dispatch in the system (as is the case in most power systems). This is because, in such cases, the generated RES power will be fully integrated as far as this maintains the power quality and stability at the required/standard levels. In this work, only Eq. (1b) is considered. The objective function in Eq. (1b) is composed of net present value (NPV) of five cost terms, each weighted by a certain relevance factor $\alpha_j; \forall j \in \{1, 2, \ldots, 5\}$. Note that, in this work, all cost terms are assumed to be equally important; hence, these factors are set to 1; however, depending on the relative importance of the considered costs, different weights can be adopted in the objective function. The first term in Eq. (1b) TInvC, represents the total investment costs under the assumption of perpetual planning horizon [42]. In other words, "the investment cost is amortized in annual installments throughout the lifetime of the installed component," as is done in Ref. [43]. Here, the total investment cost is the sum of investment costs of new and existing DGs, feeders, ESS, and capacitor banks, as in Eq. (2).

The second term TMC, in Eq. (1b) denotes the total maintenance costs, which is given by the sum of individual maintenance costs of new and existing DGs, feeders, ESS, and capacitor banks in the system at each stage and the corresponding costs incurred after the last planning stage, as in Eq. (3). Note that the latter costs depend on the maintenance costs of the last planning stage. Here, a perpetual planning horizon is assumed. The third term TEC, in Eq. (1b), refers to the total cost of energy in the system, which is the sum of the cost of power produced by new and existing DGs, purchased from upstream, and supplied by ESS at each stage, as in Eq. (4). Eq. (4) also includes the total energy costs incurred after the last planning stage under a perpetual planning horizon. These depend on the energy costs of the last planning stage. The fourth term TENSC, represents the total cost of unserved power in the system and is calculated as in Eq. (5). The last term TImiC, gathers the total emission costs in the system, given by the sum of emission costs for the existing and new DGs, as well that of power purchased from the grid at the substations.

$$\text{Maximize total RES\_Energy} = \sum_{t\in\Omega^t}\sum_{s\in\Omega^s}\rho_s \sum_{w\in\Omega^w}\pi_w \sum_{g\in\Omega^{\text{RES}}}\sum_{i\in\Omega^i}\left(P^{\text{N}}_{g,i,s,w,t}\right) \tag{1a}$$

$$\text{Minimize TC} = \alpha_1 \times \text{TInvC} + \alpha_2 \times \text{TMC} + \alpha_3 \times \text{TEC} + \alpha_4 \times \text{TENSC} + \alpha_5 \times \text{TImiC} \tag{1b}$$

$$\text{TInvC} = \underbrace{\sum_{t\in\Omega^t}\frac{(1+r)^{-t}}{r}\left(\text{InvC}_t^{\text{DG}} + \text{InvC}_t^{\text{LN}} + \text{InvC}_t^{\text{ES}} + \text{InvC}_t^{\text{CAP}}\right)}_{\text{NPV of investment cost}} \tag{2}$$

$$\text{TMC} = \underbrace{\sum_{t\in\Omega^t}(1+r)^{-t}\left(\text{MC}_t^{\text{DG}} + \text{MC}_t^{\text{LN}} + \text{MC}_t^{\text{ES}} + \text{MC}_t^{\text{Cap}}\right)}_{\text{NPV of maintenance cost}} + \underbrace{\frac{(1+r)^{-\text{T}}}{r}\left(\text{MC}_t^{\text{DG}} + \text{MC}_t^{\text{LN}} + \text{MC}_t^{\text{ES}} + \text{MC}_t^{\text{Cap}}\right)}_{\text{NPV of maintenance costs incurred after stage T}} \tag{3}$$

$$\text{TEC} = \underbrace{\sum_{t\in\Omega^t}(1+r)^{-t}\left(\text{EC}_t^{\text{DG}} + \text{EC}_t^{\text{ES}} + \text{EC}_t^{\text{SS}}\right)}_{\text{NPV of operation costs}} + \underbrace{\frac{(1+r)^{-\text{T}}}{r}\left(\text{EC}_\text{T}^{\text{DG}} + \text{EC}_\text{T}^{\text{ES}} + \text{EC}_\text{T}^{\text{SS}}\right)}_{\text{NPV operation costs incurred after stage T}} \tag{4}$$

$$\text{TENSC} = \underbrace{\sum_{t\in\Omega^t}(1+r)^{-t}\text{ENSC}_t}_{\text{NPV of reliability costs}} + \underbrace{\frac{(1+r)^{-\text{T}}}{r}\text{ENSC}_\text{T}}_{\text{NPV reliability costs incurred after stage T}} \tag{5}$$

$$\text{TEmiC} = \underbrace{\frac{\sum_{t\in\Omega^t}(1+r)^{-t}\left(\text{EmiC}_t^{\text{DG}}+\text{EmiC}_t^{\text{SS}}\right)}{}}_{\text{NPV emission costs}} + \underbrace{\frac{(1+r)^{-\text{T}}}{r}\left(\text{EmiC}_{\text{T}}^{\text{DG}}+\text{EmiC}_{\text{T}}^{\text{SS}}\right)}_{\text{NPV emission costs incurred after stage T}} \quad (6)$$

The individual cost components in Eqs. (2)–(6) are computed by the following expressions. Eqs. (7)–(10) represent the investment costs of DGs, feeders, ESS, and capacitor banks, respectively. Notice that all investment costs are weighted by the capital recovery factor, $\frac{r(1+r)^{\text{LT}}}{(1+r)^{\text{LT}}-1}$. The formulations in Eqs. (7)–(10) ensure that the investment cost of each component added to the system is considered only once in the summation. For example, suppose an investment in a particular feeder k is made in the second year of a 3-year planning horizon. This means that the feeder should be available for utilization after the second year. Hence, the binary variable associated with this feeder will be 1 after the second year while 0 otherwise, that is, $x_{k,t}=\{0,1,1\}$. In this particular case, only the difference $(x_{k,2}-x_{k,1})$ equals 1, implying that the investment cost is considered only once. It should be noted that this works regardless of the type of investment variables. Suppose instead of only binary, $x_{k,t}$ is allowed to have integer values. Assume the optimal solution is $x_{k,t}=\{0,1,2\}$. In this case, the corresponding difference $(x_{k,t}-x_{k,t-1})$ becomes $\{0,1,1\}$, indicating that the investment costs of only those components added at each stage are considered in the summation. Eq. (11) stands for the maintenance costs of new and existing DGs at each time stage. The maintenance cost of a new/existing feeder is included only when its corresponding investment/utilization variable is different from zero. Similarly, the maintenance costs of new and existing feeders at each stage are given by Eqs. (12)–(14) and are related to the maintenance costs at each stage of energy storage and capacitor banks, respectively.

$$\text{InvC}_t^{\text{DG}} = \sum_{g\in\Omega^g}\sum_{i\in\Omega^i}\frac{r(1+r)^{\text{LT}_g}}{(1+r)^{\text{LT}_{g-1}}}\text{IC}_{g,i}\left(x_{g,i,t}-x_{g,i,t-1}\right); \text{ where } x_{g,i,0}=0 \quad (7)$$

$$\text{InvC}_t^{\text{LN}} = \sum_{k\in\Omega^l}\frac{r(1+r)^{\text{LT}_k}}{(1+r)^{\text{LT}_{k-1}}}\text{IC}_k(x_{k,t}-x_{k,t-1}); \text{ where } x_{k,0}=0 \quad (8)$$

$$\text{InvC}_t^{\text{ES}} = \sum_{\text{es}\in\Omega^{\text{es}}}\sum_{i\in\Omega^i}\frac{r(1+r)^{\text{LT}_{\text{es}}}}{(1+r)^{\text{LT}_{\text{es}-1}}}\text{IC}_{\text{es}}(x_{\text{es},i,t}-x_{\text{es},i,t-1}); \text{ where } x_{\text{es},i,0}=0 \quad (9)$$

$$\text{InvC}_t^{\text{CAP}} = \sum_{c\in\Omega^c}\sum_{i\in\Omega^i}\frac{r(1+r)^{\text{LT}_c}}{(1+r)^{\text{LT}_{c-1}}}\text{IC}_c(x_{c,i,t}-x_{c,i,t-1}); \text{ where } x_{c,i,0}=0 \quad (10)$$

$$\text{MC}_t^{\text{DG}} = \sum_{g\in\Omega^g}\sum_{i\in\Omega^i}\text{MC}_g^{\text{N}}x_{g,i,t} + \sum_{g\in\Omega^g}\sum_{i\in\Omega^i}\text{MC}_g^{\text{E}}u_{g,i,t} \quad (11)$$

$$\mathrm{MC}_t^{\mathrm{LN}} = \sum_{k \in \Omega^{el}} \mathrm{MC}_k^{\mathrm{E}} u_{k,t} + \sum_{k \in \Omega^{nl}} \mathrm{MC}_k^{N} x_{k,t} \tag{12}$$

$$\mathrm{MC}_t^{\mathrm{ES}} = \sum_{es \in \Omega^{es}} \sum_{i \in \Omega^{i}} \mathrm{MC}_{es} x_{es,i,t} \tag{13}$$

$$\mathrm{MC}_t^{\mathrm{Cap}} = \sum_{c \in \Omega^{c}} \sum_{i \in \Omega^{i}} \mathrm{MC}_{c} x_{c,i,t} \tag{14}$$

The total cost of power produced by new and existing DGs is given by Eq. (15). Note that these costs depend on the amount of power generated at each scenario, snapshot, and stage. Therefore, these costs represent the expected costs of operation. Similarly, Eqs. (16), (17), respectively, account for the expected costs of energy supplied by the ESS, and that purchased from upstream (ie, transmission grid). The penalty for the unserved power, given by Eq. (18), is also dependent on the scenarios, snapshots, and time stages. Eq. (18) therefore gives the expected cost of unserved energy in the system. The expected emission costs of power generated by new and existing DGs are given by Eqs. (19)–(21), and that of energy purchased from the grid is calculated using Eq. (22). Note that, for the sake of simplicity, a linear emission cost function is assumed here. In reality, the emission cost function is highly nonlinear and nonconvex, as in Ref. [44].

$$\mathrm{EC}_t^{\mathrm{DG}} = \sum_{s \in \Omega^{s}} \rho_s \sum_{w \in \Omega^{w}} \pi_w \sum_{g \in \Omega^{g}} \sum_{i \in \Omega^{i}} \left(\mathrm{OC}^{\mathrm{N}}_{g,i,s,w,t} P^{\mathrm{N}}_{g,i,s,w,t} + \mathrm{OC}^{\mathrm{E}}_{g,i,s,w,t} P^{\mathrm{E}}_{g,i,s,w,t} \right) \tag{15}$$

$$\mathrm{EC}_t^{\mathrm{ES}} = \sum_{s \in \Omega^{s}} \rho_s \sum_{w \in \Omega^{w}} \pi_w \sum_{c \in \Omega^{c}} \sum_{i \in \Omega^{i}} \gamma^{\mathrm{dch}}_{es,i,s,w,t} P^{\mathrm{dch}}_{es,i,s,w,t} \tag{16}$$

$$\mathrm{EC}_t^{\mathrm{SS}} = \sum_{s \in \Omega^{s}} \rho_s \sum_{w \in \Omega^{w}} \pi_w \sum_{\varsigma, \Omega^{\varsigma}} \sigma_{\varsigma,s,w,t} P^{\mathrm{SS}}_{\varsigma,s,w,t} \tag{17}$$

$$\mathrm{ENSC}_t = \sum_{s \in \Omega^{s}} \rho_s \sum_{w \in \Omega^{w}} \sum_{i \in \Omega^{i}} \pi_w \upsilon_{s,w,t} \delta_{i,s,w,t} \tag{18}$$

$$\mathrm{EmiC}_t^{\mathrm{DG}} = \mathrm{EmiC}_t^{\mathrm{N}} + \mathrm{EmiC}_t^{\mathrm{E}} \tag{19}$$

$$\mathrm{EmiC}_t^{\mathrm{N}} = \sum_{s \in \Omega^{s}} \rho_s \sum_{w \in \Omega^{w}} \pi_w \sum_{g \in \Omega^{g}} \sum_{i \in \Omega^{i}} \lambda^{\mathrm{CO_2e}}_{s,w,t} \mathrm{ER}^{\mathrm{N}}_{g} P^{\mathrm{N}}_{g,i,s,w,t} \tag{20}$$

$$\mathrm{EmiC}_t^{\mathrm{E}} = \sum_{s \in \Omega^{s}} \rho_s \sum_{w \in \Omega^{w}} \pi_w \sum_{g \in \Omega^{g}} \sum_{i \in \Omega^{i}} \lambda^{\mathrm{CO_2e}}_{s,w,t} \mathrm{ER}^{\mathrm{E}}_{g} P^{\mathrm{E}}_{g,i,s,w,t} \tag{21}$$

$$\mathrm{EmiC}_t^{\mathrm{SS}} = \sum_{s \in \Omega^{s}} \rho_s \sum_{w \in \Omega^{w}} \pi_w \sum_{\varsigma \in \Omega^{\varsigma}} \lambda^{\mathrm{CO_2e}}_{s,w,t} \mathrm{ER}^{\mathrm{SS}}_{\varsigma} P^{\mathrm{SS}}_{\varsigma,s,w,t} \tag{22}$$

3.4.3 CONSTRAINTS

Kirchhoff's voltage law

The customary AC power flow equations, given by Eqs. (23), (24), are highly nonlinear and nonconvex. Understandably, using these flow expressions in power system planning applications is increasingly difficult. Because of this, Eqs. (23), (24) are often linearized by considering two practical assumptions. The first assumption is concerning the bus voltage magnitudes, which in distribution systems are expected to be close to the nominal value V_{nom}. The second assumption is in relation to the voltage angle difference θ_k across a line, which is practically small, leading to the trigonometric approximations $\sin\theta_k \approx \theta_k$ and $\cos\theta_k \approx 1$. Note that this assumption is valid in distribution systems, where the active power flow dominates the total apparent power in lines. Furthermore, the voltage magnitude at bus i can be expressed as the sum of the nominal voltage and a small deviation ΔV_i, as in Eq. (25).

$$P_k = V_i^2 g_k - V_i V_j (g_k \cos\theta_k + b_k \sin\theta_k) \tag{23}$$

$$Q_k = -V_i^2 b_k + V_i V_j (b_k \cos\theta_k - g_k \sin\theta_k) \tag{24}$$

$$V_i = V_{\text{nom}} + \Delta V_i, \quad \text{where } \Delta V^{\min} \le \Delta V_i \le \Delta V^{\max} \tag{25}$$

Note that the voltage deviations at each node ΔV_i are expected to be very small. Substituting Eq. (25) in Eqs. (23), (24) and neglecting higher order terms, we get

$$P_k \approx \left(V_{\text{nom}}^2 + 2V_{\text{nom}}\Delta V_i\right) g_k - \left(V_{\text{nom}}^2 + V_{\text{nom}}\Delta V_i + V_{\text{nom}}\Delta V_j\right)(g_k + b_k\theta_k) \tag{26}$$

$$Q_k \approx -\left(V_{\text{nom}}^2 + 2V_{\text{nom}}\Delta V_i\right) b_k + \left(V_{\text{nom}}^2 + V_{\text{nom}}\Delta V_i + V_{\text{nom}}\Delta V_j\right)(b_k - g_k\theta_k) \tag{27}$$

Note that Eqs. (26), (27) still contain nonlinearities because of the products of two continuous variables: voltage deviations and angle differences. Since these variables (ΔV_i, ΔV_j, and θ_k) are very small, however, their products can be neglected. Hence, the above flow equations become

$$P_k \approx V_{\text{nom}}\left(\Delta V_i - \Delta V_j\right) g_k - V_{\text{nom}}^2 b_k \theta_k \tag{28}$$

$$Q_k \approx -V_{\text{nom}}\left(\Delta V_i - \Delta V_j\right) b_k - V_{\text{nom}}^2 g_k \theta_k \tag{29}$$

The linear planning model proposed here is based on the above linearized flow equations. This linearization approach was first introduced in Ref. [45] in the context of transmission expansion planning problem.

When the investment planning includes switching and expansion of the DNS, Eqs. (28), (29) must be multiplied by the corresponding binary variables as in Eqs. (30)–(33). This is to make sure the flow through an existing/new feeder is zero when its switching/investment variable is zero; otherwise, the flow in that feeder should obey Kirchhoff's voltage law.

$$P_k \approx u_{k,t}\left\{V_{\text{nom}}\left(\Delta V_i - \Delta V_j\right)g_k - V_{\text{nom}}^2 b_k \theta_k\right\} \tag{30}$$

$$Q_k \approx u_{k,t}\left\{-V_{\text{nom}}\left(\Delta V_i - \Delta V_j\right)b_k - V_{\text{nom}}^2 g_k \theta_k\right\} \tag{31}$$

$$P_k \approx x_{k,t}\left\{V_{\text{nom}}\left(\Delta V_i - \Delta V_j\right)g_k - V_{\text{nom}}^2 b_k \theta_k\right\} \tag{32}$$

$$Q_k \approx x_{k,t}\left\{-V_{\text{nom}}\left(\Delta V_i - \Delta V_j\right)b_k - V_{\text{nom}}^2 g_k \theta_k\right\} \tag{33}$$

The bilinear products, involving binary with voltage deviation and angle difference variables, introduce undesirable nonlinearity to the problem. This nonlinearity can be avoided using the big-M formulation that is, by reformulating the above equations into their respective disjunctive equivalents as in Eqs. (34)–(37). As a rule of thumb, the big-M parameter often set to the maximum transfer capacity in the system.

$$MP_k(u_{k,t}-1) \le P_{k,s,w,t} - \left\{V_{\text{nom}}\left(\Delta V_{i,s,w,t} - \Delta V_{j,s,w,t}\right)g_k - V_{\text{nom}}^2 b_k \theta_{k,s,w,t}\right\} \le MP_k(1-u_{k,t}) \tag{34}$$

$$MQ_k(u_{k,t}-1) \le Q_{k,s,w,t} - \left\{-V_{\text{nom}}\left(\Delta V_{i,s,w,t} - \Delta V_{j,s,w,t}\right)b_k - V_{\text{nom}}^2 g_k \theta_{k,s,w,t}\right\} \le MQ_k(1-u_{k,t}) \tag{35}$$

$$MP_k(x_{k,t}-1) \le P_{k,s,w,t} - \left\{V_{\text{nom}}\left(\Delta V_{i,s,w,t} - \Delta V_{j,s,w,t}\right)g_k - V_{\text{nom}}^2 b_k \theta_{k,s,w,t}\right\} \le MP_k(1-x_{k,t}) \tag{36}$$

$$MQ_k(x_{k,t}-1) \le Q_{k,s,w,t} - \left\{-V_{\text{nom}}\left(\Delta V_{i,s,w,t} - \Delta V_{j,s,w,t}\right)b_k - V_{\text{nom}}^2 g_k \theta_{k,s,w,t}\right\} \le MQ_k(1-x_{k,t}) \tag{37}$$

Flow limits

The apparent power flow through a line S_k is given by $\sqrt{P_k^2+Q_k^2}$ and this has to be less than or equal to the rated value, which is denoted as

$$P_k^2 + Q_k^2 \le \left(S_k^{\max}\right)^2 \tag{38}$$

Considering line switching and investment, Eq. (38) can be rewritten as

$$P_{k,s,w,t}^2 + Q_{k,s,w,t}^2 \le u_{k,t}\left(S_k^{\max}\right)^2 \tag{39}$$

$$P_{k,s,w,t}^2 + Q_{k,s,w,t}^2 \le x_{k,t}\left(S_k^{\max}\right)^2 \tag{40}$$

The quadratic expressions of active and reactive power flows in Eqs. (39), (40) can be easily linearized using piecewise linearization, considering a sufficiently large number of linear segments, L. There are a number of ways of linearizing such functions, such as incremental, multiple choice, convex combination, and other approaches in the literature [46]. Here, the first approach (which is based on first-order approximation of the nonlinear curve) is used because of its relatively simple formulation. To this end, two nonnegative auxiliary variables are introduced for each of the flows P_k and Q_k, such that $P_k = P_k^+ - P_k^-$ and $Q_k = Q_k^+ - Q_k^-$. Note that these auxiliary variables (ie, P_k^+, P_k^-, Q_k^+, and Q_k^-) represent the positive and negative flows of P_k and Q_k, respectively. This helps one to consider only the positive quadrant of the nonlinear curve, resulting in a significant reduction in the mathematical complexity, and by implication, the computational burden. In this case, the associated linear constraints are

$$P_{k,s,w,t}^2 \approx \sum_{l=1}^{L} \alpha_{k,l} p_{k,s,w,t,l} \tag{41}$$

$$Q_{k,s,w,t}^2 \approx \sum_{l=1}^{L} \beta_{k,l} q_{k,s,w,t,l} \tag{42}$$

$$P_{k,s,w,t}^+ + P_{k,s,w,t}^- = \sum_{l=1}^{L} p_{k,s,w,t,l} \tag{43}$$

$$Q_{k,s,w,t}^+ + Q_{k,s,w,t}^- = \sum_{l=1}^{L} q_{k,s,w,t,l} \tag{44}$$

where $p_{k,s,w,t,l} \leq \frac{P_k^{\max}}{L}$ and $q_{k,s,w,t,l} \leq Q_k^{\max}/L$.

Line losses

The active and reactive power losses in line k can be approximated as follows

$$PL_k = P_{k,ij} + P_{k,ji} \approx 2V_{\text{nom}}^2 g_k (1 - \cos\theta_k) \approx V_{\text{nom}}^2 g_k \theta_k^2 \tag{45}$$

$$QL_k = Q_{k,ij} + Q_{k,ji} \approx -2V_{\text{nom}}^2 b_k (1 - \cos\theta_k) \approx -b_k V_{\text{nom}}^2 \theta_k^2 \tag{46}$$

Clearly, Eqs. (45) and (46) are nonlinear and nonconvex functions, making the problem more complex to solve. This can be overcome by having the quadratic angle differences piecewise-linearized, as it is done for the quadratic flows in the above; however, instead of doing this, the expressions in Eqs. (45) and (46) can be expressed in terms of the active and the reactive power flows, respectively, by substituting θ_k from Eqs. (28) and (29) in Eqs. (45) and (46) and neglecting higher order terms. This leads to Eqs. (47) and (48).

$$PL_{k,s,w,t} = \frac{g_k\left\{P_{k,s,w,t}^2 - 2P_{k,s,w,t} V_{\text{nom}} \left(\Delta V_{i,s,w,t} - \Delta V_{j,s,w,t}\right) g_k\right\}}{\left(V_{\text{nom}} b_k\right)^2} \tag{47}$$

$$QL_{k,s,w,t} = -\frac{b_k\left\{Q_{k,s,w,t}^2 + 2Q_{k,s,w,t} V_{\text{nom}} \left(\Delta V_{i,s,w,t} - \Delta V_{j,s,w,t}\right) b_k\right\}}{\left(V_{\text{nom}} g_k\right)^2} \tag{48}$$

Note that expressing the losses as a function of flows has two advantages. First, doing so reduces the number of nonlinear terms that has to be linearized, which in turn results in a model with a reduced number of equations and variables. For example, if Eqs. (45) and (46) are used instead, in addition to the quadratic power flow terms P_k^2 and Q_k^2, the quadratic angle differences θ_k^2 should also be linearized to make the problem linear and convex. On the contrary, if Eqs. (47) and (48) are used, we are only required to linearize P_k^2 and Q_k^2. Second, it avoids unnecessary constraints on the angle differences when a line between two nodes is not connected or remains not selected for investment. In fact, this is often avoided by introducing binary variables and using a so-called big-M formulation [45]; however, this adds extra complexity to the problem.

Note that, in addition to the quadratic flow, Eqs. (47) and (48) contain products of two continuous variables: flow and voltage magnitude deviations, which make the function nonseparable. These products can be neglected, however, because in reality, the voltage deviation variables are expected to be very small, leading to the simplified Eqs. (49) and (50), each having only the quadratic flow expressions.

$$PL_{k,s,w,t} = \frac{g_k P_{k,s,w,t}^2}{(V_{\text{nom}} b_k)^2} \tag{49}$$

$$QL_{k,s,w,t} = -\frac{b_k Q_{k,s,w,t}^2}{(V_{\text{nom}} g_k)^2} \tag{50}$$

Kirchhoff's current law (active and reactive load balances)

Load balance should be respected all the time at each node that is, the sum of all injections should be equal to the sum of all withdrawals at each node. This is enforced by adding the following two constraints

$$\begin{aligned}
&\sum_{g\in\Omega^{\text{DG}}}\left(P^{\text{E}}_{g,i,s,w,t}+P^{\text{N}}_{g,i,s,w,t}\right)+\sum_{\text{es}\in\Omega^{\text{es}}}\left(P^{\text{dch}}_{\text{es},i,s,w,t}+P^{\text{ch}}_{\text{es},i,s,w,t}\right)+P^{\text{SS}}_{\varsigma,s,w,t}+\sum_{\text{in},k\in i}P_{k,s,w,t}\\
&-\sum_{\text{out},k\in i}P_{k,s,w,t}+\delta_{i,s,w,t}=D_{i,s,w,t}+PL_{\delta,s,w,t}+\sum_{k\in i}\frac{1}{2}PL_{k,s,w,t}\quad;\ \forall\varsigma,\forall\varsigma\in i
\end{aligned} \tag{51}$$

$$\begin{aligned}
&\sum_{g\in\Omega^{\text{DG}}}\left(Q^{\text{E}}_{g,i,s,w,t}+Q^{\text{N}}_{g,i,s,w,t}\right)+\sum_{c\in\Omega^{c}}Q_{c,i,s,w,t}+Q^{\text{SS}}_{\varsigma,s,w,t}+\sum_{\text{in},k\in i}Q_{k,s,w,t}-\sum_{\text{out},k\in i}Q_{k,s,w,t}=\\
&Q_{i,s,w,t}+QL_{\varsigma,s,w,t}+\sum_{\text{in},k\in i}\frac{1}{2}QL_{k,s,w,t}+\sum_{out,k\in i}\frac{1}{2}QL_{k,s,w,t}\quad;\ \forall\varsigma,\forall\varsigma\in i
\end{aligned} \tag{52}$$

Eqs. (47), (48) stand for the active and the reactive power balances at each node, respectively.

Bulk energy storage model constraints

The generic bulk energy storage (ES) is modeled by Eqs. (53)–(59).

$$0 \le P^{\text{ch}}_{\text{es},i,s,w,t} \le I^{\text{ch}}_{\text{es},i,s,w,t} x_{\text{es},i,t} P^{\text{ch,max}}_{\text{es},i} \tag{53}$$

$$0 \le P^{\text{dch}}_{\text{es},i,s,w,t} \le I^{\text{dch}}_{\text{es},i,s,w,t} x_{\text{es},i,t} P^{\text{dch,max}}_{\text{es},i} \tag{54}$$

$$I^{\mathrm{ch}}_{\mathrm{es},i,s,w,t} + I^{\mathrm{dch}}_{\mathrm{es},i,s,w,t} \leq 1 \tag{55}$$

$$E_{\mathrm{es},i,s,w,t} = E_{\mathrm{es},i,s,w-1,t} + \eta_{\mathrm{ch,es}} P^{\mathrm{ch}}_{\mathrm{es},i,s,w,t} - \eta_{\mathrm{dch,es}} P^{\mathrm{dch}}_{\mathrm{es},i,s,w,t} \tag{56}$$

$$E^{\min}_{\mathrm{es},i} x_{\mathrm{es},i,t} \leq E_{\mathrm{es},i,s,w,t} \leq x_{\mathrm{es},i,t} E^{\max}_{\mathrm{es},i} \tag{57}$$

$$E_{\mathrm{es},i,s,w_0,T1} = \mu_{\mathrm{es}} x_{\mathrm{es},i,T1} E^{\max}_{\mathrm{es},i} \tag{58}$$

$$E_{\mathrm{es},i,s,w_1,t+1} = E_{\mathrm{es},i,s,W,t} \tag{59}$$

The limits on the capacity of ES while being charged and discharged are considered in Eqs. (53) and (54), respectively. Inequality (55) prevents simultaneous charging and discharging operation of ES at the same operational time w. The amount of stored energy within the reservoir of bulk ES at the operational time w as a function of energy stored until $w-1$ is given by Eq. (56). The maximum and minimum levels of storages in operational time w are also considered through inequality (57). Eq. (58) shows the initial level of stored energy in the bulk ES as a function of its maximum reservoir capacity. In a multistage planning approach, Eq. (59) ensures that the initial level of energy in the bulk ES at a given year is equal to the final level of energy in the ES in the preceding year. Here, $\eta_{\mathrm{dch,es}}$ is assumed to be $1/\eta_{\mathrm{ch,es}}$.

Notice that inequalities (53) and (54) involve products of charging/discharging binary variables and investment variable. In order to linearize this, new continuous positive variables $z^{\mathrm{ch}}_{\mathrm{es},i,s,w,t}$, and $z^{\mathrm{dch}}_{\mathrm{es},i,s,w,t}$, which replace the bilinear products in each constraint, are introduced such that the set of linear constraints in Eqs. (60) and (61) hold. For instance, the product $I^{\mathrm{dch}}_{\mathrm{es},i,s,w,t} x_{\mathrm{es},i,t}$ is replaced by the positive variable $z^{\mathrm{dch}}_{\mathrm{es},i,s,w,t}$. Then, the bilinear product is decoupled by introducing the set of constraints in Eq. (60) [47].

$$z^{\mathrm{dch}}_{\mathrm{es},i,s,w,t} \leq x^{\max}_{\mathrm{es}} I^{\mathrm{dch}}_{\mathrm{es},i,s,w,t}; \quad z^{\mathrm{dch}}_{\mathrm{es},i,s,w,t} \leq x_{\mathrm{es},i,t}; \quad z^{\mathrm{dch}}_{\mathrm{es},i,s,w,t} \geq x_{\mathrm{es},i,t} - \left(1 - I^{\mathrm{dch}}_{\mathrm{es},i,s,w,t}\right) x^{\max}_{\mathrm{es}} \tag{60}$$

Similarly, the product $I^{\mathrm{ch}}_{\mathrm{es},i,s,w,t} x_{\mathrm{es},i,t}$ is decoupled by including the following set of constraints

$$z^{\mathrm{ch}}_{\mathrm{es},i,s,w,t} \leq x^{\max}_{\mathrm{es}} I^{\mathrm{ch}}_{\mathrm{es},i,s,w,t}; \quad z^{\mathrm{ch}}_{\mathrm{es},i,s,w,t} \leq x_{\mathrm{es},i,t}; \quad z^{\mathrm{ch}}_{\mathrm{es},i,s,w,t} \geq x_{\mathrm{es},i,t} - \left(1 - I^{\mathrm{ch}}_{\mathrm{es},i,s,w,t}\right) x^{\max}_{\mathrm{es}} \tag{61}$$

Active and reactive power limits of DGs

The active and reactive capacity limits of existing generators are given by Eqs. (62) and (63), respectively. In the case of candidate generators, the corresponding constraints are Eqs. (64) and (65). Note that the binary variables also appear here and multiply the minimum and the maximum generation capacities of a given generator. This is to make sure that the power generation variable is zero when the generator remains either unutilized or unselected for investment.

$$P^{\mathrm{E,min}}_{g,i} u_{g,i,t} \le P^{\mathrm{E}}_{g,i,s,w,t} \le P^{\mathrm{E,max}}_{g,i} u_{g,i,t} \quad (62)$$

$$Q^{\mathrm{E,min}}_{g,i} u_{g,i,t} \le Q^{\mathrm{E}}_{g,i,s,w,t} \le Q^{\mathrm{E,max}}_{g,i} u_{g,i,t} \quad (63)$$

$$P^{\mathrm{N,min}}_{g,i} x_{g,i,t} \le P^{\mathrm{N}}_{g,i,s,w,t} \le P^{\mathrm{N,max}}_{g,i} x_{g,i,t} \quad (64)$$

$$Q^{\mathrm{N,min}}_{g,i} x_{g,i,t} \le Q^{\mathrm{N}}_{g,i,s,w,t} \le Q^{\mathrm{N,max}}_{g,i} x_{g,i,t} \quad (65)$$

Reactive power limit of capacitor banks

Inequality (66) ensures that the reactive power produced by the capacitor banks is bounded between zero and the maximum capacity.

$$0 \le Q_{c,i,s,w,t} \le x_{c,i,t} Q^{0}_{c} \quad (66)$$

Active and reactive power limits of power purchased

For technical reasons, the power that can be purchased from the transmission grid could have minimum and maximum limits, which is enforced by Eqs. (67) and (68); however, it is understood that setting the maximum and minimum limits is difficult. These constraints are included here for the sake of completeness. In this work, these limits are set to 1.5 times the minimum and maximum levels of total load in the system. Note that the multiplier is higher than one because the system has losses, which need to be covered by generating extra power.

$$P^{\mathrm{SS,min}}_{\varsigma,s,w,t} \le P^{\mathrm{SS}}_{\varsigma,s,w,t} \le P^{\mathrm{SS,max}}_{\varsigma,s,w,t} \quad (67)$$

$$Q^{\mathrm{SS,min}}_{\varsigma,s,w,t} \le Q^{\mathrm{SS}}_{\varsigma,s,w,t} \le Q^{\mathrm{SS,max}}_{\varsigma,s,w,t} \quad (68)$$

Logical constraints

The following logical constraints ensure that an investment decision cannot be reversed that is, an investment already made cannot be divested.

$$x_{k,t} \ge x_{k,t-1} \quad (69)$$

$$x_{g,i,t} \ge x_{g,i,t-1} \quad (70)$$

$$x_{\mathrm{es},i,t} \ge x_{\mathrm{es},i,t-1} \quad (71)$$

$$x_{c,i,t} \ge x_{c,i,t-1} \quad (72)$$

Radiality constraints

There are two conditions that must be fulfilled in order for a DNS to be radial. First, the solution must have $N_i - N_{SS}$ circuits. Second, the final topology should be connected. Eq. (73) represents the first necessary condition for maintaining the radial topology of DNs.

$$\sum_{k \in \Omega^{ij}} OR(x_{k,t}, u_{k,t}) = N_i - N_{SS}; \ \forall t \tag{73}$$

Note that the above equation assumes line investment is possible in all corridors. Hence, in a given corridor, we can have either an existing branch or a new one, or both connected in parallel, depending on the economic benefits the final setup (solution) brings about to the system. The radiality constraint in Eq. (73) then has to accommodate this condition. One way to do this is by using the Boolean logic operation, as in Eq. (73). Unfortunately, this introduces nonlinearity. We show how this logic can be linearized using an additional auxiliary variable $z_{k,t}$ and the binary variables associated with existing and new branches that is, $u_{k,t}$ and $x_{k,t}$, respectively. Given $z_{k,t} := OR(x_{k,t}, u_{k,t})$, this Boolean operation can be expressed using the following set of linear constraints

$$z_{k,t} \leq x_{k,t} + u_{k,t}; \ z_{k,t} \geq x_{k,t}; \ z_{k,t} \geq u_{k,t}; \ 0 \leq z_{k,t} \leq 1; \ \forall t \tag{74}$$

Note that the auxiliary variable $z_{k,t}$ is automatically constrained to be binary. Hence, it is not necessary to explicitly define $z_{k,t}$ as a binary variable; instead, defining it as a continuous positive variable is sufficient. Alternatively, if $z_{k,t}$ is defined to be binary variable from the outset, then, Eq. (73) can be converted into a single range constraint as

$$0 \leq 2z_{k,t} - x_{k,t} - u_{k,t} \leq 1; \ \forall t \tag{75}$$

Then, the radiality constraints in Eq. (73) can be reformulated using the $z_{k,t}$ variables as

$$\sum_{k \in \Omega^{ij}} z_{k,t} = N_i - N_{SS}; \ \forall t \tag{76}$$

When all loads in the DNS are only fed by power from substations, the final solution obtained automatically satisfies the two aforementioned conditions; hence, no additional constraints are required that is Eq. (74) or Eq. (75) along with Eq. (76) are sufficient to guarantee radiality; however, it should be noted that in the presence of DGs and reactive power sources, these constraints alone may not ensure the radiality of the distribution network, as pointed out in Ref. [48] and further discussed in Ref. [49]. This is, however, out of the scope of this work. If this is indeed found out to be a critical issue, additional constraints need to be added to guarantee that all buses are linked, as proposed in Refs. [43,49–51].

3.5 UNCERTAINTY AND VARIABILITY MANAGEMENT

There are various sources of uncertainty and variability in a distribution system's planning problem, particularly with intermittent renewable sources. These are related to the variability in time and the randomness of operational situations [52]. In addition, there are other uncertainties mostly related to the long-term electricity, carbon and fuel prices, rules, regulations, and policies, etc. Exhaustive modeling of all sources of uncertainty and variability is out of the scope of this work; however, variabilities due to intermittent DG power outputs (mainly, wind and solar) and demand are captured by considering a sufficiently large number of operational states, also known here as "snapshots." To ensure tractability, a standard clustering technique (k-means) is used to reduce the number of snapshots to 200. Here, each cluster represents a group of similar operational situations. A representative snapshot, the medoid in this case, is then selected from each cluster. A weight is then assigned to each representative snapshot, which is proportional to the number of operational situations in its group.

The hourly demand at each node (which is largely predictable) is assumed to be available. Here, an hourly demand series of a real-life distribution network is considered. Wind speed is assumed to follow a Weibull probability distribution. A total of 8760 samples (corresponding to the number of hours in a year) are generated randomly from this probability distribution. Similarly, the hourly solar radiation is assumed to follow beta probability distribution, and the same number of samples is generated accordingly. Note that these generated random samples cannot be used, as they are in the planning process. They should be readjusted to reflect the temporal correlations that naturally exist among demand, solar radiation, and wind speed series. To this end, the correlation between wind and solar sources is considered to be −0.3, while that of wind and demand is 0.28, which is in line with the results in Ref. [53]. A correlation of 0.5 is assumed between solar and demand, according to Ref. [54]. Using these correlations and the demand series as a reference, the wind speed and solar radiation time series are readjusted by making use of Cholesky factorization, which can easily be implemented in MATLAB. This way, new wind speed and solar radiation series are obtained that meet the mentioned intercorrelations. Then, the hourly wind and solar power output are determined by plugging in these readjusted series into their corresponding power curves given by Eqs. (77) and (78).

$$P_{\mathrm{wnd},h} = \begin{cases} 0; \ 0 \le v_h \le v_{\mathrm{ci}} \\ P_r\left(A + Bv_h^3\right); \ v_{\mathrm{ci}} \le v_h \le v_r \\ P_r; \ v_r \le v_h \le v_{\mathrm{co}} \\ 0; \ v_h \le v_{\mathrm{co}} \end{cases} \tag{77}$$

In the above equation, A and B are parameters represented by the expressions in Refs. [55,56]. Similarly, the hourly solar power output $P_{\mathrm{sol},h}$ is determined by plugging in the hourly solar radiation levels in the solar power output expression given in Eq. (78) [57].

$$P_{\mathrm{sol},h} = \begin{cases} \dfrac{P_r R_h^2}{R_{\mathrm{std}} \times R_c}; \ 0 \le R_h \le R_c \\ \dfrac{P_r R_h}{R_{\mathrm{std}}}; \ R_c \le R_h \le R_{\mathrm{std}} \\ P_r; \ R_h \ge R_{\mathrm{std}} \end{cases} \tag{78}$$

3.6 CASE STUDY

3.6.1 SYSTEM DATA AND ASSUMPTIONS

The DNS, shown in Fig. 1, is used to test the developed planning model. Information regarding network and maximum demand data is provided in Table A1 [58]. The total active and reactive loads in the system are 4.635 MW and 3.25 MVAr, respectively. The nominal voltage of the system is 12.66 kV. The following assumptions are made when carrying out the simulation:

- A 3-year planning horizon is considered, which is divided into yearly decision stages.
- Interest rate is set to 7%.
- For the sake of simplicity, maintenance costs are taken to be 2% of the corresponding investment costs.
- The lifetime of capacitor banks and ESSs is assumed to be 15 years, while that of DGs and feeders is 25.
- A 5% voltage deviation is considered to be the maximum allowable deviation in the system.
- The power transfer capacity of all feeders is assumed to be 6.986 MVA.
- All big-*M* parameters are set to 10, which is higher than the power transfer capacity of all feeders.
- The number of piecewise linear segments is limited to 5; this balances well the accuracy with computation burden, as concluded in Ref. [59].
- The efficiency of the bulk ES is assumed to be 90%.
- The unit cost of capacitor banks is assumed to be €25/kVAr.
- The size of the minimum deployable capacitor bank is considered to be 0.1 MVAr.

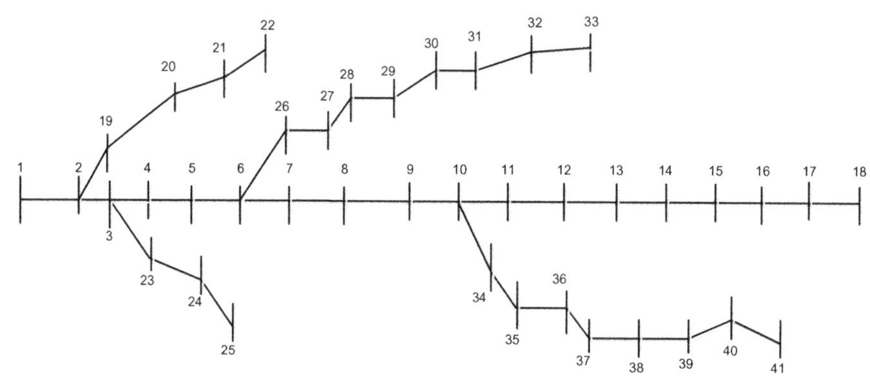

FIG. 1

Single-line diagram of the IEEE 41-bus distribution network system.

- The investment cost of a 1.0 MW bulk ES, whose energy reservoir is 5 MWh, is considered to be 1.0 M€.
- The emission rate of power purchased is arbitrarily set to 0.4 tCO_2e/MWh.
- The investment cost of a given feeder is assumed to be directly proportional to its impedance that is, $C_{ij} = \text{constant} \times Z_{ij}$ where the proportionality constant is 10,000€/Ω.
- Wind- and solar-type DGs, each with 1 MW installed capacity, are considered as potential candidates to be deployed in the system; the investment costs of these generators are assumed to be 2.64 and 3 M€, respectively.
- Yearly demand growths of 5%, 10%, and 15% are assumed for the planning stages.
- The emission prices in the first, second, and third stages are set to 25, 45, and 60€/tCO_2e, respectively.
- Variable power generation sources (wind and solar, in particular) are assumed to be available in every node; this assumption emanates from the fact that distribution networks span over a small geographical area, hence, the distribution of resources in this area can be considered to be the same.
- The substation node (node 1) is considered as a reference; hence, its voltage magnitude and angle are set to $1.02 \times V_{\text{nom}}$ and 0, respectively.
- The cost of unserved energy is set to 3000€/MWh.

3.6.2 RESULTS AND DISCUSSION

Intermittent power generation sources such as wind and solar PV-type DGs normally operate with a fixed lagging power factor [60]. In other words, such generators absorb reactive power, instead of producing and contributing to the voltage regulation in the system (also known as reactive power support). In power systems, voltage regulation has been traditionally supported by conventional (synchronous) generators; however, this is likely to change in the near future, given the upward trend of integrating such resources in power systems. These generators will be equipped with reactive power support devices, which are predominantly based on power electronics, to enhance their capability to provide reactive power when it is needed in the system. Here, we have carried out the system expansion considering without reactive power support, and the results of the simulation are discussed as follows.

The power factor of wind and solar PV-type DGs is set to 0.95 lagging [60]. This means such DGs consume reactive power all the time. The system is expanded considering this case, and the expansion results are discussed in the following section.

The optimal solution for capacitor banks, DGs, and bulk ES in the system are shown in Tables 1–3, respectively. In general, the majority of the investments are made in the first stage. This is because the NPV of operation and emission costs are higher in the first stage than those in any of the subsequent stages. This makes it attractive to invest more in renewables in the first stage than in the other stages so that these costs are drastically reduced.

Table 1 Optimal Investment Solution for Capacitor Banks at Each Stage

Location (Bus)	Time Stages		
	T1	T2	T3
	$x_{c,i,t}$		
7	1	1	1
8	4	4	4
14	4	4	4
24	0	0	2
25	1	1	3
29	1	1	1
30	8	9	9
31	1	1	1
32	2	2	2
37	1	1	1
38	8	8	9
39	1	1	1
40	2	2	2

Table 2 Optimal Investment Solution for DGs at Each Stage

DG Type	Location (Bus)	Time Stages		
		T1	T2	T3
		$x_{g,i,t}$		
Solar	30	1	1	1
Wind	7	0	0	1
Wind	14	2	2	2
Wind	18	1	1	1
Wind	30	1	1	1
Wind	31	1	1	1
Wind	37	1	1	2
Wind	38	1	1	1

Table 3 Optimal Investment Solution for Bulk Energy Storage at Each Stage

Location (Bus)	Time Stages		
	T1	T2	T3
	$x_{es,i,t}$		
14	2	2	2
30	1	1	1
39	2	2	2

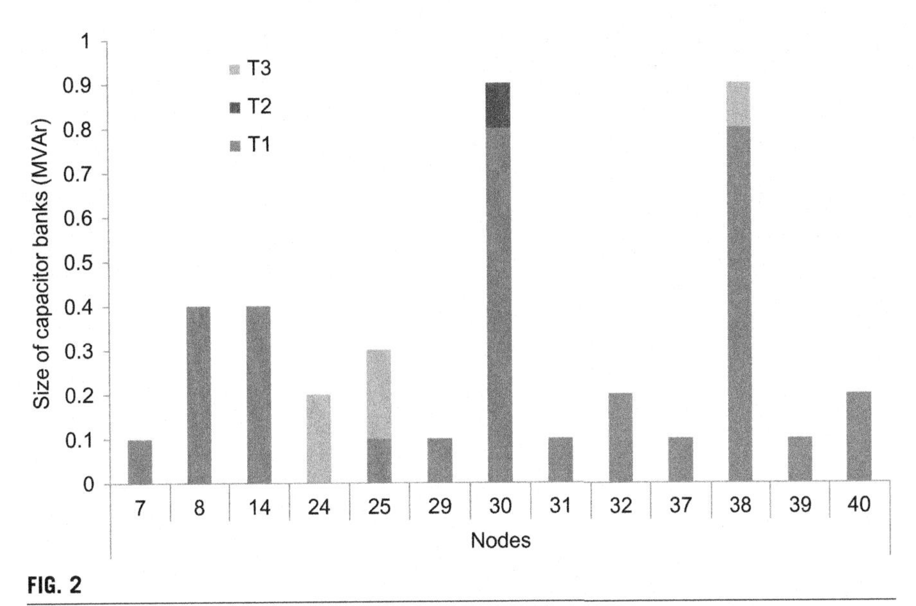

FIG. 2

Optimal location and size of capacitor banks at each stage.

As we can see in Table 1, the optimal location of capacitor banks mostly coincides with high-load connection points (nodes) as well as with those closer to the end nodes. This is expected from the system operation point of view, because capacitor banks are required at such nodes to meet the reactive power requirements and thus keep the corresponding voltages within allowable operational limits. Otherwise, the voltages are expected to drop at these nodes without a compensation mechanism put in place. As shown in Fig. 2, the total size of investment in capacitor banks required throughout the planning horizon is 4.0 MVAr, out of which investments in 3.4, 0.1, and 0.5 MVAr are made in the first, second, and third stages, respectively.

Table 2 shows that more investments are made in wind than in solar PV-type DGs. This is because of the higher capacity factor of potential wind power generators compared to solar PV ones. In general, the total MW of DG power installed at each node and stage in the system is shown in Fig. 2. Here, the optimal size of DGs integrated in the system is 8 and 2 MW in the first and the third stages, respectively.

The results in Tables 1–3 (also conveniently shown in Fig. 3) demonstrate the strong complementarity of variable generation, ESSs, and compensators. Based on the results here, the bulk ES systems and DGs in particular are optimally located close to one another.

It is well known that bulk ES can bring significant benefits, such as load following, power stability improvements, and enhancing the dispatchability of RESs from the system operator's point of view according to their operation modes. Likewise, the optimal deployment of capacitor banks also brings substantial benefits to the system. The combination of all these entirely helps one to dramatically increase the size of RESs that can be integrated into the system without violating system constraints. The optimal size of RESs would otherwise be limited to only 3 MW. It is interesting to

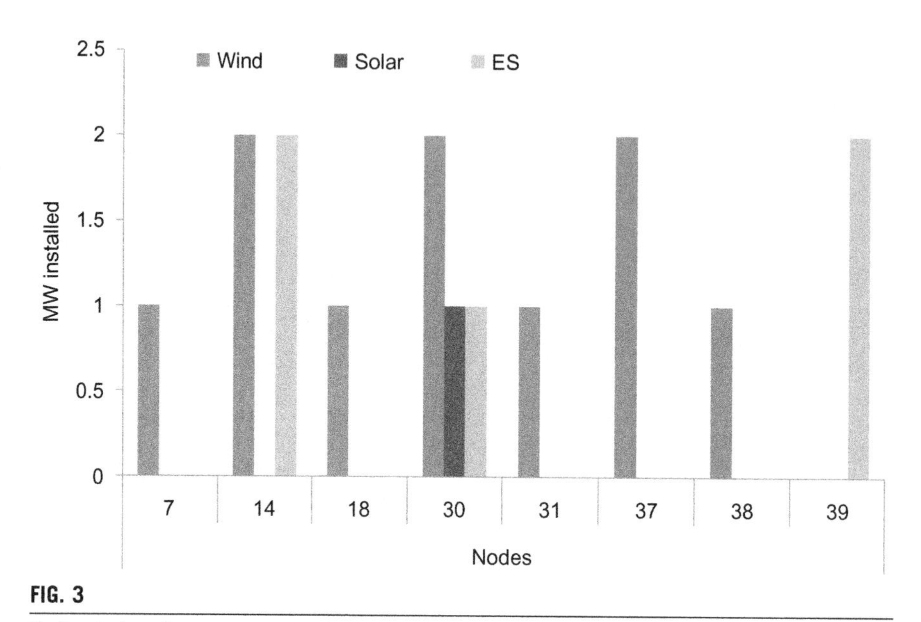

FIG. 3

Optimal size of energy storage, installed solar, and wind power throughout the planning horizon.

see here that the integration of ESs and capacitor banks has such a dramatic impact on the level of DG integration. This is due to the fact that ESSs and capacitor banks bring about significant flexibility and control mechanism to the system. Substantial improvements in voltage controllability are also clearly visible in Figs. 4 and 5.

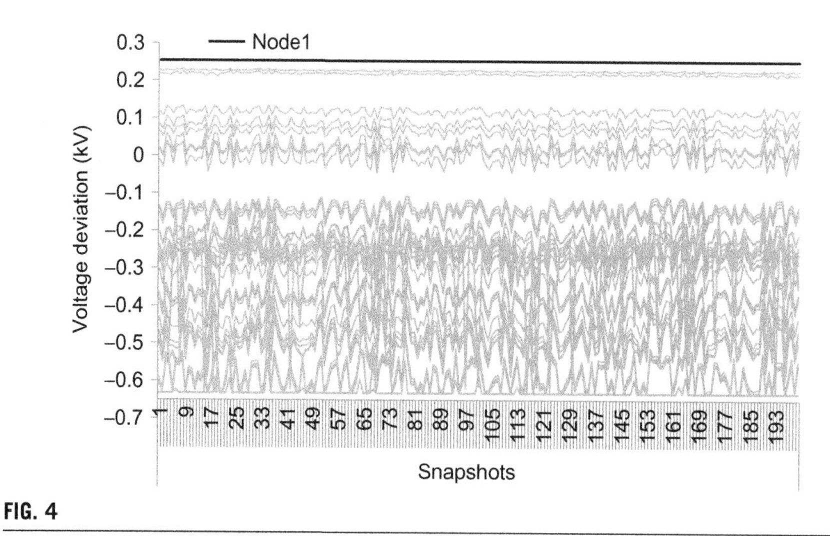

FIG. 4

Profiles of voltage deviations without system expansion in the first stage.

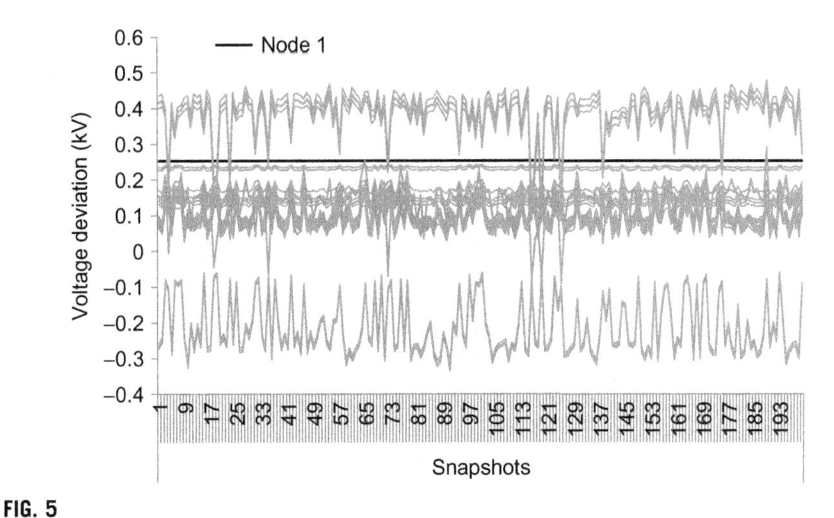

FIG. 5

Profiles of voltage deviations at each node after expansion in the first stage.

These figures show the voltage deviation profiles at each node with the selected operational situations (which can alternatively be understood as "long hours") without and with system expansion, respectively. In the base case (shown in Fig. 4), one can see that some of the node voltage deviations (especially those at the extreme nodes) tend to be very close to the minimum allowable limit. On the contrary, all node voltages largely stay very close to the nominal one (with an average deviation of approximately 1.5%), leaving significant margins to the operational limits. Alternatively, Fig. 6 conveniently shows the variance of the voltage deviations at each node. It is also evident to see here that the variance of most of the deviations is very low. The highest variances at nodes 20–22 are due to high impedance of feeder connected between nodes 19 and 20 (see Table A1). The same reasoning explains the relatively high variances in nodal voltage deviations between nodes 13 and 18; however, these variances are negligible when put into perspective with the square of maximum deviation, that is, $(\Delta V^{\max})^2$, which in this case is approximately $(0.05 \times 12.66\,\text{kV})^2 \approx 400{,}000\,\text{V}^2$. In general, such a substantial improvement in voltage controllability has come from the combined effect of expansion decisions in DG, ES, and capacitor banks.

Other important aspects in this expansion analysis are related to the impact of system expansion on the network losses and investments. Fig. 7 shows a comparison of the network losses in the base case and with expansion for every operational state. We can see a significant reduction in network losses (by nearly 50% on average) in the system after the expansion planning is carried out. This is one of the major benefits of integrating DGs in the system. Concerning investments in lines, in this particular case study, not a single feeder is selected for reinforcements. This clearly indicates line investments are deferred when DGs are integrated in the system.

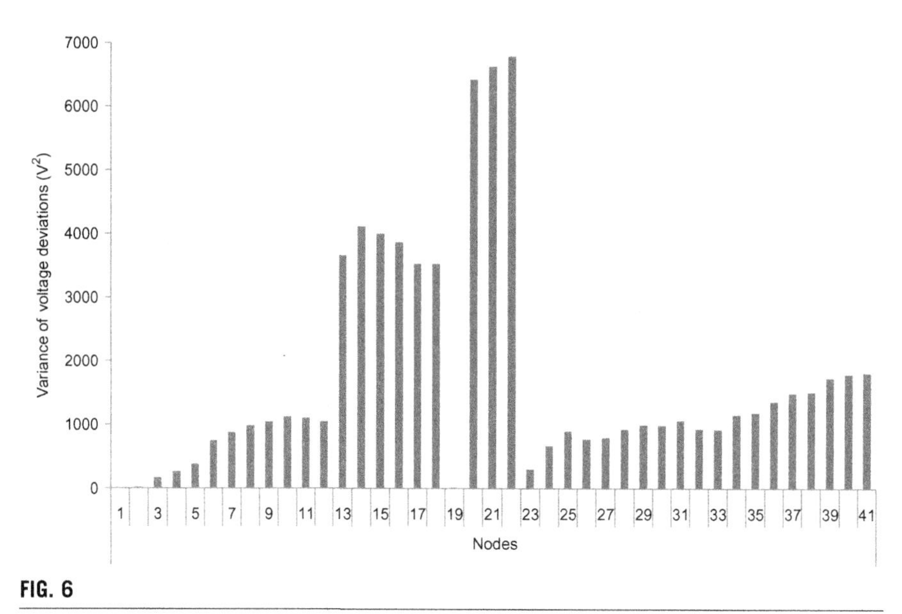

FIG. 6

Variance of voltage deviations as a result of variations in system operational states.

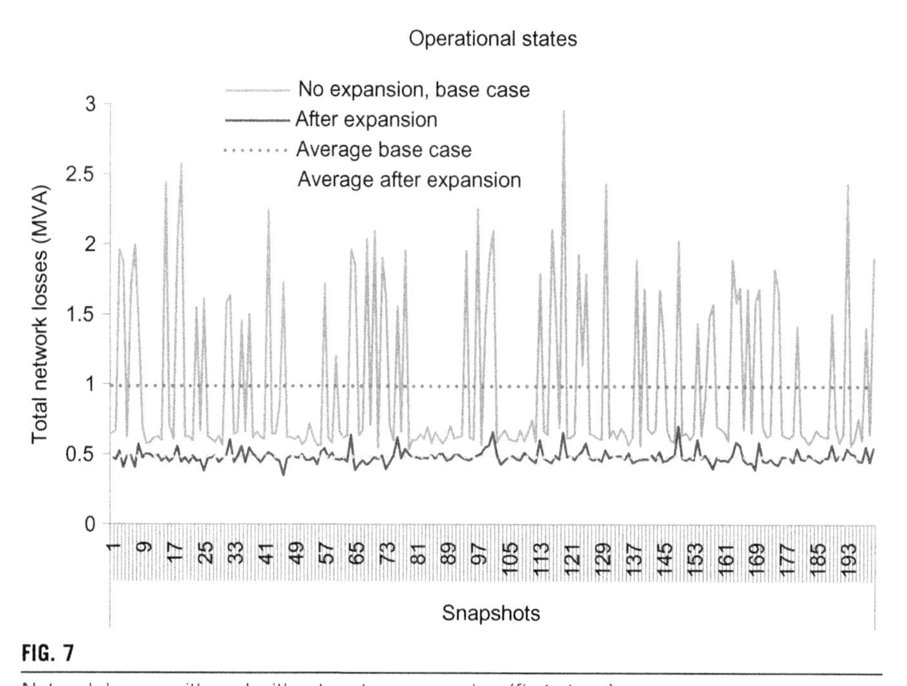

FIG. 7

Network losses with and without system expansion (first stage).

For this particular case study, the total NPV investment costs for the three stages are 30.427, 0.003, and 5.294 M€, respectively, bringing the total investment costs to 37.724 M€. And the NPV cost of energy, emissions, maintenance, and unserved power throughout the planning horizon for the corresponding stages are 27.105, 8.089, 9.442, and 0.868 M€, respectively. The overall NPV cost in this case is 83.228 M€.

3.6.3 A STRATEGY FOR REDUCING COMBINATORIAL SEARCH SPACE

In the case study presented earlier, all nodes in the system are assumed to be candidates for the placement of DGs, ESSs, and capacitor banks; however, this is not possible when the planning work is carried out on large-scale DNSs, because the size of the problem becomes huge as a result of combinatorial explosion, rendering difficulty in solving the problem to optimality. Owing to this fact, the potential candidate nodes are often predetermined either arbitrarily or using some criteria for the selection, such as the level of load, availability of resources, etc. For example, the possible connection points of RES-based DGs are often known *a priori* based on the availability of primary energy sources (such as wind speed and solar radiation). In fact, the variation in the availability of wind speed and solar radiation among the connection points in the DNS is not expected to be significant because it normally spans over a geographically small area.

Here, we show how the combinatorial search space can be substantially reduced using a simple heuristic method. The method is based on solving a relaxed version of the original problem. This is done by treating all (normally integer) investment variables except the line reinforcement variables as continuous ones. This effectively means fractional investment decisions are allowed. The method here works by first establishing a threshold for each fractional investment solution (ie, corresponding to DGs, ESS, and capacitor banks). Then, those nodes whose corresponding values of investment solutions are lower than the preset thresholds are neglected. For instance, consider the investment solution of the relaxed problem corresponding to ESS at each node, as shown in Fig. 8. In this case, the threshold is set to 0.15. As we can see, for most of the nodes, the investment values corresponding to ESS fall below this threshold. Only those values at the following nodes are significant: {14, 18, 29, 30, 31, 32, 37, 38, 39, 40}. This set of nodes is considered the most likely locations in the system for ESS placements in the "brute force" planning model (ie, the full MILP version). It should be noted that such a reduction in possible connection points (from 41 to 10) results in a substantial reduction of the combinatorial search space, by implication of the computational burden. Note that the procedure/criterion for setting the threshold is an open question.

Similarly, the reduced set of nodes for possible capacitor and DG connections are obtained by using 1 and 0.2 as thresholds, respectively, as shown in Figs. 9 and 10. In this case, {7, 8, 14, 24, 25, 29, 30, 31, 32, 37, 38, 39, 40} is the reduced set of nodes for capacitor bank connections, and that of DGs is {7, 8, 14, 18, 25, 29, 30, 31, 32, 37, 38, 39, 40}.

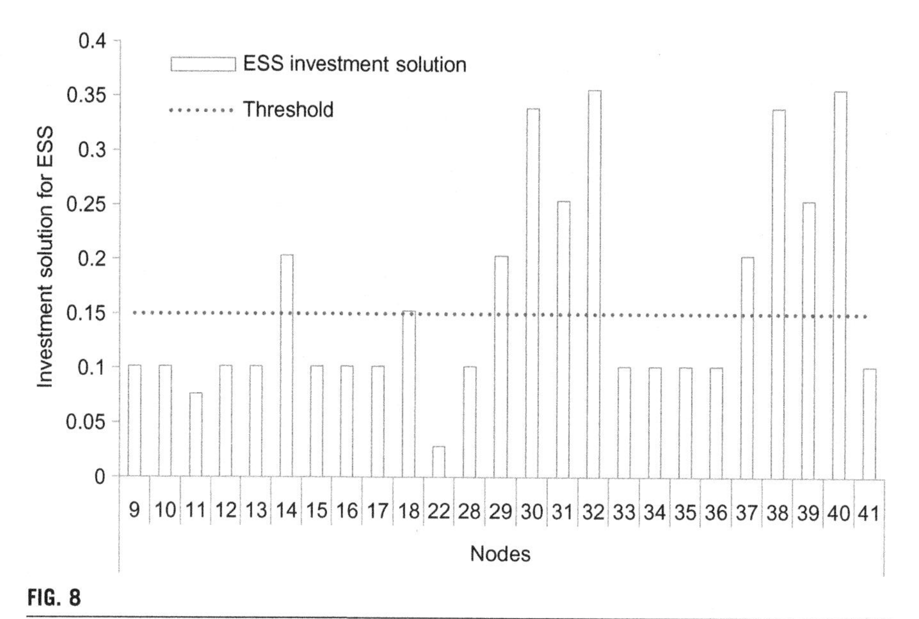

FIG. 8

Decision variable for ESS (last stage).

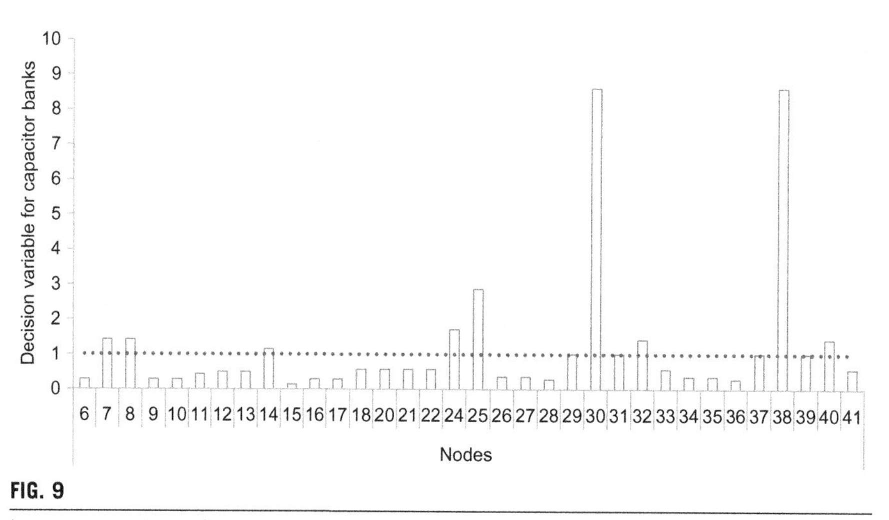

FIG. 9

Investment solution for capacitor banks (last stage).

The heuristic method, proposed here, has been applied in the case study, and the results are compared with that of the "brute force" model. The investment decisions remain the same in both cases, but the computational requirements substantially differ from one another. This heuristic method has significantly reduced the combinatorial solution search space and thus the computational effort by more than sevenfold.

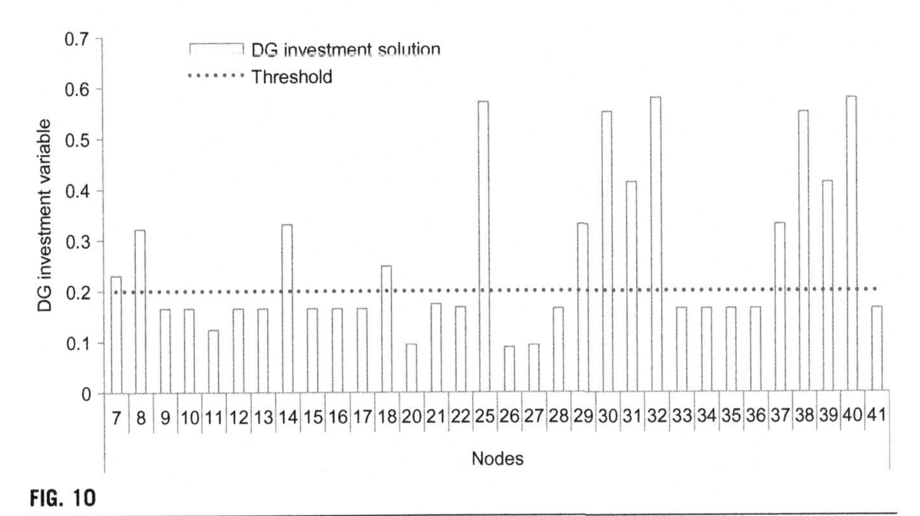

FIG. 10

Investment solution for DGs (last stage).

3.7 CONCLUSIONS

This work has developed a new joint multistage mathematical optimization model considering smart-grid-enabling technologies such as ESS, compensators, and network switching and/or expansion to support large-scale DG integration. The integrated planning model simultaneously determines the optimal sizing, time, and placement of ESSs and compensators, as well as that of RESs in distribution networks. The ultimate goal of this optimization work is to maximize the RES power absorbed by the system while maintaining the power quality and stability at the required/standard levels at the minimum cost possible. The model, formulated as an MILP optimization, employs a linearized AC network model that better captures the inherent characteristics of power network systems and balances accuracy with computational burden. The standard IEEE 41-bus distribution system is used to test the developed model and carry out the required analysis from the standpoint of the objectives set in this work.

The results of the case study show that the integration of ESSs and compensators helps to significantly increase the size of variable generation (wind and solar) in the system. For the case study, a total of 10 MW demand wind and solar power has been added to the system. One can put this into perspective with the peak load 4.635 MW in the system. This means it has been possible to integrate RES power more than twice the peak demand in the base case. It has been demonstrated that the joint planning of DGs, compensators, and ES systems, proposed in this work, brings about significant improvements to the system, such as reduction of losses, cost of electricity and emissions, voltage support, and many others.

The expansion planning model proposed here can be considered a major leap forward toward developing controllable grids, which support large-scale integration of RESs (as opposed to the conventional "fit and forget" approach). It can also be a handy tool to speed up the integration of more RESs until smart-grids are realized in the future.

APPENDIX INPUT DATA

Table A1.

Table A1 Load and Network Data for the IEEE 41-Bus Distribution Network System

	Demand Data		Network Data					
Node	Active Power (kW)	Reactive Power (kVAr)	From Node	To Node	Resistance (Ω)	Reactance (Ω)	Capacity (MVA)	Investment Cost (×1000 €)
2	100	60	1	2	0.0992	0.0470	6.9860	0.9920
3	90	40	2	3	0.4930	0.2511	6.9860	4.9300
4	120	80	3	4	0.3660	0.1864	6.9860	3.6600
5	60	30	4	5	0.3811	0.1941	6.9860	3.8110
6	60	20	5	6	0.8190	0.7070	6.9860	8.1900
7	200	100	6	7	0.1872	0.6188	6.9860	1.8720
8	200	100	7	8	0.7114	0.2351	6.9860	7.1140
9	60	20	8	9	10.300	0.7400	6.9860	103.00
10	60	20	9	10	10.440	0.7400	6.9860	104.40
11	45	30	10	11	0.1966	0.0650	6.9860	1.9660
12	60	35	11	12	0.3744	0.1238	6.9860	3.7440
13	60	35	12	13	14.680	11.550	6.9860	146.80
14	120	80	13	14	0.5416	0.7129	6.9860	5.4160
15	60	10	14	15	0.5910	0.5260	6.9860	5.9100
16	60	20	15	16	0.7463	0.5450	6.9860	7.4630
17	60	20	16	17	12.890	17.210	6.9860	128.90
18	90	40	17	18	0.7320	0.5470	6.9860	7.3200
19	90	40	2	19	0.1640	0.1565	6.9860	1.6400
20	90	40	19	20	15.042	13.554	6.9860	150.42
21	90	40	20	21	0.4095	0.4784	6.9860	4.0950
22	90	40	21	22	0.7089	0.9373	6.9860	7.0890
23	90	50	3	23	0.4512	0.3083	6.9860	4.5120
24	420	200	23	24	0.8980	0.7091	6.9860	8.9800
25	420	200	24	25	0.8960	0.7011	6.9860	8.9600
26	60	25	6	26	0.2030	0.1034	6.9860	2.0300
27	60	25	26	27	0.2842	0.1447	6.9860	2.8420
28	60	20	27	28	10.590	0.9337	6.9860	105.90
29	120	70	28	29	0.8042	0.7006	6.9860	8.0420
30	200	600	29	30	0.5075	0.2585	6.9860	5.0750
31	150	70	30	31	0.9744	0.9630	6.9860	9.7440
32	210	100	31	32	0.3105	0.3619	6.9860	3.1050
33	60	40	32	33	0.3410	0.5302	6.9860	3.4100
34	60	25	10	34	0.2030	0.1034	6.9860	2.0300
35	60	25	34	35	0.2842	0.1447	6.9860	2.8420
36	60	20	35	36	10.590	0.9337	6.9860	105.90

Continued

Table A1 Load and Network Data for the IEEE 41-Bus Distribution Network System—*Cont'd*

Node	Demand Data		Network Data					
	Active Power (kW)	Reactive Power (kVAr)	From Node	To Node	Resistance (Ω)	Reactance (Ω)	Capacity (MVA)	Investment Cost (×1000 €)
37	120	70	36	37	0.8042	0.7006	6.9860	8.0420
38	200	600	37	38	0.5075	0.2585	6.9860	5.0750
39	150	70	38	39	0.9744	0.9630	6.9860	9.7440
40	210	100	39	40	0.3105	0.3619	6.9860	3.1050
41	60	40	40	41	0.3410	0.5302	6.9860	3.4100

NOMENCLATURE

SETS/INDICES

c/Ω^c index/set of capacitor banks
es/Ω^{es} index/set of energy storages
$g/\Omega^g/\Omega^{DG}$ index/set of DGs
h index for hours
i/Ω^i index/set of buses
k/Ω^ℓ index/set of branches
s/Ω^s index/set of scenarios
t/Ω^t index/set of time stages
w/Ω^w index/set of snapshots
$\varsigma/\Omega^\varsigma$ index/set of substations

PARAMETERS

$E_{es,i}^{min}, E_{es,i}^{max}$ energy storage limits
$ER_g^N, ER_g^E, ER_\varsigma^{SS}$ emission rates of new and existing DGs, and energy purchased at substations, respectively
g_k, b_k, S_k^{max} conductance, susceptance, and flow limit of branch k
$IC_{g,i}, IC_k, IC_{es,i}, IC_{c,i}$ investment cost of DG, line, energy storage system, and capacitor banks, respectively
L total number of linear segments
$LT_g, LT_k, LT_{es}, LT_c$ lifetimes of DG, line, energy storage system, and capacitor banks, respectively
MC_c, MC_{es} maintenance cost of capacitor bank and energy storage system per year
MC_g^N, MC_g^E maintenance costs of new and existing DGs per year
MC_k^N, MC_k^E maintenance costs of new and existing branch k per year
MP_k, MQ_k big-M parameters associated with active and reactive power flows through link k, respectively

N_i, N_{SS}	number of buses and substations, respectively
$OC^N_{g,i,s,w,t}, OC^E_{g,i,s,w,t}$	cost of unit energy production by new and existing DGs
$P^{ch,max}_{es,i}, P^{dch,max}_{es,i}$	charging and discharging power limits of a storage system
P_r	rated power a DG unit
$P_{sol,h}$	hourly solar PV output
$P_{wnd,h}$	hourly wind power output
Q^0_c	rating of minimum capacitor bank
r	interest rate
R_c	a certain radiation point (often taken to be 150 W/m^2)
R_h	hourly solar radiation
R_{std}	solar radiation in standard condition (usually set to 1000 W/m^2)
α_l, β_l	slopes of linear segments
$\gamma^{dch}_{es,i,s,w,t}$	cost of energy discharged from storage system
$\eta_{ch,es}, \eta_{dch,es}$	charging and discharging efficiencies of a storage system
$\lambda^{CO_2e}_{s,w,t}$	price of emissions (€/tons of CO_2 equivalent—€/tCO_2e)
v_{ci}	cut-in wind speed
v_{co}	cut-out wind speed
v_h	observed/sampled hourly wind speed
v_r	rated wind speed
ρ_s, π_w	probability of scenario s and weight (in hours) of snapshot group w
$\sigma_{\varsigma,s,w,t}$	price of electricity purchased from upstream
$\upsilon_{s,w,t}$	penalty for unserved power

VARIABLES

$D_{i,s,w,t}, Q_{i,s,w,t}$	active and reactive power demand at node i
$E_{es,i,s,w,t}$	stored energy
$I^{ch}_{es,i,s,w,t}, I^{dch}_{es,i,s,w,t}$	charge-discharge indicator variables
$p_{k,s,w,t,l}, q_{k,s,w,t,l}$	step variables used in linearization of quadratic flows
$P^{ch}_{es,i,s,w,t}, P^{dch}_{es,i,s,w,t}$	power charged to and discharged from storage system
P_k, Q_k, θ_k	active and reactive power flows, and voltage angle difference of link k, respectively
$P^N_{g,i,s,w,t}, P^E_{g,i,s,w,t}$	active power produced by new and existing DGs
$P^{SS}_{\varsigma,s,w,t}, Q^{SS}_{\varsigma,s,w,t}$	active and reactive power imported from grid (upstream)
PL_k, QL_k	active and reactive power losses, respectively
$PL_{\varsigma,s,w,t}, QL_{\varsigma,s,w,t}$	active and reactive losses at substation ς
$Q_{c,i,s,w,t}$	reactive power injected by capacitor bank at node i
$Q^N_{g,i,s,w,t}, Q^E_{g,i,s,w,t}$	reactive power produced by new and existing DGs
$u_{g,i,t}, u_{k,t}$	utilization variables of existing DG and lines
V_i, V_j	voltage magnitudes at nodes i and j
$x_{g,i,t}, x_{es,i,t}, x_{c,i,t}, x_{k,t}$	investment variables for DG, energy storage system, capacitor banks, and distribution lines
$\delta_{i,s,w,t}$	unserved power at node i

FUNCTIONS

EC^{SS}_t	expected cost of energy purchased from upstream
$EmiC^{DG}_t$	expected emission cost of power production using DG

$\mathbf{EmiC}_t^{\mathbf{N}}, \mathbf{EmiC}_t^{\mathbf{E}}$	expected emission cost of power production using new and existing DGs, respectively
$\mathbf{EmiC}_t^{\mathbf{SS}}$	expected emission cost of purchased power
$\mathbf{ENSC}_t$	expected cost of unserved power
$\mathbf{InvC}_t^{\mathbf{CAP}}, \mathbf{MC}_t^{\mathbf{Cap}}$	NPV investment/maintenance cost of capacitor banks
$\mathbf{InvC}_t^{\mathbf{DG}}, \mathbf{MC}_t^{\mathbf{DG}}, \mathbf{EC}_t^{\mathbf{DG}}$	NPV investment/maintenance/expected energy cost of DGs, respectively
$\mathbf{InvC}_t^{\mathbf{ES}}, \mathbf{MC}_t^{\mathbf{ES}}, \mathbf{EC}_t^{\mathbf{ES}}$	NPV investment/maintenance/expected energy cost of an energy storage system, respectively
$\mathbf{InvC}_t^{\mathbf{LN}}, \mathbf{MC}_t^{\mathbf{LN}}$	NPV investment/maintenance cost of a distribution line

REFERENCES

[1] Energy Information Administration. International energy outlook 2014. Washington, DC: US Department of Energy; 2014.

[2] National Research Council. Advancing the science of climate change. Washington, DC: The National Academies Press; 2010.

[3] Jordehi AR. Optimisation of electric distribution systems: a review. Renew Sustain Energy Rev 2015;51:1088–100. http://dx.doi.org/10.1016/j.rser.2015.07.004.

[4] Soares T, Pereira F, Morais H, Vale Z. Cost allocation model for distribution networks considering high penetration of distributed energy resources. Electr Power Syst Res 2015;124:120–32. http://dx.doi.org/10.1016/j.epsr.2015.03.008.

[5] Ray PK, Mohanty SR, Kishor N. Classification of power quality disturbances due to environmental characteristics in distributed generation system. IEEE Trans Sustainable Energy 2013;4:302–13. http://dx.doi.org/10.1109/TSTE.2012.2224678.

[6] Martinez IC, Chen C-Y, Teng J-H. Utilising energy storage systems to mitigate power system vulnerability. IET Gener Transm Distrib 2013;7:790–8. http://dx.doi.org/10.1049/iet-gtd.2012.0694.

[7] Georgilakis PS, Hatziargyriou ND. Optimal distributed generation placement in power distribution networks: models, methods, and future research. IEEE Trans Power Syst 2013;28:3420–8. http://dx.doi.org/10.1109/TPWRS.2012.2237043.

[8] Paliwal P, Patidar NP, Nema RK. Planning of grid integrated distributed generators: a review of technology, objectives and techniques. Renew Sustain Energy Rev 2014;40:557–70. http://dx.doi.org/10.1016/j.rser.2014.07.200.

[9] Hung DQ, Mithulananthan N, Bansal RC. An optimal investment planning framework for multiple distributed generation units in industrial distribution systems. Appl Energy 2014;124:62–72. http://dx.doi.org/10.1016/j.apenergy.2014.03.005.

[10] Abu-Mouti FS, El-Hawary ME. Heuristic curve-fitted technique for distributed generation optimisation in radial distribution feeder systems. IET Gener Transm Distrib 2011;5:172. http://dx.doi.org/10.1049/iet-gtd.2009.0739.

[11] Pal BC, Jabr RA. Ordinal optimisation approach for locating and sizing of distributed generation. IET Gener Transm Distrib 2009;3:713–23. http://dx.doi.org/10.1049/iet-gtd.2009.0019.

[12] Maciel RS, Rosa M, Miranda V, Padilha-Feltrin A. Multi-objective evolutionary particle swarm optimization in the assessment of the impact of distributed generation. Electr Power Syst Res 2012;89:100–8. http://dx.doi.org/10.1016/j.epsr.2012.02.018.

[13] Vinothkumar K, Selvan MP. Fuzzy embedded genetic algorithm method for distributed generation planning. Electr Power Compon Syst 2011;39:346–66. http://dx.doi.org/10.1080/15325008.2010.528533.

[14] Ugranlı F, Karatepe E. Convergence of rule-of-thumb sizing and allocating rules of distributed generation in meshed power networks. Renew Sustain Energy Rev 2012;16:582–90. http://dx.doi.org/10.1016/j.rser.2011.08.024.

[15] Murty VVSN, Kumar A. Optimal placement of DG in radial distribution systems based on new voltage stability index under load growth. Int J Electr Power Energy Syst 2015;69:246–56. http://dx.doi.org/10.1016/j.ijepes.2014.12.080.

[16] Bagheri A, Monsef H, Lesani H. Integrated distribution network expansion planning incorporating distributed generation considering uncertainties, reliability, and operational conditions. Int J Electr Power Energy Syst 2015;73:56–70. http://dx.doi.org/10.1016/j.ijepes.2015.03.010.

[17] Sheng W, Liu K, Liu Y, Meng X, Li Y. Optimal placement and sizing of distributed generation via an improved nondominated sorting genetic algorithm II. IEEE Trans Power Delivery 2015;30:569–78. http://dx.doi.org/10.1109/TPWRD.2014.2325938.

[18] Mena R, Hennebel M, Li Y-F, Ruiz C, Zio E. A risk-based simulation and multi-objective optimization framework for the integration of distributed renewable generation and storage. Renew Sustain Energy Rev 2014;37:778–93. http://dx.doi.org/10.1016/j.rser.2014.05.046.

[19] Foster JD, Berry AM, Boland N, Waterer H. Comparison of mixed-integer programming and genetic algorithm methods for distributed generation planning. IEEE Trans Power Syst 2014;29:833–43. http://dx.doi.org/10.1109/TPWRS.2013.2287880.

[20] Ameli A, Bahrami S, Khazaeli F, Haghifam M-R. A multiobjective particle swarm optimization for sizing and placement of DGs from DG Owner's and distribution company's viewpoints. IEEE Trans Power Delivery 2014;29:1831–40. http://dx.doi.org/10.1109/TPWRD.2014.2300845.

[21] Wen S, Lan H, Fu Q, Yu DC, Zhang L. Economic allocation for energy storage system considering wind power distribution. IEEE Trans Power Syst 2015;30:644–52. http://dx.doi.org/10.1109/TPWRS.2014.2337936.

[22] Sedghi M, Aliakbar-Golkar M, Haghifam M-R. Distribution network expansion considering distributed generation and storage units using modified PSO algorithm. Int J Electr Power Energy Syst 2013;52:221–30. http://dx.doi.org/10.1016/j.ijepes.2013.03.041.

[23] Saboori H, Hemmati R, Abbasi V. Multistage distribution network expansion planning considering the emerging energy storage systems. Energy Convers Manag 2015;105:938–45. http://dx.doi.org/10.1016/j.enconman.2015.08.055.

[24] Aman MM, Jasmon GB, Bakar AHA, Mokhlis H. A new approach for optimum simultaneous multi-DG distributed generation Units placement and sizing based on maximization of system loadability using HPSO (hybrid particle swarm optimization) algorithm. Energy 2014;66:202–15. http://dx.doi.org/10.1016/j.energy.2013.12.037.

[25] Rao RS, Ravindra K, Satish K, Narasimham SVL. Power loss minimization in distribution system using network reconfiguration in the presence of distributed generation. IEEE Trans Power Syst 2013;28:317–25. http://dx.doi.org/10.1109/TPWRS.2012.2197227.

[26] Nekooei K, Farsangi MM, Nezamabadi-Pour H, Lee KY. An improved multi-objective harmony search for optimal placement of DGs in distribution systems. IEEE Trans Smart Grid 2013;4:557–67. http://dx.doi.org/10.1109/TSG.2012.2237420.

[27] Sedighizadeh M, Esmaili M, Esmaeili M. Application of the hybrid Big Bang-Big Crunch algorithm to optimal reconfiguration and distributed generation power allocation in distribution systems. Energy 2014;76:920–30. http://dx.doi.org/10.1016/j.energy.2014.09.004.

[28] Abdelaziz AY, Hegazy YG, El-Khattam W, Othman MM. Optimal allocation of stochastically dependent renewable energy based distributed generators in unbalanced distribution networks. Electr Power Syst Res 2015;119:34–44. http://dx.doi.org/10.1016/j.epsr.2014.09.005.

[29] Othman MM, El-Khattam W, Hegazy YG, Abdelaziz AY. Optimal placement and sizing of distributed generators in unbalanced distribution systems using supervised Big Bang-Big Crunch method. IEEE Trans Power Syst 2015;30:911–9. http://dx.doi.org/10.1109/TPWRS.2014.2331364.

[30] Rama Prabha D, Jayabarathi T, Umamageswari R, Saranya S. Optimal location and sizing of distributed generation unit using intelligent water drop algorithm. Sustain Energy Technol Assess 2015;11:106–13. http://dx.doi.org/10.1016/j.seta.2015.07.003.

[31] Mohamed Imran A, Kowsalya M, Kothari DP. A novel integration technique for optimal network reconfiguration and distributed generation placement in power distribution networks. Int J Electr Power Energy Syst 2014;63:461–72. http://dx.doi.org/10.1016/j.ijepes.2014.06.011.

[32] Singh AK, Parida SK. Allocation of distributed generation using proposed DMSP approach based on utility and customers aspects under deregulated environment. Int J Electr Power Energy Syst 2015;68:159–69. http://dx.doi.org/10.1016/j.ijepes.2014.12.076.

[33] Al Abri RS, El-Saadany EF, Atwa YM. Optimal placement and sizing method to improve the voltage stability margin in a distribution system using distributed generation. IEEE Trans Power Syst 2013;28:326–34. http://dx.doi.org/10.1109/TPWRS.2012.2200049.

[34] Atwa YM, El-Saadany EF, Salama MMA, Seethapathy R. Optimal renewable resources mix for distribution system energy loss minimization. IEEE Trans Power Syst 2010;25:360–70. http://dx.doi.org/10.1109/TPWRS.2009.2030276.

[35] Yang Y, Zhang S, Xiao Y. An MILP (mixed integer linear programming) model for optimal design of district-scale distributed energy resource systems. Energy 2015;90:1901–15. http://dx.doi.org/10.1016/j.energy.2015.07.013.

[36] Montoya-Bueno S, Munoz JI, Contreras J. A stochastic investment model for renewable generation in distribution systems. IEEE Trans Sustainable Energy 2015;6:1466–74. http://dx.doi.org/10.1109/TSTE.2015.2444438.

[37] Sfikas EE, Katsigiannis YA, Georgilakis PS. Simultaneous capacity optimization of distributed generation and storage in medium voltage microgrids. Int J Electr Power Energy Syst 2015;67:101–13. http://dx.doi.org/10.1016/j.ijepes.2014.11.009.

[38] Mahmoud K, Yorino N, Ahmed A. Optimal distributed generation allocation in distribution systems for loss minimization. IEEE Trans Power Syst 2015;1–10:http://dx.doi.org/10.1109/TPWRS.2015.2418333.

[39] Vinothkumar K, Selvan MP. Hierarchical agglomerative clustering algorithm method for distributed generation planning. Int J Electr Power Energy Syst 2014;56:259–69. http://dx.doi.org/10.1016/j.ijepes.2013.11.021.

[40] Atwa YM, El-Saadany EF. Probabilistic approach for optimal allocation of wind-based distributed generation in distribution systems. IET Renew Power Gener 2011;5:79. http://dx.doi.org/10.1049/iet-rpg.2009.0011.

[41] Murty VVS, Kumar A. Mesh distribution system analysis in presence of distributed generation with time varying load model. Int J Electr Power Energy Syst 2014;62:836–54. http://dx.doi.org/10.1016/j.ijepes.2014.05.034.

[42] Blank L, Tarquin A. Engineering economy. 7th ed. New York: McGraw-Hill Science/Engineering/Math; 2011.

[43] Munoz-Delgado G, Contreras J, Arroyo JM. Joint expansion planning of distributed generation and distribution networks. IEEE Trans Power Syst 2015;30:2579–90. http://dx.doi.org/10.1109/TPWRS.2014.2364960.

[44] Phonrattanasak P. Optimal placement of DG using multiobjective particle swarm optimization. In: 2nd international conference on mechanical and electrical technology (ICMET); 2010. p. 342–6. http://dx.doi.org/10.1109/ICMET.2010.5598377.

[45] Zhang H, Heydt GT, Vittal V, Quintero J. An improved network model for transmission expansion planning considering reactive power and network losses. IEEE Trans Power Syst 2013;28:3471–9. http://dx.doi.org/10.1109/TPWRS.2013.2250318.

[46] Vielma JP, Ahmed S, Nemhauser G. Mixed-integer models for nonseparable piecewise-linear optimization: unifying framework and extensions. Oper Res 2009;58:303–15. http://dx.doi.org/10.1287/opre.1090.0721.

[47] Williams HP. Model building in mathematical programming. 4th ed. New York: Wiley; 1999.

[48] Romero-Ramos E, Riquelme-Santos J, Reyes J. A simpler and exact mathematical model for the computation of the minimal power losses tree. Electr Power Syst Res 2010;80:562–71. http://dx.doi.org/10.1016/j.epsr.2009.10.016.

[49] Lavorato M, Franco JF, Rider MJ, Romero R. Imposing radiality constraints in distribution system optimization problems. IEEE Trans Power Syst 2012;27:172–80. http://dx.doi.org/10.1109/TPWRS.2011.2161349.

[50] Zhang J, Yuan X, Yuan Y. A novel genetic algorithm based on all spanning trees of undirected graph for distribution network reconfiguration. J Mod Power Syst Clean Energy 2014;2:143–9. http://dx.doi.org/10.1007/s40565-014-0056-0.

[51] Ahmadi H, Martí JR. Mathematical representation of radiality constraint in distribution system reconfiguration problem. Int J Electr Power Energy Syst 2015;64:293–9. http://dx.doi.org/10.1016/j.ijepes.2014.06.076.

[52] Georgilakis PS, Hatziargyriou ND. A review of power distribution planning in the modern power systems era models, methods and future research. Electr Power Syst Res 2015;121:89–100. http://dx.doi.org/10.1016/j.epsr.2014.12.010.

[53] Sinden G. Characteristics of the UK wind resource: long-term patterns and relationship to electricity demand. Energy Pol 2007;35:112–27. http://dx.doi.org/10.1016/j.enpol.2005.10.003.

[54] Subhadarshi S. Minding the P's and Q's: real and reactive power assessment of hybrid energy conversion systems with wind and solar resources. Iowa State University; 2013.

[55] Li YZ, Wu QH, Li MS, Zhan JP. Mean-variance model for power system economic dispatch with wind power integrated. Energy 2014;72:510–20. http://dx.doi.org/10.1016/j.energy.2014.05.073.

[56] Zhao M, Chen Z, Blaabjerg F. Probabilistic capacity of a grid connected wind farm based on optimization method. Renew Energy 2006;31:2171–87. http://dx.doi.org/10.1016/j.renene.2005.10.010.

[57] Aien M, Rashidinejad M, Fotuhi-Firuzabad M. On possibilistic and probabilistic uncertainty assessment of power flow problem: a review and a new approach. Renew Sustain Energy Rev 2014;37:883–95. http://dx.doi.org/10.1016/j.rser.2014.05.063.

[58] Wang M, Zhong J. Islanding of systems of distributed generation using optimization methodology. In: 2012 IEEE Power and Energy Society general meeting; 2012. p. 1–7. http://dx.doi.org/10.1109/PESGM.2012.6345009.

[59] Fitiwi DZ, Olmos L, Rivier M, de Cuadra F, Pérez-Arriaga IJ. Finding a representative network losses model for large-scale transmission expansion planning with renewable energy sources. Energy 2016;101:343–58. http://dx.doi.org/10.1016/j.energy.2016.02.015.

[60] Ellis A, Nelson R, Von Engeln E, Walling R, MacDowell J, Casey L, et al. Reactive power performance requirements for wind and solar plants. In: 2012 IEEE Power and Energy Society general meeting; 2012. p. 1–8. http://dx.doi.org/10.1109/PESGM.2012.6345568.

CHAPTER

Scheduling interconnected micro energy grids with multiple fuel options

4

A. Zidan*,†, H.A. Gabbar*

University of Ontario Institute of Technology (UOIT), Oshawa, ON, Canada Assiut University, Assiut, Egypt†*

4.1 MICRO ENERGY GRIDS AND NETWORKS

The continuous growth in energy demands, the development of generation technologies, and the environmental impacts of power generation highlight the urgent need for smart energy grids. In response, governments and utilities have created incentives for transforming distribution grids from passive to active networks. This goal could be achieved through the creation of small networks, called micro energy grids (MEGs), that can be autonomous and independent from the backbone grid.

4.1.1 MICRO ENERGY GRID STRUCTURE

An MEG is a small local grid at the low voltage or medium voltage distribution level, which combines different types of loads and distributed energy resources (DERs) [1]. The MEG can autonomously meet the power, energy, and quality requirements of the customers in its area. As shown in Fig. 1, MEG is connected to the main grid through a switch. MEGs can be integrated with wind turbines (WT), photovoltaics (PV), battery energy storage systems (ESSs), boilers, and combined heat and power (CHP) generators with different types and technologies (eg, microturbines (MTs), fuel cells (FCs), etc.). CHPs generate heat and power simultaneously. All of the energy sources coordinate to satisfy the power and heat load in the microgrid energy grid. The MEG can be connected to the main grid, and it supplies/absorbs excess/deficient power to/from the main grid.

4.1.2 MICRO ENERGY GRID CONTROL LOOPS

As with conventional power systems, control is a key technology for the deployment of an MEG. An MEG has a hierarchical control structure with different layers. A general control scheme for operating an MEG is shown in Fig. 2 [2]. For instance, during the islanded mode, the MEG must cope with the variations such as load

Smart Energy Grid Engineering.

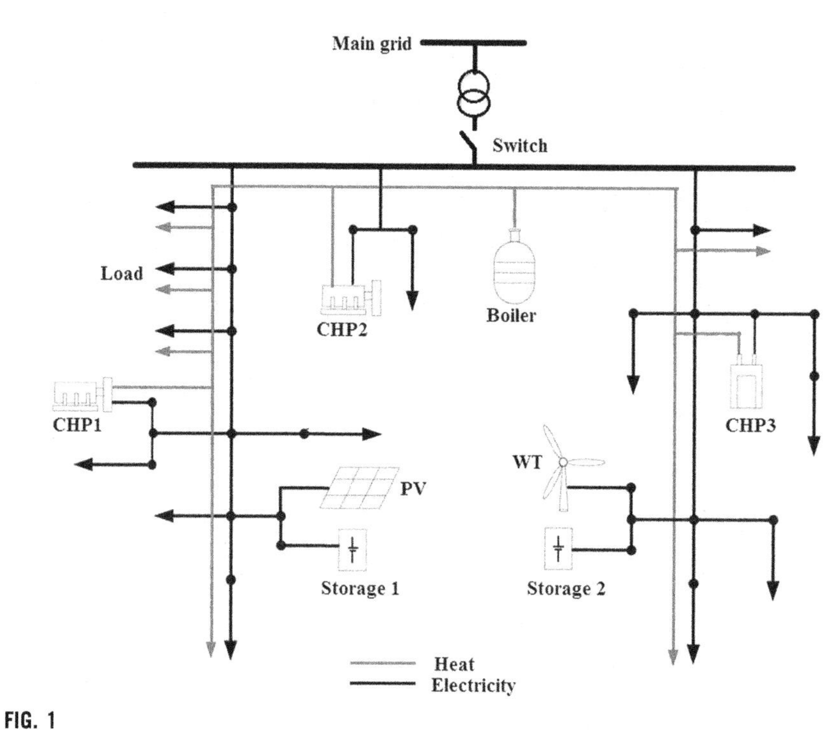

FIG. 1

Basic structure of a micro energy grid.

disturbances. So, it needs to perform active power/frequency regulation and reactive power/voltage regulation (Q/V) by using proper control loops. The control loops in MEGs have four levels: local, secondary, central/emergency, and global controls [2]. The local control level deals with initial primary controls such as current and voltage control loops for distributed energy sources. The secondary control level is responsible for inside ancillary services and for regulating the frequency and voltage deviations of the MEG toward zero after any change in load and/or supply. The central/emergency control level is performed by the MEG central controller (CC), which interfaces between its own MEG and other microgrids, as well as the main grid. This control level is responsible for all possible emergency control schemes and protection plans, in order to maintain the stability and availability of its MEG. The global control coordinates the CCs for interconnected MEGs. The central global control provides optimal economical operation for each MEG and organizes the energy flow among interconnected MEGs and the main grid.

4.1.3 GLOBAL CONTROL AND POWER DISPATCHING

Global control is accomplished by a market operator acting in an economy-based energy management level for power exchanges with the main grid and/or other neighboring MEGs. This global control objective requires wide area monitoring and estimation for operating indices such as fuel and device storage conditions,

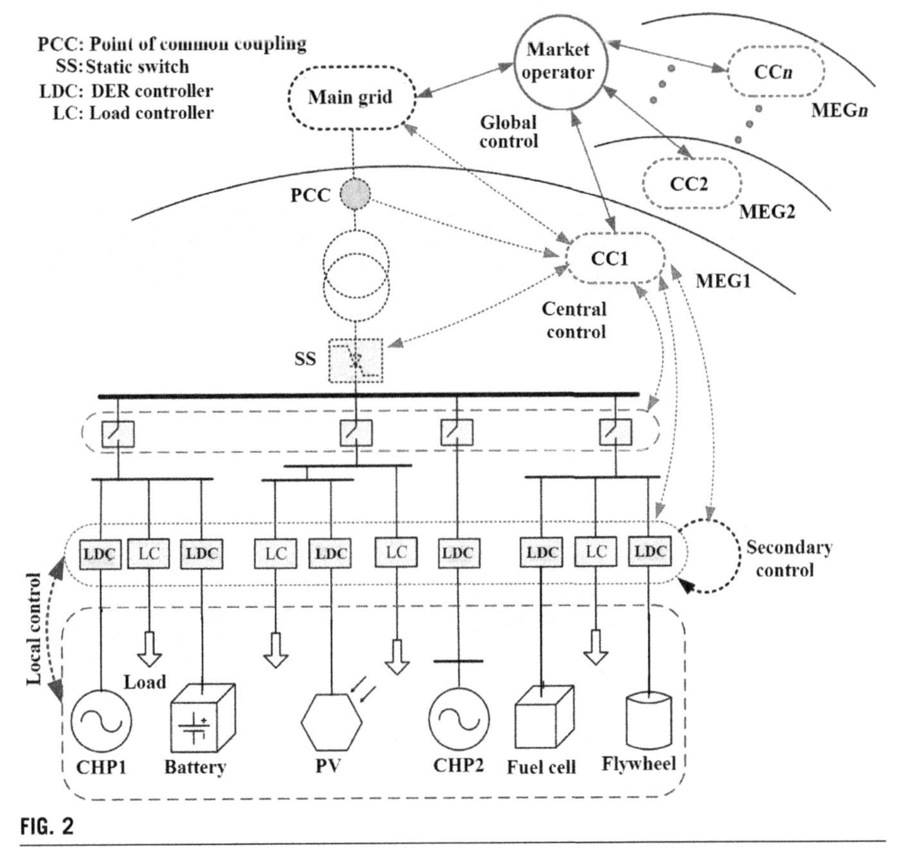

FIG. 2

A general scheme for MEG control levels [2].

commercial power cost and demand charge tariffs, MEG reliability, real/reactive power components, predicted weather, system constraints, and demand pattern. The global control center (market operator) for interconnected MEGs is responsible for determining the optimal generation schedule (ie, output power of the DERs) in order to minimize production costs and to balance the demand and generation from both MEGs and the main grid.

Global control and MEG CCs supervise the MEGs' market activities, such as buying and selling active and reactive power to the main grid and transferring energy from an MEG to nearby feeders. They perform energy management for the MEGs through a subset of basic functions, including load and weather forecasting, economic scheduling, overall security assessment, and demand-side management. The global control for interconnected MEGs is implemented through the cooperation of various MEG CCs located in all MEGs, based on two-way communication, DERs data, and control commands. The global control can be deployed by optimizing the power exchanged between MEGs, as well as the main grid, in order to maximize the production in response to market prices and security constraints.

4.2 MODELING MICRO ENERGY GRID NETWORKS

Accurate mathematical models of MEG components are needed to analyze the various characteristics of the system and the power/heat flow among them. For instance, the efficiency of each device, such as the CHP prime mover, is not a constant value, because it varies with the capacity, the load ratio, and other system factors. However, the intermittent or climate-based characteristics of renewable energy sources; the couplings between electricity and heating and cooling networks; and the diverse characteristics of the MEG components make it very difficult to model these systems. As shown in Fig. 3, a typical MEG consists of four main elements: the CHP prime mover, electrical/heat storage systems, loads, and renewable energy sources. All energy flows (ie, electricity and heat) depend on the efficiency parameters of each of the components.

4.2.1 MODELING OF CHP PRIME MOVERS

CHP prime movers play a vital role in the cogeneration system because they generate electricity and heat simultaneously. Thus, the precision of the prime mover's modeling affects the accuracy of the whole system model. There are several types of CHP prime movers in the energy market, such as steam turbines, combustion turbines, MTs, and FCs. MTs have great flexibility, as they are available in different scales. Given that they are fed by natural gas, MTs have the environmental advantage of low emissions compared to other oil-based generators. Similarly, FCs are environmentally friendly and produce high energy efficiencies under varying load rates. There

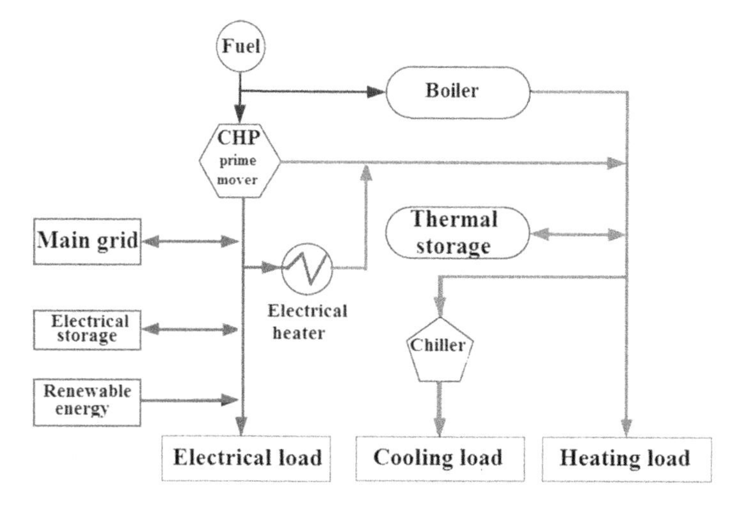

FIG. 3

Energy flow for MEGs.

are five FC technologies available in the market, but the proton exchange membrane fuel cell and the solid oxide fuel cell are most widely used [3].

The electrical efficiency, thermal efficiency, and the pollutant gas emission characteristics of MTs depend on their capacities and their load factors [4,5]. Bando et al. [5] provided empirical formulas for the MT's electrical thermal efficiencies. Presently, it is difficult to create a generalized model; however, the model could consider a specific MT with a given capacity. A generalized DER model can be represented by the following equations:

$$f_{\mathrm{fue},i}(t) = \frac{P_i(t)}{\eta_{i,P}} \forall t \in T, \quad P_i(t) \leq P_{\mathrm{r},i}, \ i \in G \tag{1}$$

$$H_i(t) = P_i(t) \frac{\eta_{i,H}}{\eta_{i,P}} \quad \forall t \in T, \ i \in G \tag{2}$$

$$E_i(t) = K_i P_i(t) \forall t \in T, \quad i \in G \tag{3}$$

where G is the set of DERs, $f_{\mathrm{fue},i}$ is the fuel consumed by DER i at hour t, P_i is the output power from DER i at hour t, T is the set of hourly periods in the next day, $\eta_{i,P}$ is the power efficiency of DER i, $P_{\mathrm{r},i}$ is the rated power for DER i, H_i is heat power generated by DER i at hour t, $\eta_{i,H}$ is the heat efficiency of DER i, E_i is the CO_2 emissions from DER i at hour t, and K_i is the carbon footprint for the energy produced by DER i in kg CO_2/kW h.

4.2.2 MODELING ENERGY STORAGE SYSTEMS

The economic evaluation of ESSs (electrical, thermal, or fuel) is a challenging task because of difficulties such as the intertemporal nature of storage and the need to make decisions despite the great uncertainty that surrounds future circumstances. However, fluctuations in renewable energy sources and loads lead to dynamic changes in the power, heat, and cooling balance between generation and loading. To deal with this situation, ESSs are vital components. ESSs can smooth out intermittent renewable power generation, flatten the load (by charging when the load is low and discharging during peak load times), and exploit time-varying electricity prices for arbitrage. Therefore, any local shortage in supplying the loads could be met by discharging the ESS or by purchasing power/heat from one of the MEG neighbors. The size of ESS units depends on the load size and the renewable energy (wind and PV) penetration. More importantly, when an MEG is isolated, the ESS can provide voltage and frequency references for the microgrid, alone or together with other sources [6]. For instance, for solar generation systems at night and wind generation systems during periods of stillness, ESS can supply energy to consumers. Therefore, the rational use of an ESS is a vital factor for providing the safe, economical, and reliable operation of MEG systems.

ESSs are based on different principles, and they take different forms, which can be divided into chemical energy storage and physical energy storage. For example, battery storage and electrochemical capacitors provide the chemical energy storage, and pumped hydroelectric storage, flywheel storage, compressed air energy storage, and superconducting magnetic energy storage provide the physical energy storage [7]. The ESS plays an important role in MEGs, especially given the high penetration of renewable sources, but their high cost restricts their widespread application. Batteries are the oldest and most mature storage technology. Thus, they are the most common storage devices in MEGs [8].

Many battery models have been developed in the literature for different applications, and they can be divided into three groups:

(1) Chemical mechanism modeling based on the internal battery [9]: This model is the most accurate one to date. However, it contains many coupled differential equations, and it involves great computational complexity.
(2) Circuit modeling based on the equivalent circuit of the battery system [10,11]: This model is easy to understand, because it has a simple structure. So, it is used in most studies. However, this model has some defects and needs a large volume of experimental data in order to estimate the circuit parameters.
(3) A numerical model [12]: In this model the battery cell is assumed to have two parts: the bound charge and the available charge. The available charge provides electrical power to the load, and the bound charge is the charge available at a certain rate [13,14]. The model is expressed as a set of two differential equations. Hence, through appropriate discretization, these two equations can be transformed into five submodels, each of which has explicit meanings and is easy to calculate.

In addition, thermal storage tanks can be used for space heating and cooling, and hot water tanks can be used to supply hot water. Heat loss depends on the heat radiation from tanks. Their corresponding parameters are always set on the basis of the manufacturer's specifications.

4.2.3 MODELING RENEWABLE ENERGY SOURCES

Renewable energy refers to energy that is collected from resources that are naturally replenished on a human timescale, such as sunlight, wind, rain, tides, waves, and geothermal heat. Renewable energy can be provided in the form of electricity generation, air and water heating and cooling, transportation, and rural energy services. The use of renewable energy systems, especially PV and wind power, is increasing because of their inexhaustible and environmentally friendly nature. Renewable energy systems have intermittent output power, however, because they depend on climatic conditions (eg, PV output power depends on solar irradiance, and the output power from wind generators depends on wind speed). Thus, many studies have focussed on building forecasting models for PV and wind power output. For instance, in day-ahead energy management studies, the output power of PV and wind generators is modeled using two main methods: (1) the deterministic model, which is the most common model, giving the forecast output power over a certain period; and (2) the probabilistic model, which represents the random nature of the PV and wind power sources by using chance-constrained programming or an expectancy model.

Several published studies have focussed on the prediction of PV power with good accuracy. Solar irradiance can be forecasted on different timescales, using many methods such as artificial neural networks (ANNs) [15], auto-regression [16], and hybrid methods [17]. Furthermore, a deterministic model based on the Hottel-Liu-Jordan formula can be applied to calculate the solar radiation [18]. Then, the forecasted irradiance and temperature data can be used as inputs for commercial PV simulation software such as TRNSYS [19], PVFORM [20], and HOMER [21]. Other methods can directly predict PV power based on prior information or other readily accessible data [22].

The output power of a wind generator can be determined by the wind speed. Thus, many methods, including ANNs [23], Kalman filtering methods [24], and time-series methods [25], have been proposed for forecasting wind speed. In addition, wind speed can be forecasted using a hybrid forecasting approach, which typically consists of an autoregressive integrated moving average prediction model for the linear component of a time series and a nonlinear prediction model for the nonlinear component [26]. The hybrid forecasting approach offers another approach to short-term forecasting, but it does not always produce better performance under differing conditions [26].

4.2.4 MODELING LOADS

The load for each MEG can be divided into heat, cold, and electricity loads, which are important input parameters for both planning and energy management studies. Load forecasting techniques help the designer make important decisions for MEGs, such as purchasing or selling power associated with the MEG, charging or discharging the ESS, and purchasing or selling power via the main grid. In order to provide accurate load forecasting, historical data on energy consumption and appropriate weather data, such as temperature and humidity, must be provided by smart meters and sensors available at different locations in the MEG. At each time step, these measured data on previous periods can be used to predict the loads for a future predefined horizon.

Load modeling can be established by either physical or statistical methods [27]. Physical models use the analysis of physical processes such as heat transfer, and statistical models depend on surveys of the dominant end effects without much emphasis on the details of the processes. For instance, a statistical analysis of residential, commercial, and industrial load profiles was presented in [28]. Furthermore, methods such neural networks, time series, and fuzzy logic can be used [29,30]. In the future smart grid, energy consumption will be affected by real-time energy prices, which will be included in load modeling. Moreover, customers may own renewable energy generators such as PVs on their house roofs. So, the fluctuations in renewable energy availability will also affect the customers' behavior, which is another factor to be considered when modeling loads.

4.3 INTERCONNECTION OF MICRO ENERGY GRIDS

Future power systems are expected to include interconnected micro energy grids combined with solar, wind, and plugin electrical vehicles. MEGs with excessive energy generation can trade with other MEGs with power deficits for their mutual benefit. The local energy demand for each MEG can be served by its local generation and storage discharging, as well as by using power purchased from the main power grid and/or neighboring MEGs. As shown in Fig. 4, each MEG operator is responsible for the energy scheduling in its MEG and for its energy trading with the main grid and other interconnected MEGs.

4.3.1 INTERCONNECTED MEGs: INTERACTION AND COOPERATIVE POWER DISPATCHING

The management of energy flows is a challenge in interconnected MEGs. MEGs in different locations have different distributed energy sources, renewable generation patterns, and local load profiles. Hence, through trading energy with each other, interconnected MEGs can exploit the diversities of supply and demand patterns, thus improving their operational performance. MEGs are independent players, because each MEG interacts with other MEGs to optimize its own performance. To maximize mutual benefits among interconnected MEGs, they can interact with each other based on the Nash bargaining theory, which is a cooperative game theoretical framework that helps decision makers achieve fair and Pareto-optimal benefit sharing [31]. Each MEG bargains with other interconnected MEGs to determine the amount of energy trading and the associated payment. The energy trading and payments between MEGs should satisfy the market clearing constraints. Unlike a single

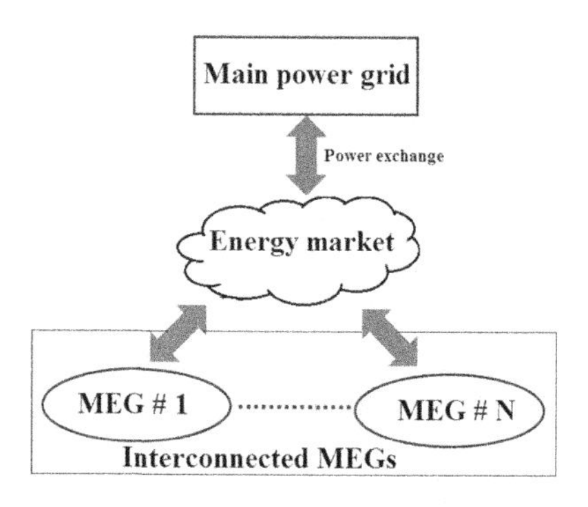

FIG. 4

Energy trading among multiple interconnected MEGs.

MEG operation, interconnected MEGs not only coordinate their local energy supplies and demands, but they also conduct energy trading with each other. Each MEG aims to optimize its own performance, in terms of minimizing its operating cost through energy trading.

4.3.2 ADVANTAGES OF INTERCONNECTED MEGs

A single MEG can only generate and distribute energy within a localized area. Also, the intermittent power generated by renewable energy sources leads to variations in the output power, voltage, and frequency. Additionally, if one or more local generators in the single MEG fail or become otherwise unavailable while it operates in islanded mode, the single MEG is not secure enough to meet its load demand, due to its restricted energy generation capability. In such cases, the single MEG is forced to apply a load-shedding technique to ensure the supply-demand balance, which is inconvenient and economically undesirable because it causes interruption to customers' operations.

To overcome this challenge, MEGs in neighboring locations can be interconnected in order to exchange power with each other and hence to ensure the improved security of supply. The feasibility and benefits of interconnecting MEGs are discussed in the literature. For example, improved reliability, high penetration of renewable energy, and improved generation efficiency can be achieved by using interconnected MEGs. Interconnecting MEGs has several additional benefits: (1) better utilization of assets, optimization, and efficient operation; (2) accommodation of all generation and storage options; (3) supply of the energy quality needed by today's industry; (4) operation secured against attacks and natural disasters; and (5) introduction of new services, products, and markets.

4.4 COMPUTATIONAL OPTIMIZATION TECHNIQUES: A BRIEF INTRODUCTION

Computational optimization is a group of mathematical techniques used to select an optimal solution based on a specific criterion from a set of available alternatives. Optimization includes finding the best available values of an objective function given a set of constraints. Diwerak [32] described the optimization process as an iterative procedure composed of an optimizer and a model. The model is the process of identifying objectives, variables, and constraints for a given problem [33]. The optimizer invokes the model with values for the decision variables, while the model calculates the objective function and constraints. The optimizer uses these calculated values to generate a new set of decision variables. This iterative sequence continues until the optimization criterion is satisfied [32].

Different optimization algorithms can be used, depending on the type of optimization problem (ie, types of decision variables, objective functions, and constraints). Authors in [34,35] defined different categories of optimization problems, including continuous and discrete, constrained and unconstrained, global and local, stochastic and deterministic, multimodal and multiobjective, heuristic, and metaheuristic optimization. In some optimization problems, the optimizer cannot always find the optimum solution. Also, some optimization problems can be unfeasible due to the given characteristics. For example, nondeterministic polynomial-time hard problems might need exponential computation time in order to obtain an optimal solution; hence, they require high computational times for practical purposes [36].

Heuristic methods can find a good solution among a large set of feasible solutions, with less computational effort than optimization techniques [37]. Therefore, they are useful approaches for optimization problems, when applying classic optimization techniques is not practical. Metaheuristic methods are another category of optimization techniques that can find an optimal solution in a discrete search-space. Metaheuristics can combine many heuristic methods because the first heuristic method can find a primary or initial solution, and the other one can be used to find a better solution.

Currently, hybrid and parallel metaheuristics have been applied in some applications. The hybrid metaheuristic combines other optimization approaches with the metaheuristic one. The parallel metaheuristics model runs multiple metaheuristic searches in parallel by using parallel computing techniques. In parallel computing, a large problem can be divided into smaller ones, with their calculations carried out simultaneously. Common types of problems found in parallel computing for MEG applications are the Monte Carlo simulation [38] and dynamic programming [39].

4.5 OPTIMIZATION TECHNIQUES APPLIED TO MICRO ENERGY GRID PROBLEMS

MEG planning and operational scheduling problems have largely addressed cost minimization objectives. Beyond economic goals, other goals can also be considered, including environmental impact, power quality, and reliability.

4.5.1 GENERATION MIX SELECTION AND SIZING

Vafaei and Kazerani [40] used traditional optimization techniques to select and size different generation technologies and storage devices for a microgrid, in order to minimize operational costs. This optimization model was formulated as a mixed-integer programming problem in the general algebraic modeling system environment. In [41] the reduced-gradient method was used for the power mix selection of microgrids, and Matlab software was used to simulate the system. Heuristics are widely used in sizing and power generation mix selection. Erdinc and Uzunoglu [37] highlights some heuristic optimization techniques for the sizing of hybrid renewable energy systems, such as genetic algorithm, particle swarm optimization, and ant colony. Chen et al. [42] proposes a cost-benefit analysis for the optimal sizing of an ESS in a microgrid. Time-series and feed-forward neural network techniques are used for forecasting wind speed and solar radiance, respectively. The planning problem is formulated as a mixed-integer linear programming model, which is solved using a modeling language for mathematical programming (AMPL). Presently, some mathematical programming methods are implemented by commercial software tools that are widely used in microgrid planning. ETAP, HOMER, and WEBOPT are introduced as examples of these tools [43]. HOMER software is widely used for microgrid modeling purposes. For instance, the study in [44] used it to define the layout of power plants for a hybrid microgrid in remote islands of the Republic of Maldives.

4.5.2 SITING

The literature has many studies on the allocation of energy resources for community energy systems. There are two main approaches: power-line layout and equipment siting (ie, power and storage). Both are focussed on power-loss minimization and quality maintenance goals. The study in [45] used dynamic programming as a traditional approach for designing cost-optimized microgrid architectures subject to reliability constraints. This method determines the optimal power-line layout between local generators and load points, given their locations and the rights of way for possible interconnections. Regarding microgrid siting problems, some multiobjective optimization algorithms are combined with sensitivity analysis. Buayai et al. [46] proposes a two-stage multiobjective optimization process for microgrid planning. In the first stage, the loss sensitivity factor is used to identify the microgrid area in a primary distribution system. In the second stage, a Pareto-based nondenominated sorting genetic algorithm (NSGA-II) finds the locations and sizes of a specified number of generators within the microgrids. The multiobjective functions include power loss, load voltage deviation, and annualized investment cost.

4.5.3 OPERATION-SCHEDULING PROBLEM

The operation-scheduling problem is usually solved using mathematical computing techniques and computer software. Final scheduling must fulfill microgrid goals, given demand, the operational and system constraints of the available resources, and the corresponding transmission capabilities. Niknam et al. [47] propose a probabilistic approach for the economic or emission management of microgrids, which includes uncertainties and a modified multiobjective algorithm based on the gravitational search algorithm, used to find the Pareto-optimal front of the operation management problem. Furthermore, some authors look for new approaches to power-source scheduling in microgrids. For instance, a calculation method for microgrid surplus load is proposed in [48]. A simulation model for a microgrid with wind generation, MTs, and FCs is also established. Similarly, Palma-Behnke et al. [49] proposes an energy management system that includes demand-side management to minimize operational costs while supplying load demands.

4.6 INTERCONNECTED MICRO ENERGY GRIDS: SCHEDULING OPTIMIZATION

The problem of scheduling optimization for interconnected MEGs is similar to the economic dispatch with traditional generation units. The purpose of the problem is to minimize total operational cost and emissions, while considering equality and inequality constraints.

4.6.1 OBJECTIVE (FITNESS) FUNCTION

The objectives for the interconnected micro energy grids scheduling problem are represented as follows:

$$\text{Minimize}(OF_1, OF_2) \tag{4}$$

$$OF_1 = \sum_{t \in T} \left(C_{\text{ope}}(X_t) + C_{\text{pur}}(X_t) + C_{\text{losses}}(X_t) \right) \tag{5}$$

$$OF_2 = \sum_{t \in T} \left(E_{\text{grid}}(X_t) + E_{\text{DER}}(X_t) + E_{\text{bo}}(X_t) \right) \tag{6}$$

$$C_{\text{ope}}(X_t) = \sum_{i \in G} \left(C_{\text{gas},i} f_{\text{fue},i}(t) + C_{\text{m},i} P_i(t) + C_{\text{su},i} SU_i(t) \right) \quad \forall t \in T \tag{7}$$

$$C_{\text{pur}}(X_t) = \sum_{j \in n_{\text{MG}}} \left(C_{\text{buy}}(t) P_{\text{buy},j}(t) - C_{\text{sell}}(t) P_{\text{sell},j}(t) + H_{\text{bo},j}(t) (C_{\text{heat}} + C_{\text{m,bo}}) \right) \quad \forall t \in T \tag{8}$$

$$C_{\text{losses}}(X_t) = \sum_{j \in n_{\text{MG}}} C_{\text{buy}}(t) P_{\text{losses},j}(t) \quad \forall t \in T \tag{9}$$

$$E_{\text{grid}}(t) = K_{\text{grid}} P_{\text{grid}+}(t) \quad \forall t \in T \tag{10}$$

$$E_{\text{DER}}(t) = \sum_{i \in G} E_i(t) \quad \forall t \in T \tag{11}$$

$$E_{\text{bo}}(t) = \sum_{j \in n_{\text{MG}}} K_{\text{bo}} H_{\text{bo},j}(t) \quad \forall t \in T \tag{12}$$

where OF_1(cost) and OF_2(emission) are the objectives required to be minimized, X_t is the decision variable vector at hour t, C_{ope} is the operational cost, C_{pur} is the energy purchase cost, C_{losses} is the energy losses cost, E_{grid} is the CO_2 emissions from the main grid, E_{DER} is the total CO_2 emissions from all DERs in the interconnected-MEG system, E_{bo} is the CO_2 emissions from all boilers in the interconnected-MEG system, $C_{\text{gas},i}$ is the gas price required for DER i, $C_{\text{m},i}$ is the maintenance cost of DER i, $C_{\text{su},i}$ is the start-up cost of DER i, SU_i is the start-up status of DER i at hour t, n_{MG} is the total number of MEGs in the system, C_{buy}/C_{sell} is the buying/selling price of electricity at hour t, P_{grid+}/P_{grid-} is the power bought/sold from/to the main grid at hour t, C_{heat} is the heat price, $H_{bo,j}$ is the heat supplied by the boiler of MEG j at hour t, $C_{m,bo}$ is the maintenance cost of the boilers, $P_{losses,j}$ is the energy losses of MEG j at hour t, K_{grid} is the carbon footprint for the energy purchased from the grid in kg CO_2/kW h, and K_{bo} is the emission from the boiler in kg CO_2/kW h.

4.6.2 CONSTRAINTS

Power and heat balance constraints

The operational constraints for the interconnected micro energy grids scheduling problem are represented as follows:

$$\sum_{i\in G} P_i(t) + \sum_{j\in n_{MG}} \left(P_{buy,j}(t) - P_{sell,j}(t)\right) = \sum_{j\in n_{MG}} \left(P_{ld,j}(t) + P_{losses,j}(t)\right) \quad \forall t\in T \tag{13}$$

$$0 \le P_{grid+}(t), P_{grid-}(t) \le P_{grid,max} \quad \forall t\in T \tag{14}$$

$$P_{i,min} \le P_i(t) \le P_{i,max} \quad \forall i\in G,\ t\in T \tag{15}$$

$$Q_{i,min} \le Q_i(t) \le Q_{i,max} \quad \forall i\in G,\ t\in T \tag{16}$$

$$\sum_{i\in G_j} H_i(t) + H_{bo,j}(t) \ge H_{ld,j} \quad \forall t\in T,\ j\in n_{MG} \tag{17}$$

where $P_{ld,j}$ and $H_{ld,j}$ are the power and heat demand of MEG j at hour t; $P_{grid,max}$ is the upper capacity of the main grid; $P_{i,min}$ and $P_{i,max}$ are the lower and upper active power generation of DER i, respectively; Q_i is the output reactive power from DER i at hour t; $Q_{i,min}$ and $Q_{i,max}$ are the lower and upper reactive power generation of DER i, respectively; and G_j is the DERs of MEG j.

Voltage limits constraints

$$V_{min} \le V_k(t) \le V_{max} \quad \forall k\in N_{bus}, t\in T \tag{18}$$

where V_k is the voltage magnitude in p.u. (voltage in *per-unit system* as fraction of a defined base unit quantity) at bus k and hour t; N_{bus} is the total number of buses for the interconnected-MEG system; and V_{min} and V_{max} are the minimum and maximum voltage limits, respectively (ie, 0.9 and 1.05 p.u.).

Current limits constraints

$$I_{l,min} \le I_l(t) \le I_{l,max} \quad \forall l\in N_l,\ t\in T \tag{19}$$

where I_l is the current magnitude flowing in line l during hour t, N_l is the total number of lines for the interconnected-MEG system, and $I_{l,\min}$ and $I_{l,\max}$ are the minimum and maximum current limits for line l.

Finally, the decision variables (optimal active and reactive output powers for each generator) can be determined on an hourly basis:

$$X_t = [P_1(t)P_2(t)\cdots P_G(t)Q_1(t)Q_2(t)\cdots Q_G(t)] \quad \forall t \in T \tag{20}$$

4.6.3 CHALLENGES

Modern energy systems display a high degree of complexity and uncertainty due to rising factors such as aging and constrained infrastructure, the penetration of distributed generation, new loads such as electric vehicles, the emergence of energy storage devices, and the integration of renewable energy sources. Therefore, new technical challenges have arisen in microgrid planning, operation, and control. Intermittent power generation from renewable sources and the variation in load demands should be considered when modeling energy systems. For instance, renewable power generation may deviate from its forecasted values; hence, predefined microgrid operation schedules may be violated. Furthermore, the seasonal and yearly variation of weather conditions may affect the operation of microgrids, and, thus, these variations should be investigated during microgrid planning. In addition, the mobility of dynamic electric vehicles leads to uncertainty as to their numbers at a specific location at a given time, leading to randomness in the capacity of vehicle-to-grid power.

In order to address these challenges, stochastic modeling and optimization tools should be used for microgrid planning, operation, and control. These methods can also be used to investigate the following issues:

- The two operation modes (ie, islanded mode and grid-connected mode) of a microgrid have two different objectives. During islanded mode, each microgrid minimizes its own generation cost. On the other hand, the grid-connected mode allows energy transactions between the microgrid and the main grid and/or neighboring microgrids. Therefore, in order to accommodate both modes of operation, appropriate stochastic models should be developed for each mode of operation, and the trade-off between the operation objectives of the main grid and microgrids needs to be investigated.
- Each microgrid is designed to supply its local electrical and heat loads in a small geographical area. So, weather conditions, such as wind speeds and solar radiation, are relatively similar within the service area. Consequently, renewable power generation and electrical and heat loads may have substantial spatial correlations. This possible spatial correlation should be investigated to improve the accuracy of the modeling.

- Unlike traditional electric distribution systems, which only supply electrical loads, microgrids have both electrical and heat loads, due to the implementation of CHP plants. Therefore, a two-dimensional model of electricity and heat flows should be developed for microgrids. Furthermore, the differences in the storage and charging/discharging characteristics of electricity and heat should be investigated in microgrid modeling and optimization.

REFERENCES

[1] Wu X, Wang X, Qu C. A hierarchical framework for generation scheduling of microgrids. IEEE Trans Power Delivery 2014;29(6):2448–57.

[2] Fathi M, Bevrani H. Statistical cooperative power dispatching in interconnected microgrids. IEEE Trans Sustainable Energy 2013;4(3):586–93.

[3] Wu DW, Wang RZ. Combined cooling, heating and power: a review. Prog Energy Combust Sci 2006;32(5–6):459–95.

[4] Campanari S. Full load and part-load performance prediction for integrated SOFC and microturbine systems. ASME J Eng Gas Turbines Power 2000;122(2):239–46.

[5] Bando S, Watanabe H, Asano H, Tsujita S. Impact of various characteristics of electricity and heat demand on the optimal configuration of a microgrid. Electr Eng Jpn 2009;169:6–13.

[6] Xiao J, Bai L, Li F, Liang H, Wang C. Sizing of energy storage and diesel generators in an isolated microgrid using discrete Fourier transform (DFT). IEEE Trans Sustainable Energy 2014;5(3):907–16.

[7] Roberts B. Capturing grid power. IEEE Power Energ Mag 2009;7(4):32–41.

[8] Strzelecki RM, Benysek G. Power electronics in smart electrical energy networks; 2008. Springer-Verlag.

[9] Ye Y, Shi Y, Cai N, Lee J, He X. Electro-thermal modeling and experimental validation for lithium ion battery. J Power Sources 2012;199:227–38.

[10] Chao K-H, Chen J-W. State-of-health estimator based-on extension theory with a learning mechanism for lead-acid batteries. Expert Syst Appl 2011;38(12):15183–93.

[11] Srivastava AK, Kumar AA, Schulz NN. Impact of distributed generations with energy storage devices on the electric grid. IEEE Syst J 2012;6(1):110–7.

[12] Achaibou N, Haddadi M, Malek A. Modeling of lead acid batteries in PV systems. Energy Procedia 2012;18:538–44.

[13] Jongerden MR, Haverkort BR. Which battery model to use? IET Softw 2009;3(6):445–57.

[14] Ceraolo M. New dynamical models of lead-acid batteries. IEEE Trans Power Syst 2000;15(4):1184–90.

[15] Yona A, Senjyu T, Funabashi T. Application of recurrent neural network to short-term-ahead generating power forecasting for photovoltaic system, In: IEEE power engineering society general meeting; 2007. p. 1–6.

[16] Bacher P, Madsen H, Aalborg Nielsen H. Online short-term solar power forecasting. Sol Energy 2009;83(10):1772–83.

[17] Cao JC, Cao SH. Study of forecasting solar irradiance using neural networks with preprocessing sample data by wavelet analysis. Energy 2006;31(15):3435–45.

[18] Rahman MH, Nakamura K, Yamashiro S. A grid-connected PV-ECS system with load leveling function taking into account solar energy estimation. In: IEEE international conference on electric utility deregulation, restructuring and power technologies, vol.1, pp.405–10; 2004.
[19] Alamsyah TMI, Sopian K, Shahrir A. Predicting average energy conversion of photovoltaic system in Malaysia using a simplified method. Renew Energy 2004;29(3):403–11.
[20] Ropp ME, Begovic M, Rohatgi A, Long R. Design considerations for large roof-integrated photovoltaic arrays. Prog Photovolt Res Appl 1997;5:55–67.
[21] Dalton GJ, Lockington DA, Baldock TE. Feasibility analysis of renewable energy supply options for a grid-connected large hotel. Renew Energy 2009;34(4):955–64.
[22] Chakraborty S, Weiss MD, Simoes MG. Distributed intelligent energy management system for a single-phase high-frequency AC microgrid. IEEE Trans Ind Electron 2007;54(1):97–109.
[23] Varshney K, Poddar K. Prediction of wind properties in urban environments using artificial neural network. Theor Appl Climatol 2012;107(3):579–90.
[24] Louka P, Galanis G, Siebert N, Kariniotakis G, Katsafados P, Pytharoulis I, et al. Improvements in wind speed forecasts for wind power prediction purposes using Kalman filtering. J Wind Eng Ind Aerodyn 2008;96(12):2348–62.
[25] An X, Jiang D, Liu C, Zhao M. Wind farm power prediction based on wavelet decomposition and chaotic time series. Expert Syst Appl 2011;38(9):11280–5.
[26] Shi J, Guo J, Zheng S. Evaluation of hybrid forecasting approaches for wind speed and power generation time series. Renew Sustain Energy Rev 2012;16(5):3471–80.
[27] Chung M, Park H-C. Building energy demand patterns for department stores in Korea. Appl Energy 2012;90(1):241–9.
[28] Jardini JA, Tahan CMV, Gouvea MR, Ahn SU, Figueiredo FM. Daily load profiles for residential, commercial and industrial low voltage consumers. IEEE Trans Power Delivery 2000;15(1):375–80.
[29] Amjady N, Keynia F, Zareipour H. Short-term load forecast of microgrids by a new bilevel prediction strategy. IEEE Trans Smart Grid 2010;1(3).
[30] Kasbekar GS, Sarkar S. Pricing games among interconnected microgrids. In: IEEE power and energy society general meeting; 2012. p. 1–8.
[31] Osborne MJ, Rubinstein A. Bargaining and markets. Academic Press; 1990.
[32] Diwerak U. Introduction to applied optimization. 2nd ed. Springer; 2008.
[33] Nocedal J, Wright SJ. Numerical optimization. 2nd ed. New York: Springer; 2006.
[34] Fouskakis D, Draper D. Stochastic optimization: a review. Int Stat Rev 2002;70:315–49.
[35] Bianchi L, Dorigo M, Maria Gambardella L, Gutjahr WJ. A survey on metaheuristics for stochastic combinatorial optimization. Nat Comput 2009;8(2):239–87.
[36] Kogan K, Nikolenko S, Keshav S, Lopez-Ortiz A. Efficient demand assignment in multi-connected microgrids with a shared central grid. In: Sustainable Internet and ICT for Sustainability, pp. 1–5, Oct. 2013; 2013.
[37] Erdinc O, Uzunoglu M. Optimum design of hybrid renewable energy systems: overview of different approaches. Renew Sustain Energy Rev 2012;16(3):1412–25.
[38] Rueda-Medina AC, Padilha-Feltrin A. Pricing of reactive power support provided by distributed generators in transmission systems. In: IEEE Trondheim Power Tech, pp. 1–7, 19–23 June; 2011.
[39] Kanchev H, Francois B, Lazarov V. Unit commitment by dynamic programming for microgrid operational planning optimization and emission reduction. In: International

Aegean conference on electrical machines and power electronics and electromotion, pp. 502–507, 8–10 Sept.; 2011.
[40] Vafaei M, Kazerani M. Optimal unit-sizing of a wind-hydrogen-diesel microgrid system for a remote community. In: IEEE Trondheim Power Tech, pp. 1–7, June; 2011.
[41] Augustine N, Suresh S, Moghe P, Sheikh K. Economic dispatch for a microgrid considering renewable energy cost functions. In: IEEE innovative Smart Grid technologies, pp. 1–7, Jan.; 2012.
[42] Chen SX, Gooi HB, Wang MQ. Sizing of energy storage for microgrids. IEEE Trans Smart Grid 2012;3(1):142–51.
[43] Farahmand F, Khandelwal T, Dai JJ, Shokooh F. An enterprise approach to the interactive objectives and constraints of Smart Grids. In: IEEE Conference on Innovative Smart Grid Technologies—Middle East, pp. 1–6, Dec.; 2011.
[44] Nayar C, Tang M, Suponthana W. Wind/PV/diesel micro grid system implemented in remote islands in the Republic of Maldives. In: IEEE International Conference on sustainable energy technologies, pp. 1076–80, 24–27 Nov.; 2008.
[45] Khodaei A, Shahidehpour M. Microgrid-based co-optimization of generation and transmission planning in power systems. IEEE Trans Power Syst 2013;28(2):1582–90.
[46] Buayai K, Ongsakul W, Mithulananthan N. Multi-objective micro-grid planning by NSGA-II in primary distribution system. Eur Trans Electr Power 2012;22:170–87.
[47] Niknam T, Golestaneh F, Reza Malekpour A. Probabilistic model of polymer exchange fuel cell power plants for hydrogen, thermal and electrical energy management. J Power Sources 2013;229:285–98.
[48] Chen M, Zhu B, Xu R, Xu X. Ultra-short-term forecasting of microgrid surplus load based on hybrid intelligence techniques. Electr Power Autom Equip 2012;32:13–28.
[49] Palma-Behnke R, Benavides C, Aranda E, Llanos J, Saez D. Energy management system for a renewable based microgrid with a demand side management mechanism. In: IEEE symposium on computational intelligence applications in Smart Grid, pp. 1–8, April; 2011.

CHAPTER

Safety design of resilient micro energy grids

5

H.A. Gabbar, Y. Koraz

University of Ontario Institute of Technology (UOIT), Oshawa, ON, Canada

5.1 INTRODUCTION

A resilient micro energy grid (MEG) can be defined as a system that comprises intelligent energy sources and distribution systems, automated metering, and a specialized computing system. Where MEG also integrates with renewable energy sources such as solar, wind power, small hydro, geothermal, waste-to-energy, and combined heat and power systems.

The MEG has played a significant role in achieving renewable energy utilization, and improving the resiliency of energy sources and distribution grids. A MEG fundamentally reduces energy consumption and increases self-healing and flexibility with great capability [1]. Also, the MEG is seen as an integrated energy system, which offers highly dynamic energy supplies for the area, such as electricity, cooling, and heating energy [2]. Therefore, it is a necessity to design a MEG with higher safety fault tolerances against variant types of risks and hazards, than with the various discrete systems used earlier. A MEG is a small or medium scale autonomous energy supplying system, which is comprised of smart meters and sensors, automated controls, and advanced software. The advanced software utilizes real-time distribution data to detect and isolate faults by reconfiguring the energy flow in order to minimize the customer's impact and to increase the system resiliency. It may include load-distributed generators, energy storage devices, and predictive energy management to achieve the simultaneous goals of electricity costs, energy reliability, and emissions [3]. From the systems point of view, the MEG, as one controllable unit which combines energy sources, loads, and energy storage units, can supply electricity, cooling, and heating energy to the end users. The MEG performs dynamic control over energy sources, enabling autonomous and automatic self-healing operations. During normal or peak usage, or during a capital energy grid failure, a MEG can operate independently of the larger grid, and isolate its generation nodes and energy loads from disturbance without affecting the larger grid's integrity.

Smart Energy Grid Engineering.

Effective design for the management of fault-tolerant control systems can help in realizing the full capability of energy resiliency and eco-friendly energy provided by the MEG. The MEG is comprised of a complex system with varied dynamic response characteristics at numerous time-scales. A hierarchical pattern is usually used for the control of such systems [4,5]. It includes an overall supervisory control independent protection layer (IPL), which defines the set-points for the significant MEG operation parameters based on energy demand, such as which distributed energy resources (DERs) should be operating (on/off states), and at what conditions they must be operating at (energy levels, power level, temperatures, pressures, mass flow rates, and so on).

The inner IPL, consisting of component level feedback controllers, seeks to attain these set-points by fine-tuning the essential actuators, such as control valves, engine, compressors, and pumps [6].

Defining the optimal set-points needs the development of accurate models which can precisely characterize the effects on fault-tolerant vital variables, such as the amount of the load demand, energy source efficiency, operating costs, and emissions, which can be realized by optimizing the operational parameters, or set-points, of the MEG system controllers.

The advantage of resilient MEG can be summarized as follows:

1. enhancing the reliability of the performance system,
2. enhancing customer awareness and choice,
3. encouraging greater efficiency of decisions from the utility provider, and
4. providing a closer proximity between energy generation and energy use.

Where resilient MEG technology is applied over a city, the city is called a "Smart Green City," which is remarked over the world, such as at Canada's dockside or UAE Masdar.

This study will assess the impact of utilizing thermal energy storage (TES) tanks on the resiliency of cooling dispatch for nonuniform thermal cooling units. Hence, TES has accompanied the facility of time-shifting electricity demand from on-peak demand periods, where energy storage can be an auxiliary source for additional on-peak generation in order to prevent the hazard of on-peak demand blackouts, which are caused by dynamic energy demand and/or irregular renewable generation.

Renewable energy generation requires MEG technologies to balance generation intermittency, and to reduce the need for capital utility [7]. This study uses 2 MW of solar power (PV) and 2 MW of wind farms (WT) as renewable resources of energy because of its direct benefit in the elimination of gas emissions.

Detailed models for MEG system component dynamics are extensively available [8]. Nevertheless, integration of such detailed models for complete MEG optimization would actually result in undesirably huge computation times and other related challenges [6]. Therefore, reduced order models are necessary. Without losing important dependencies expressed by detailed models, these models will be integrated in an optimization framework.

5.2 MEG INFRASTRUCTURE

5.2.1 SYSTEM DESCRIPTION

MEGs consist of localized energy generation equipment. It may consist of microturbines, solar panels, wind turbines, fuel cells, etc., which can provide energy to a local area in a cleaner way. MEGs can operate either in a main grid-connected mode or in an islanded mode [9]. In a main grid-connected mode, MEGs can supply energy generated by renewable sources to the capital grids. In the case of an energy outage on the main grid, MEGs can take charge and provide energy to the load. However, when islanding occurs, there is intermittency in the energy flow. In a simulation it was observed that the dynamic performance of the MEGs during and after islanding was better when supplementary storage devices supported the MEGs, as compared to those without energy storage. Therefore, it is a better option to have MEGs equipped with storage devices for better overall dynamic performance.

The proposed MEG model shown in Fig. 1 has the ability for self-sufficiency in its electricity, cooling, and heating demands most of the year by utilizing distributed generator, PV, WT, and district heating/cooling units with TES and super-capacitor bank for swift and dynamic power backup. Despite that the MEG has the ability of operating in islanded mode, it is interconnected with the capital grid to ensure operation resiliency in hazard scenarios, and offers backup sources at uncertain increasing demands. A set of six electric thermal cooling units of varied size and performance characteristics, shown in Table 1, produces cold water to supply the cooling demand, and/or is stored in a 175 MWh TES tank for future use. An on-site 15 MW cogeneration gas (CG) turbine is the essential source of electrical power for the facility. Furthermore, exhaust gas from the CG is used to provide steam in a heat recovery steam generator (HRSG). Where the steam is used for driving a 3 MW steam turbine in order to produce additional electrical power, and to produce heat energy in order to meet the majority of the facility-heating load shown in Fig. 2 [6].

5.2.2 DETAILED MODEL FOR MEG COMPONENTS

Cogenerator gas (CG) turbine

The gas turbine is one of the most effective power generation technologies, which operates on the thermodynamic cycle or Brayton cycle. This turbine is mainly composed of three stages: a compressor, a combustor, and a turbine. The compressor increases atmospheric pressure and compresses it into the combustor. The combustor merges this air with fuel then burns the mixture. Then the hot exhaust gases are sent into the turbine to convert the energy into mechanical work [10]. Fig. 3 shows the principal components of a simple-cycle gas turbine. The gas turbine is used in the MEG to produce electrical power as a conversion of the turbine mechanical work with an electrical efficiencies range from about 20% to 25%, as well as produce hot exhausted gases which can be 800–1100°F, depending on the type of turbine. These high exhaust temperatures are a ground for several studies and research for restoring the wasted thermal energy [11].

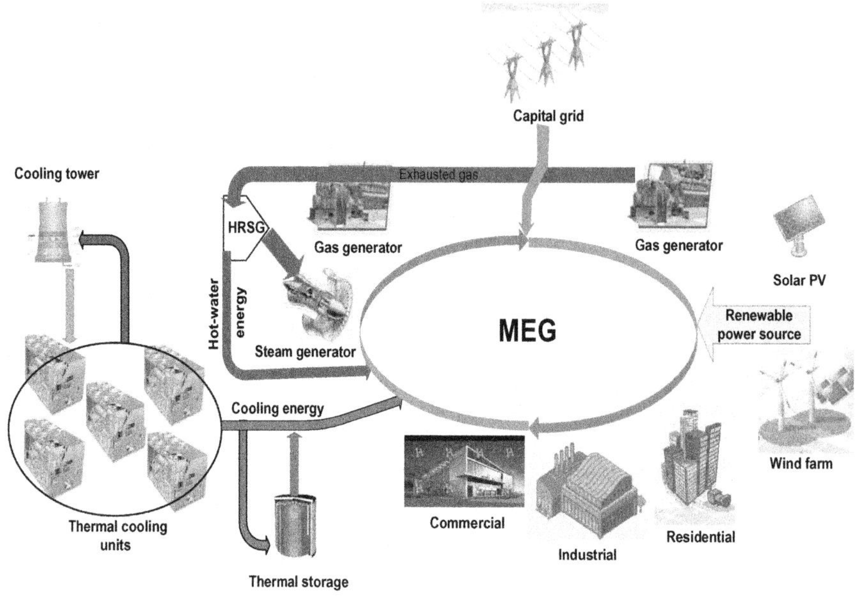

FIG. 1

Proposed MEG model.

Table 1 Thermal Cooling Units Rated Size and Coefficient of Performance (COP)

Thermal Cooling	Thermal Cooling Unit-1	Thermal Cooling Unit-2	Thermal Cooling Unit-3	Thermal Cooling Unit-4, 5, and 6
Size (tons)	900	1000	2800	3500
Size (kWe)	600	620	2000	2100
COP	5.0	5.5	4.5	6.0

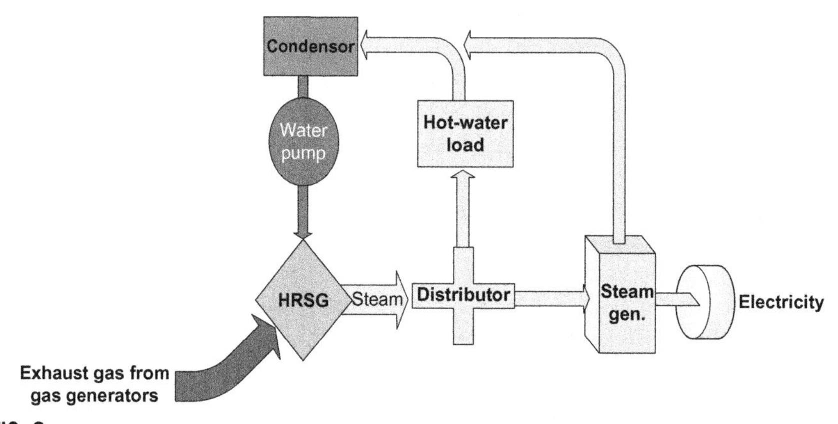

FIG. 2

Steam loop schematic diagram.

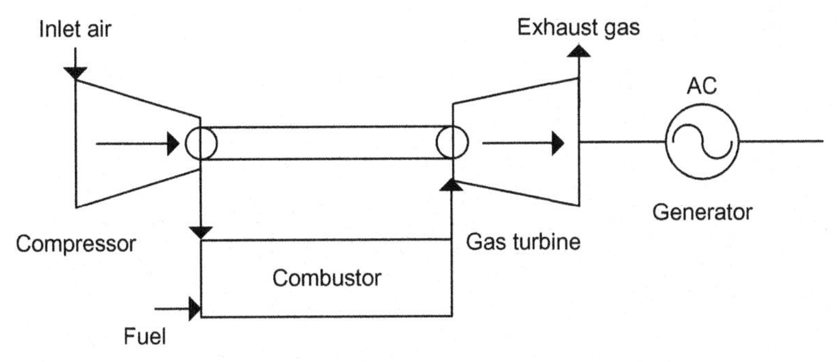

FIG. 3

Gas turbine model.

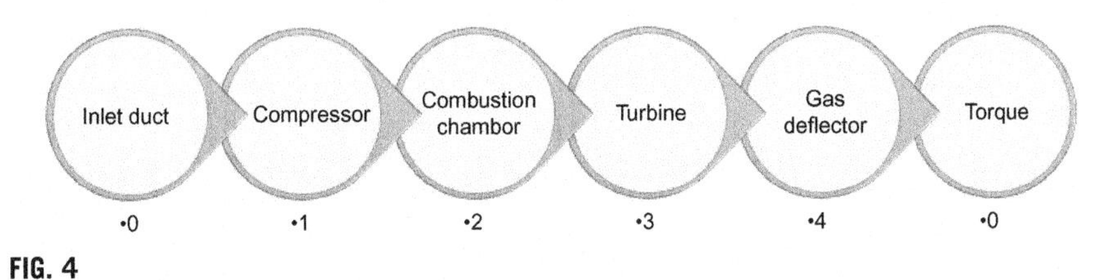

FIG. 4

Schematic diagram for gas turbine.

The dynamic behavior of the gas generator can be simulated by identify the nonlinear form of mass and energy conservation equations for each component. Besides, applying some static equations to complete the linear model. Schematic of the gas turbine is shown in Fig. 4.

The static model for a CG is obtained using thermodynamic equations and maps of the components in order to determine the off-design performance of the CG with constant output power [12]. The linearization of the dynamic nonlinear equations is realized using finite difference method. The final linear equations are developed into the state-space form as follows:

$$\begin{aligned} \dot{x} &= Ax + Bu \\ y &= Cx + Du \end{aligned} \tag{1}$$

Compressor linearized equations:

$$\begin{aligned} \Delta w_1 - \Delta w_2 &= \frac{v_{\mathrm{com}}}{R}\left(\frac{1}{T_2}\Delta\dot{P}_2 - \frac{P_2}{T_2^2}\Delta\dot{T}_2\right) \\ & w_1\Delta h_1 + h_2\Delta w_1 - w_2\Delta h_2 - h_2\Delta w_2 - \Delta pow_{\mathrm{C}} \\ &= \frac{v_{\mathrm{C}}}{R}\left(h_2\left(\frac{1}{T_2}\Delta P_2 - \frac{P_2}{T_2^2}\Delta\dot{T}_2\right) + \frac{P_2}{T_2}\left(\frac{\partial h_2}{\partial p_2}\Delta\dot{P}_2 + \frac{\partial h_2}{\partial T_2}\Delta\dot{T}_2\right)\right) \end{aligned} \tag{2}$$

Combustion chamber linearized equations:

$$\Delta w_2 - \Delta w_{\mathrm{f}} - \Delta w_3 = \frac{v_{\mathrm{com}}}{R}\left(\frac{1}{T_3}\Delta \dot{P}_3 - \frac{P_3}{T_3^2}\Delta \dot{T}_3\right)$$

$$w_2\Delta h_2 + h_2\Delta w_2 - \dot{Q}_{\mathrm{f}}\eta_{\mathrm{com}}\Delta w_{\mathrm{f}} - w_3\Delta h_3 - h_3\Delta w_3 \tag{3}$$

$$= \frac{v_{\mathrm{com}}}{R}\left(h_3\left(\frac{1}{T_3}\Delta P_3 - \frac{P_3}{T_3^2}\Delta \dot{T}_3\right) + \frac{P_3}{T_3}\left(\frac{\partial h_3}{\partial p_3}\Delta \dot{P}_3 + \frac{\partial h_3}{\partial T_3}\Delta \dot{T}_3\right)\right)$$

Turbine linearized equation:

$$\Delta w_3 - \Delta w_4 = \frac{v_{\mathrm{T}}}{R}\left(\frac{1}{T_4}\Delta \dot{P}_4 - \frac{P_4}{T_4^2}\Delta \dot{T}_4\right)$$

$$w_3\Delta h_3 + h_3\Delta w_3 - w_4\Delta h_4 - h_4\Delta w_4 - \Delta pow_{\mathrm{T}} \tag{4}$$

$$= \frac{v_{\mathrm{T}}}{R}\left(h_4\left(\frac{1}{T_4}\Delta \dot{P}_4 - \frac{P_3}{T_4^2}\Delta \dot{T}_4\right) + \frac{P_4}{T_4}\left(\frac{\partial h_4}{\partial p_4}\Delta \dot{P}_4 + \frac{\partial h_4}{\partial T_4}\Delta \dot{T}_4\right)\right)$$

Shaft mechanical energy linearized equation:

$$\Delta \dot{N} = \frac{1}{4\pi^2 IN}\left(\eta_{\mathrm{mech}}\Delta pow_{\mathrm{T}} - \Delta pow_{\mathrm{C}} - \frac{6\pi}{50}N\Delta\tau - \frac{6\pi}{50}\tau\Delta N\right) \tag{5}$$

where W is the mass flow rate (kg s^{-1}), T is the temperature (K), ρ is the density (kg m^{-3}), *pow* is the power, N is the speed (rpm), η is the efficiency, σ is the pressure loss coefficient, τ is the torque (Nm), P is the pressure (Pa), f is the fuel

Heat recovery steam generator

The HRSG is defined as an energy recovery heat exchanger that recaptures heat from a hot gas stream. The steam is generated to drive a steam turbine. A combined-cycle power station (CC) is a common application for an HRSG, where the hot exhaust gas produced from a gas turbine is fed to an HRSG to produce steam to drive a steam turbine (Fig. 5). The CC produces electricity more efficiently than either the gas turbine or steam turbine individually, where the electrical efficiency range from about 25% to 45% and overall CC efficiency of 65% to 80% for combined electrical and heat energies [11]. The HRSG consists of four major components: Evaporator, Super heater, Economizer, and Drum. The different components are combined to meet the operating requirements of the unit [13]. The high quality heat from the gas turbine exhaust allows for the utilization of the thermal energy to generate electricity by a steam turbine along with the gas turbine in a combined cycle system process that also allows the thermal energy to be restored and used for heating or cooling of the premises and used to provide domestic hot water [14].

The following equation evaluates the volume of steam that HRSG is able to generate [15]:

$$W_{\mathrm{s}} = \frac{W_{\mathrm{g}}C_{\mathrm{p}}(T_1 - T_3)eLf}{h_{\mathrm{sh}} - h_{\mathrm{saf}}} \tag{6}$$

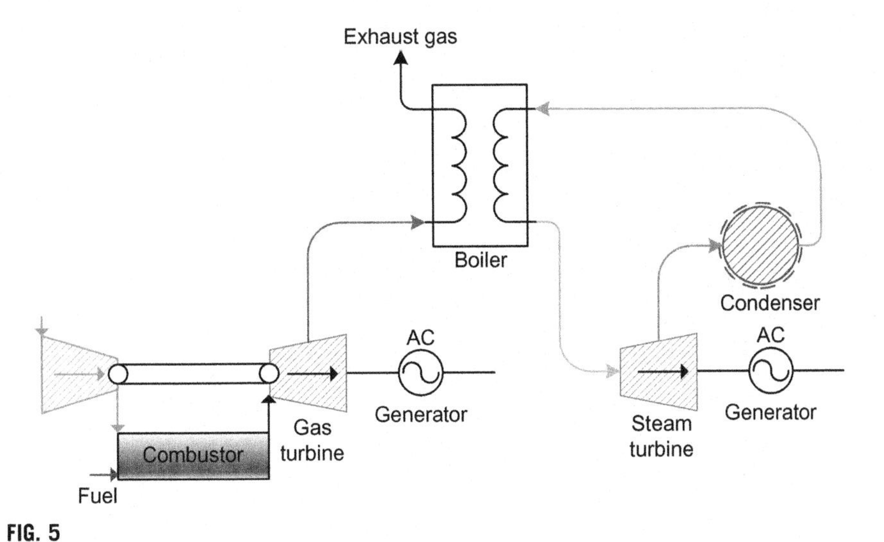

FIG. 5

Combined cycle power plant.

where W_S is the steam flow rate; W_g is the exhaust flowrate to HRSG; C_p is the specific heat of products of combustion; T_1 is the gas temperature after burner; T_3 is the saturation temperature in steam drum; L a factor to account radiation and other losses, 0.985; h_{sh} is the enthalpy of steam leaving superheater; h_{saf} is the saturated liquid enthalpy in steam drum; e is the HRSG effectiveness $= \frac{(T_1 - T_2)}{(T_1 - T_3)}$; and f is the fuel factor, 1.0 for fuel oil, 1.015 for gas.

Thermal energy storage (TES)

The principle idea behind TES is to provide a buffer to balance fluctuations in supply and demand of energy [16]. Energy demand in the residential, commercial, and industrial regions fluctuates in course of day periods, intermediate periods (e.g., 7 days), and seasons (spring, summer, autumn, and winter). Consequently, various TES systems are utilized to match the demand as well as to reshape the actual demand on the energy sources. TES has been used for decades in different forms for space and process heating/cooling applications. Different types of materials such as latent or phase change materials (PCM) and sensible heat materials have been applied as prospective heat transfer media for energy storage applications. For the latent, the thermal energy is absorbed and released by a phase change of the storage media by fusion. However, sensible heat storage materials were utilized based on their ability to raise or lower the temperature of storage media without a phase change [17].

A stratified cylindrical tank operates both on transfer and retrieval mode so, consequently, the concluding model is hybrid. The relationship between the number of

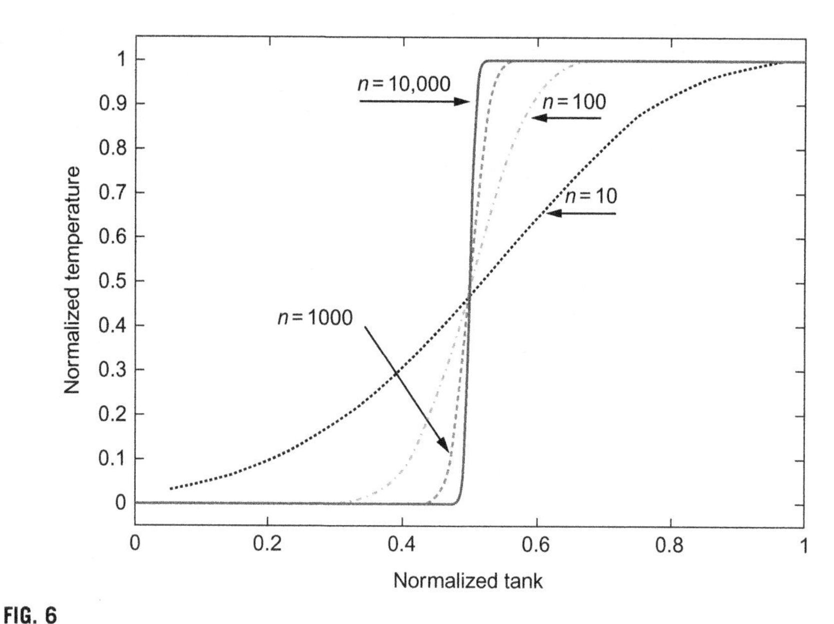

FIG. 6

Simulated TES profile.

nodes used in simulation and the degree of stratification, which the model predicts are shown in Fig. 6. The tank operates either on a charging or discharging mode, therefore, the resulting model is a hybrid. The TES can be modeled using a dynamic finite element based, which divides the tank into 100 control volumes along its height. Energy and mass conservation laws are applied to each control volume [6,12].

The approximate dynamic tempreture profile of a TES system can be simplified as follows:

$$\rho C_p A_{xs} \Delta x \frac{dT_i}{dt} = C_p \dot{m}_{\sin k}(T_{i-1} - T_i) + C_p \dot{m}_{\text{source}}(T_{i+1} - T_i) \\ - UP\Delta x(T_i - T_{\text{amb}}) + \frac{\varepsilon A_{xs}}{\Delta x}(T_{i+1} - 2T_i + T_{i-1}) \tag{7}$$

where ρ is the storage fluid density, C_p is the storage fluid heat capcity, Δx is the length of node, $\dot{m}$ is the mass flow rate, T is the time, U is the tank fluid to ambient overall heat transfer coefficient, P is the tanks perimeter, and A_{xs} is the tank cross-sectional area.

District thermal cooling unit

The district thermal cooling unit is modeled using the standard approach for an integration of static models for essential components, such as evaporators, condensers, compressors, cooling towers, and pumps [3]. Where the ideal compressor equation:

$$W_{\text{comp-s}} = \left(\frac{n}{n-1}\right)\dot{m}_{\text{ref}} P_2 V_2 \left[\left(\frac{P_2}{P_1}\right)^{(n-1/n)} - 1\right] \tag{8}$$

The compressor polytropic efficiency is evaluated by

$$\eta_{pol} = \frac{1}{CPR}\left[\frac{W_{comp-s}}{\eta_m W_{comp-design}}\right] \tag{9}$$

And the actual compressor work is defined by:

$$W_{comp} = \frac{W_{comp-s}}{\eta_m \eta_{pol}} \tag{10}$$

5.3 MEG PERFORMANCE MODELING

In order to design the future MEGs infrastructure, it is important to provide a modeling and simulation environment in order to evaluate different design and operational scenarios, with consideration toward different grid process data and performance indicators. Grid process data can be expressed using state variables that can be classified as: process variables (PV) (such as voltage, current, phase, and pressure temperature); equipment variables (such as parameters related to grid equipment such as transformers, generators, lines, pump, boiler, and pipeline); environment variables (such as humidity, temperature, and dust); and human variables (such as operator related parameters, consumer related parameters, and so on). Other parameters include organizational or external parameters, such as transportation and societal parameters.

5.3.1 PROCESS VARIABLES (PV)

Phasor measurement unit (*PMU*): voltage, current, phase, power factor, and noise.
Transformer/energy convertor: voltage, current, phase, power factor, efficiency, flow rate, and capacity.
Thermal storage: temperature, pressure, energy density, leakage, efficiency, and capacity.
Photo voltaic cell (*PVC*): incident sunlight, angle of installation, and efficiency.
Wind turbine: wind velocity, wind direction, and efficiency.
Communication channel: distance, frequency, mode, communication technology, and network topology.
Microgrid (*PMU*): number of users, voltage-in form grid, voltage-out to other microgrid, current-in form grid, current-out to other microgrid, and system losses.

Performance indicators:

(i) Total cost of production, distribution, and storage
(ii) Efficiency:

$$\frac{\text{Total energy production} - \text{Total system loss}}{\text{Total energy production}} \quad (11)$$

(iii) Risk index

Computational intelligent algorithms (genetic algorithms, learning/reasoning) were developed to perform multidimensional optimization for different performance indicators for each simulation scenario.

Each component should be linked with a knowledgebase for different model components, that is, business processes, regulations, power, energy, asset integrity, physical topology, and geographical information.

5.3.2 CRITICAL PERFORMANCE PARAMETERS

The following major critical parameters should be considered when studying the performance of the MEGs.

Energy measurement accuracy: Since the energy is produced from distributed unit sources, it is important to accurately consider every source of energy generation in the network. For example, if a consumer has installed solar panel in his or her house, he will be contributing the energy generated by the solar panels to the grid and, simultaneously, he will be consuming energy from the grid through any of his home appliances. This might result in some inaccuracy in power measurement.

Weather conditions: The energy generation by renewable sources such as solar panels and wind turbines, vary with different weather conditions. There may be a case in which their energy production does not follow a particular trend in some climatic conditions. Therefore, it is important to consider the weather conditions during energy calculation.

Delay: Since there will be many energy generation sources, there will be more computations and subsequently, more computational delay. Besides computational delay, there will also be transmission delays, since all the computed values will be transferred to a common station [18].

5.3.3 COST ANALYSIS

It is important to quantify the related cost reductions in MEG simulation models for different grid lines. However, there are uncertainties in energy demand and energy generation capacity, for example, wind output. This will have an impact on the grid cost in view of the different types of generation. A projection of marginal generation cost, including carbon cost for the future year, could be used to reflect the complete cost analysis scenarios of the grid. The energy generation cost shows that $/kWh for wind is the lowest, and gas is the highest. Similarly, the cost of different types of energy storage systems should be analyzed and compared to identify potential energy storage technology. It is relatively difficult to compare technologies, especially with different parameters involved in the performance evaluation. However, super-capacitor technologies are probably one of the best in respect to life cycle, while lead-acid energy storage technologies are relatively better in terms of $/kWh. Some considerations use a 12-year lifetime for the life cycle cost analysis. However, some memory has an effect on batteries, which has a negative impact over time. Research shows the storage cost of lead-acid technologies is around 400 $/kWh, whereas the generation cost for wind is 0.0001 $/kWh, which reflects the potential reduced cost of generation over storage [19].

5.4 RISK MANAGEMENT FOR MEGs

This section explores the available methodologies in realizing a resilient MEG system with immunity protection layers against various types of hazards. A resilient MEG has the ability to handle a risk assessment approach, including hazards, faults, and accidents, and identify the critical requirements of security and safety within the demand scenarios, and then automatically map the IPL in real time for online self-healing verification.

5.4.1 RISK ASSESSMENT FRAMEWORK FOR MEGs

Risk assessment is important for increasing the profitability and the return on investment (ROI) of the MEGs infrastructures, and to support energy supply operations with minimum risk. Fig. 7 shows the proposed risk assessment framework of the MEG. The starting point is to identify the main building blocks of the MEGs systems: topology, dynamics, control system, and communications.

Risk probability could be used in the *risk assessment* process; it allows the severity of the risk of an event occurring to be determined. The risk of any particular hazard, R_i, can be defined as its probability, P_i, multiplied by its consequence severity, S_i. In layman's terms: how likely it is to happen and how bad it would be, if it happens.

$$R_i = P_i \times S_i \tag{12}$$

Therefore, the total risk, R, of an event, e, is the sum of the n potential risk that would result in that event:

$$R_e = \sum_{i=0}^{n} R_i \tag{13}$$

The first step of risk assessment is to identify the hazard event from the network through various sensors and monitoring systems. In energy system protection, this consists of many components, such as instrument transducers (sensors), relays, valves, tripping, and shutdown devices. In modern energy systems, many intelligent electronic devices (IEDs) are available in substations. Data from IEDs need to be merged with data from supervisory control and data acquisition (SCADA) to enhance monitoring capabilities [20].

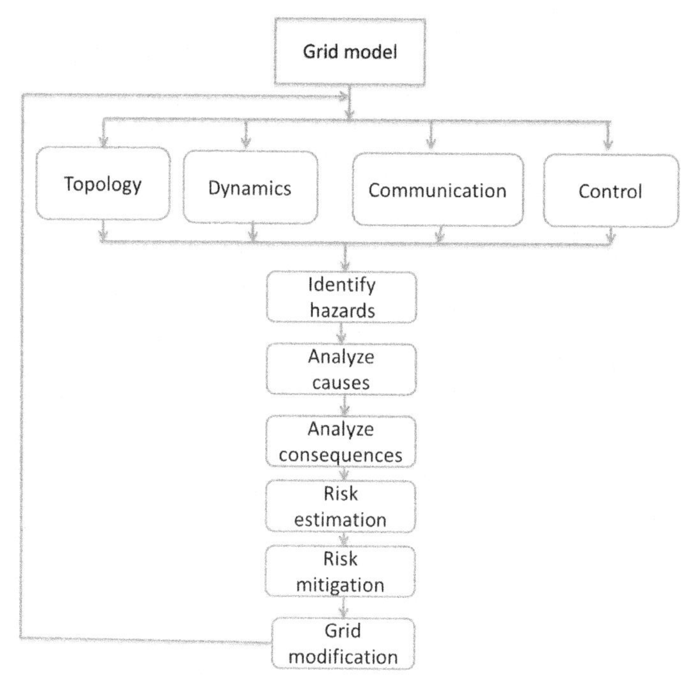

FIG. 7

MEG risk assessment framework.

The next step is to investigate/analyze the root cause of the fault that occurred, calculate the consequence of the fault, and to estimate the risks. An optimal fault location algorithm that uses data from substation IEDs, as well as data from SCADA, PI Historian, and simulation data, is implemented, which is shown in Fig. 8 [20]. This illustrates the concept of detecting and mitigating cascading events, which demonstrates local and system-wide monitoring of data. Condition-based data from substation IEDs need to be utilized for failure rate assessment and risk estimate. A failure mode can be described using symptoms, enablers, PV, causes, and consequences [18].

In order to mitigate risks associated with the MEGs, one approach is to provide real-time and risk-based accident forecasting mechanisms and tools. A fault semantic network (FSN) is one mechanism/tool that works with process deviations and links with possible accident scenarios. A FSN includes a knowledge structure for a fault model that links grid structure, behavior, operation, and associate process/equipment variables.

The FSN is utilized to contrast flexible fault knowledge structures in a qualitative manner. Also the FSN is associated with fault propagation and is linked with risk quantitative, deterministic, and probabilistic estimation, and identifies protection layers with each fault propagation scenario. By comparing estimated PV with a FSN we can learn the possible cause and consequence, which is then used to define possible corrective actions.

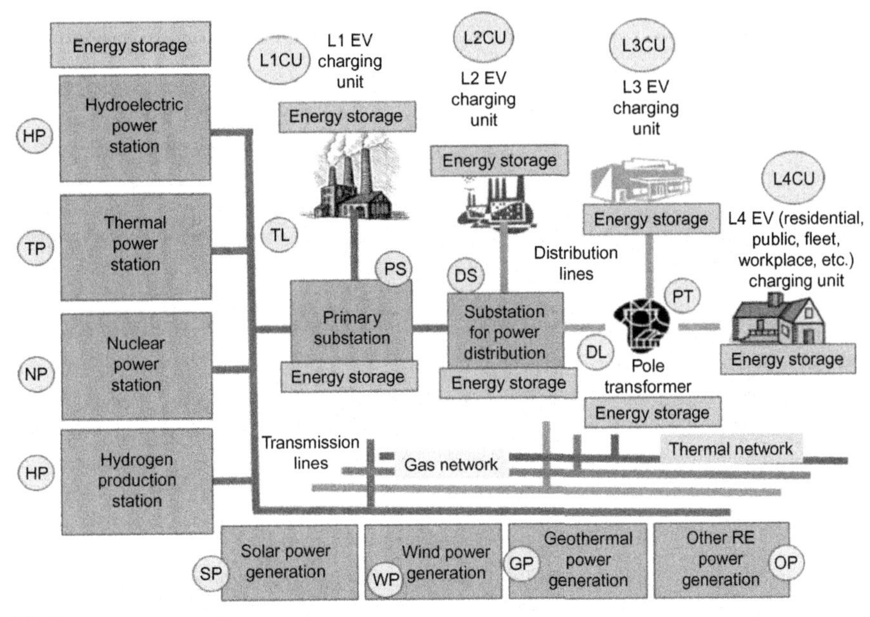

FIG. 8

Hybrid energy generation and storage within smart grid.

In order to be able to perform a risk assessment, some key performance indicators (KPI) must be specified. The performance of the smart grid under varying conditions can then be compared with a defined performance indicator or requirement. These KPIs can be calculated by a formula using different PV. Based on these KPIs, corrective measurement action can be taken.

Hazards to people and the environment are a serious concern, and the Occupational Safety and Health Administration (OSHA) has given this topic much attention. Protection schemes, which achieve fast fault clearing times for bus faults and feeder faults, will be reviewed. The different technical solutions will be evaluated against economic considerations.

The fast fault clearing/risk mitigation technique can be illustrated as follows [19]:

- *Using different setting groups*: Modern numerical relays/actuators normally offer several setting groups, which can be activated at any time. The goal would be to archive instantaneous tripping/action in the case of a fault. After the work is finished the original settings, which were coordinated with all other relays/ actuators, would be reactivated.
- *Normal reverse interlocking scheme*: The reverse interlock scheme provides an instantaneous tripping/action bus protection for substations with radial feeders.
- *Reverse interlocking scheme with several infeeds*: Bus systems with several infeed feeders are not uncommon, and are more likely at higher upstream levels. Even in distribution and industrial applications, the so-called main-tie application

can be considered as a bus (actually two busses) with at least two infeeds. This is because the tie-bus connection can be used as a load as well as a source.

- *Advanced reverse interlocking scheme*: The goal of the advance reverse interlocking scheme is to reduce the tripping time for faults on the bus, and for faults on feeder locations close to the bus. By reducing the tripping time, the incident energy will be reduced.

5.5 HAZARD AND RISK ANALYSIS TECHNIQUES IN MEGs

Foundation design is not an inherent safeguard protection layers, Table 2 shows the major hazards that threaten the MEG system in electrical, cooling, and heating grids, and the possible remedy action for overcoming the related consequences, and for avoiding the risk of failure or blackout.

The hazard level can be calculated with the following formula [21]:

$$\text{Hazard level}\,(H_L) = S_i \times C_i = (P_i + F_i + A_i) \times C_i \tag{14}$$

where S_i is the consequence severity of the hazard event, C_i is the class hazard event likelihood, P_i is the probability, F_i is the frequency, and A_i is the ability for failure avoidance.

5.5.1 ELECTRICAL MEG HAZARDS

The following points summarize the main hazard events in electrical MEG:

1. *Over load* (*above the grid capability*): Electrical demand can be suddenly increased for a limited period of time due to different reasons, such as extremely hot and cold weather, which may lead to several negative impacts:
   I. *Impacts on humans*: Demand not served (DNS).
   II. *Impact on facility*: Overheated transmission and distribution cables, asset damages, fire, and power blackout.
   III. *Impacts on the environment*: Fire causes CO_2 emission.

   This can be prevented by several remedial actions and/or IPLs such as:
   - upgrade grid capacity, which costs time and money;
   - shift on-peak power demand, by using an intelligent energy storage system, such as super capacitor, fly wheel, TES and pumped hydro, or hydrogen storage; and
   - dynamic grid mapping based on the load demand and priority.
2. *Lack of DER at the MEG*: DER can be out of service due to routine maintenance, or due to system breakdown or failure. The negative impact could be as follows:
   I. *On humans*: Interruption on service.
   II. *On the facility*: Could lead to a risk of losing the electricity power for a wide region, or of a general blackout.

III. *On the environment*: Lack of DER means an increase in the demand on fossil fuel generators, which cause a dramatic increase of emissions.
Preventing IPL action can be through high-dynamic performance from the distributed power and energy system by:
- storing off-peak power production for use for an emergency or for on-peak demand;
- utilize a cogenerator; and
- connect to the capital grid (utility).

3. *Utilization of on-site renewable sources*: Despite the fact that renewable resources are known as eco-friendly energy sources, it has an accompanying hazard of intermittency and noncoincidence in energy production, which leads to a lack in providing the required power demand. This can be prevented by utilizing the IPLs mentioned in point 2.
4. *Integration of multisources DERs*: Has negative impacts on grid parameters, such as active power (P), reactive power (Q), voltage (V), phase shift (α), and frequency (f). Remedial action and IPL as follows:
   - full utilization of DERs to increase energy efficiency;
   - improve power quality; and
   - enhance system stability.

5.5.2 THERMAL COOLING MEG HAZARDS

Cooling MEG resiliency is affected by the following hazards:

1. *High correlation between cooling demand and electricity demand*: This relation has negative effects on the MEG resiliency as follows:
   I. Impacts on humans: Uncomfortable condition (temperature and humidity).
   II. Impacts on the facility: The on-peak demand for both electricity and cooling grids are accrued at the same time, and this subsequently leads to an increase in the actual electricity of on-peak demand, which might cause interruption and/or blackout for both services.
   III. Impacts on the environment: Increases the demand on fossil fuel generation, and its consequences on emissions.

This hazard can be avoided by shifting on-peak cooling demand to off-peak demand by utilizing TES and intelligent management control.

2. *MEG contingency with lack of chiller unit*: This might cause an inability to meet the cooling on-peak demand, which has the following undesirable influences:
   I. Impacts on humans: Uncomfortable condition due to DNS.
   II. Impacts on the facility: Shortage on cooling production leads to lack of service.
   III. Impacts on the environment: Individual A/C units are a solution in overcoming the lack of service, and A/C unit usage has an impact on electricity demand and global warming.

Table 2 List of Hazards at the MEG System

No. #	Grid Type	Hazard Event	Severity	Frequency	Probability	Avoidance
		SCORE	1 = Negligible 2 = Marginal 3 = Critical 4 = Catastrophic	1 = Less 2 = yearly 3 = Monthly 4 = Weekly 5 = Daily	1 = Negligible 2 = Rarely 3 = Possible 4 = Likely 5 = Common	1 = Likely 3 = Possible 5 = Impossible
1	Electrical MEG	Over load (above the grid Capability)	4	3	4	1
2		MEG has lack of DER	4	3	3	3
3		Varity of on-site renewable sources	3	5	5	3
4		Integration of multisources DERs	2	5	5	3
5	Cooling MEG	High correlation of cooling demand with electricity demand	4	5	5	3

Class	Hazard Level	Adverse Effects		
3–4 = Very Low 5–7 = Low 8–10 = Moderate 11–13 = High 14–15 = Extremely High	1 = Low 2 = Moderate 3 = High	(1) On Human (2) On Facility (3) On Environment	Action (Remedial or Prevention)	Required IPLs
8	H	(1) Demand not served (DNS) (2) Overheated transmission and distribution cables, asset damage, fire and power blackout (3) Fire cause CO_2 emission	1. Upgrade grid capacity 2. Shift on-peak power demand 3. Dynamic grid mapping based on load demand and priority	1. Cost money and time 2. Intelligent energy storage system (super capacitor, fly wheel, TES and pumped hydro, or hydrogen storage 3. Intelligent fault tolerant controller 4. Ranking the loads as per its prioritization level
9	H	(1) Interruption on service (2) Power interruption and/or blackout (3) Lack of DER = more demand on fossil fuel generators which cause Emission	High dynamic performance from the distributed power and energy system by: • Store off-peak power production for using at on-peak demand • Utilize gas generator • Connect to capital grid (utility)	1. Intelligent Energy Storage System (super capacitor, fly wheel, TES and pumped hydro, or hydrogen storage 2. Following generator (fuel cells, microgas turbines, and hybrid fuel cell gas turbine systems) 3. Higher level Self-Healing management controller
13	H	(1) Disturbance on service (2) Intermittency and noncoincidence of power production (3) Lack of DER = more demand on fossil fuel generators		
13	H	(1) Operation Failure of Sensitive Devices (2) Negative impacts on grid parameters such as active power (P), reactive power (Q), voltage (V), phase shift (α) and frequency (f). On other word bad power quality (3) Excessive on energy resources and emission	1. Full utilization of DERs to increase energy efficiency 2. Improve power quality 3. Enhance system stability	1. Advanced D-FACTS system on AC/DC MEG to achieve resilient MEG 2. Create Robust KPI parameters able to optimize feedback control coefficients
13	H	(1) Demand not served (2) Increase on-peak electricity demand could cause interruption and/or blackout (3) Increase demand on fossil fuel generation	Shift on-peak cooling demand to off-peak demand	1. Utilize TES tanks 2. Predictive energy management

Continued

Table 2 List of Hazards at the MEG System—*Cont'd*

No. #	Grid Type	Hazard Event	Severity	Frequency	Probability	Avoidance
		SCORE	1 = Negligible 2 = Marginal 3 = Critical 4 = Catastrophic	1 = Less 2 = yearly 3 = Monthly 4 = Weekly 5 = Daily	1 = Negligible 2 = Rarely 3 = Possible 4 = Likely 5 = Common	1 = Likely 3 = Possible 5 = Impossible
6		MEG contingency with lack of chiller unit	4	3	3	3
7	Heating MEG	Irregular hot-water demand	3	5	4	3
8	Transportation	Transportation energy demand	4	5	5	1
9	All	Earth quake	4	2	2	2
10	All	Water flood	4	2	2	2

Class	Hazard Level	Adverse Effects	Action (Remedial or Prevention)	Required IPLs
3–4 = Very Low 5–7 = Low 8–10 = Moderate 11–13 = High 14–15 = Extremely High	1 = Low 2 = Moderate 3 = High	(1) On Human (2) On Facility (3) On Environment		
9	H	(1) Uncomfortable condition for human (2) Can't meet the on-peak cooling demand (3) Using individual A/C units lead to increase Global Worming	1. Store off-peak cooling production for using at on-peak demand	1. Utilize TES tanks 2. Intelligent contingency energy management (for emergency procedure)
12	H	(1) Uncomfortable condition for human (2) Failure to meet the Hot water on-peak demand (3) Alternative heat sources like furnace produce emission	1. Store off-peak Hot water production for using at on-peak demand	1. Utilize TES tanks 2. Predictive energy management
11	H	(1) Loss of life's, injury and delay (2) Failure in energy threaten the safety for Properties and the public (3) Back-up Engines works using fossil fuel which increase Emission	1. Achieve energy management balance between transportation units and MEG for more reliability and security enhancement, reduced emissions and improved energy quality	1. Energy Storage System (super capacitor, fly wheel, TES and pumped hydro, or hydrogen storage 2. Following Generator (fuel cells, microgas turbines, and hybrid fuel cell gas turbine systems) 3. Intelligent management controller
6	M	(1) Loss of life's, injury and delay (2) Failure in energy threaten the safety for properties and the public (3) Spreading the damages and may initiate new hazards	Isolate the affected area from the service	Intelligent management controller
6	M	(1) Loss of life's, injury and delay (2) Failure in energy threaten the safety for properties and the public (3) Spreading the damages and may initiate new hazards	Isolate the affected area from the service	Intelligent management controller

This hazard can be evaded by storing off-peak cooling production for use at on-peak demand, using TES and management control to ensure higher MEG reliability levels.

5.5.3 THERMAL HEATING MEG HAZARDS

From the historical data on heating demand, it can be clearly defined that there is an irregular heating demand with a low correlation with electrical demand, which might lead to a failure to meet the on-peak heating demand. The consequence of the failure has several negative impacts:

I. Impacts on humans: Uncomfortable condition (temperature and humidity).
II. Impacts on the facility: Failure to meet the heating on-peak demand.
III. Impacts on the environment: Increases the requirement for alternative heat sources such as furnaces, which increase the gas emissions.

In order to prevent the hazard of heating failure, a strategy to storing off-peak heating production should be utilized, in order for it to be available when required at on-peak demand.

5.5.4 TRANSPORTATION MEG HAZARDS

Transportation is a vital service for the society and the public; therefore, the energy demand conjugated with it is essential for its resiliency. Any interruption might have harmful impacts as follows:

I. Impacts on humans: Loss of life, injury, and delay.
II. Impact on the facility: Failure in energy threatens the safety of properties and the public.
III. Impacts on the environment: Back-up engines work by using fossil fuel, which increases emissions.

Achieving an energy management balance between transportation units and MEG is one of the main solutions for more reliability, security enhancement, emissions reduction, and energy quality improvement.

5.6 SAFETY DESIGN AND PROTECTION LAYERS FOR MEGs

The safety design for the MEG is aimed to improve the stability of the energy system during abnormal conditions, and to prevent fault and damage propagation. This can be achieved by interrupting and isolating faulted or failed components from the system, as well as providing safety strategies for properties and public safeguards.

The dynamic structure of the MEGs, and their various operating conditions, required the development of adaptive protection strategies using intelligent control and monitoring units based on safety design criteria.

Table 3 SIS Operating Condition

SIS Operating Condition	Process	Protection Available	Failure Indication
Normal	Operating normally	Yes	N/A
Fail-safe	Falsely operating	N/A	Yes
Fail-danger	Operating normally	No	Without diagnosis

The ANSI/ISA-84.00.01-2004 (IEC 61511) standard defines a safety instrumented system (SIS) as an instrumented system used to implement one or more safety instrumented functions (SIFs). A SIS is a combination of sensor(s), logic solver(s), and final element(s). IEC 61508 uses the term "safety-related system" instead of the term SIS. This term describes the same principle but with a different language context that can be broadly applied to many industries [22].

The main purpose of a control loop, in a basic process control system (BPCS) is generally to maintain process parameters within prescribed limits. A SIS monitors process parameters and interferes when required [23].

Risk analysis teams generally assess the hazardous situations that occur during the operation stream. The average time period between hazardous events is commonly evaluated to be over 10 years when the process design is successfully oriented toward safety. In consequence, the SIS is activated only once every 10 years or more in some cases and it is passive during normal operation. A chart of SIS operating conditions is shown in Table 3. The essential problem is the fail-danger mode. Hence, the process is operating normally under this condition but without the automatic protection of the SIS and without indication that something has failed [24].

5.6.1 SAFETY INSTRUMENTED SYSTEM ENGINEERING REQUIREMENTS

Nevertheless a SIS is similar to a BPCS in numerous ways, the differences found in the unique design, maintenance, and automated integrity requirements. These can be listed as follow [22]:

- design to fail-safe,
- design diagnostics to automatically detect fail-danger,
- design manual test procedures to detect fail-danger, and
- design to meet international and local standards.

In addition to the normal functional and performance requirements correlated with control system design.

5.6.2 SAFETY INSTRUMENTED FUNCTION

A SIF is defined, in ANSI/ISA-84.00.01-2004 (IEC 61511 Mod), 3.2.71, as a "safety function with a specified safety integrity level which is necessary to achieve functional safety." A safety function can be defined as a "function to be implemented by a SIS, other technology safety-related system, or external risk reduction facilities, which is intended to achieve or maintain a safe state for the process, with respect to a specific hazardous event" [10].

5.6.3 INDEPENDENT PROTECTION LAYERS

The IPL can be defined as a device, system, or action that has the capability of preventing a scenario from proceeding to its undesired consequence, independent from the initiating event or the action of any other layer of protection associated with the scenario. The main characteristics of IPL are

- its ability in blocking fault consequence when the IPL functions as designed and
- it should be auditable where the assumed effectiveness in terms of consequence prevention and probability of failure on demand (PFD) must be capable of validation (by documentation, review, testing, and so on).

The layer of protection analysis (LOPA) is utilized to determine if there are sufficient IPLs that can tolerate the risk, and supress the consequences of an accident scenario [25]. Each IPL has its own PFD.

$$\mathrm{PFD} = p_n \tag{15}$$

where n is the indicate layer level.

Where its value has direct impact on system resiliency, as declared on the LOPA path equation:

$$\text{LOPA path} = f_n = \left(\prod_{i=1}^{i=n-1} p_i \right) \times f_0 \tag{16}$$

Fig. 9 shows the proposed IPL layers required to tolerate the hazard of losing energy dispatch for a MEG system, where cogenerators, TES, and supervisory fault-tolerant predictive energy management control has been utilized to achieve the simultaneous goals of energy dispatch resiliency, energy production quality/cost, and emissions:

I. *IPL-1* cogenerators such as fuel cells, microgas turbines, and hybrid turbine systems are able to cover the lack of power demand and to respond to renewable interruptions.

II. *IPL-2* TES is an effective solution for MEG applications due to the listed below points:

   A Centralized infrastructure where large thermal reservoirs provide flexibility to manage cooling dynamics, as well as lower emissions and energy failure risks.

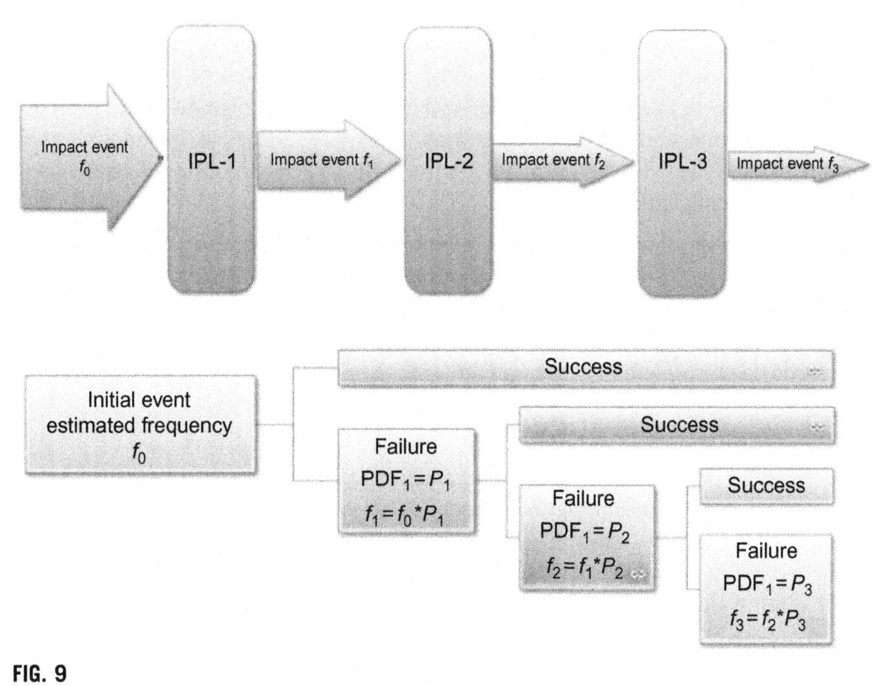

FIG. 9

LOPA path diagram for MEG system.

B Shifts on-peak production demand to off-peak hours in order to mitigate congested demand-subsequent hazards.

III. *IPL-3* Supervisory fault-tolerant energy management (FTEM) controller where the management of distributed resources near the renewable energy source is seen as the most effective means for increasing renewable penetration.

Different types of IPLs that can be utilized to increase MEG resiliency are shown in Table 1 and can be summarized as follows:

1. MEG storage system (E/T/C),
2. intelligent control system at normal operation to ensure rigid performance,
3. smart energy asset management for both sources and load within MEG boundary,
4. emergency control for resilient systems on abnormal cases,
5. risk assessment platform,
6. MEG safety shutoff and restoration system, and
7. upper-level centralized/decentralized MEGs management (capital grid).

These IPLs can be implemented and studied in later research for comparison between different techniques and their performance on MEG resiliency, and for enhancing MEG self-healing characteristics.

5.6.4 FAULT TREE FOR THE MEG

In reliability engineering, the primary statistical variable of interest is Time to Failure (T). The time to failure measurement can be analyzed to generate another important measurement, failure rate. Instantaneous failure rate is a commonly used measure of reliability that gives the number of failures per unit of time from a quantity of components exposed to failure.

$$\lambda(t) = \text{Failure rate} = \frac{\text{Failures per unit time}}{\text{quantity exposed}} \tag{17}$$

$$R(t) = e^{-\lambda t}$$

$$F(t) = 1 - e^{-\lambda t} \approx \lambda t$$

$$\text{MTTF} = 1/\lambda$$

Fault tree is a common technique to illustrate probability combinations. This technique begins with the definition of an "undesirable event," usually a system failure of some type. The analyst continues by identifying all events and combinations of events that result in the identified undesirable event. The fault tree is therefore quite useful when modeling failures in a specific failure mode. These different failure modes can be identified as different undesirable events in different fault trees. Fig. 10 shows a top event which is defined as PFD for a proposed MEG SIF. This method provides a clear way to express the reality of multiple failure modes. And the PFD for a MEG can be estimated by using the following equation:

$$\begin{aligned} F(\text{MEG}) &= F(\text{Capital Grid}) \times F(\text{Distribution Energy}) \\ &= [F(\text{Capital Grid}) + F(\text{Chiller}) + F(\text{MT})] \\ &\quad + [(F(\text{Co}-\text{gen}) \times F(\text{Renw})) + (F(\text{Co}-\text{gen}) \times F(\text{TES})) \\ &\quad + (F(\text{Co}-\text{gen}) \times F(\text{Manag})) + (F(\text{Renw}) \times F(\text{TES})) \\ &\quad + (F(\text{Renew}) \times F(\text{Manag})) + F(\text{TES})] \end{aligned} \tag{18}$$

where the PFD for the individual systems can be illustrated from historical database and experience. Individual PDF were demonstrated in Tables 4 and 5.

Compensation of individual failure rate for each device in Eq. (16) illustrates that the probability of energy blackout for the MEG was reduced nine times by utilizing proposed IPLs and distribution energy sources where the PFD became 78.7e − 3 while it was 649.2e − 3 for the conventional energy grid.

In the same way LOPA shows a reduction on the system average interruption frequency index (SAIFI) from 0.4893 for the conventional energy grid [26,28] to 4.03e − 6 with the proposed IPLs, Fig. 11 shows the results and proposed protection layers.

Where LOPA path can be defined using Eq. (16); $\text{LPOA} = f_3 = 0.4893 \times 0.1647 \times 0.05 \times 0.001 = 4.03e^{-6}$.

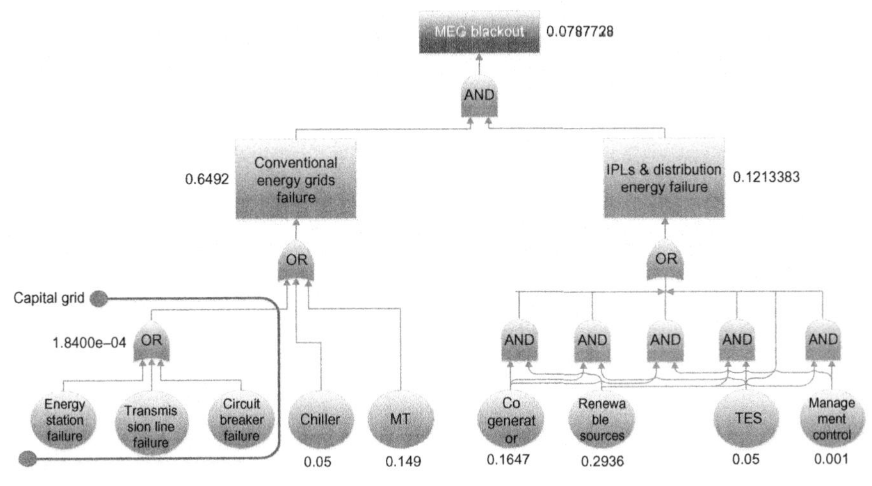

FIG. 10

A MEG fault tree.

Table 4 Failure Rate and Repair Time [26]

Type	Failure Rate (f/yr)	Reliability $e^{-\lambda T}$	Probability of Failure $1-e^{-\lambda T}$	Repair Time (h)
PV	0.11	0.8958	0.1042	72
WT	0.21	0.8106	0.1894	60
Cogenerator	0.18	0.8353	0.1647	12
Capital grid	0.000184	0.9998	0.000184	
Chiller		0.95 [12]	0.05	
FC	0.11	0.8958	0.1042	72
Battery	0.22	0.8025	0.1975	60
HRSG	0.16	0.8521	0.1479	16

Table 5 Typical Outage Rate for a Consumer [27]

Contributor	Minutes/Year	%
Generation/transmission	0.5	0.5
132 kV	2.3	2.4
66 and 33 kV	8	8.3
11 and 6.6 kV	58.8	60.7
Low voltage	11.5	11.9
Scheduled shutdown	15.7	16.2
Total	96.8	100

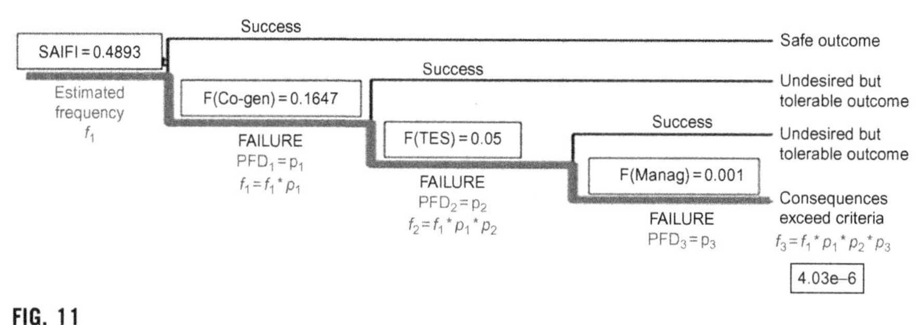

FIG. 11

LPOA diagram for the proposed resilient MEG.

5.7 CONTROL TYPES FOR MEGs

Numerous control methodologies for MEG have been proposed and studied. The centralized, decentralized, and multilevel hierarchical control of MEG have been discovered in previous theoretical and laboratory experimental research [29–34]. An illustration of these different control structures will be discussed in this section.

5.7.1 DECENTRALIZED CONTROL FOR MEGs

Decentralized control methodology for the MEG can be summarized as follows, the individual energy sources has the right to share the demand as per their specific capacity and local control characteristics, see Fig. 12. Those are fixed during installation and planning phases. Consequently, it is difficult to make any re-scheduling for instantaneous energy production, or for each source to achieve optimum generation cost and emission conditions. This fact led to the underutilization of the energy sources, although they may have high efficiency and lower operating rates [35].

5.7.2 CENTRALIZED CONTROL FOR MEGs

The centralized control methodology for the MEG mainly consists of a central control system for the remote control of all energy sources in the MEG boundary, see Fig. 13. Optimal performance can be achieved by using a centralized control system, but it has a significant disadvantage on the reliability of an energy system where the central controller fails, most likely the overall energy system will collapse. The centralized control methodology relies on a communication network, where the speed and reliability of the communication system has direct impact on the MEG performance, reliability, and resiliency.

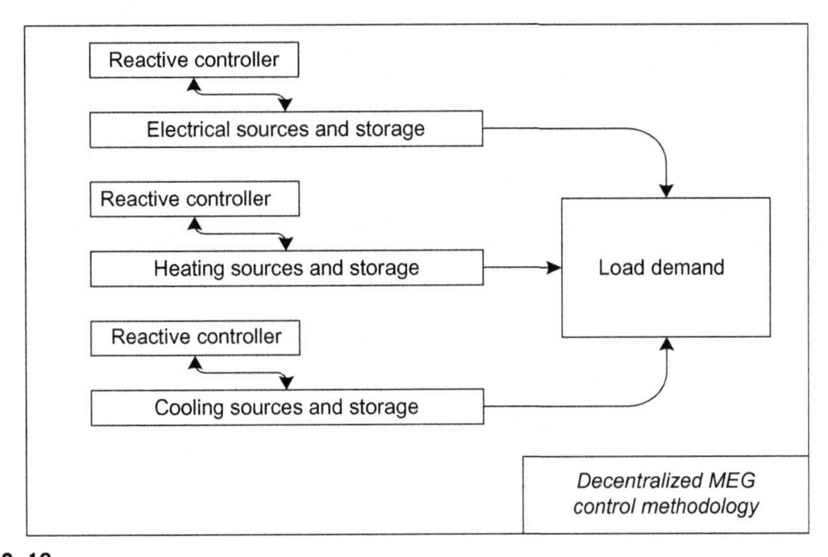

FIG. 12

Decentralized MEG control methodology.

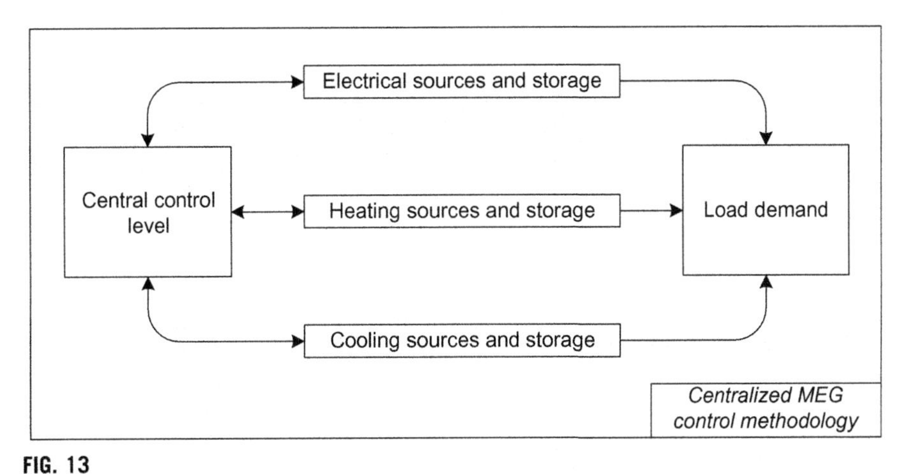

FIG. 13

Centralized MEG control methodology.

5.7.3 MULTILEVEL HIERARCHICAL CONTROL FOR MEGs

Multilevel hierarchical control of the MEG provides a better methodology to overcome most of the obstacles that accompanying the centralized and decentralized control methodologies [36]. This control type has a significant role in achieving the optimum operation of the MEG system that is similar to a centralized control

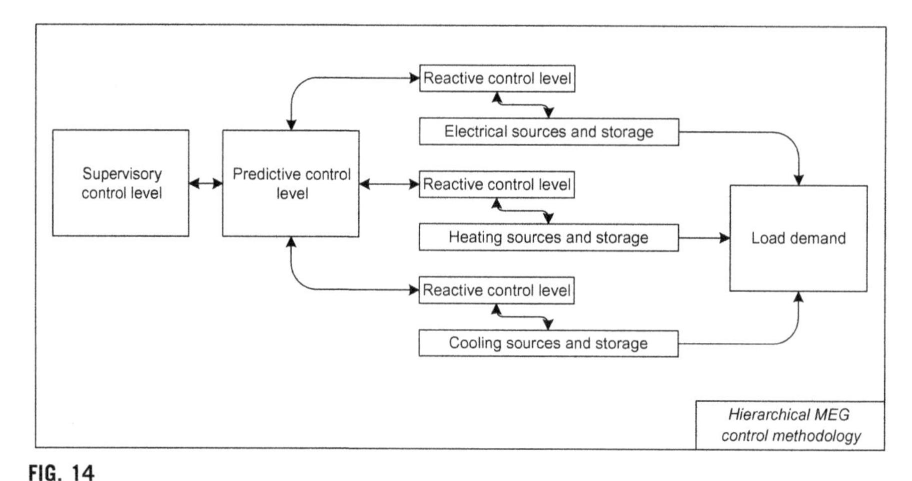

FIG. 14

Hierarchical MEG control methodology.

methodology but under lower speed and the reliability level of the communication network requirements, see Fig. 14. However, the main challenge of the hierarchical control methodology is the necessity for clear boundaries of control range and domain-based control levels [37]. In the hierarchical control the supervisory and predictive control levels generally depend on a communication network to achieve the MEG system optimization operation that is the same as a centralized methodology. But hierarchical has the advantage of a decentralized methodology, where the reactive control level is not dependent on the communication network. This feature immunizes the MEG from the loss of operation once a failure occurs in a higher-level control and/or network. Whilst the hierarchical may lose optimal performance during such hazardous event.

The MEG performs dynamic control over energy sources, enabling autonomous and automatic self-healing operations. During normal or peak usage, or at times of the capital energy grid failure, an MEG can operate independently of the capital grid and isolate its generation nodes and energy loads from disturbance without affecting the capital grid's integrity. A multilevel hierarchical control is proposed to provide autonomous self-healing supervisory control for the MEG system. The control architecture consists of three levels working together to achieve the overall operational goal.

Hierarchical control architecture

Hierarchical control architecture was realized in order to operate energy sources proficiently as well as to employ the MEG components efficiently. It embraces three levels, comprises an autonomous decision making level, a predictive control level, and a reactive control level. Each level has its own local objectives and they work together to realize a resilient operational performance. The higher-level controller involves a fault tolerant control formulation in order to deal with the uncertainty of hazardous conditions and to determine the best action for each subsystem. The predictive control level utilizes prescheduled operational timing to handle the district cooling units (DC) operation. The predictive control aims to shift the operation of the DC units to off-demand periods by charging the TES for further uses at on-demand peak periods. The lower-level controller is a load following control for the demand, which needs fast response. The quality of any control depends on the assumed information and control structures [36]. A central decision maker defines values of control based on available information collected from all subsystems. Nevertheless a centralized method might be difficult to realize in a large-scale system, where a process of transmission and transformation of information are more complicated. The decentralization of information and control structures is a reasonable solution to overcome this problem. Control problems with decentralized measurement information are the main element for hierarchical control. The decomposition of a large system into subsystems is mainly aimed at minimizing the required computations to further to reduce the amount of information required for a decision making level [38].

Previously the MEG was classified either as an islanded or grid-connected approach. Whereas, it is essential for a resilient energy system to modify an elastic MEG configuration capable of operating in both grid-connected and islanded modes [39]. This system opens the door on great challenges where establishing such systems requires an integration of different technologies, energy sources, energy storage, and energy management systems. In addition to safety issues, such as fault monitoring, predictive maintenance, or protection, which are fundamental principles for a MEG with a high level of self-healing capability.

This section's emphasis is on hierarchical control design with the employment of an adaptive neuro-fuzzy system for energy balancing and optimization purpose and to improve the resiliency of the MEG system. Hierarchical control design is also aimed at providing real-time backup methodologies in case of a hazard event in cooling generation and/or cooling load profiles in order to minimize its negative impact on the electrical energy system.

Design of adaptive neuro-fuzzy decision making method

Fig. 15 shows an adaptive-network-based fuzzy inference system (ANFIS) that has an optimized structure of 5 layers organized as follow 2:10:25:25:1.

This structure was created from initial data using a MATLAB/anfiseditor environment. The Takagi-Sugeno-Kang fuzzy model-based ANFIS has been used with architecture of two inputs and one output, which is tuned online using a combination of least-squares estimation and back-propagation methods. The error between reference chillers operation and actual chillers operation is used to tune the neuro-fuzzy model parameters. The functions of each layer in the ANFIS architecture are formalized as follows [40,41]:

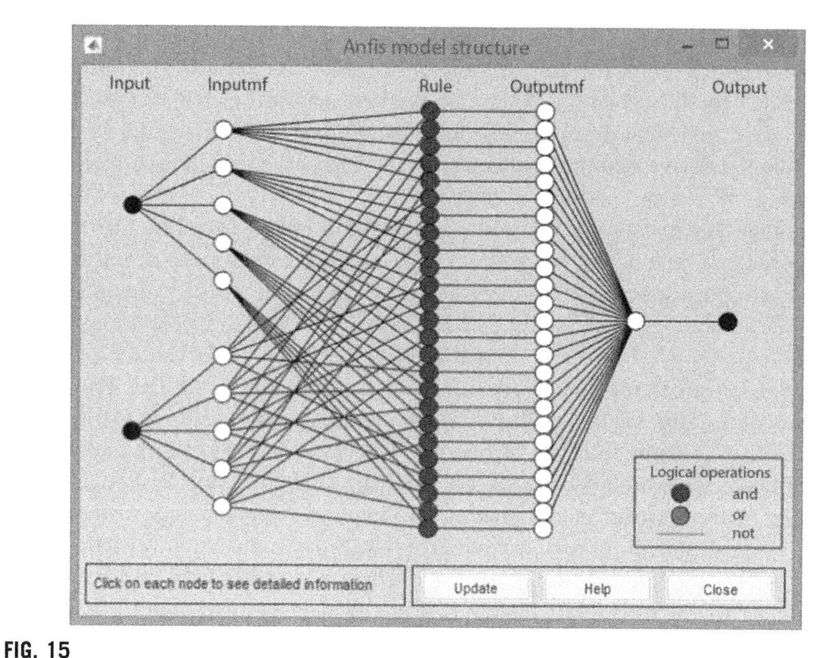

FIG. 15

Optimized ANFIS architecture.

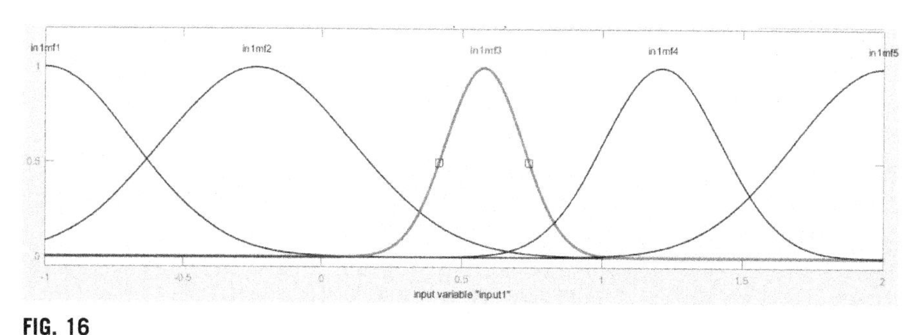

FIG. 16

Gaussian curve fuzzy membership.

Layer 1: It is a fuzzification layer where each node is symbolized by a membership. Five Gaussian curve membership functions are designated to each input as shown in Fig. 16, and its node equations is given as follows:

$$\text{Gaussian}(\xi; c, \sigma) = e^{-\frac{1}{2}\left(\frac{\xi - c}{\sigma}\right)^2} \tag{19}$$

where c is MFs center and σ is MFs width.

Layer 2: Each node in this layer is a multiplier that multiplies the input signals and forwards the result to the 3rd layer

$$\mu_{i,j} = \mu_{A_i(\xi_1)} \cdot \mu_{B_j(\xi_2)}, \quad i = j = 1, 2,3,4,5 \tag{20}$$

This equation characterizes the firing strength of a rule.

Layer 3: Each node in this layer calculates the normalized firing strength of each rule as given in the following:

$$\bar{\mu}_{i,j} = \frac{\mu_{i,j}}{\sum_{i=1}^{5}\left(\sum_{j=1}^{5}(\mu_{i,j})\right)} \tag{21}$$

Layer 4: In this layer each node is multiplied by tuned variable weights $(a_0^{i,j}, a_1^{i,j})$ as shown in the following equation:

$$O_{i,j} = \bar{\mu}_{i,j} \cdot f_{i,j} = \bar{\mu}_{i,j}\left(a_0^{i,j} + a_1^{i,j} \cdot \xi\right), \quad i,j = 1,2,3,4,5 \tag{22}$$

Layer 5: Is the final output layer of fuzzy system. Output of the system is the summation of all incoming signals from layer 3, computed as follows:

$$Y = \sum_{i=1}^{5}\left(\sum_{j=1}^{5}(O_{i,j})\right), \quad i,j = 1,2,3,4,5 \tag{23}$$

Energy system procedure

Fig. 17 summarizes the procedure for energy production and flow for electrical, cooling and heating energy systems within a resilient MEG.

5.8 SIMULATION SCENARIOS

In order to evaluate the IPL performance and MEG resiliency, the following scenarios will be studied:

Scenario-1: Study operation and performance of MEG (without TES and cogeneration)

This scenario examines the natural resiliency of the MEG, before applying proposed IPLs, in order to define the initial event estimated frequency, f_1, which depends on the reliability of capital grid and renewable resources.

Scenario-2: Study operation and performance of MEG (with cogeneration)

Here, we applied IPL-1 in order to define the improvement level of MEG resiliency by interconnecting the gas generator in order to eliminate renewable resource penetration, and to cover around 60% of the energy demand requirements. In other words, this means a reduction of the severity risk of energy failure to 60% of the capital grid total failure.

Scenario-3: Study operation and performance of MEG (with TES)

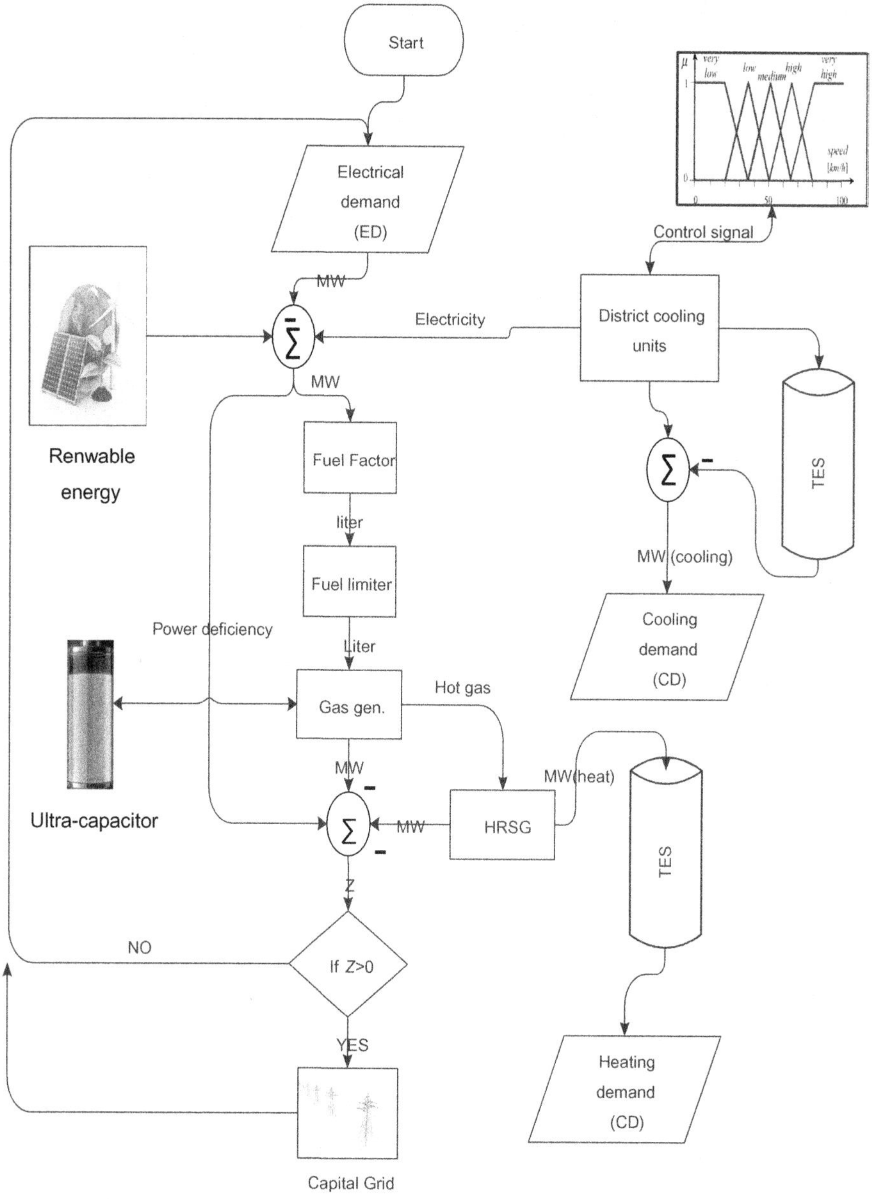

FIG. 17

MEG energy production flow chart.

IPL-2 was applied alone to define its capability of MEG self-healing by partially shifting the energy demand from the on-peak period to other off-peak timing; and it can be shown that it improves more than 17% of the total energy demand.

Scenario-4: Study operation and performance of MEG with TES and cogeneration)

Here IPL-1, IPL-2, and IPL-3 were provided to the MEG in order to increase its capability, and to operate on the islanded mode, which means that the IPL-3 is providing the reminder of the 23% of the total energy demand by applying optimum prediction and management control for the MEG system.

5.9 CASE STUDIES AND DISCUSSIONS

In order to demonstrate the dynamic behavior of the MEG system when utilizing different levels of the IPLs, as mentioned earlier in this chapter, a case study for the proposed MEG system is presented in this section using a Simulink environment. Modeling and simulating of dynamic time-varying systems can be implemented in the Simulink environment, with support of the Matlab software environment. Simulink has the ability to convert mathematical equations, which describe the behavior of a model into an interactive graphical shape, where the interactive graphical environment supports the creation of understandable models.

Here, the proposed MEG system is implemented in the Simulink environment (Fig. 18) in order to study the system performance in different operational scenarios to examine the MEG system's resiliency for prescribed cooling, heating, and electricity energy demands.

Data for 1 week in the summer has been analyzed to evaluate and improve the MEG system operation. The interaction between cogenerators, thermal cooling units, TES storage, and the capital power grid are explored to increase safety, resiliency, and self-healing for the MEG system. Two baseline strategies are used in this section.

- In the first baseline strategy, no IPLs were utilized.
- In the second baseline strategy, two potentially valuable structures, that is, the TES and cogeneration are used.
- In the third baseline strategy, a heuristic, rule-based methodology using physical anticipation is used to determine the operating parameters of the MEG without installing additional MEG hardware.
- In the fourth baseline strategy, study operation and performance of the MEG under a fault hazard event (four district cooling units out of service).

The objective of the proposed strategy is to reduce the hazards of system failure through optimal application of TES by shifting cooling demand to off-peak hours. Measurements of the MEG generation, capital's power imports, and the thermal cooling unit operations have been processed for optimum management of the electricity, heating, and cooling energy demands.

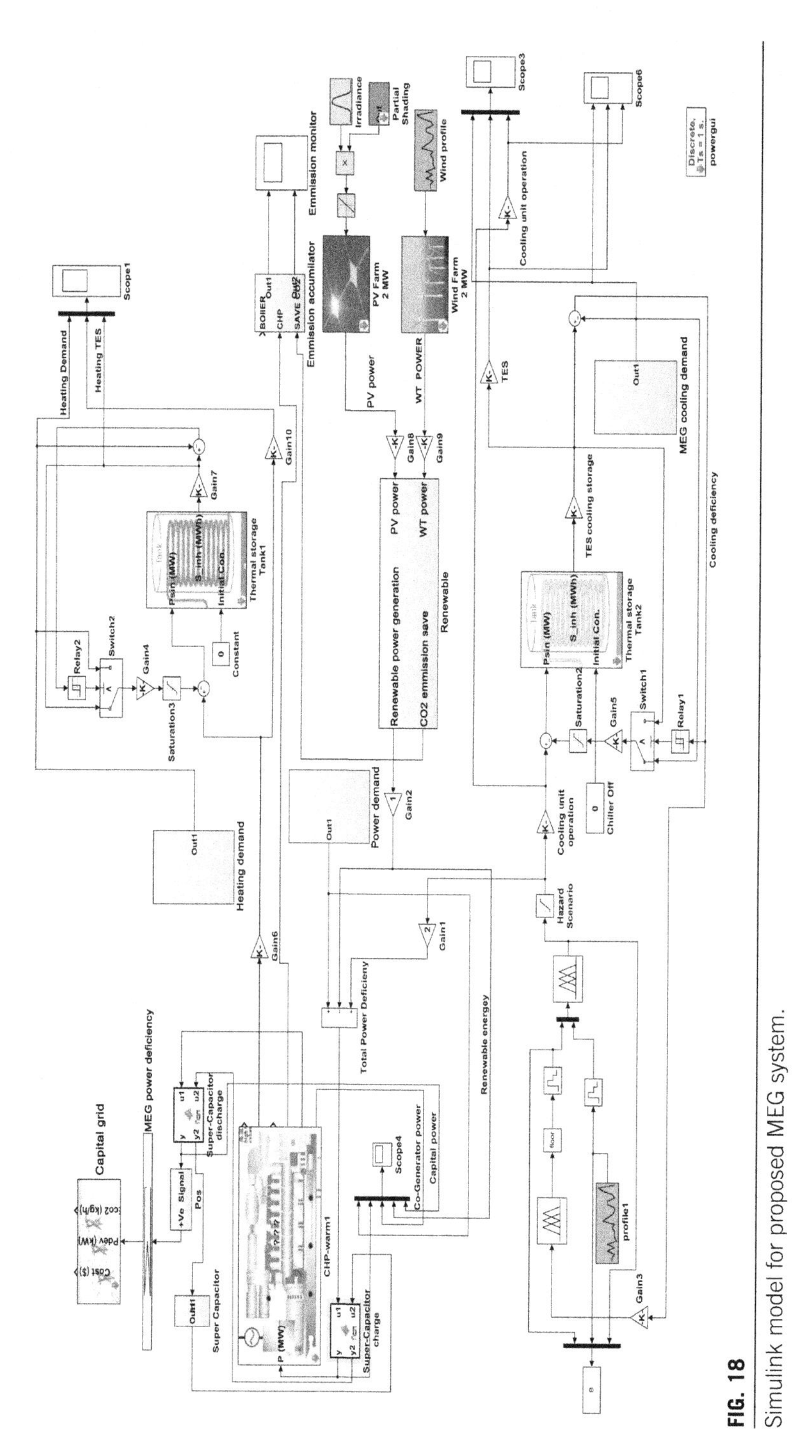

FIG. 18

Simulink model for proposed MEG system.

5.9.1 FIRST BASELINE STRATEGY

Fig. 19 illustrates the power demand profile for 1 week in the summer for a foundation MEG system without the inherent safeguards of the proposed IPLs. The figure clearly defines that the combination of cogenerator and renewable sources can't handle the power demand alone; therefore, the capital power is required to cover the power deficiency caused by sudden raise on the power demand. The power deficiency caused by two reasons, first due to limited capacity of DERs and secondly due to the dynamic behavior of cogenerator, which led to a delay in responding to the rapid change on the demand profile.

Cooling profile in Fig. 20 shows the MEG cooling demand during 1 week in summer without utilizing the cogenerator and TES, the figure illustrates a high frequency of on- and off-district cooling units (DC) operation during a course of the day. This leads to a dramatic increase on the inrush current during DC units' start-up, which might cross beyond a double of the chiller-rated current. On other hand DC units are in duty most of the day with an increasing number of operated units during on-demand period. Keep in consideration the fact of high correlation for cooling demand with electricity.

5.9.2 SECOND BASELINE STRATEGY

Fig. 21 presents a sample of the resulting power demand profile for 1 week in summer using cogenerator and TES IPLs only. From the figure above it can be clearly defined that the cogenerator is completely covering the power demand for the first 4 days with the support of renewable power sources, while in the last 3 days the capital power is partially required to cover the power deficiency caused by a sudden raise on the power demand. The power deficiency occurred 2 h a day with maximum 4 MW while the cogenerator serves an average of 14 MW with maximum production of 18 MW.

Cooling profile in Fig.22 shows the MEG cooling demand within 1 week in the summer while the cogenerator and TES were utilized, the figure also illustrate that the TES improved the imports of cooling with less operational hours for the thermal cooling units. However, there is the fact of a high correlation for cooling demand with electricity demand.

Fig. 23 presents a sample of the heating demand profile for 1 week in the summer. The figure shows extensive coverage of the heating demand by the heat generated by the cogenerator unit. Furthermore, there is a comparatively low correlation between the electrical demand and heating demand, particularly during the summer season.

5.9.3 THIRD BASELINE STRATEGY

Utilizing the three IPLs, proposed earlier, the MEG system operates in an islanded mode under most of the operation scenarios without the need for capital grid imports. Fig. 24 shows that the cogenerator is completely covering the power demand for the first 5 days with the support of renewable power sources, while in the last 2 days the capital power was slightly required to cover the power deficiency caused by a sudden rise in power demand. The power deficiency occurred twice within the test week, for

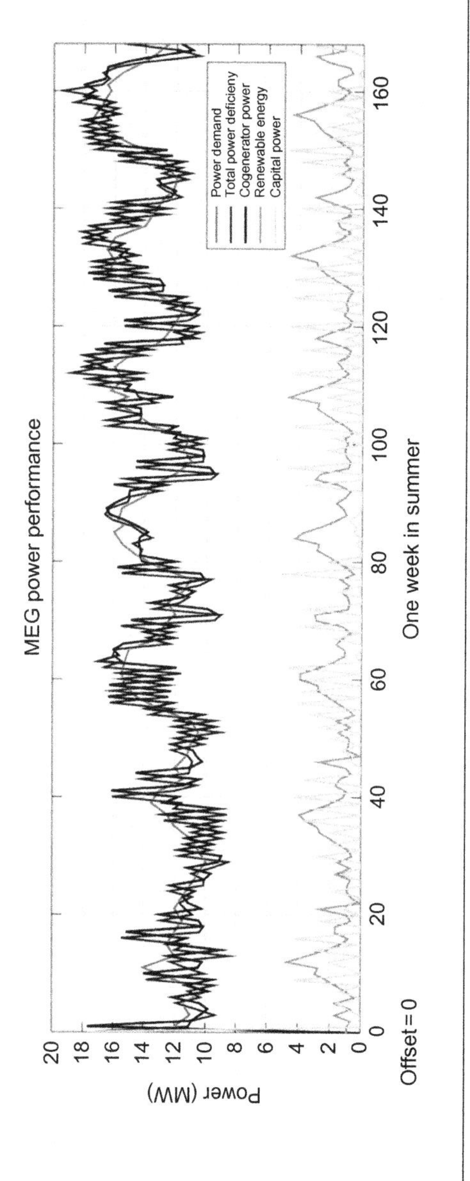

FIG. 19
Power profile for foundation MEG.

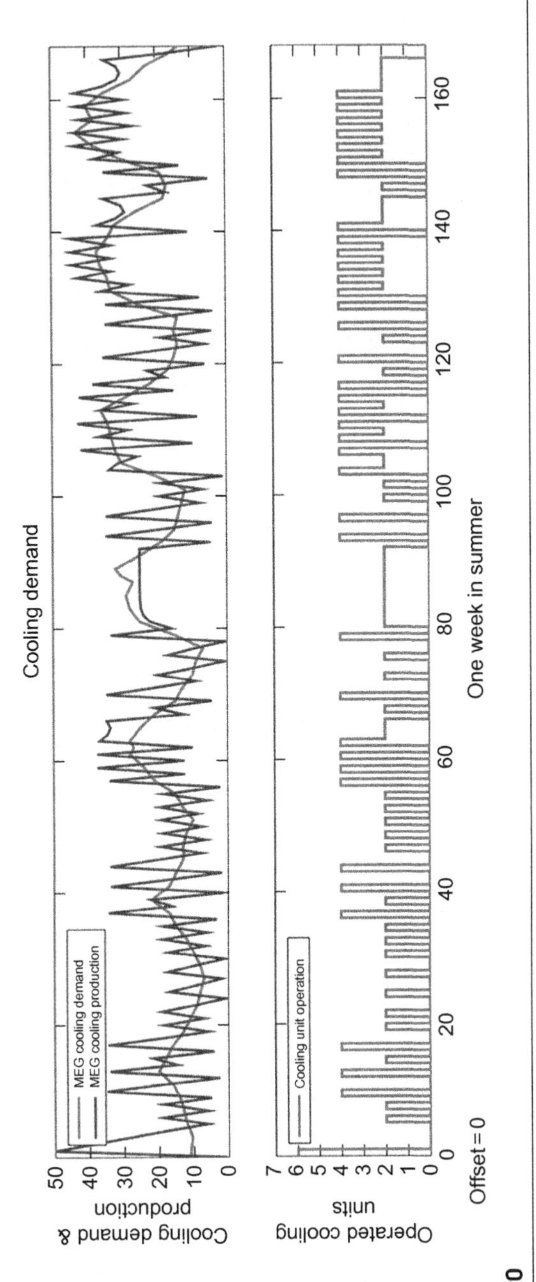

FIG. 20

Cooling profile for foundation MEG.

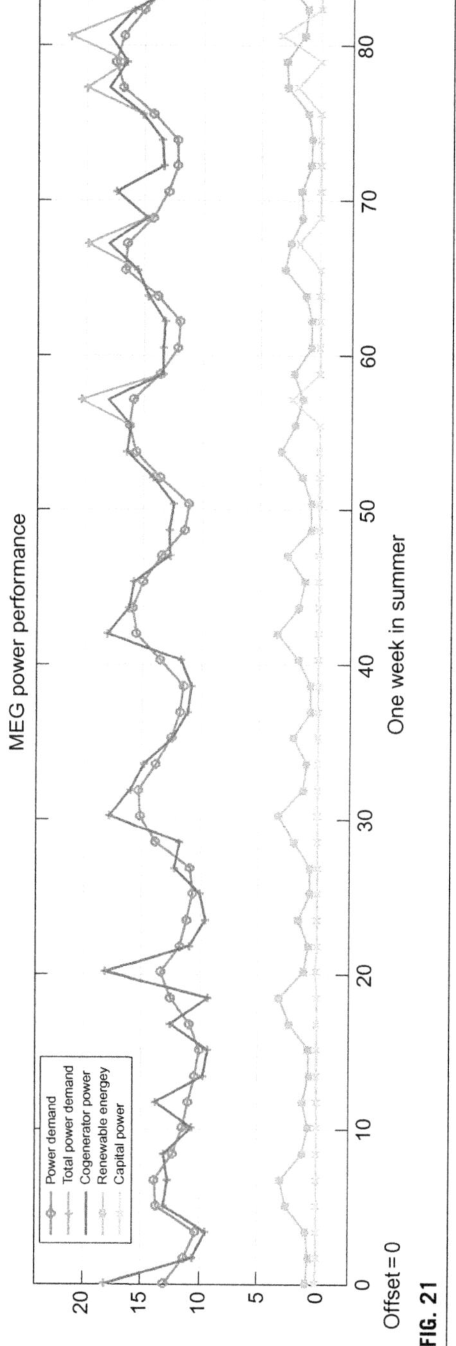

FIG. 21

MEG power profile with utilizing cogeneraation and TES IPLs.

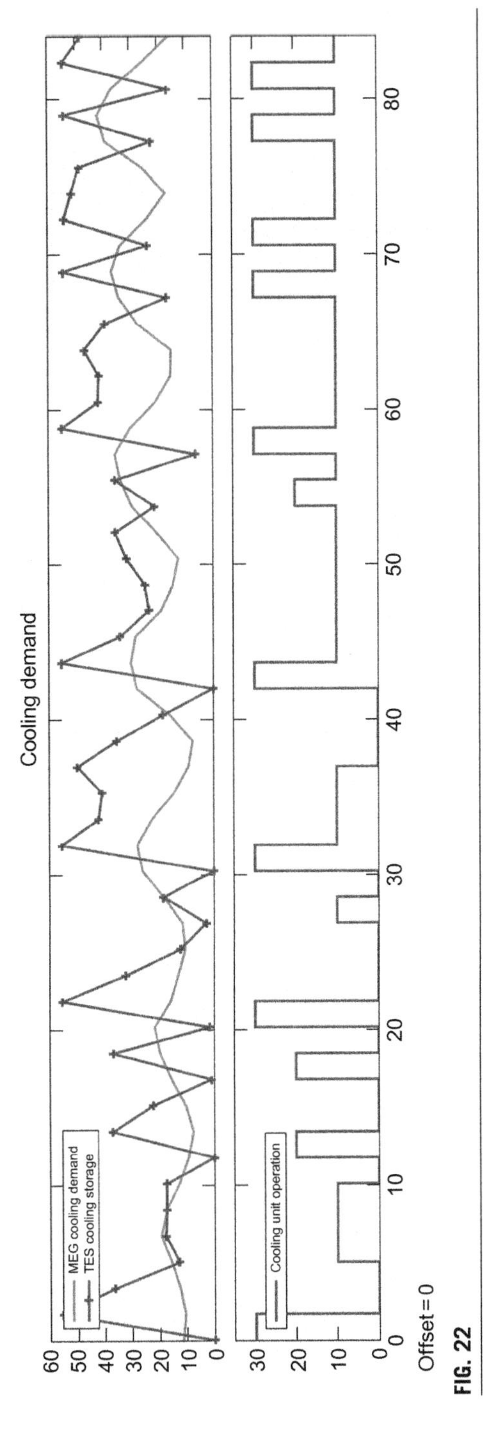

FIG. 22

MEG cooling profile with utilizing cogeneraation and TES IPLs.

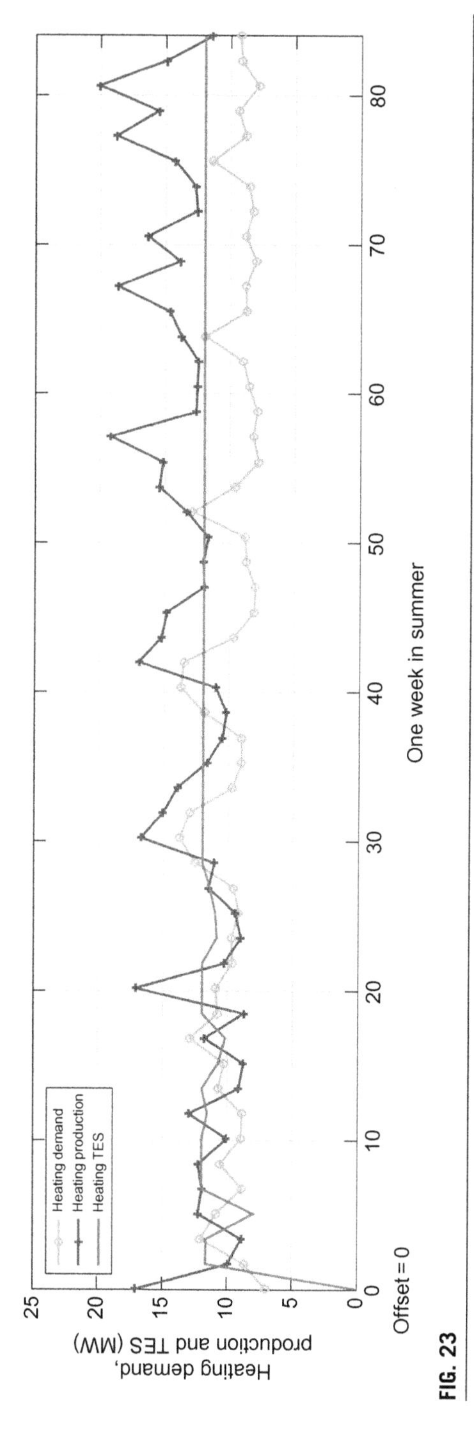

FIG. 23

MEG heating profile with utilizing cogeneraation and TES IPLs.

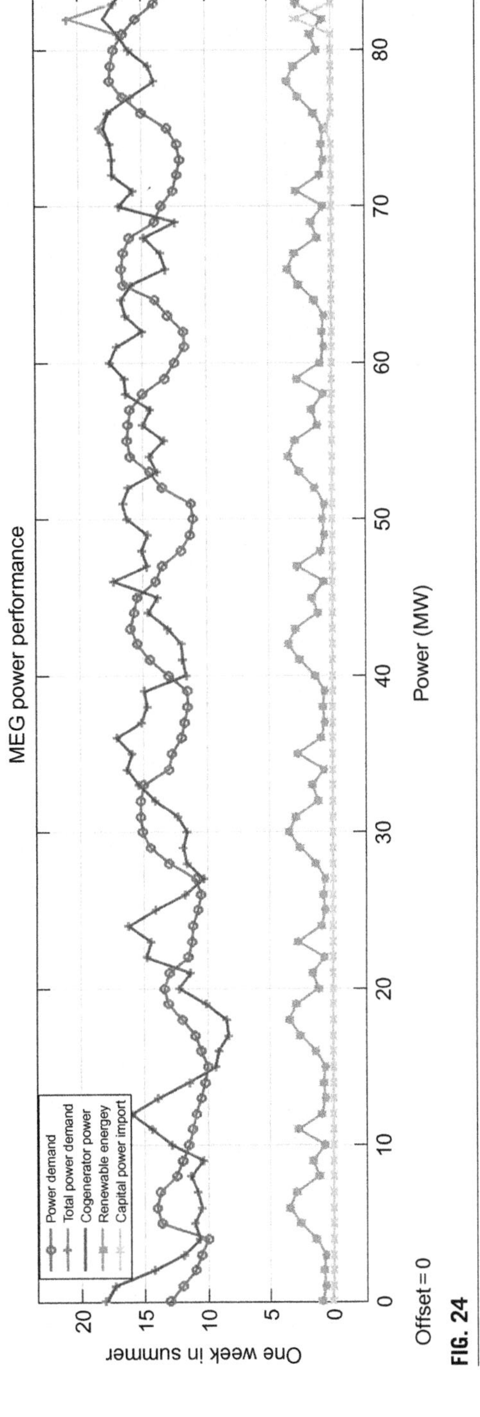

FIG. 24

MEG power profile with utilizing IPL-1, IPL-2, and IPL3.

a period of 1 h in each, with maximum 3 MW while the cogenerator serves an average of 12 MW with maximum production of 18 MW.

The cooling profile in Fig. 25 shows an improvement in the thermal cooling unit operations, where cooling on demand was shifted completely to the off-demand period, by rescheduling the operation of the thermal cooling units. The shifting of cooling to on-demand has a major positive impact on both power and cooling profiles, subsequently it is increasing the MEG capabilities without additional physical hardware upgrades for the MEG system, to further increase the MEG resiliency and self-healing competency.

A sample of the heating demand profile for 1 week in the summer was demonstrated in Fig. 26. Widespread coverage of the heating demand can be achieved by the heat generated from the cogenerator unit, but the low correlation of the electricity demand and heating demand, particularly during the summer season, makes it asynchronous between the heating on-demand and the heating generated by the cogenerators, in case the heuristic methodology is not applied. However, by comparing Figs. 23 and 26 it can be defined that heating production becomes more correlated to the heating demand when the heuristic methodology is employed.

5.9.4 FOURTH BASELINE STRATEGY

In order to observe the behavior of a hierarchical control on the MEG resiliency, when four out of six cooling units are out of order. In this case the prescheduled chiller operation has failed to produce the required cooling energy within the off-demand period, therefore district cooling units have to fulfill the requirements of the cooling demand during on-demand period as illustrated in Fig. 27. Obviously, the controller reaction helps to maintain serving cooling energy whilst a fault event has occurred. Nevertheless, the MEG has lost the optimal flat profile for cogenerator power production and it is following the energy demand, as shown in Fig. 28.

5.10 CONCLUSION

A safety design for a MEG system is proposed in this chapter in order to mitigate major hazards, which threaten the current MEG, by increasing the resiliency of the system using three IPLs. Cogeneration, TES, and a FTEM controller are used as IPLs to enable the MEG to work in an islanded mode for normal energy demands during all four seasons. A hierarchical control in the three-level structure was implemented using an adaptive-network-based fuzzy inference. Coordination between control levels has been realized in order to achieve the high system resiliency and to optimize the energy production profile based on aggregated information collected from local subsystems, which determine some "directions" to reactive controllers as well as the decision maker controller level to plan the overall energy procedure.

Those IPLs increase the MEG reliability to more than double its normal capability, while the cogenerator, TES, and FTEM offer significant reduction on the capital

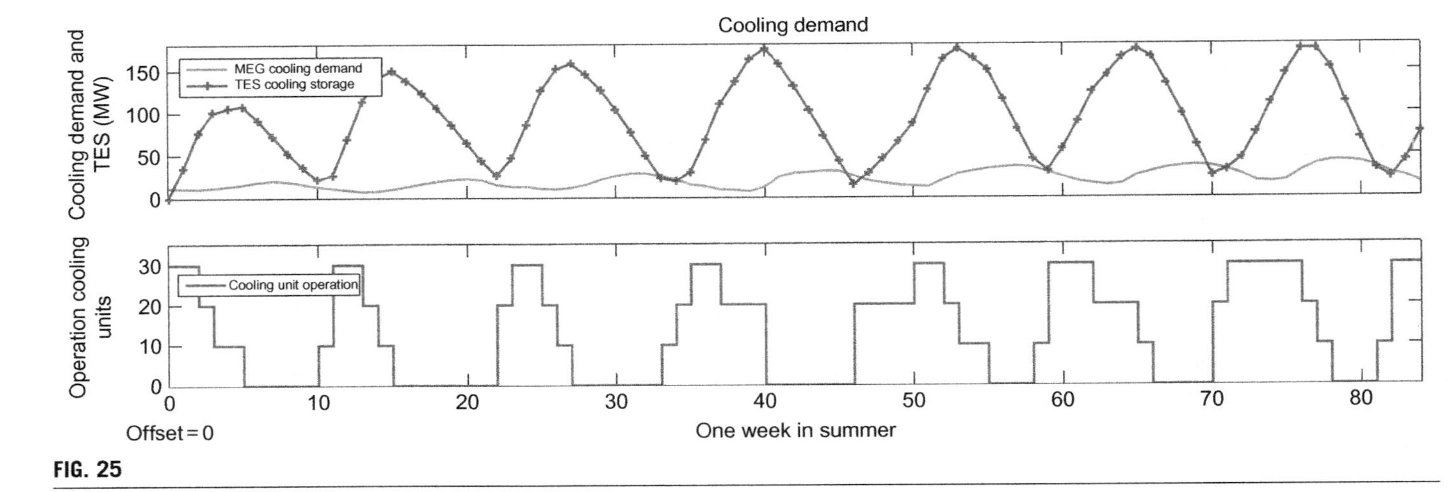

FIG. 25
MEG cooling profile with utilizing IPL-1, IPL-2, and IPL3.

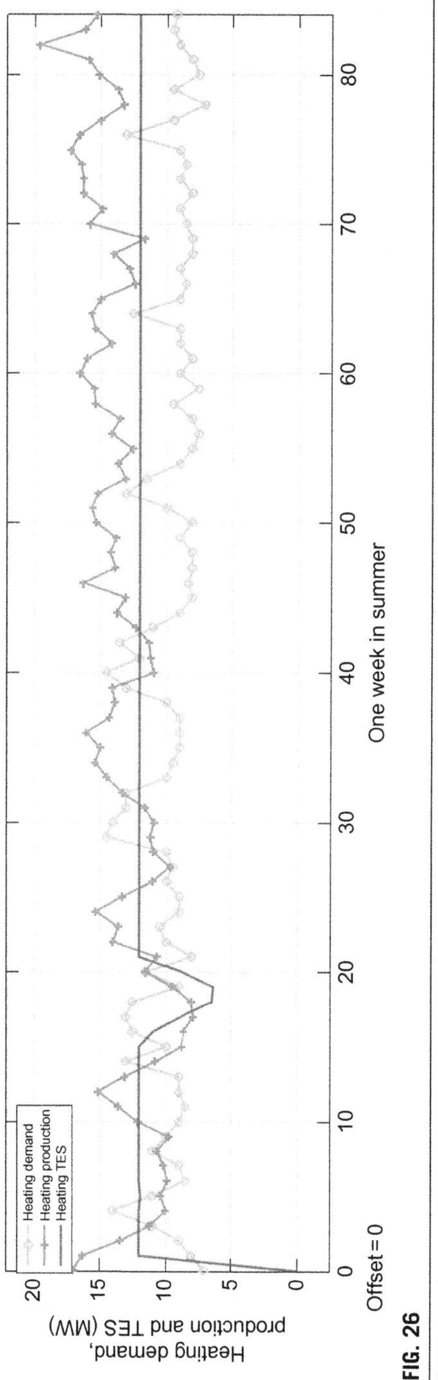

FIG. 26

MEG heating profile with utilizing IPL-1, IPL-2, and IPL3.

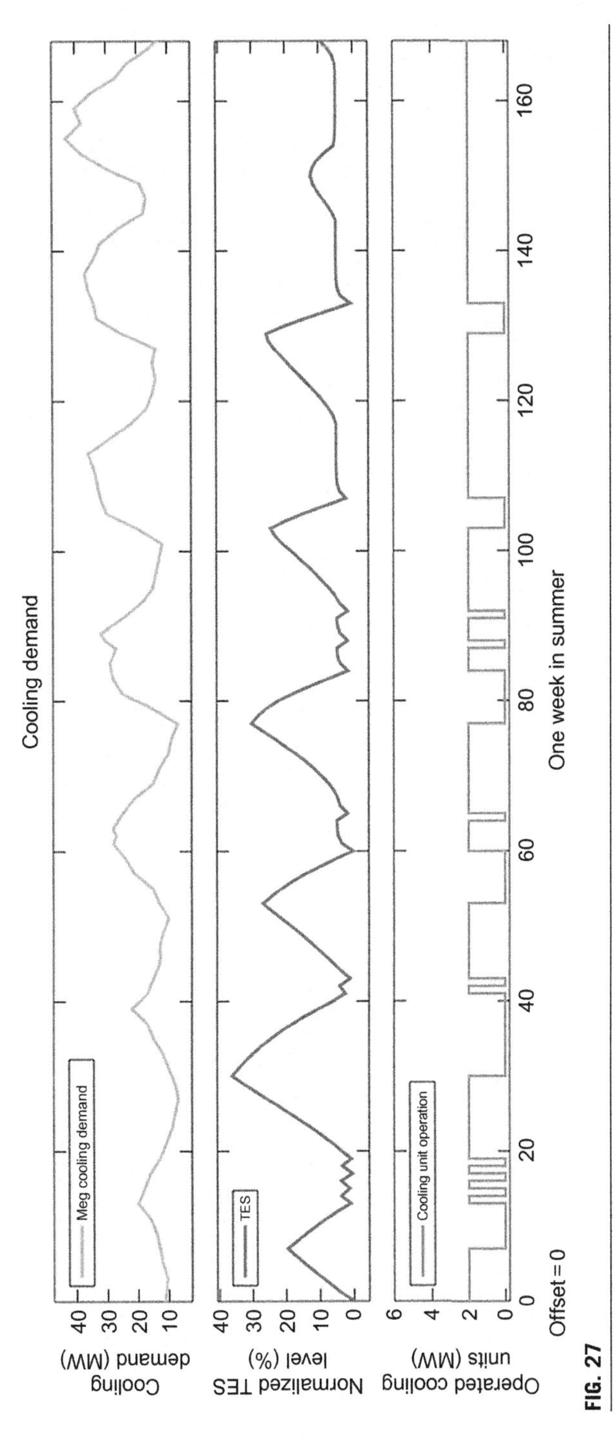

FIG. 27
Cooling profile for a resilient MEG at hazard event.

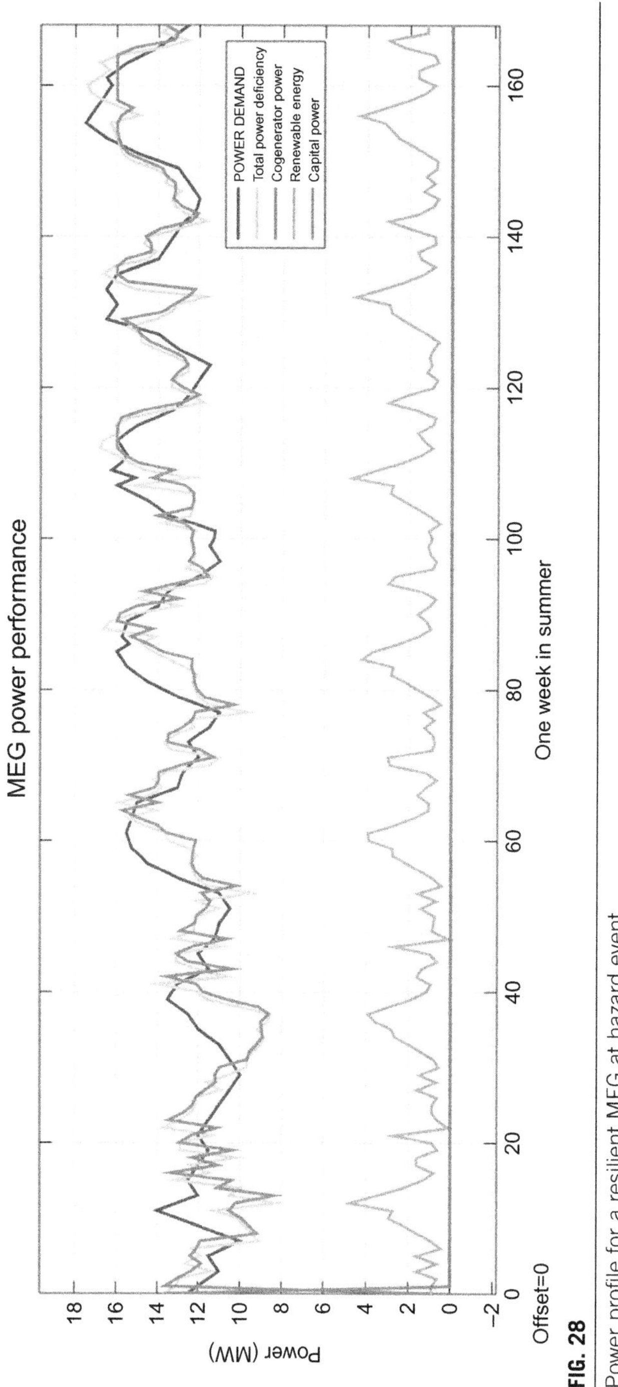

FIG. 28

Power profile for a resilient MEG at hazard event.

grid risk severity. Subsequently, the IPLs enable the improvement of MEG performance with practical everyday considerations, such as equipment maintenance and variation in energy demand that affect MEG operation and load distribution. Predicting future load profiles from historical data can provide tolerable approximates in scheduling the dispatch of MEG resources, which can be easily adapted using the real-time energy dispatch control for effective management of MEG resources and energy flow mapping.

REFERENCES

[1] Gabbar HA. Design and planning support tool for interconnected micro energy grids. Br J Appl Sci Technol 2016;12(6):1–15. Article no. BJAST.21968.

[2] Zhang J, Zhou Q, Yang Z. Reliability assessment for micro-grid with multi-energy demand. In: China international conference on electricity distribution (CICED 2014); 2014.

[3] McLarty D, Sabate C, Brouwer J, Jabbari F. Micro-grid energy dispatch optimization and predictive control algorithms; a UC Irvine case study. Electr Power Energy Syst 2015;65:179–90.

[4] Dieck-Assad G, Masada GY, Flake RH. Optimal set-point scheduling in a boiler-turbine system. IEEE Trans Energy Convers 1987;16:388–95.

[5] Garduno-Ramirez R, Masada GY, Lee KY. Multiobjective optimal power plant operation through coordinate control with pressure set-point scheduling. IEEE Trans Energy Convers 2001;I6:115–22.

[6] Chandan V, Do A-T, Jin B, Jabbari F, Brouwer J, Akrotirianakis I, et al. Modeling and optimization of a combined cooling, heating and power plant system, In: American control conference, Fairmont queen Elizabeth, Montreal, Canada, June 27–29, 2012; 2012.

[7] Maarten W. The research agenda on social acceptance of distributed generation in smart grids: renewable as common pool resources. Renew Sustain Energy Rev 2012; 16(1):822–35. http://dx.doi.org/10.1016/j.rser.2011.09.006.

[8] Costamagna P, Magistri L, Massardo AF. Design and part-load performance of a hybrid system based on a solid oxide fuel cell reactor and a micro gas turbine. J Power Sci 2001;96:352–68.

[9] Girgis AA, Mathure S. Application of active power sensitivity to frequency and voltage variations on load shedding. Electr Power Syst Res 2010;80(3):306–10.

[10] Schultz CC, p. e. carl c. schultz. Closing the cogeneration gap. Engineered Systems 2010;27(6):38.

[11] Frankland JH. A model-based feasibility study of combined heat and power systems for use in urban environments. Atlanta, GA: 2013.

[12] Hosseinalipour SM, Abdolahi E, Razaghi M. Static and dynamic mathematical modeling of a micro gas turbine. J Mech 2013;29(2):327–36.

[13] Abobghala A. Modelling, simulation and optimisation for the operation of heat recovery steam generators. Huddersfield: The University of Huddersfield; 2010.

[14] Ghaffari A, Chaibakhsh A. Neuro-fuzzy modeling of heat recovery steam generator. Int J Mach Learn Comput 2011;2:479–83.

[15] Turner WC, Doty S. Energy management handbook. 8th ed. Lilburn, GA: The Fairmont Press Inc.; 2013.
[16] Lee KS. Underground thermal energy storage. London, Heidelberg, New York, Dordrecht: Springer; 2013.
[17] Petchers N. Combined heating, cooling & power handbook: technologies & applications an integrated approach to energy resource optimization. Lilburn, GA: Fairmont Press; 2012.
[18] Gabbar HA, Islam R, Isham MU, Trivedi V. Risk-based performance analysis of microgrid topology with distributed energy generation. Int J Electr Power Energy Syst 2012; 43(1):1363–75.
[19] Schroeder A. Modeling storage and demand management in power distribution grids. Appl Energy 2011;88(12):4700–12.
[20] Wang Y, Li W, Lu J, Liu H. Evaluating multiple reliability indices of regional networks in wide area measurement system. Electr Power Syst Res 2009;79(10):1353–9.
[21] McManus TN. Management of hazardous energy: deactivation, de-energization, isolation and lock-out. Boca Raton, London, New York: CRC Press; 2012. ISBN 9781439878361. CAT# K13638.
[22] Goble WM, Cheddie H. Safety instrumented systems verification: practical probabilistic calculations. Research Triangle Park, NC: ISA—The Instrumentation, Systems, and Automation Society; 2004.
[23] Wolter K. Stochastic models for fault tolerance: restart, rejuvenation and checkpointing. Heidelberg, Dordrecht, London, New York: Springer; 2010.
[24] Dubrova E. Fault-tolerant design. New York, Heidelberg, Dordrecht, London: Springer Science and Business Media; 2013.
[25] Islam MdR, Gabba HA. Study of micro grid safety & protection strategies with control system infrastructures. Smart Grid Renew Energy 2012;3:1–9.
[26] Daneshi H, Khorashadi-Zadeh H. Microgrid energy management system: a study of reliability and economic issues, In: IEEE power and energy society general meeting; 2012. p. 1–5.
[27] eeh power systems laboratory. UCTE system adequacy forecast; 2007.
[28] Abbasi AR, Seifi AR. Considering cost and reliability in electrical and thermal distribution networks reinforcement planning. Energy 2015;84:25–35.
[29] Momoh JA. Smart grid fundamentals of design and analysis, vol. 53(9). Hoboken, NJ: John Wiley & Sons; 2012.
[30] Hwang T, Choi M, Kang S, Lee I. Design of application-level reference models for micro energy grid in IT perspective, In: International conference on computing and networking technology; 2012.
[31] Telecom K, Korea S. Korea micro energy grid technology. The use case of the First-Town in Sejong. In: Network operations and management symposium (APNOMS), 2013 15th Asia-Pacific; 2013.
[32] Barati M, Lotfi S, Rahmati A. A fault tolerance algorithm for resource discovery in semantic grid computing using task agents. J Softw Eng Appl 2014;7:256–63.
[33] Gabbar HA, Bower L, Pandya D, Agarwal A, Tomal MU, Islam FR. Resilient micro energy grids with gas-power and renewable technologies. In: Proc—ICPERE 2014 2nd IEEE conference on power engineering and renewable energy 2014; 2014. p. 1–6.
[34] Tomal MU, Gabbar HA. Key performance assessment of fuel cell based distributed energy generation system in resilient micro energy grid. In: IEEE international conference on smart energy grid engineering; 2015. p. 1–6.

[35] Lipan LC, Pislaru-Danescu L, Dasu DC, Tomescu L, Pisica I, Comanescu G. Design and implementation of a combined cooling, heating and power plant to support data centre supply reliability. In: 2012 47th international universities power engineering conference; 2012. p. 1–6.

[36] Li Y, Mavris D. A hierarchical control architecture for resource allocation, vol. 2012; 2012. p. 5–7.

[37] Wandhare RG, Thale S, Agarwal V. Reconfigurable hierarchical control of a microgrid developed with PV, wind, micro-hydro, fuel cell and ultra-capacitor. In: Conference proceedings—Applied Power Electronics Conference and Exposition (APEC); 2013. p. 2799–806.

[38] Duda Z. Two-level hierarchical control in a large scale stochastic system, In: Proceedings of the European control conference; 2001. p. 2193–7.

[39] Guerrero JM, Vasquez JC, Matas J, De Vicuña LG, Castilla M. Hierarchical control of droop-controlled AC and DC microgrids—a general approach toward standardization. IEEE Trans Ind Electron 2011;58(1):158–72.

[40] Chmielowski WZ. Fuzzy control in environmental engineering, vol. 31. Cham: Springer International Publishing; 2016.

[41] Kurano M, Yasuda M, Nakagami J, Yoshida Y. A fuzzy approach to Markov decision processes with uncertain transition probabilities. Fuzzy Sets Syst 2006;157(19):2674–82.

CHAPTER

Regional transportation with smart energy grids and hybrid fuel options

H.A. Gabbar*, A.M. Othman*,†, N. Ayoub‡,§

University of Ontario Institute of Technology (UOIT), Oshawa, ON, Canada Zagazig University, Zagazig, Egypt† The University of Missouri, Columbia, MO, United States‡ Helwan University, Cairo, Egypt§*

6.1 INTRODUCTION

This chapter discusses the development and design of smart energy grids (SEGs) with effective strategies and hybrid fuel options. One potential application is the integration with the regional transportation and railway infrastructures as a new green technology. This integration will enable interconnected SEGs to work transparently with the recently stored energy. Railway transportation with an SEG model proposes to balance energy flows between the train's moving and braking energy, energy storage system, and a main power utility network. As an energy optimization tool for the interconnected railway, SEGs can be used for energy management application and for achieving economical costs during operation.

This chapter also discusses and presents modeling concepts for hybrid transportation techniques (HTTs), which can contain sustainable, environment friendly vehicles—including electric and hydrogen vehicles. The HTTs has their integrality with the modeling that supports them, such as the infrastructure technical system with respect to SEGs, fuel chain modeling, environmental gain model, cost model, and the market-predicating model that feeds back to the HTTs with the database. The forecasting and predicating can be on medium-term or on long-term market forecasting, based on the autoregressive integrated moving average model. The engineering model for infrastructure and layers of energy grids with HTT has been given. The fuel supply chain model is discussed and developments regarding the most significant techniques for implementing regional green transportation are presented.

6.2 SEGs FOR TRANSPORTATION

The SEG has recently received a great deal of interest due to its applicability and effectiveness. SEGs provide electricity with enhanced reliability, reduced emissions, improved power quality by supporting voltage and reducing voltage dips, and lead to

Smart Energy Grid Engineering.

lower energy supply costs. The purpose of evaluating the performance is to ensure resilient operation, and it is an important trend for many of the presented projects and publications. Developed grids enhance methods for automatic and seamless transitions between grid-connected and islanding modes of operation. An SEG should have control strategies that achieve voltage and frequency stability under grid and islanding operating conditions. There has been a global movement in the direction of adoption and deployment of distributed and renewable resources in SEGs. Hybrid distributed energy resources (DERs) provide clean energy supply alternatives that differ from conventional fossil fuel sources. The integration of such time-variable, distributed, or embedded sources into an SEG requires special consideration. Energy storage devices, such as batteries, energy capacitors, and flywheels, play an important role in SEG operation. Several researches have been performed to optimize the operation, load dispatch, and management of the energy storage system of the SEGs. There are some important drawbacks; one of them is that it does not consider all uncertainties of the problem.

In general, electric railway systems are more effective than the diesel ones due to their minimal losses and low emissions. Also, electric railway systems have the ability to regenerate energy from braking power. Application of an SEG on railway infrastructure will save the energy by utilization of the braking energy and renewable powers in the presence of recent storage systems such as flywheel and supercapacitor. This procedure will lead to improved power quality and system reliability, where, at this moment, the railway traction system will participate in the supplying of power. In addition, this will confirm the SEG as an effective solution for normal operation, and, in the case of an emergency, as more reliable and secure. This will decrease the failure of the energy supply, increase energy storage capabilities, and reduce the demand not survived (DNS). Moreover, the reduction in pollution and greenhouse gas emissions will be actualized.

In view of some research challenges, it is important to build and demonstrate a working model for an SEG for transportation, especially railways with high-performance criteria using state-of-the-art control systems and technologies that will evaluate and confirm the technical benefits. It is important to support railway infrastructure with decision making to modernize power grid systems and introduce SEG concepts with integrated renewable energy generation and energy storage technologies. SEGs can be applied to manage the power flowing between trains as a unit of railway systems and the utility grid, and can also manage the operation of storage elements. The exchanged power from and to the power utility grid will be affected by the price tariff on power and energy. Along the railway system, there will be SEGs that can achieve these objectives with minimal costs. The SEG energy flow optimization problem is defined as a scheme, which will be simulated in any scenario as time slots.

6.2.1 INTERCONNECTION FORMULATION

Interconnected SEGs can be proposed with integrated energy storage and hybrid DERs with distribution lines. The railway transportation SEG model proposes to balance energy flows between the train's moving and braking energy, energy storage system, and a main power utility network as shown in Fig. 1.

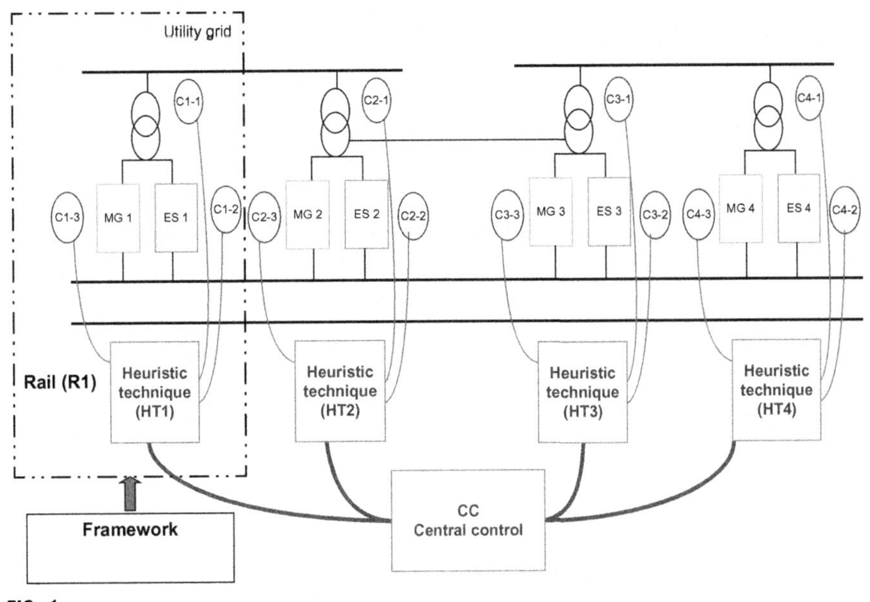

FIG. 1

General layout of interconnected SEG with railway.

The description of the optimization problem will depend on getting the optimal values of power of the SEG to supply the railway, and also the storage rate of the storage system that minimizes the operating cost of the SEG. This should be achieved by keeping the lines loading, and the voltage profile that will act as an energy optimization tool for the interconnected railway-SEG system.

Performance optimization

The task is concerned with finding the optimal power supply of an SEG and the optimal size of the energy storage system to minimize operational costs.

The main general description of the equations

$$\text{Min.Fitness}\, F_t(X, U) \tag{1}$$

Subject to:

$$G_t(X, U) = 0.0,\ H_t(X, U) \leq 0.0 \tag{2}$$

where

$F_t(X,U)$ represents the fitness function to be minimized;
$G_t(X,U)$ represents the vector of the equality constraints corresponding to active and reactive power balance equations;
$H_t(X,U)$ represents the vector of the inequality constraints corresponding to system parameter bound limits, active and reactive power generation limits, bus voltage limits, and phase angle limits;

X represents the vector of the states consisting of voltage magnitude and phase angles and the other dependent elements; and
U represents the vector of control variables to be optimized, which will be the output of the optimization process, optimal power supply of SEGG, and the optimal size of the storage.

The fitness function will depend on minimizing the operational cost of the power drawn from the utility grid with keeping the lines loading and the voltage profile within acceptable ranges:

$$Ob = F_t(X, U) = \sum_{k=1} C_k P_{\text{utility}\,k} \tag{3}$$

With constraint:

$$\sum_{i=1}^{NB} \begin{cases} 0, 0.95 \le V_{i\text{operating}} \le 1.05 \\ \log\left(W_i * \left|\frac{V_{i\text{nominal}} - V_{i\text{operating}}}{V_{i\text{operating}}}\right|^{K_{VV}}\right), \text{otherwise} \end{cases} + \sum_{j=1}^{NL} \begin{cases} 0, P_{j\text{operating}} \le P_{j\text{thermal capacity}} \\ \log\left(W_j * \left(\frac{P_{j\text{operating}}}{P_{j\text{thermal capacity}}}\right)^{K_{OL}}\right), \text{otherwise} \end{cases} \tag{4}$$

where

$P_{\text{utility}k}$ represents the power drawn from the utility power
C_k represents electricity price profiles
$V_{i\text{ operating}}$ represents the operating voltage magnitude at buses
$V_{i\text{ nominal}}$ represents the buses nominal voltage
W_i represents the weight and is determined in order to have a certain index value for various percentages of voltage differences, also used to adjust the slope of the logarithm
K_{VV} represents the coefficient used to penalize more or less voltage variations
NB represents the number of the buses in the system
$S_{j\text{ operating}}$ represents current Volt-Ampere power in lines
$S_{j\text{ thermal capacity}}$ represents the Volt-Ampere power rate of lines
W_j represents the weight and is determined in order to have a certain index value for various percentages of branch loading, also used to adjust the slope of the logarithm
K_{OL} represents the coefficient used to penalize more or less line overloads
NL represents the number of lines in the system

System constraints

Equality constraints

The active and reactive power balance equations at each bus in the network, which are described as

$$P_{Gi} - P_{Di} - P_{\text{Lines}\Rightarrow\text{bus}i}(V, \theta) = 0.0 \tag{5}$$

$$Q_{Gi} - Q_{Di} - Q_{\text{Lines}\Rightarrow\text{bus}i}(V,\theta) = 0.0 \tag{6}$$

Inequality constraints

– Generation Power Limits:

$$P_{Gi}^{\min} \leq P_{Gi} \leq P_{Gi}^{\max}, \quad i = 1,\ldots,nG \tag{7}$$

$$Q_{Gi}^{\min} \leq Q_{Gi} \leq Q_{Gi}^{\max}, \quad i = 1,\ldots,nG \tag{8}$$

– Bus Voltage Limits:

$$V_i^{\min} \leq V_i \leq V_i^{\max}, \quad i = 1,\ldots,NB \tag{9}$$

– Phase Angles Limits:

$$\delta_i^{\min} \leq \delta_i \leq \delta_i^{\max}, \quad i = 1,\ldots,NB \tag{10}$$

– Capacitor Constraints:

$$Q_c \leq Q_c^{\max}, \tag{11}$$

– Power Lines Limits

$$P_{ij} \leq P_{ij}^{\max}, \quad i = 1,\ldots,NL \tag{12}$$

SEG design and demonstrations

SEG configuration has different AC- and DC-distributed generating (DG) units that are supplying different AC and DC sides. Based on the output of the planning stage, the AC sources are the DFIG wind turbine generator and the microgas turbine generator. The DC sources are the battery, and fuel cell stack based on hydrogen and PV arrays. For full utilization operation, there are boost converters. AC/DC, DC/AC, and DC/DC converters are connected to the microgrid. Various types of DC and AC loads can be represented, such as resistive loads, motorized DC series motor loads, linear AC loads, nonlinear AC loads, and three-phase induction motorized loads.

6.3 HYBRID TRANSPORTATION MODEL

The hybrid transportation system contains different conventional vehicles that operate by fuelling with diesel and gasoline. This system contains sustainable, environmentally friendly vehicles like hydrogen vehicles, electric trains, electric vehicles, and others. There is an increase in the use of these vehicles in the current

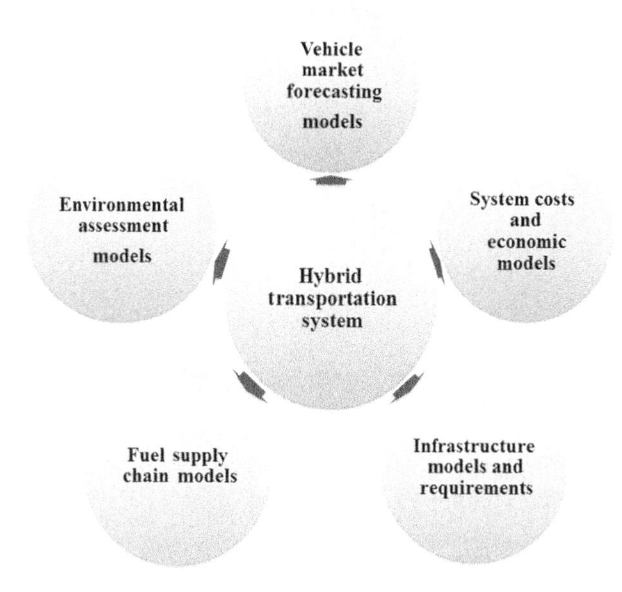

FIG. 2

Hybrid transportation system modeling.

transportation systems in various cities and countries throughout the world, showing promising trends for the future. See Fig. 2. There are different models and procedures are required to be concerned to make the transportation system work properly and effectively.

6.3.1 INFRASTRUCTURE MODELS

The penetration of electric vehicles, electric trains, hydrogen vehicles, and hydrogen-enriched natural gas (HENG) will affect and lead to evolutionary developments in the current infrastructure and predicting models needed to sustain the recently required operation criteria. It is important to facilitate supply chain infrastructure, modern communication channels, and sensor systems to effectively apply the integrated SEG. There are few survey works that talk about infrastructure predication for hybrid transportation systems. Based on the SEG framework and National Institute for Standard and Technology (NIST) research, the power electricity grid is integrated to natural gas grid. That integration can be in many ways for possible integration of the power electricity grid and natural gas grid. The energy significance of a general charging infrastructure on PHEVs is tested to emphasize that producing a widespread general charging service will decrease PHEV gasoline demand to 30%. Also, this grid facility helps to decrease the costs accrued by electric vehicle drivers. The significance of supplied energy infrastructure and supplied chain depends on CO_2 emissions and energy utilization. With comparing the traditional and electrical vehicles, the electric vehicles are better in carbon and energy consumption of supplied fuel than traditional ones. That conclusion was without the fact that off peak durations re-charging facility will lead to EVs operate as energy storage element that can operate as shaved load during the peak durations by vehicle to grid (V2G) theory to present a valued framework for charging electric vehicles as introduction of two recent electricity market models which are development of infrastructure and provider of services. There are various research survey concerns with the environmental and economical impacts of transportation systems and infrastructures. It is less difficult to get research references concerning the economic analysis and engineering analysis points of view. For example, Schroeder has analyzed the Return on Investment (ROI) of activating a large scale of rapid charging infrastructure based on the most risky parameters of market penetration of electric vehicle. Optimization methodology for the cost effective design of fuel units can be presented for infrastructure and alternative fuel vehicles. The mathematical modeling can estimate the best places to establish backup transportation fuel units in charge of their locations where the locations can maximize the amount of vehicles while keeping budget limitations are not violated.

6.3.2 INFRASTRUCTURE TECHNICAL MODELING

There are various developed techniques for the possible integration of the power electricity grid and natural gas grid. The power generated by renewable sources, like wind turbines and solar PV cells, is known to have fluctuations when transferred to a gas network by the electrolysis system, which are then directed to the gas grid system to produce hydrogen-enriched natural gas (Fig. 3).

6.3.3 HYBRID FUEL OPTIONS

Hybrid fuel options are available by new NG, hydrogen, and electricity fuel options while keeping the current fuel options. Based on the amount of vehicles within each assigned type, the distribution map of the managed, upcoming stations will be primarily planned. The distribution map of the fuel stations will depend on GIS systems, with respect to the distribution and the spread of vehicles, and the space from the nearest stations and their closest suppliers. That space is considered an index of the fuel selection for minimizing the cost; the most popular—referring to fuel applications—is using the nearest fuel station for the demand so the distribution map of fuel stations should be close to the demand areas.

There are many economical gains to be had using hybrid fuel options, such as:

CO_2 reduction

There will be reduction in CO_2 emissions, that gain can be estimated by the following formula:

$$B_1 = \alpha_{CO_2} \sum N^{veh}$$

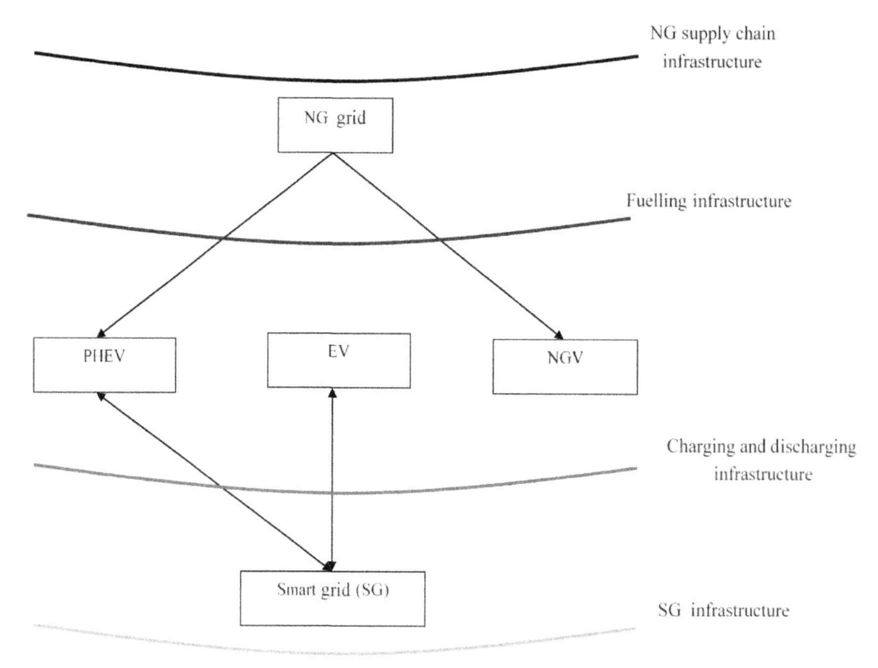

FIG. 3

Schematic representation for the infrastructure engineering model.

where

B_1 is gain the reduction in CO_2 emissions,
α_{CO_2} is the gain parameter by the average vehicle emissions, CO_2/km,
N^{veh} is the amount of vehicles with the average travel distance of 16,000 km per year.

Life time costs

The gain of the total savings comes from comparing new vehicles to the conventional vehicles. That gain can be estimated by the following formula, based on the Difference in Life Time Costs (DLTC) and the amount of vehicles for each kind N^{veh}:

$$B_2 = \sum \text{DLTC}\ N^{veh}$$

Infrastructure costs

The alternative fuel vehicles will need a resilient infrastructure of fuel stations to achieve high impact, so it is important to investors that extra hydrogen, electricity, and CNG centers or stations are installed. The cost of installing extra stations to the infrastructure includes the fixed cost for additional fuel station units, as well as the cost of integrating each station within the utility network.

$$B_3 = (1 + \alpha_{\text{Share}}) * \sum_j N_j\ \alpha_{j\text{build}} + \sum_{ij} \alpha_{ij\text{conn}}\ d_{ij\text{conn}}$$

where,

N_j is the amount of fuel stations per each vehicle j,
$\alpha_{j\text{build}}$ is the cost of installing fuel stations j per year,
α_{share} is the percentage of share when installing a station in the space of a current gas station,
$\alpha_{ij\text{conn}}$ is the cost of linking the fuel station i with the fuel station of vehicle j,
$d_{ij\text{conn}}$ is the distance from fuel station i to fuel j and the utility network km.

6.4 FUELS SUPPLY CHAINS

Based on the fuel kind, there are various supply chains corresponding to each fuel that can be utilized for the HTTs. The comparison between these chains depends on several parameters such as the economical situation, the natural of the applications and target consumers, the fuel characteristics, and others.

6.4.1 COMPRESSED NATURAL GAS

Oil fields are locations where we can find natural gas. Generally, gas resources and many other resources (such as waste treatment locations, landfills, and shale) contain NG traps inside their geologic space. As shown in Fig. 4, the handling of NG starts in the extraction field. First, condensate and water are eliminated, and then transport makes a processing facility. Within that processing facility, there are many chemical processes (such as filters, heating, and cooling treatments) that are applied to make sure the NG is pure, free of impurities. Later on in the NG processing, it is transported by pipelines (in the short spaces). There are two types of NG for long distance, which are liquefied natural gas (LNG) and compressed natural gas (CNG).

For ocean crossings, LNG is better where NG is liquefied at the departure location and returns back to the normal form at the destination. Another way is for it to be loaded on LNG trucks for further utilization. The transportation of CNG requires less capital cost in some of the small installations. The storage of natural gas can be done

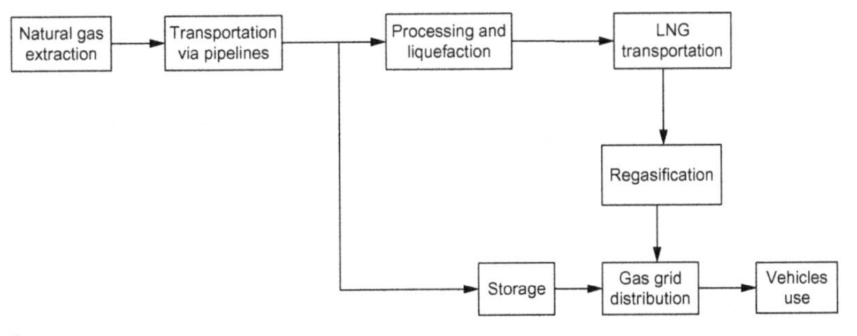

FIG. 4

Chain of NG resource.

in basins of depleted gas, aquifer reservoirs, or salt domes. The best economical option is the depleted gas reservoir, but its reliability is less because of there is one round of injection and withdrawal annually while the other two methods are higher in the cost.

6.4.2 ELECTRICITY SUPPLY

Fossil and renewable resources can be used in the production of electricity. There are great trends in the research on the supplied chains of electricity generation. Fig. 5 indicates the two forms as typical traditional and renewable supplied chains.

6.4.3 CONVENTIONAL GASOLINE

The gasoline and diesel supply chains can be considered the same, as they both appear at the extraction site when crude oil is pulled out. The extracted crude oil goes flowing through pipelines to refineries, and then on to manufacture the gasoline. Then, for refinery processing, gasoline flows inside pipelines, tanks, and trucks to fuel stations or to distribution centers for storage or utilizations. The conventional supplied chain of the gasoline is shown in Fig. 6.

6.4.4 PURE HYDROGEN FUEL

There are many requirements to make hydrogen available for utilization, such as infrastructure, installation, and storage. Also, there are strategic essential factors (such as the stock selection, the technology of conversion, and the technology of transportation between production locations and consumer centers) that should be emphasized economically and technically to impact performance. Fig. 7 indicates the supplied chains of hydrogen production where the second supplied chain is applied to infrastructure calculations.

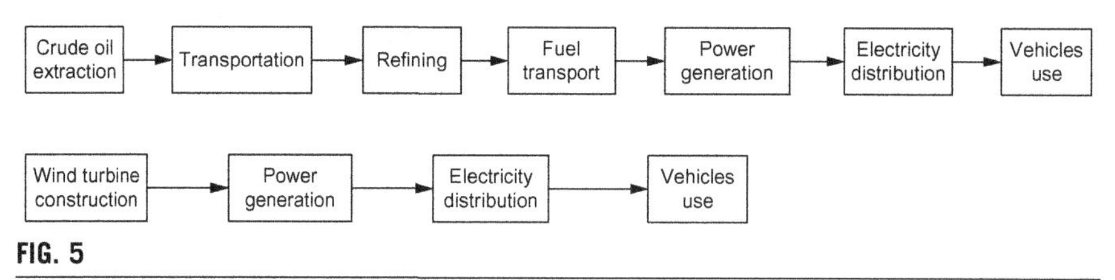

FIG. 5

Supplied chains for electricity generations.

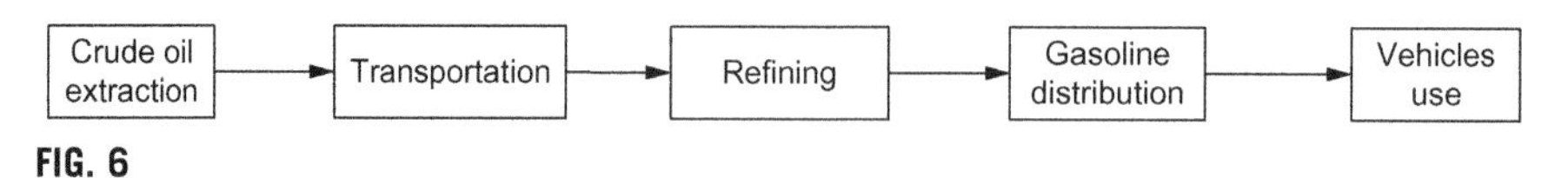

FIG. 6

Conventional supplied chain of the diesel and gasoline.

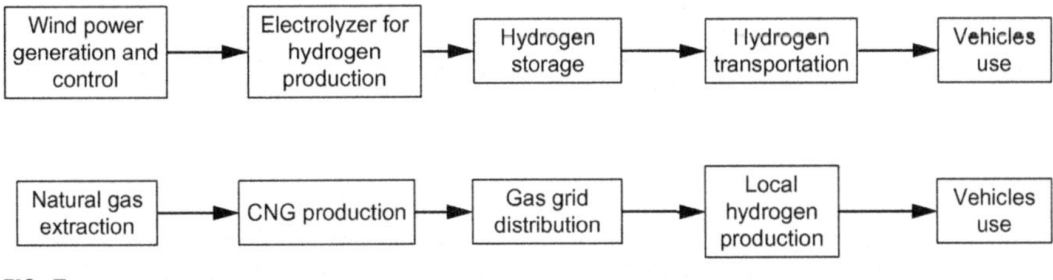

FIG. 7

Supplied chains of for hydrogen production.

6.4.5 HYDROGEN-ENRICHED NATURAL GAS (HENG)

Hydrogen-enriched natural gas is considered a combination of NG and hydrogen with,10–20% as hydrogen volume. The HENG is the most promising and cost-effective trend for reducing GHGs. HENG can mainly operate on the existing NG infrastructure—transmission, distribution, and consumption equipment. The supplied HENG chain is a combination of gas and hydrogen together in the pipelines of the distribution network.

6.4.6 THE ADVANCED SUPPLIED MODEL

The supplied fuel chain grid structure can be represented with a single output product and multiple-stage processes, which take into consideration the physical characteristics of each fuel kind, as indicated in Fig. 8. It is assumed that the supplied fuel is

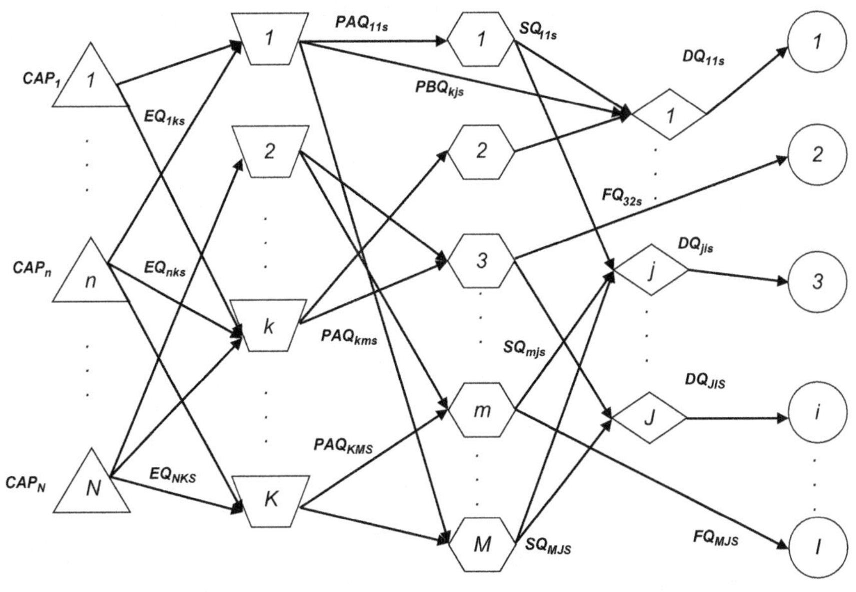

FIG. 8

Supplied fuel chain grid structure.

provided in large-scale fuel spots located in the heart of consumption points by selecting the optimal locations for installing fuel distribution centers.

The objective function of the fuel supply chain can be proposed as a mixed-integer linear programming model. The objective is to minimize the total cost of the supply chain with the following concerns: (1) the amount of fuel stations and their needs; (2) the maximum amount of refineries, storage spots, and fuel distribution center's strength and their structures; (3) the distance to the extraction site is not considered; (4) fuel stations are fed with fuel from a single source or storage space according to the nearest distance.

The mathematical representation of the supplied fuel supply chain model is explained as follows:

Model decision variables

$$w_k = \begin{cases} 1 & \text{if processing plant is used for fuel production} \\ 0 & \text{Otherwise} \end{cases}$$

$$r_j = \begin{cases} 1 & \text{if FDC is used for receiving and distributing fuels} \\ 0 & \text{Otherwise} \end{cases}$$

$$v_m = \begin{cases} 1 & \text{if storage place is used} \\ 0 & \text{Otherwise} \end{cases}$$

$$g_{ji} = \begin{cases} 1 & \text{if FDC } j \text{ is used to supply fueling station } i \\ 0 & \text{Otherwise} \end{cases}$$

$$u_{mi} = \begin{cases} 1 & \text{if storage place } m \text{ is used to supply fueling station } i \\ 0 & \text{Otherwise} \end{cases}$$

$$c_{mi} = \begin{cases} 1 & \text{if storage place } m \text{ is used to supply FDC } j \\ 0 & \text{Otherwise} \end{cases}$$

$$a_{km} = \begin{cases} 1 & \text{if processing plant } k \text{ is used to supply storage place } m \\ 0 & \text{Otherwise} \end{cases}$$

The objective function is defined by the total cost of the supplied fuel chain for transportation system. It is concerned with the transportation costs of oil and gas from extraction sites, and their later transportation fuel cost of shipping from refineries to consumers after storage.

Model mathematical indices

n is an index for fuel extraction process ($n \in N$),
s is an index for transportation method ($s \in S$)
k is an index for refineries ($k \in K$)
m is an index for storage place ($m \in M$)
j is an index for fuel distribution center, FDC, ($j \in J$)
i is an index for fuelling station type ($i \in I$)

Variables of the model

EQ_{nks}: the amount of extracted oil and gas transported from extraction site n to the processing plant k by transportation mean s,
PAQ_{kms}, PBQ_{kjs}: the amount of the fuel flowed from plant k to storage space m and fuel distribution center j, respectively, by transportation mean s,
SQ_{mjs}: the fuel amount transported from storage site m to fuel distribution center j by transportation mean s,
FQ_{mis}: amount of fuel transported from storage space m to fuel station i by transportation mean s,
DQ_{jis}: amount of fuel moved from fuel distribution center j to station i by transportation mean s.

Parameters of the model

W: the maximum amount of fuel distribution centers,
P: the maximum amount of refineries,
L: the maximum amount of storage spaces,
CAP_n: the maximum rate of extraction sites n,
d_i: the fuel demand required at fuel station i,
DTH_j: the yearly throughput of fuel distribution centers j,
STH_m: the yearly throughput of storage space m,
PTH_k: the yearly throughput of running a site k,
ETC_{nks}: transportation cost of fuel from extraction location n to site k by transportation mean s,
$PATC_{kms}$: the transportation cost (in units) of fuel from site k to storage space m by transportation mean s,
$PBTC_{kjs}$: the transportation cost (in units) of fuel from site k to fuel distribution centers j by transportation mean s,
STC_{mjs}: the transportation cost (in units) of fuel from storage apace m to fuel distribution centers j by transportation mean s,
FTC_{mis}: the transportation cost (in units) of fuel from storage space m to fuel station i by transportation mean s,
DTC_{jis}: the unit transportation cost of fuel from fuel distribution centers j to fuel station i by transportation mean s.

6.5 MATHEMATICAL MODELING

$$\min z = \sum_n \sum_k \sum_s ETC_{nks}\, EQ_{nks} + \sum_k \sum_m \sum_s PATC_{kms}\, PAQ_{kms} + \sum_k \sum_j \sum_s PBTC_{kjs}\, PBQ_{kjs} + \sum_m \sum_j \sum_s SQ_{mjs}\, STC_{mjs} + \sum_j \sum_i \sum_s DTC_{jis}\, DQ_{jis} + \sum_m \sum_i \sum_s FTC_{mis}\, FQ_{mis}$$

$$\sum_j g_{ji} = 1, \ \forall\, i, \ \sum_m u_{mi} = 0$$

$$\sum_m u_{mi} = 1, \ \forall\, i, \ \sum_j g_{ji} = 0$$

$$\sum_m c_{mj} = 1, \ \forall\, j, \ \sum_m b_{kj} = 0$$

$$\sum_j d_i\, g_{ji} \le DTH_j\, r_j \ \ \forall\, i,j$$

$$\sum_m d_i\, u_{mi} + \sum_m DTH_j\, c_{mj} \le STH_m\, v_m \ \ \forall\, m,i,j$$

$$\sum_m a_{km} = 1, \ \forall\, m$$

$$\sum_k b_{kj} = 1, \ \forall\, j, \ \sum_m c_{mj} = 0$$

$$\sum_k DTH_j b_{kj} + \sum_m STH_m a_{km} \le PTH_k\, w_k \ \ \forall\, k,j,m$$

$$\sum_k w_k \le P$$

$$\sum_j r_j \le W$$

$$\sum_m v_m \le L$$

$$\sum_s DQ_{jis} = d_i\, g_{ji} \ \ \forall i,j$$

The mathematical model equations indicate the objective function for the total SCN cost, represent the significant assignment of fuel distribution centers to fuel stations, fuel distribution centers to storage spaces, and fuel stations to storage spaces, respectively. Next, the capacity constraint of fuel distribution centers and the storage capacity constraints to meet all sources and chains are presented. The significant assignment of storage space m from processing location k and of fuel distribution centers j to processing location m, respectively, is presented.

There are descriptions for the capacity limitation of processing location m, which is dependent on the number of available refineries, storage spaces, and fuel distribution centers that have to have less than the maximum allowed amounts of P, L, and W.

There is another constraint that all the values of shipped fuel amount to fuel station have to be less than their maximum rates, with keeping the significance of supplied fuel station. Also, constraint over refineries and extraction locations should be observed. The model supposes that the integrality limitations are on the decision parameters and variables, which should not be negative natural. There are several solution techniques to solve the mixed-integer linear programming system. The genetic algorithm method is perfectly poised to get a set of possible populations and solutions in place of performing a succession of multiple trials using conventional mathematical algorithms and procedures.

FURTHER READING

[1] Liu X, Wang P, Loh C. A hybrid AC/DC microgrid and its coordination control. IEEE Trans Smart Grid 2011;2(2):278–86.

[2] Fadaee M, Radzi M. Multi-objective optimization of a stand-alone hybrid renewable energy sys. By using evolutionary algorithms: a review. Renew Sust Energ Rev 2012;16(5):3364–9.

[3] Salas V, Olias E, Alonsob M, Chenlo F. Overview of the legislation of DC injection in the network for low voltage small grid-connected PV systems in Spain and others. Renew Sustain Energy Rev 2008;12(2):575–83.

[4] Cossent R, Gomez T. Towards a future with large penetration of distributed generation: is the current regulation of electricity distribution ready? regulatory recommendations under a European perspective. Energy Policy 2009;37(3):1145–55.

[5] Falvo MC, Lamedica R, Bartoni R, Maranzano G. Energy management in metro-transit systems: an innovative proposal toward an integrated and sustainable urban mobility system including plug-in vehicles. Electr Power Syst Res 2011;81.

[6] Pankovits P, Pouget J, Robyns B, Delhaye F, Brisset S. Towards railway-smartgrid: Energy management optimization for hybrid railway power substations. In: IEEE PES Innovative Smart Grid Technologies Conference Europe (ISGT-Europe), pp. 12–15 Oct. 2014; 2014.

[7] Cornic D. Efficient recovery of braking energy through a reversible dc substation. IEEE Alstom Transport; 2010.

[8] Okui A, Hase S, Shigeeda H, Konishi T, Yoshi T. Application of energy storage system for railway transportation in Japan. In: International Power Electronics Conference (IPEC), pp. 3117–3123, June 2010; 2010.
[9] Leung PCM, Lee EWM. Estimation of electrical power consumption in subway station design by intelligent approach. Appl Energy 2013;101.
[10] Gulin M, Vašak M, Baotić M. Analysis of microgrid power flow optimization with consideration of residual storages state. In: Proceedings of the 2015 European control conference, Austria, pp. 3131–3136, 2015; 2015.
[11] Yiqiao C, Jiahai W. Differential evolution with neighborhood and direction information for numerical optimization. IEEE Trans Cybernet 2013;43(6):2202–15.
[12] Xinggao L, Yunqing H, Jianghua F, Kean L. A novel penalty approach for nonlinear dynamic optimization problems with inequality path constraints. IEEE Trans Autom Control 2014;59(10):2863–7.
[13] Azizipanah-Abarghooee R. A new hybrid bacterial foraging and simplified swarm optimization algorithm for practical optimal dynamic load dispatch. Electr Power Energy Syst 2013;49:414–29.
[14] Rajabi A, Fotuhi M, Othman M. Optimal unified power flow controller application to enhance total transfer capability. IET Generation, Transmission & Distribution 2015;9(4):358–68.
[15] Hong Y, Wei Z, Chengzhi L. Optimal design and techno-economic analysis of a hybrid solar-wind power generation system. J Appl Energy 2009;86(2):163–9.
[16] Anglani N, Muliere G. Analyzing the impact of renewable energy technologies by means of optimal energy planning. In: 9th International conference on environment and electrical engineering (EEEIC), Prague, 2010, pp. 1–5; 2010.
[17] Patnaik SS, Panda AK. Particle swarm optimization and bacterial foraging optimization techniques for optimal current harmonic mitigation by employing active power filter. J Appl Comput Intell Soft Comput 2012;2012:1–10.
[18] Acharya DP, Panda G, Lakshmi YVS. Effects of finite register length on fast ICA, bacterial foraging optimization based ICA and constrained genetic algorithm based ICA algorithm. J Digital Signal Process 2011;20(3):964–75.
[19] Biswas A, Das S, Abraham A, Dasgupta S. Stability analysis of the reproduction operator in bacterial foraging optimization. J Theoret Computer Sci 2010;411:2127–39.
[20] Al-Alawi BM, Bradley TH. Review of hybrid, plug-in hybrid, and electric vehicle market modeling studies. Renew Sust Energ Rev 2013;21:190–203.
[21] Al-Alawi BM, Bradley TH. Total cost of ownership, payback, and consumer preference modeling of plug-in hybrid electric vehicles. Appl Energy 2013;103:488–506.
[22] Altiparmak F, Gen M, Lin L, Paksoy T. A genetic algorithm approach for multi-objective optimization of supply chain networks. Comput Ind Eng 2006;51:196–215.
[23] An H, Wilhelm WE, Searcy SW. Biofuel and petroleum-based fuel supply chain research: a literature review. Biomass Bioenergy 2011;35:3763–74.
[24] Andress D, Nguyen TD, Das S. Reducing GHG emissions in the United States' transportation sector. Energy Sustain Develop 2011;15:117–36.
[25] Ayoub N. A multilevel decision making strategy for designing and evaluating sustainable bioenergy supply chains. Process system engineering. Yokohama: Tokyo Institute of Technology; 2007. p. 150.
[26] Ayoub N, Elmoshi E, Seki H, Naka Y. Evolutionary algorithms approach for integrated bioenergy supply chains optimization. Energy Convers Manag 2009;50:2944–55.

CHAPTER

High-performance large microgrid

7

H.A. Gabbar*, A.M. Othman*,†

University of Ontario Institute of Technology (UOIT), Oshawa, ON, Canada Zagazig University, Zagazig, Egypt†*

7.1 INTRODUCTION

This chapter is aimed at the design and development of a high-performance large microgrid (LMG). The LMG can be defined as a microgrid with a large generation capacity of AC/DC power, using distributed generations (DGs) technologies in the form of DG farms (DGF) or large generation units, such as small modular reactors (SMRs). Advanced, intelligent protection and control schemes can be designed and evaluated according to IEEE, IEC, and other standards, as part of electric engineering practices. For decision support on LMG operation and automation, an integrated modeling and simulation environment can be developed with quantitative and qualitative knowledge. LMG protection and safety concepts can be analyzed based on the risk and reliability of engineering fundamentals. Multivariate data analysis approaches are employed to analyze real-time monitoring data, which are synchronized with the simulation environment. In view of these research challenges, it is important to build and demonstrate a working model of the LMG with state-of-the-art protection and control systems, and technologies that will evaluate and confirm the benefits to the power grid, in terms of reduced operation costs and environmental stresses within the integrated DG and energy storage (ES) technologies.

7.2 LMG DESIGN AND CONFIGURATION

In the last decade, the increase in demand for electricity, as well as climate changes, provoked governments and industries to invest more in renewable energy technologies as alternative energy sources and power supply systems. LMGs can provide a robust power network that can maintain a sustainable energy supply that covers local power demands and needs in terms of power, while supplying excess power back to the grid in a cost-effective manner.

LMGs can be considered to be a microgrid with a large generation capacity of AC/DC power, using DG technologies in the form of DGF or large distributed

Smart Energy Grid Engineering.

generation (LDG) units, such as SMRs. SMRs have many technical merits, such as flexibility, reliability, and a cost-effective electric energy supply for future applications. SMRs could allow for a stabilized output of energy as required. SMRs and other DGs can be analyzed by their power performance, geographic profile, and environmental effects, using modeling and simulation results. The power of SMRs could also be analyzed using safety, risk management, and special real cases. SMRs are classified as DG and are used mainly in the recent emergency applications. Based on the International Atomic Energy Agency (IAEA) standards, a unit having an electrical supplied outlet of less than 300 MW can be considered a SMR. An SMR can have a vital contribution as a DG within LMGs. Small reactors can be divided into functional modes that include a researching reactor, prototyping reactor, testing reactor, and a trading efficient reactor [1–4].

Within any nuclear reactor, there is a sustaining fission caused by the reacting of chains that is desired for the supply of electrical power. Identically, uranium oxide, known as UO_2, is applied to generate fuel output aligned inside the core of the reactor. The supplied energy of the fission reaction is directed as heat by the coolant, which is moved to a steam source for steam production. The steam is subjected to a high-pressure steam from the source that leads the turbines to generate electricity. The essential parts contain heat transfer units, fuel chains, the core of the reactor, a steam source, a turbine system, and a generating unit. Table 1 indicates the main variables of the different SMR systems. SMRs should be designed considering various parameters in order to allow for flexible and reliable high-performance criteria. An SMR can be constructed in many environmental places and sites. There are many

Table 1 Different SMR Technology with Particulars

SMR Name	Technology	Thermal Capacity	Electrical Capacity	Fuel	Refueling Cycle
Westinghouse	PWR	800 MWt	225 MWe	<5% enriched ^{235}U	2 years
SMR					
mPower	PWR	530 MWt	150–180 MWe	<5% enriched ^{235}U	4+ years
NuScale	PWR	160 MWt	45 MWe	4.95% enriched	2 years
IRIS	PWR	300–1000 MWt	100–335 MWe	5% enriched ^{235}U	5 years
Gen4 (hyperion)	FNR	75 MWt	25 MWe	Uranium nitride	10 years (replaced)
4S	LMR	30 MWt	10 MWe	19.9% enrichment	10–30 years

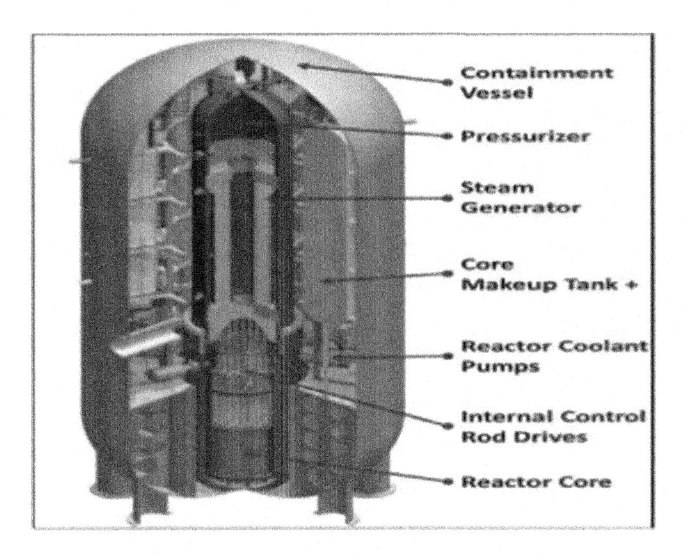

FIG. 1

Westinghouse small module reactor [1].

significant merits of SMRs, such as simple maintenance procedures, a long lifetime, and higher performance of the protection and operation system (Fig. 1).

There are many aspects related to the rate and technical structure of the SMRs that could be a backup guide for integrated LMGs. Various types of SMRs, as per the rate, can change from 10 to 300 MW, such as the graphite moderator reactor (GMR), fast neutron reactor (FNR), and the light water reactor (LWR). With respect to their electrical profile, many SMRs can be applied in one location within an LMG. Because SMRs can save stable energy, the output energy could be exchanged with the utility network. SMRs may not be considered distributed, such as solar PVs or wind turbines, due to safety considerations. In addition, SMRs could provide more rates than normal traditional plants [5, 6].

SMR technical innovations and design procedures are in progress all over the world with high developments. FNR, GMR, and LWR are vital types of SMRs. The LWR has the least risk rates, while the FNR has the simplest and longest refueling option. The GMR generates much more concern about installation. This leads to the FNR and LWR being a better design for SMRs. The SMR can be flexibly designed as per rates and types. The main concern of SMR design includes plant installation within the location, less required time, and more reasonable with higher effective process. The previous statement indicates that SMR can be allowed, with flexibility and simplicity, as power backup. Proper designing and safety concerns allow for the least core damage frequency (CDF) between the different kinds of nuclear units. A pressurized water reactor (PWR) has an annual CDF value of 10^{-4} to 10^{-6}. The LWR has a CDF per year that equals 10^{-6} to 10^{-7}. SMRs provide the least CDF among the others, annually equaling 10^{-8}.

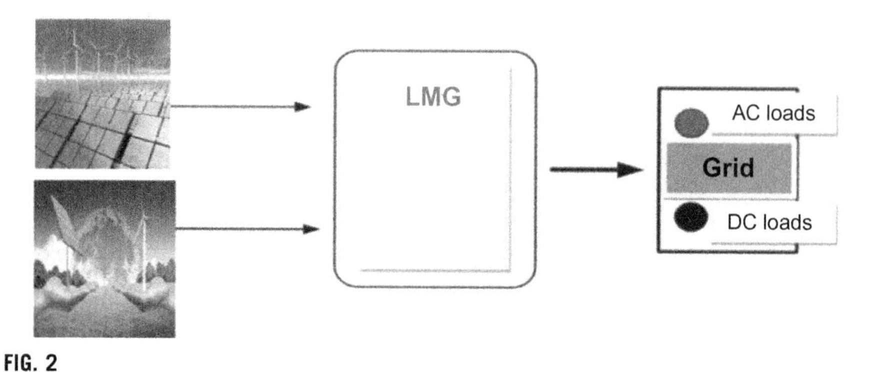

FIG. 2

LMG using the DGF with the integration of WFs with solar PV.

Another form of LMG, which could be in the form of a DGF, is the integration between wind farms (WFs) and Solar PV farms. The integration of WFs with electricity grids can lead to overall improved performance of energy supply grids. Wind technology investors are seeking potential means for enhancing the performance of energy generation, and integrating wind farms to support their expansion. An important objective, in order to achieve the integration of WFs with PV solar energy, is to support AC/DC circuits by enhancing the quality and controllability of MGs with new control design techniques. LMGs can emphasize advanced engineering practices for the improved integration of WFs and PVs with new design features and evaluation of overall energy generation performance, which increases the return on investment, and reduces the gap in generation between the grid and the loads. Also, there are enhanced protection and stability aspects [7–10]. LMGs can integrate the ES issue of MGs with the presence of WF and PV integration. Evaluation of overall energy generation performance, and using simulations based on key performance indexes, should be incorporated to study weather forecasting, and investigate adaptive wind design features with PV integration (Fig. 2).

Over the past two decades, there have been many studies that have worked on directing sustainable energy sources and smart grid systems, as a solution for managing integrated energy solutions. These studies have provided important contributions that focus on developing new intelligent methodologies that make use of the hybrid renewable energy resources of the WFs and PVs in an integrated applicable power system. Traditional power grids are characterized by their simplicity in centralizing energy production and consumption, but are undergoing much more complex conditions with interconnected models. However, MGs allow energy production, especially for solar PV panels and WF turbines. They are advantageous in the reduction of pollution and greenhouse gas emissions, and they reduce the national dependence on oil suppliers [11, 12].

Therefore, the integration of microgrid resources will be characterized by an increased utilization of renewable energies, and a widespread application of new power electronic devices that apply intelligent control strategies. MGs must also have a robust power control system that controls, and allows its connection to, the electric grid.

Hybrid energy integration enables the use and exploitation of on- and off-shore wind energy and PV solar energy, but this requires the development of an appropriate power-intelligent control strategy for a grid-connected hybrid generation in order to make a consistent power transfer to the system. Hence, LMGs can focus on the implementation of flexible solutions that investigate economic operations, the minimization of environmental impact, and the efficient utilization of resources, while reducing failures in the energy supply, and allowing transfers of unconsumed energy to the utility power network, thereby, reducing the necessity for increasing power plant numbers, or their power capacity. Recently there have been many artificial intelligent controllers applied in the design of the controller. And, in order to achieve optimal performance, many recent innovative heuristics optimization methods have been presented and applied.

LMGs could be intelligent and smooth for energy production, especially with solar PV panels and WF turbines. The operation of WFs faces some technical problems, such as the instability in some operating conditions due to dynamic responses. Also, the power quality becomes poor at the connection with the grid. To support the rate of the WF, integration with solar PV can be a solution; and this will face the challenge of AC/DC integration and operation. The arrangement, alignment, and optimal size of the maximum power-point tracking mode (MPPT) should be considered and well selected. The type and size of the storage elements are very important factors for normal operation and the stability aspect, as well as for protection, especially during the faults profile.

Power networks are seeking higher-performance energy-power supplies in the presence of MGs with DGs. There is an increasing need for the design of intelligent control systems in order to achieve efficient MGs, while also applying heuristic optimization techniques to optimize the overall MG performance. Flexible AC transmission system devices will be applied as a tool in order for the power system control to overcome the limitations of the AC/DC circuit integration.

LMGs can consolidate performance improvements of WFs when integrated with the utility grid, which includes power quality and cost. In addition, it can investigate the design and control of integrated WFs with recent ES technologies, and the optimization of the type and size of the ES system. Moreover, it can study and optimize the integration between WFs, ES, and PV systems, as an MG, with the utility grid.

Researchers can develop an intelligent control design framework in order to achieve resilient and high-performance MGs with DGs, and examine control designs for integrated controllers as part of the MGs, and implement them in power grids. Artificial intelligence techniques can be adopted to synthesize design and control scenarios, and optimize MG performance. Case studies of MGs could be examined for residential and commercial buildings with different load profiles.

Some of the challenges of implementing an efficient MG include:

(a) LMG design based on available resources, existing grid topology, and load and energy demand;
(b) dynamic protection with adequate fault propagation analysis based on the dynamic nature of LMGs in islanded and grid-connected modes;
(c) power quality and performance maximization in view of different operation parameters;
(d) cost–benefit analysis for different energy technologies with realistic profit planning; and
(e) considerations of international standards and national regulations for energy systems with a multiview analysis based on detailed business models (Fig. 3).

One important feature of the LMG is its ability to deal with low or medium voltage as part of the distribution networks, which minimizes the complexity and cost of hierarchically structured topologies and control systems traditionally used at high-voltage levels.

The integration of microgrids with existing electric utilities is most likely to impact their operation, in the sense of operation, control, protection, and stability. Researchers have found that many of the protection, control, and stability challenges need more analysis. Microgrid protective and control approaches should be subjected to harsh conditions that emulate site situations. The protective digital relay should not be vulnerable to malfunction due to switching, transformer energizing, saturation of current transducers, or communication channel failure.

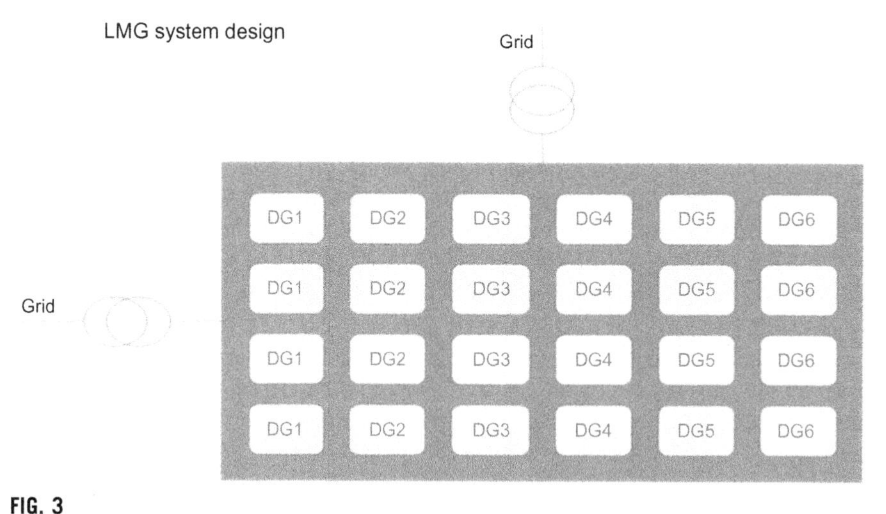

FIG. 3

LMG using distributed generation farms (DGF).

7.3 LMG CONTROL

The advanced control strategy will depend on the design and integration of the grid-connected LMG with reliability monitoring, control, protection, and management. Accurate technology and configuration of the LMG system with the main grid should be modeled. It has a tool for optimal rating, sizing, and arrangement of the LMG for supplying the demand side. Application of the smart distributed flexible AC transmission system (DFACTS) for both AC/DC sides will maximize energy utilization. Dynamic real-time monitoring will stabilize the connection, and save its ability for proper operation between the LMG and the utility grid.

A detailed design of the microgrid integrated controller (MGIC) for the LMG will consist of various levels of control and administration, the hierarchical architecture of MGIC for each level, along with its components, is shown in Fig. 4. The section of (C-Opt) has a role to determine the optimal rating, while the section of (C-Con) is responsible for defining the best configuration based on the number, size, and topology of the DGs based on optimal scheduling and planning.

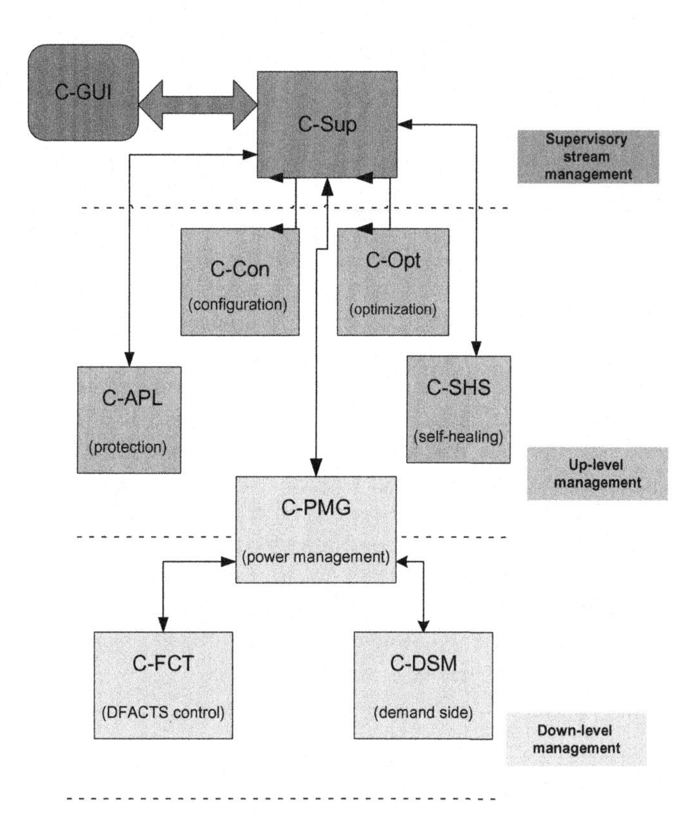

FIG. 4

Hierarchical architecture of MGIC for the LMG control.

Some sections have roles that control and enhance performance by controlling actions internally and externally for energy efficient technologies, and also achieve demand/supply management, as per demand type, in order to adapt the voltage and consumption profile.

In reference to Fig. 4, a certain section of the MGIC is based on load control, where load, here, refers to AC and DC loads. Depending on the voltage and current measurements across or through the AC/DC bus and each load, the LMG's capability could be determined in the moment. As it varies, loads could either be fully supplied by the AC/DC bus, or partially shut off to match the LMG's capability. Loads can be supplied at all times by the sources: wind turbine, battery bank, or others. Design of LMG can cover AC/DC loads fully supplied overweighed loads, in the control algorithm, LMG's capability is defined as the first thing to be determined if it's capable of supplying AC/DC load demand. All loads should be switched on and be fully supplied by the buses when the LMG's capability is beyond load demand; otherwise, loads can be selectively shut off to enable the LMG capable of supplying the rest of the loads. Meanwhile, the LMG's capability, which refers to the power generation over the AC/DC bus, should be continuously monitored and detected by the voltage and current sensors. The loads that had been shut off earlier should again be switched back on and powered up by the AC/DC bus when the LMG's capability has been detected beyond the load demand. In terms of loads, the precondition of being powered up by the buses is that all loads must be fully supplied in the moment. Excess power available on the AC/DC bus that has not been drawn by the load should be checked against the power consumption by other loads. Taking the DC to AC source conversion efficiency into account, AC loads would only be powered up by the DC bus if the extra power were determined to be over the AC load demand; otherwise, the connection of the AC loads with the DC bus would be cut off. The 120 V AC voltage source would come into use, instead, to fully supply the AC loads, which would be connected back to the DC bus when the LMG's capability has been detected as over the AC and DC load demand. Battery charging mode control only happens when excess power, not being drawn by the AC and DC loads, is available on the DC bus, and the state of charge is below 80%. While having the battery charged, its state of charge is monitored. When it reaches beyond 80%, the charging mode is switched back to the discharge mode in order to be prepared to supply the DC bus at all times. Overall, the power generated on the DC bus should be always checked against the load demand in order to determine whether the loads should be fully supplied or partially shut off to match the LMG's capability; and the battery state of charge should also be monitored so as to control its charge and discharge mode, and to ensure that the battery is charged in time.

7.4 LMG OPERATION AND MANAGEMENT

The LMG of DGF, or LDG, will present an innovative architecture in the application of intelligent synchronizing units for connecting the LMG with the main grid management. There will be innovative schemes for energy efficiency and system stability

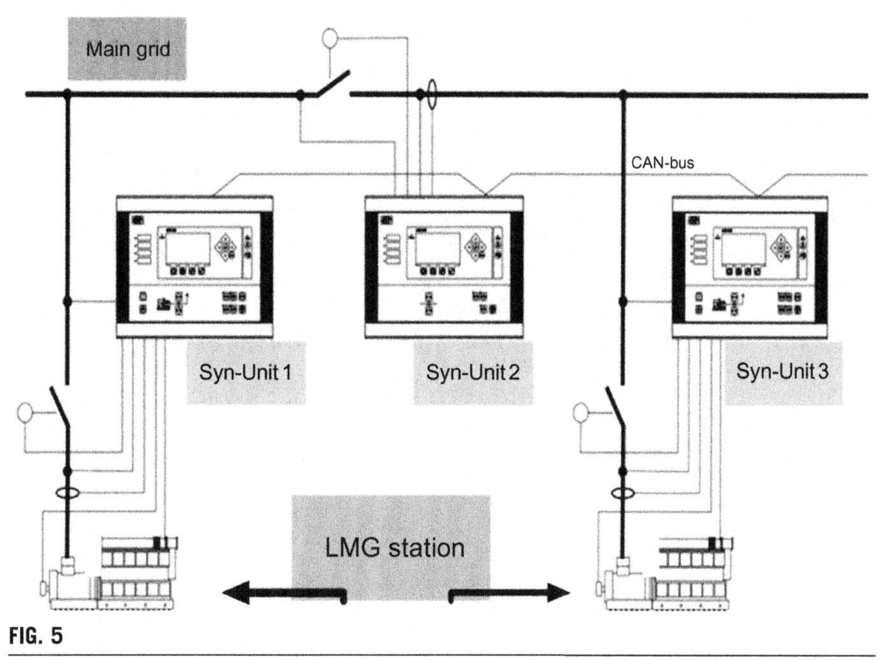

FIG. 5

Synchronizing connection between the LMG and the main grid.

with monitoring and control. Sustainable technological solutions should be merged for the balancing of electricity supply and demand. The operation of the LMG will provide a comprehensive framework for evaluating the effects of voltage and volt-ampere reactive control methods for local and coordinated control schemes applied to the LMG, considering uncertainties related to the variation of the load demand, and others. It proposes solutions for the optimum system configurations of the AC and DC sides to increase the efficiency, power availability, flexibility, and dispatch ability (Figs. 5 and 6).

DGs provide vital contributions in the planning of LMGs and their operations. The main objective for proposing an LMG is to present a larger renewable and distributed energy to utility networks, and to raise the sustainability levels of modern power systems. The main DG categories are wind turbine, solar PVs, microturbine, fuel cells, and others. This chapter introduces the SMR as a DG within LMGs. In order to achieve a sustainable LMG, all DGs should support the demands within the LMGs. Solar PV cells, as renewable sources, are from distinguished systems. It is a clean energy source that has the least greenhouse gas emissions [13–15]. The generated power of solar PV cells will rely on solar radiation, and its duration, which changes from one condition to another.

Concerning wind power and its operation: the power generated will vary depending on wind speeds, which conditionally change with time. The supplied electrical power will depend on the wind speed which should be more than minimum threshold speed. Gas emissions are the main consideration for environmental effects and global

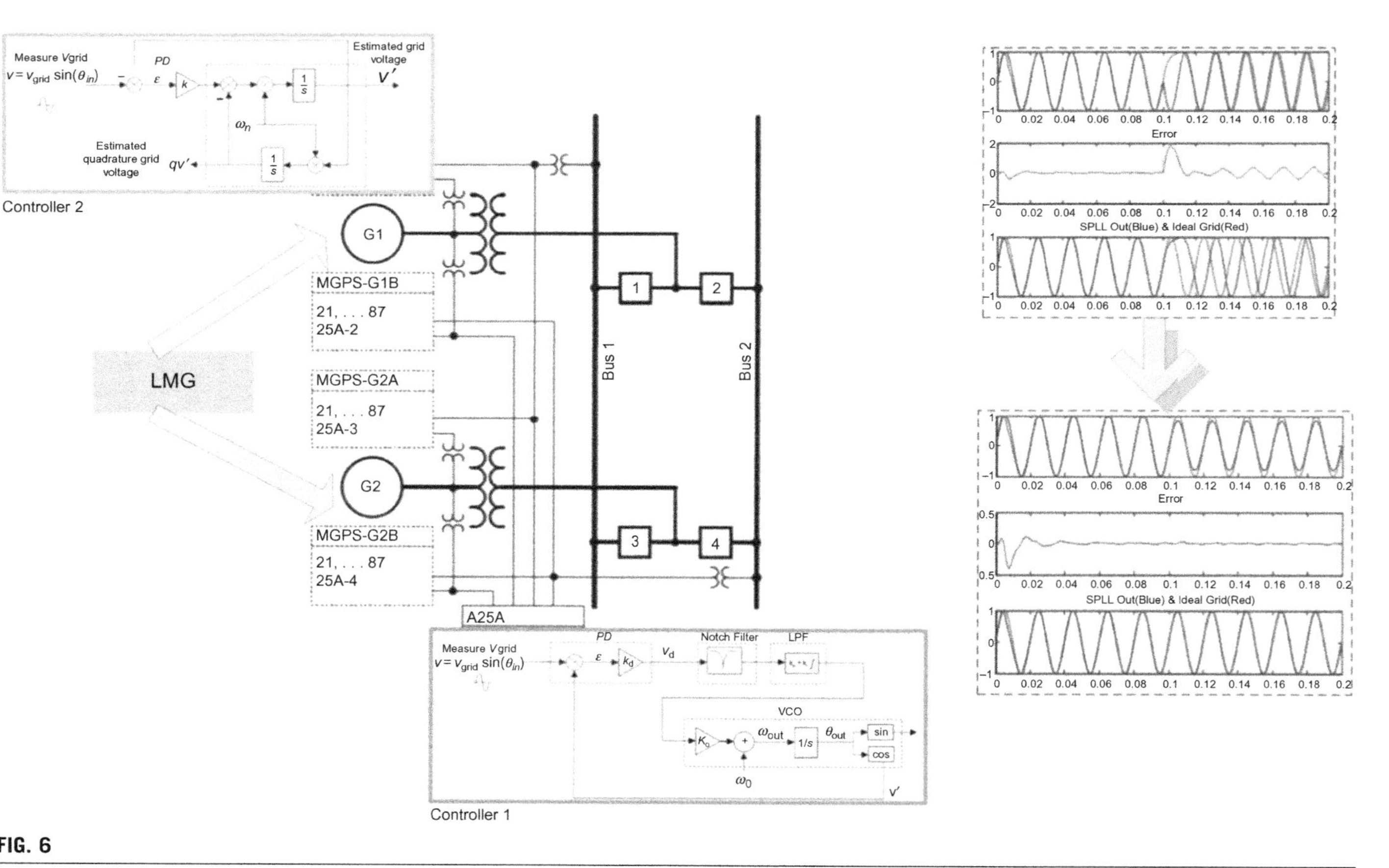

FIG. 6

Details for the monitoring and frequency adaptation between the LMG and main grid. Squares 1, 2, 3, and 4 are relays. Bus 1 and Bus 2 are *out-bound* power busses (redundant). MGPS units are GPS-synchronized clock sources for timing the generators. A25A is the measurement and control unit (contains a microprocessor).

warming. Greenhouse gases are produced from various human applications, especially the combustion of various resources, such as coal and oil, in electrical productions, and transportation processes. Some of the DGs have an environmental impact during their integration with the power grids or islanded operations.

Traditional power units have significant clear emissions during burn processes, while hydro units, solar plants, wind farms, and nuclear plants have zero clear emissions, but there are some other emissions based on the life cycle. The available options are to present more renewable power and nuclear sources to the network in order to provide future loads. Also, traditional power plants can be changed through carbon dioxide-free resources like SMR.

Though utility grids are needed in society, they possess several disadvantages. A majority of the disadvantages come from the high initial cost for its construction due to: (1) the requirement of step up and step down transformers, (2) the cost associated with high-voltage towers and lines, (3) the area that the transmission lines occupy, (4) the high cost of regular maintenance and reconstruction in the case of infrastructure damage due to bad weather conditions (storms, floods, blizzards), (5) the risk of power outage for long periods (days or weeks) in large areas in the case of destruction of large sections of the transmission line system (caused by bad weather conditions, or war or military attack to the infrastructure system), and (6) the loss of power during transmission.

Functional decomposition of the LMG shows the flow of energy and interaction of individual components within the subfunction. The power sources used for the LMG are the programmable DC source, the wind turbine, the battery storage unit, and others, and are used primarily to store surplus power from the buses for use as a backup power supply. The load demand is controlled by the load control module, which makes decisions based on the control algorithm with the aid of current sensors that control relay switches that switch resistive loads on and off to match the LMG's power supply capability.

7.5 HIGH-PERFORMANCE LMG

The following potential aspects are presented for enhanced research and operations.

7.5.1 LMG DISTRIBUTED PROTECTION

LMG protection is a growing area of research. There are two major challenges that mandate new methods and technologies for the LMG protective systems: (a) LMG topologies are moving toward intelligent devices and equipment, which require smarter protective systems; and (b) interconnected LMGs with both operation modes (i.e., islanded and grid-connected) will require dynamic protection strategies and systems based on the understanding of the associated risks, and the dynamic fault propagation analysis. This can offer innovative techniques for real-time fault propagation analysis based on advanced safety design methods of the LMG, which will be the basis of LMG and smart grid protection systems technologies.

7.5.2 ADVANCED LMG MONITORING TECHNIQUES

Phasor monitoring units (PMUs) are widely used in the modern grid systems. The accurate measurement of phase shifts across LMGs should be identified as an improved way to monitor the LMGs for better performance and control, as well as for timely protection capabilities. A PMU is introduced to monitor the transmission line/distribution line and their integration with the LMG. In addition, advanced frequency monitoring methods are also proposed, where it may lead to dynamic and high-performance LMGs with reliable protection mechanisms.

7.5.3 ADVANCED LMG MODELING AND SIMULATION

Innovative LMG modeling and simulation can be proposed, where integrated simulation is proposed and includes grid physical elements, control and instrumentation, communication and security, and the application layer. It is essential to accurately evaluate different LMG designs, load forecasting, control, and protection scenarios. In addition, qualitative and quantitative LMG knowledge should be constructed and maintained for effective lifecycle decision making.

7.5.4 SMART BUILDING INTEGRATION WITHIN LMG

LMGs are intelligent grid structures that provide efficient and reliable energy that covers regional needs. One main constituent of any region is the buildings and the societal and industrial facilities. Advanced methods and technologies should be developed to construct reliable and effective integration between building energy management systems with LMGs in the form of smart metering, two-way control, and building energy conservation and management. This will pave the way toward efficient building and facilities within the LMG with intelligent control and protection strategies.

7.5.5 LMG DISTRIBUTED ADAPTIVE CONTROL

Advanced methods for microgrid distributed control should be investigated based on DG, ES, and load forecasting in the selected region. The proposed adaptive control methods should be based on neuro-fuzzy and genetic algorithms with adaptive rules that will enable the dynamic modeling of LMG systems, fine tuning with the fuzzy rules, and constructing the distributed controllers accordingly. These methods are significant, as they will provide scalable, distributed controllers with intelligent capabilities that control the operation of the LMG, while optimizing its overall performance.

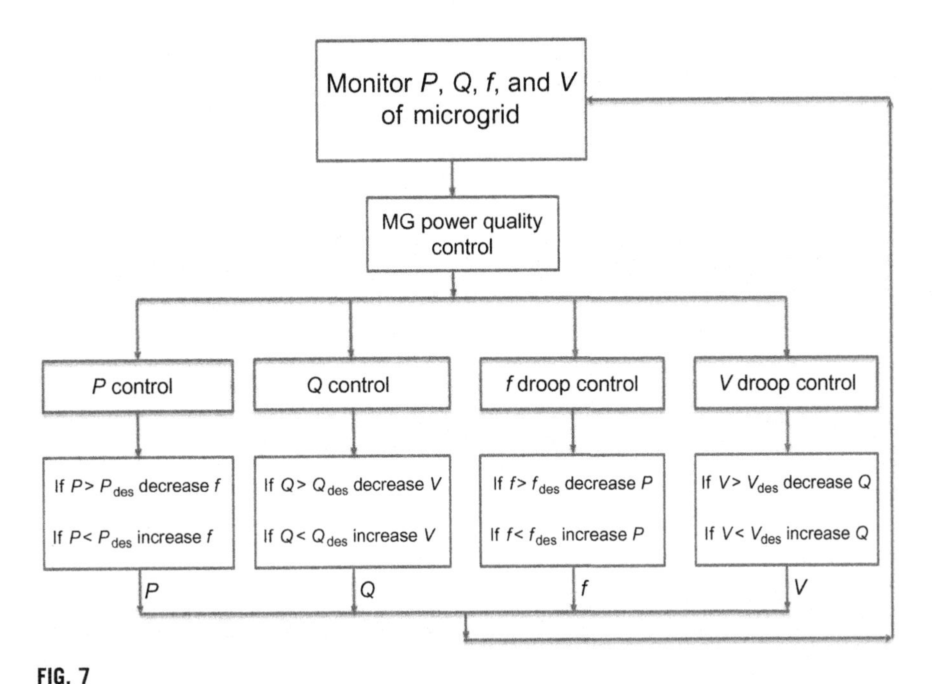

FIG. 7

LMG power control analysis.

The adaptive control system is the core of the LMG design and operation, and it can be structured as three groups: (1) peer-to-peer controller, (2) two-layer controller, and (3) multiagent controller. The peer-to-peer controller main functions are the power control of the DGs or load without coordinating with the central controller or communication unit. The main control strategies are *PQ* control, droop control, and frequency/voltage control. Fig. 7 shows the preliminary LMG control model.

PQ control: *PQ* control is used so that the output active and reactive power is kept constant. *PQ* control consists of the *P* and *Q* controller. During *P* control, the frequency droop characteristic is adjusted to keep the active output power at the desired value (P_{des}) while the frequency is changed. During *Q* control, the voltage droop characteristic is adjusted to keep the reactive output power at the desired value (Q_{des}) while the voltage is changed.

Droop control: Typical droop control includes a frequency droop controller and a voltage controller. During islanded mode the control strategy of ES should be changed to enable local frequency control. The power of the ES is regulated according to the predefined droop characteristics.

Frequency/voltage control: Load changes in the LMG result in a deviation of frequency and voltage from the steady state within the droop control action. Depending on the voltage/frequency sensitivity of the load and the droop characteristics, the energy resource should contribute in changing the overall generation in order to restore the voltage/frequency of the LMG to their desired values.

In a two-layer controller, the control strategy consists of two layers: the distributed controller and the LMG central controller. The main functions of the central controller are the startup and shut down operation of the distributed energy resource, to confirm an optimal schedule for DGs within economical dispatching strategy, to minimize total power loss by sharing the load among DG units, to maintain power quality including the voltage profile and harmonic distortions, to set a specific limit for each DG unit considering the type of the DG unit (cost of generation, time dependency, maintenance interval, and environmental impact), and to improve the stability margin and dynamic response.

7.5.6 INTELLIGENT CONTROL

A fuzzy learning controller has adaptive capabilities, and it was an idea by Mamdani. The variables of the fuzzy control system should be adapted within the operation to cover certain desired performances based on the system parameter tuning. The fuzzy model reference learning control (FMRLC) applies the reference model that indicates the desired response for adapting the procedure. The terminology learning refers to the memory ability of the fuzzy control system to enhance the response, which could be subject to the plant by the dynamic operating conditions affected by the state variables.

The technique consists of the following modules: (1) the reference model, (2) the inverse model, (3) the adaptation mechanism, and (4) the fuzzy controller. The reference model is defined by the first- or second-order linear system that matches the desired performance. In this work, a second-order system is used as a reference model given by:

$$\frac{Y_{\text{ref}}(s)}{R_{\text{ref}}(s)} = \frac{\omega_{\text{n}}^2}{s^2 + 2\eta\omega_{\text{n}}s + \omega_{\text{n}}^2} \tag{1}$$

where $R_{\text{ref}}(s)$ is the reference input, $Y_{\text{ref}}(s)$ is the output of the reference model, ω_{n} is the desired natural frequency, and η is the desired damping ratio. ω_{n} is set to 10 radian per seconds, while η is set to one (no overshoot is desired).

The reverse modeling consists of the first-order fuzzy structure of rule-based groups showing the reverse dynamic states of the plant. It is responsible for the calculation of the proper level of control actions by getting the fuzziness output, which enforces the operator in tracking the reference model. With no prior data on the plant dynamic states, fuzzy data on the system reverse dynamic states could be coded to the rule sets of the reverse modeling.

The parameters to the reverse modeling are indicated by the equations:

$$y_{\text{e}}(k) = y_{\text{m}}(k) - y(k) \tag{2}$$

$$y_{\text{I}}(k) = T\left(y_{\text{e}}(k) + \sum_{i=1}^{k-1} y_{\text{e}}(i)\right) \tag{3}$$

with y, y_{m}, y_{e}, and y_{I} referring to the outcomes of the procedure, the outcome of the reference modeling, the error values of the reference modeling and the procedure, and the integrated error, respectively. The third section of the control technique depends on learning and adapting to the process that encodes the outcome of the reverse modeling to the rule sets of the fuzzy control system. These sets show the nonlinear response, which solves and analyzes the overall performance in order to cover the reference-modeling trend. It has a responsibility to adapt and control the response in order to manage the tuned variables of the demands connected to the utility grid.

The adaptation mechanism updates the centers of the output membership function as follows:

$$b_i(k) = b_i(k-1) + \mu g_p p(k) \tag{4}$$

where b_i is the center of the output membership function, μ is the learn parameter, g_p is the adapting parameter, and $p(k)$ is the outcome of the reverse modeling. The learning factor, μ, should be small to avoid learning instability. In this work, μ was set to 0.0004. In order to avoid generating control actions exceeding the limits of the process input, the following two equations were added:

$$b_i(k) = b_{max}, \quad b_i(k) \geq b_{max} \tag{5}$$

$$b_i(k) = b_{min}, \quad b_i(k) \leq b_{min} \tag{6}$$

where b_{min} and b_{max} are the minimum and maximum controlling action variables. Eq. (4) refers to the FMRLC that allows for adaption and learning. The adapting procedure has direct dependency on the inputs of the fuzzy control system. Also, it is searching for a specific performance, which is termed by the reference model, without keeping the variation of the procedure variables. The mismatch between the adaptation mechanism and the inverse dynamics of the system results in a continuous but decaying learning process. To avoid learning oscillation that could cause the oscillation and instability of the system, learning remarked bands have been proposed for the adaptation process, as per the following equation:

$$\begin{aligned} b_i(k) &= b_i(k-1) + \mu g_p p(k), & -p_{DB} \geq g_p p(k) \geq p_{DB} \\ b_i(k) &= b_i(k-1), & \text{otherwise} \end{aligned} \tag{7}$$

where p_{DB} is the lead remarked band bound for the learning procedure. The fuzzy control system has a PID controller and various input values, which is indicated by the following equations:

$$e(k) = r(k) - y(k) \tag{8}$$

$$c(k) = \frac{e(k) - e(k-1)}{T} \tag{9}$$

$$e_I(k) = T\left(e(k) + \sum_{i=1}^{k-1} e(i)\right) \tag{10}$$

with e, c, and e_I referring, respectively, to the error variable of the referenced input, r, and measuring outcome y; the derivative; and the integral within the real sample k. PID fuzzy control systems are usually applied within the controlling loops inside the nonlinear process. It is well known that the process with the PD control blocks will have huge steady-state errors. The PID fuzzy control system will be subject to an impaired transient response. PID fuzzy control system will upgrade the size of the rule set matrix and then update the learning procedure to achieve better operation. The algorithm is designed to process only nine fired rules at the rule base, in order to reduce the computational complexity. Uniformly distributed triangle membership functions are used for this work on the controller inputs and outputs for both the main fuzzy controller and the inverse model. The dynamic performance, met by the membership curves, is stabilized at the range (−1,1). The modification of the waveform's dynamic interval met by the membership curves is obtained by the *I*/*O* scaling parameters, which is gx, where x reflects to the term of the signal of the parameter is located.

Each integrator, presented by Eq. (6) in the algorithm, is followed by a limiter in order to cancel out the accumulation of huge rates within the transient time. These huge values, which are more in the dynamic interval of the appropriate waveform, can lead to a slowdown of the control system response. The minimum and the bounds of the maximum are settled to the magnitude of the waveform outcome of the dynamic interval of the integrator.

Five membership functions are used to represent each input and output of the inverse model and fuzzy controller. The minimum and maximum methods are used for fuzzy and/or operations, the product is used for inference, and the center of area is used for the defuzzification process.

7.6 CRITICAL OPERATION FACTORS OF LMG

The next sections are some critical factors that one should be concerned with while analyzing the operation of the LMG.

7.6.1 ACCURACY OF METERING

Because power is supplied from DGs resources, it is vital to make accurate considerations of each power supply resource in the grid. In case of a demand that is loaded with a solar panel at home, it can contribute to the power supply by the solar PV cells to the utility, and it can consume power from the network through other home devices. This could lead to inaccuracy in the measuring of the power.

7.6.2 WEATHER CASES

The power supply with renewable resources, such as solar PV cells and WFs, is changeable under various weather cases. There may be conditions when the power supply varies more than a particular flow in climatic cases. Therefore, it is vital to consider weather cases during power planning.

7.6.3 TIME LAG

Because there can be various power supply resources, there may be more calculations required for processing time lag and delay. In addition to a processing delay, there may also be a transmission lag where all the controlled signals can be moved by a certain station via transmission channels.

Table 2 Summary of the LMG Performance Analysis

#	Component	Total Harmonic Distortion (THD)
1	Transformer	Transformer losses increase as current distortion is increased. This leads to overheating and shortening of the life of the transformer **1.** The *K*-factor of loads of a particular area can be measured, and a special transformer can be made based on the *K*-factor, which can withstand the distortion caused by the loads of that area
2	LMG loads	The converters in LMG and other nonlinear loads are responsible for distortion **1.** Using high-switching frequency converters and LCL filters on the converters decreases THD in microgrids **2.** Keeping the amount of thyristor rectifier loads reasonably small decreases the THD in microgrids
3	Standby generators	The generators having less subtransient reactance leads to less harmonic distortion **1.** Keeping the flux density high, by using an appropriate cage design, reduces subtransient reactance **2.** When balanced three-phase loads are used, another way of reducing subtransient reactance is by using a longer coil pitch

7.6.4 COST ANALYSIS

It is important to quantify the related cost reductions in LMG modeling cases for various grid heads. However, there are many uncertain reasons for power loads and energy supply capacity, such as wind trends. This can impact the network operation and cost as per various kinds of generations. An identical point of view on generating cost, including carbon cost for the future year, can be applied to indicate the total cost analysis case of utility. The energy generation cost shows that $/kWh for wind is the lowest and gas in the highest. Similarly, the cost of different types of ES systems should be analyzed and compared to identify potential ES technologies.

It is relatively difficult to compare technologies, especially with different parameters involved in the performance evaluation. However, super-capacitor technologies are probably one of the best with respect to life cycle, while lead-acid ES technologies are relatively better in terms of $/kWh. Some considerations use a 12-year lifetime for the life cycle cost analysis, although there is some memory effect on batteries, which has a negative impact over time. Research shows the storage cost of lead-acid technologies are around 300 $/kWh, whereas the generation cost for wind is 0.0001 $/kWh, which reflects the potential reduced cost of generation over storage. Table 2 shows a summary of the distortion performance of different components of the LMG.

7.7 DG TECHNOLOGIES

Table 3 shows an analysis of DG technologies where size, cost, advantages, and drawbacks are conducted and analyzed in order to provide a strong base for detailed protection and control of LMGs.

Table 3 Distributed Generation Technology Summary [2]

DG Technology	Reciprocating Engines	Gas Turbines	Microturbine	Fuel Cells	Photovoltaic	Wind
Size (kW)	Diesel: 20–10,000+ Gas: 50–5000+	1–20,000	30–200	50–1000+	1–20	200–3000
Efficiency (%)	36–43 28–42	21–40	25–30	35–54	N/A	N/A
Fuel	Heavy fuel oil and biodiesel Natural gas, biogas, and landfill gas	Gas, kerosene	Mainly natural gas, but also landfill and biogas	Methanol, hydrogen, or natural gas	Sun	Wind
CO_2 emissions (kg/MWh)	650 500–620	58–680	720	430–490	No direct emission	No direct emission
No_x emissions (kg/MWh)	10 0.2–1	0.3–0.5	0.1	0.005–0.01	No direct emission	No direct emission
Gen. cost (USD/KW)	Diesel: 125–300 Gas: 250–600	300–600	500–750	1500–3000	5000–7000	900–1400
O & M cost (USD/MWh)	5–10 7–15	3–8	5–10	5–10	1–4	10
Applications	Emergency, standby services, and CHP	CHP and peak power supply	Transportation sector, power generation, and CHP	Transportation sector, power generation, and CHP	Household, small-scale and off-grid applications	Central generation, more than DG
Advantages	Low capital cost, large size, fast start up, good efficiency, and high reliability	Low maintenance cost and lower NO_x emissions	High speed, less noise, low NO_x emissions	Compact, high efficiency, reliability, and low emissions	Low operating cost and no emissions	No emissions
Drawbacks	Noise, costly maintenance, and high NO_x emissions	Noise and lower efficiency	High cost and recently commercialized	High cost and recently commercialized	High capital cost and unpredictable output	High capital cost and unpredictable output

REFERENCES

[1] Shulyak N. Westinghouse small modular reactor: taking proven technology to the next level. In: Presented to the IAEA INPRO dialogue forum Vienna, October 2011; 2011. http://www.iaea.org/INPRO/3rd_Dialogue_Forum/12.SMRWestinghouse.pdf.

[2] Yeong K, Tsorng J, Chen J. Novel maximum-power-point-tracking controller for photovoltaic energy conversion system. IEEE Trans Ind Electron 2001;48(3):594–601.

[3] Pascual J, San Martin I. Implementation and control of a residential microgrid based on renewable energy sources. In: IEEE energy conversion congress & exposition (ECCE), Denver; 2013. p. 204–9.

[4] Kroposki B, Lasseter R, Ise T, Morozumi S, Papatlianassiou S, Hatziargyriou N. Making microgrids work. IEEE Power Energy Mag 2008;6(3):40–53.

[5] Gabbar H, Abdelsalam A. High performance AC/DC microgrid in grid-connected and islanded modes with SVC-based control. Energy Convers Manag 2016. In press.

[6] Buayai K, Ongsakul W, Mithulananthan N. Multi-objective MG planning by NSGA-II in primary distribution system. Eur Trans Electr Power 2012;22(2):170–87.

[7] Islam M, Gabbar H. Study of small modular reactors in modern microgrids. Int Trans Electr Energy Syst 2015;25(9):1943–51.

[8] Katiraei F, Iravani R, Hatziargyriou N, Dimeas A. Microgrids management. IEEE Power Energy Mag 2008;6(3):54–65.

[9] Surprenant M, Hiskens I, Venkataramanan G. Phase locked loop control of inverters in a microgrid. In: IEEE energy conversion congress and exposition (ECCE), 17–22 September 2011; 2011. p. 667–72.

[10] Kuo YC, Liang TJ, Chen JF. Novel maximum-power-point-tracking controller for photovoltaic energy conversion system. IEEE Trans Indus Electron 2001;48(3):594–601.

[11] Jimeno J, Anduaga J, Oyarzabal J, Muro A. Architecture of a microgrid energy management system. Euro Trans Electr Power 2011;21:1142–58.

[12] Sungwoo B, Kwasinski A. Dynamic modeling and operation strategy for a microgrid with wind and photovoltaic resources. IEEE Trans Smart Grid 2012;3(4):1867–76.

[13] Xiong L, Peng W, Poh C. A hybrid AC/DC microgrid and its coordination control. IEEE Trans Smart Grid 2011;2(2):278–86.

[14] Luis AS, Wen Y, Rubio J. Modeling and control of wind turbine. Math Probl Eng 2013;2013. http://dx.doi.org/10.1155/2013/982597.

[15] Pierce K, Jay L. Wind turbine control system modeling capabilities. In: Proceedings of the American controls conference, Philadelphia, PA, USA, June 1998; 1998. p. 24–6.

CHAPTER

Design and control of V2G

8

A. Zidan*,†, H.A. Gabbar*

*University of Ontario Institute of Technology (UOIT), Oshawa, ON, Canada**

Assiut University, Assiut, Egypt†

8.1 BACKGROUND

An electric vehicle (EV) uses electric motors or traction motors for propulsion. An electric vehicle is powered by electricity from an off-vehicle source, or may be self-contained within a battery or generator that converts fuel to electricity. The future of transportation is shifting to more efficient electric drive systems. When considering the number of EVs in the coming years, and the capacity of their energy storage systems (ESSs), there are possible additional advantages and uses for this source of stored energy. However, there are many questions, such as what additional hardware/software would be required to deliver stored energy, what communication systems would be required, can this be done without affecting the needs of the driver, what would the impact be on battery life, and what motivations exist to accomplish this, and who benefits. The above unknowns are explored in the following sections of this chapter.

8.1.1 PLUG-IN ELECTRIC VEHICLES

An EV is any vehicle whose driving torque is produced by electric motors. Three main types of EVs are currently available in the market.

(1) Hybrid EVs (HEVs): have an electric propulsion system coupled with a fossil-fuel internal combustion engine. The only source of energy is the fossil fuel. HEVs have efficiency-improving technologies such as regenerative braking to reduce fuel consumption and emissions. Therefore, they produce fewer emissions and provide greater efficiency than fossil-fuel internal combustion engine vehicles.

(2) Plug-in EVs (PEVs): two types of PEVs are available in the markets:
   - **(a)** Plug-in hybrid EVs (PHEVs): their construction is almost the same as that of HEVs, except that they use higher-capacity batteries to be recharged through a connection to an external electric power source. Therefore, they can operate as pure EVs. This type of vehicle has a higher efficiency and lower emissions than HEVs.

Smart Energy Grid Engineering.

   (b) Plug-in pure EVs: considered as a special case of the PHEVs without emissions as they completely depend on energy stored in their batteries and are not equipped with a conventional fossil-fuel internal combustion engine. The market for them is still limited due to its high capital cost and limited charging stations.

(3) Fuel cell EVs: utilize fuel cell technology to supply the motor by converting the hydrogen chemical energy to electric energy. The driving range of this type of vehicle is acceptable when compared to that of conventional fossil-fuel vehicles. The market for them is still limited due to a limited hydrogen infrastructure, which is very expensive.

PEVs are the most promising of these types of electric vehicles. PEVs have two operating modes: battery charge depletion (BCD) or battery charge sustaining (BCS) [1]. During BCD mode, the energy stored in the battery is used to supply the motor with power. When all of the energy stored in the battery has been used, the vehicle enters the BCS mode and it operates as a conventional fossil fuel–consuming vehicle.

8.1.2 BATTERY CHARGERS

An electric car charger is a device used to recharge electric vehicle batteries. It contains a plug that connects electric vehicle batteries to an electrical source to provide the batteries with electrical energy. Three main types of chargers are commonly used for PEV charging. The specifications for each type are listed in Table 1 [2,3]. A level-two charger is the most often used in Canada, as it is recommended by vehicle manufacturers.

8.1.3 COMMUNICATION AND CONTROL

Electric vehicles charging during peak load periods require capacity expansion. Communication with EVs and infrastructure could shift charging to off-peak times. Any communication between the grid and a vehicle can be executed through the charger because it is fixed in place. As shown in Table 2, vehicles' chargers are equipped

Table 1 PEV Charging Levels

Type	Specifications
Level 1R	• 110/120 V, AC, 15–20 A • Does not require installation and can use typical 120 V electrical outlet • Typical charge times: 8–12 h
Level 2R	• 208–240 V, AC, 15–30 A • Requires special installation • Typical charge times: 3–8 h
Level 3R	• DC fast charging • 440 V, DC, 125 A, 55 kW or higher • Requires special installation • Typically returns 50% of PEV battery charge in under 30 min

Table 2 Communication and Control Levels

Level	Capabilities
Level 1	• Only charges the vehicle (discharging not allowed) • Controls the charging current and voltage of the battery to match its specifications • Supplies electric energy to the vehicle battery directly when it is plugged in
Level 2	• Includes the features of level 1 • Includes feature for time delay as each vehicle owner can control the starting and ending times for charging his/her vehicle
Level 3	• Includes the features of level 2 • Enables two-way communications with the electric utility as it can receive an on or off enabling signal from the electric utility • Reports vehicle identification to the electric utility when the vehicle is plugged in
Level 4	• Includes the features of level 3 • Includes bidirectional power flow to enable vehicle to grid (V2G)

with different levels of communication and control capabilities [4]. A variety of methods for accomplishing communication between the chargers and the grid are possible. Each technology is evaluated based on how well it provides low-power, low-throughput, and high-security communication applicable to electric vehicles.

- Wireless: a transceiver is installed at each charger location with a low-rate of data to be sent (9.6–56 kb/s) [5]. Time division multiple accesses are used for sharing the communication medium among customers in each region.
- Power line carrier (PLC): sufficient for EV control and home energy management applications. However, issues, related to reliability and vulnerability, require additional investigation.
- Over Internet Protocol (IP): the cheapest and simplest. A local area network circuit is built inside the charger to be connected to the Internet. Each user can have an individual account to monitor his/her vehicle from his/her computer or mobile.

8.1.4 IMPACT OF EV CHARGING ON ELECTRICAL DISTRIBUTION SYSTEMS

Opinions on the effects of EVs on the power grid vary. Due to the required long charging time for level 1R and 2R, the majority of EVs are expected to be charged at home, in parking lots, or in public locations. Level 1R charging does not have significant impacts on distribution systems. However, level 2R charging can cause potential risk to the system even at low EV penetration (5–10%) [6]. Risk could occur if EVs have been clustered in specific geographical areas. The energy consumed by these EVs adds considerable loading to the distribution network. Therefore, if not managed well, EV charging can have negative impacts, such as increased power losses, phase imbalance, power quality problems, overloading of feeders, transformer degradation, and fuse blowouts.

8.2 VEHICLE TO GRID (V2G)

Electric vehicles, whether powered by batteries, fuel cells, or gasoline hybrids, have energy sources and power electronics capable of producing electricity. When connections are added to allow this electricity to flow from cars to a power grid, it is called a "vehicle to grid", or V2G. The key to realizing the economic value of a V2G is grid-integrated vehicle controls that dispatch according to power system needs.

8.2.1 DEFINITION

V2G technology refers to the capability of controllable and bi-directional electrical energy flow between a vehicle and the electrical grid. The electrical energy flows from the grid to the vehicle to charge its battery. The electrical energy flows from the vehicle to the grid when the grid requires the energy, for example, to provide peaking power. Studies indicate that vehicles are not in use for active transportation over 90% of the time [7]. Therefore, during these times, the batteries of EVs can be utilized to serve electricity markets without affecting their primary transportation function. V2G technology includes vehicle-to-home (ie, when the electric vehicle is at a residence) or vehicle-to-building (ie, when the electric vehicle is at a commercial building). In these cases, the battery power can be used to supply the local building's electrical load without transferring to the electrical grid.

8.2.2 BENEFITS

The electrical energy that is stored inside the batteries of EVs can be used for peak shaving and power quality applications. Peak shaving is to shave the load at peak times. Under the smart grid vision, electric vehicles can become an important part of the grid by functioning as distributed sources. EVs will provide storage, and support grid stability by giving the required amount of power with less pollution. EV owners can also get some cash back, which they probably spent for the gasoline.

- Peak load leveling: V2G helps to provide power to the grid when the demand is high (peak shaving) and charges the batteries of EVs at night when the demand is low (valley filling).
- Financial: EV owners can get the financial benefit from V2G.
- Renewable energy storage: EVs can support the increase of renewable energy penetration. EVs can store excess energy during windy and/or sunny periods. Later, EVs can return energy back to the grid when the demand is high. In this way, EVs can stabilize and smooth the intermittency of renewable power.
- Support during power outage: V2G is expected to have a vital role during power outages. For example, if an EV can produce a few kilowatts of continuous power, its owner can run his/her critical electrical appliances with that power during a power outage.

8.2.3 RISKS

- Capital and power electronics costs: a bi-directional interface is required to connect EVs to the grid, which is expensive and needs additional inverters.
- Battery life: EV batteries have a life of 1000 cycles [8]. Hence, charging and discharging cycles will shorten the battery life. Hence, each EV owner should sell the stored energy back to the grid when the demand is high enough to get the cost back. Also, it is good if a deep cycle is avoided, as a shallow cycle does not give as much pressure on the battery.
- Modeling complexity: the V2G concept is still an ongoing research, as most of the designs and control schemes that have been proposed by researches are too complex to implement. Still more studies are required to reduce the complexity and cost for V2G compatibility on EVs.
- Market: the prime market is used to buy electricity from EV owners, it is necessary to be sure that utilities are going to get these amounts of power. All the power from EVs can be aggregated before transferring to the utility.

8.2.4 GRID INTEGRATION

Before integrating EVs with the grid, in order to send power back into the system, a number of issues should be considered.

- Based on the IEEE 519 standard, total harmonics distortion should be lower than 5% as directly related to the grid pollution.
- Power factor should be close to unity.
- The vehicle needs to have, at least, a good amount of charge reserved in the vehicle, otherwise, even after hours of integration, the vehicle will not be able to deliver any power to the grid, and that would be a waste of resources.

8.3 V2G OPERATING MODES AND FUNCTIONALITIES

Three system components are involved for charging/discharging energy to/from EVs: (1) the location where the EV connects with the electrical grid; (2) the EV supply equipment to which the vehicle connects; and (3) the battery for the EV that manages the state of charge (SOC). The location where the EV connects with the electrical grid may be an owner's residence, a vehicle parking lot, or a public charging station. The EV supply equipment can be designed to provide AC or DC power to the vehicle at different power levels. Each EV can have several components to control and regulate the battery charging rates. All of these components play a role in determining operating modes and functionality.

8.3.1 EV SUPPLY EQUIPMENT DESIGN AND POWER LEVELS

There are some ongoing changes over designations of EV supply equipment power levels. The Society of Automotive Engineers (SAE) has established the AC and DC charging levels, as shown in Fig. 1 and Table 3 [9]. The main differences between AC charging and DC charging are as follows:

- AC charging requires an onboard power inverter and battery management system. However, in DC charging, AC/DC conversion is performed off-board.
- In AC charging, power management is shared between the EV and its supply equipment. However, in DC charging, EV supply equipment provides significant power management.
- AC charging has lower power transfer capabilities compared to DC charging.

Utility power is delivered as AC to the premises where the EV supply equipment is installed. The EV battery stores DC power. So, a conversion from AC to DC is

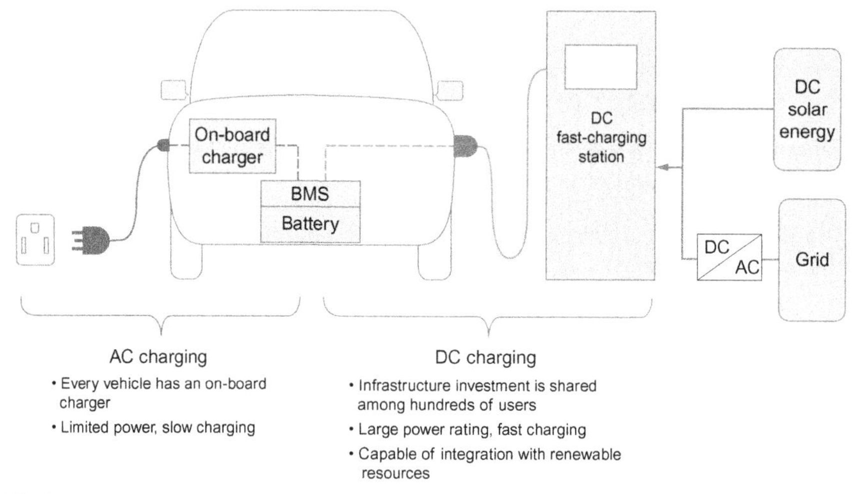

FIG. 1

Basic structure of AC and DC charging.

Table 3 AC and DC Charging Power Levels

AC charging	DC charging
Level 1: 120 V, single-phase, maximum 16 A, maximum 1.9 kW	Level 1: 200–450 V, maximum 80 A, maximum 19.2 kW
Level 2: 240 V, single-phase, maximum 80 A, maximum 19.2 kW	Level 2: 200–450 V, maximum 200 A, maximum 90 kW
Level 3: To be determined, may include AC three-phase	Level 3: To be determined, may cover 200–600 V, maximum 400 A, maximum 240 kW

required to complete charging. Conversely, when operating in the V2G mode, the DC power in the EV battery must be converted to AC in order to be delivered back to the grid.

AC charging level 1

AC charging level 1 is the most basic level of EV charging, as the public has easy access to its required electricity at home or at work. This is because the typical voltage ratings in residential and commercial buildings in North America are between 110 and 120 V with a maximum current flow of 16 A. The AC charging level 1 will be significantly longer. Thus, most EV owners will use level 2 charging. Some EV providers suggest that their level 1 cord set should be used only during unusual circumstances (ie, when level 2 is not available, such as when parked overnight at a nonowner's home, or in an emergency travel situation). Because of the very low level of power transfer capable with AC level 1, and its total lack of control and monitoring capabilities, V2G applications will not be practical or available with this unit.

AC charging level 2

AC charging level 2 is the primary and preferred method for private and public facilities. It specifies a single-phase current with voltage ratings from 220 to 240 V with a maximum rating of 19.2 kW. Higher voltages allow for faster EV charging. AC level 2 charging supports vehicle-refueling modes coincident with destination locations. The SAE (J1772) approved connector allows for a current as high as 80 A [9]. When the EV is connected, its battery management system determines the charge required, and draws the current from the EV supply equipment accordingly. AC level 2 charging is expected to be the preferred method for V2G capabilities because of its commonly available input power, and its compatibility with applications that allow users to control charging.

DC charging

DC level 2 charging (DC fast charging) can be used in commercial and public applications as it is intended to perform similarly as commercial gasoline service stations. DC fast charging can provide around 80% recharge in 30 min for an 85–100-mile range. DC fast charging uses an off-board charger to provide the AC to DC conversion. The vehicle's onboard battery management system controls the off-board charger, delivering DC to the battery. The off-board charger is served by a three-phase AC circuit at 208, 240, 380, 480, or 575 V. As battery capacities increase, it is expected that DC charger power will increase as well for shorter recharge times. Electric buses for school districts and cities will use DC levels 2 and 3.

8.3.2 CHARGING STATION ENVIRONMENT

EV owners will be able to charge their vehicles at four locations: (1) at their residence or primary overnight parking location, (2) at their place of employment, (3) at fleet vehicle charging locations, and (4) at commercial stations.

Residential charging

If the infrastructure is available, EV owners will conduct most of their charging at their residence. EVs are parked for a significant amount of time. Utility rates are low during the off-peak evening hours, early morning hours, and during weekend periods. Therefore, EV owners are encouraged to charge during these off-peak times. Some EV owners may own roof-mounted solar panels to provide power demand for the building, and to supply any surplus power back onto the distribution system. Vehicle-to-house (V2H) applications can be applied during residential blackouts. The EV battery would be used as a storage device for house backup power. V2H presents a preliminary step toward V2G, as only the house receives the power, and the EV remains isolated from the electrical grid.

Most residential charging is expected to be accomplished through AC level 2 charging, as it has shorter charge times compared to the AC level 1. Manufacturers of AC level 2 EV supply equipment will provide a variety of features (ie, basic charging, communication module, revenue-grade meter, and touch-screen functionality). Hence, based on the features provided, some versions of this equipment can be within the financial reach of EV owners. Also, utilities may provide subsidies for EV owners through advanced controllable EV supply equipment for enabling demand response (DR) programs. Although AC level 1 charging requires longer charge times, it may be sufficient for some EV owners to restore the battery capacity overnight. EV suppliers will provide AC level 1 cord sets with EVs as these AC level 1 cord sets are simple devices that deliver power only. DC level 1 (when available) causes overloading for residential service panels, and DC level 2 is impractical for family residences due to the significant cost required to upgrade the utility grid.

Employer facility charging

There is a debate about the importance of workplace charging at employer's facilities. The benefits for employers, or commercial hosts, to provide charging are to provide services for their clients, and the potential for advertising revenue. However, employers may have difficulties, such as tax implications of the benefits provided, cost of EV supply equipment and infrastructure, and managing available charge stations. Also, commercial hosts may have difficulties, such as added load to their facility, corresponding demand charges from their energy provider, and loss of general parking spaces.

EVs arriving for work will be connected to be fully charged before the area's peak demand hours. Then, these EVs, as aggregated storage, could offer a cheaper solution in satisfying peak demand requirements. Also, they can be used to supply a building (ie, vehicle-to-building, V2B), or to deliver backup power for critical business operations. Some EV batteries may be depleted during the peak, and may be unable to recharge for the home trip. Therefore, each EV owner will need to determine a minimum SOC for his/her EV to be used for travel after work. In V2B, the building's owner benefits from the reduced energy cost, and EV owners receive payments from the facility for the use of the battery.

Fleet charging

Fleet vehicles are groups of vehicles owned or leased by a business, government agency, or other organization rather than by an individual or family. Examples are vehicles operated by car rental companies, taxicab companies, public utilities, public bus companies, and police departments. Overall, fleet vehicles account for about 18.2% of passenger and light duty vehicles in the United States. Fleet charging occurs at the work environment for EVs owned by a company. These EVs would be available for V2G services on the non-business hours. Some types of fleets may have greater advantages in the area of V2G than others. For instance, electric school bus fleets may provide a significant availability of stored energy that can be resourced during off-duty times. As these buses have large battery packs, are operated on well-known routes during the weekday, and then parked for known durations overnight and on weekends, the V2G potential is large.

Commercial charging

Presently, there is a push in the commercial sector to install charging stations. The type of businesses that will install charging stations will be varied, and the version of EV supply equipment installed will vary. AC level 2 stations can be installed in locations where EV owners will stay enough time to allow their EVs to complete a significant level of charging. These locations may include restaurants, theaters, shopping malls, doctor, lawyer, and dental offices. DC level 2 charging can be used in locations such as restaurants, coffee shops, convenience stores, and gasoline stations as customers will receive a significant charge in minutes. Also, DC fast chargers can be used along freeway corridors between metropolitan areas.

8.3.3 PHYSICAL CONNECTION TO THE GRID

Using the stored energy in EVs requires actual equipment and services, which may be complex. The following subsections discuss the required equipment, onboard equipment, and V2G communications.

Equipment

The required equipment will include the EV supply equipment, and other necessary equipment, to provide the reverse current flow. The EV supply equipment represents the bridge between the EV and the rest of the premises wiring and control.

(1) V2G premises equipment: the battery for the EV, through connection with the EV supply equipment, acts as a variable load. The EV supply equipment has a role similar to the PV system. A significant difference is the ability of the EV supply equipment to provide a bi-directional power flow to allow the EVs to charge and discharge. The EV supply equipment represents the interface between the local electricity grid and the electric vehicle. Voltage and frequency controls on the EV supply equipment determine whether the current flow is into or out of the EV. If there is no onboard meter to measure electricity consumed, and output by, the EV, a smart residential electricity meter may be used to perform the required measurement and sensing.

(2) V2H premises equipment: V2H or V2B power flow is easier to implement than V2G, as a reverse power interface with the local electricity grid is not required. The EV supply equipment can provide the power from the EV to local loads. Thus, designs for intentional islanding will be required. Also, the system design must consider the equipment and procedures required to restore the electric utility grid when it is recovered.

Onboard vehicle equipment

Onboard vehicle equipment has three main parts: (1) the EV inlet, (2) the onboard charger, and (3) the ESS (ie, battery). The vehicle inlet and EV supply equipment connector provides the interface between the EV supply equipment and the EV. If the EV supply equipment is AC level 2 and the AC current is delivered to the vehicle, the reverse flow from the vehicle is AC. Consequently, the vehicle converts the battery's DC to output AC through an onboard inverter. If the EV supply equipment is DC level 2 and the DC current is delivered to the vehicle, the reverse current flow from the vehicle is DC, and the inverter is included in the off-board EV supply equipment.

8.3.4 IMPLEMENTATION ISSUES

There are many implementation issues that should be addressed before V2G systems can be widely adopted. Examples are business process, public policy, and standard domains.

Stakeholders

(1) Electric utilities: electric utilities have some barriers due to V2G applications. As a new technology, V2G barriers are interoperability requirements, evolving standards, impacts on EV battery life, and network latency requirements. Nontechnology barriers are lack of investment capital, grid reliability responsibility and accountability, and lack of certainty in market prices.

(2) Vehicle manufacturers: large investments have been made in research, development, and production of EVs. However, V2G causes uncertainty related to battery life and capacity. Thus demonstration tests and economic analysis are required to prove the concept, and to determine whether it makes sense for the manufacturers and battery suppliers. EV manufacturers are unwilling to permit the discharge of energy from batteries by any control other than the EV's control system. This is because opening up EVs to external control interfaces, such as price or regulation signals from utilities or aggregators, may cause additional risk. EV manufacturers will have additional difficulty, as they cannot know prior to vehicle delivery whether the owner will want to participate in V2G operations or not. Thus they will need to either provide all vehicles with the capabilities of long vehicle service with V2G

operations, or base warranties on other metrics, such as the number of battery cycles.

(3) EV owners: EV owners will be motivated by the benefits balanced with the risks. Benefits include monetary, environmental, and grid benefits, which should be weighed against effects of battery lifetime, vehicle availability, and ease of use. EV owners need to know how often, and how much, they will get paid.

(4) Governments: due to the uncertainty of the new electric transportation market, there is a lack of clear policy directives, standards, and market support.

Test and evaluation

Sufficient testing and evaluation are required by the industry and academia to provide supporting data that would minimize risks and mitigate the barriers in the adoption of the V2G. This is one of the primary obstacles for decision making by regulators. Test programs to evaluate V2G may include:

(1) Battery impact: impact on the battery lifecycle from frequent charge/discharge cycles, and thermal impact from rapid charge/discharge rates.

(2) Network operation: message transfer, bi-directional, and alternative transport comparison.

(3) System response: power quality, application response times, and data collection, storage, and presentation.

8.4 BI-DIRECTIONAL CHARGER SYSTEM

If EV batteries are directly connected to a DC bus, the charging/discharging current cannot be controlled. Then, once the load changes significantly, the rush current would destroy the EV battery. Therefore, a bidirectional converter needs to be inserted between the DC bus and the EV battery to control the charging and discharging current [10]. A sample topology for the V2G system has been shown in Fig. 2. The bi-directional charger is the interface between the grid and EVs; and it has two stages: a grid-connected AC/DC converter, and a DC/DC converter. This bi-directional charger can achieve three major functions: battery charger mode, V2G, and V2H, which are the main topics on the integration of EVs with the grid.

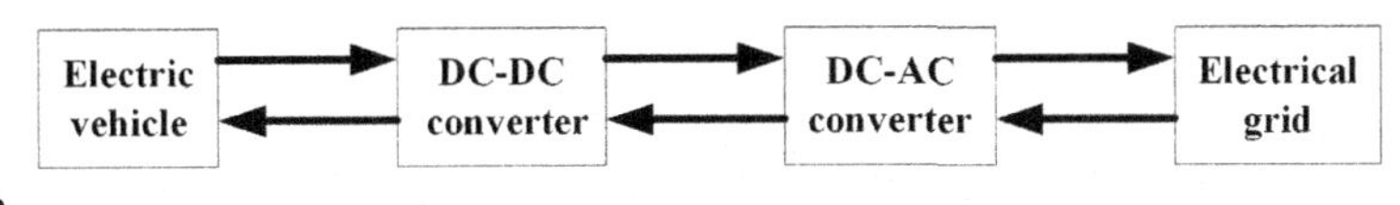

FIG. 2

Block diagram of V2G technology.

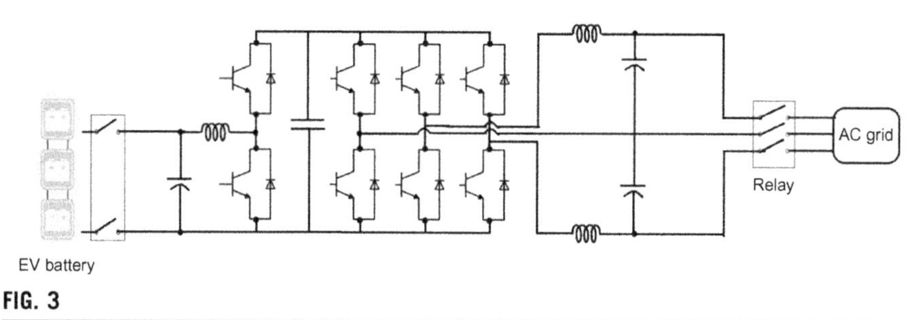

FIG. 3

Topology of a bi-directional charger.

8.4.1 TOPOLOGY SELECTION

The topology of the bi-directional battery charger is shown in Fig. 3. This bidirectional charger has two stages: stage 1 is a grid-side converter, and stage 2 is a battery-side converter. During charging mode (ie, grid to vehicle, G2V), the converter transfers the power from the grid to charge the battery. The grid-side converter uses different half-bridges to convert AC power based on different input voltages. During the discharging mode (ie, V2G or V2H), the power inside the battery is inversely fed back to either the grid or the loads. During the V2G mode, the grid-side converter operates in current-mode control, which regulates the grid current to be a low-harmonics sinusoidal current. While at V2H, the converter operates in voltage-mode control, which regulates the output voltage to be sinusoidal with any type of load. The battery-side converter in both modes regulates the DC bus voltage by operating in the boost mode.

8.4.2 POWER DESIGN OF CONVERTERS

(1) Passive components design: when designing the DC bus capacitor for a single-phase rectifier, the second-order harmonic needs a large capacitor to smooth the DC voltage. The capacitor value can be calculated and chosen based on its stored energy. Then, the filter components can be selected. The filter inductor for the input/output filter is designed based on the current ripple on that inductor.

(2) Efficiency test: after constructing the prototype, an efficiency test should be performed to test the converter reliability, and measure its efficiency.

8.4.3 CONTROL STRUCTURE OF CONVERTERS

An electric vehicle charging and discharging system consists of a three-phase power supply (AC grid) that is connected to the electric vehicle's batteries and to a control loop through a transformer, inverter module, filter circuit, and bi-directional DC/DC

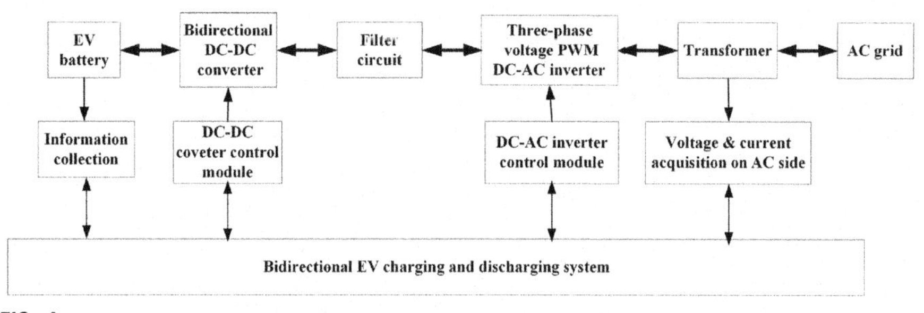

FIG. 4

Typical structural diagram of the bidirectional charge and discharge system.

converter. The basic structure is shown in Fig. 4. The inverter module adopts the voltage-type PWM converter for energy exchange between the AC grid and the DC bus. The voltage-type PWM converter works in the rectifying state when charging, and when the battery draws energy from the grid. The converter works in the inverting state when discharging its power to the grid from the EV battery. As shown in Fig. 4, the V2G control circuit includes: the voltage and current acquisition of the AC side, the PWM control circuit, the bi-directional DC/DC control circuit, and the EV battery information acquisition (ie, battery voltage, current, temperature, and state of charge). The control circuit obtains charge/discharge instructions, work mode, parameter setting, and real-time capacity.

The overall control structure for the V2G is shown in Fig. 5 [11]. It is composed of a double-loop structure for the AC/DC inverter stage, and a single loop for DC/DC converter stage. The battery management system monitors the battery parameters, such as the SOC, state of health, voltage, and temperature. Then, the controller decides how to charge this battery based on its condition. Currents and voltages are available for measurement in the AC/DC inverter control. According to the active and reactive power references, the inverter allows for bi-directional power exchange with the AC grid during the charging mode or discharging (V2G) mode. When the power system needs the EVs to operate at the V2G for a service, such as frequency regulation and load leveling, an operator will arrange a number of EVs, and send the power requirement.

8.5 POWER MANAGEMENT FOR MULTIPLE EVs

With the growing number of EVs integrated into the power grid, a large number of EV batteries need to be charged/discharged via the infrastructure, such as charging stations and parking lots. The existing power grid structure is insufficiently considered for EV integration. Therefore, a framework for integrating EVs is required. The management strategy for V2G operation can be based on a hierarchical control network.

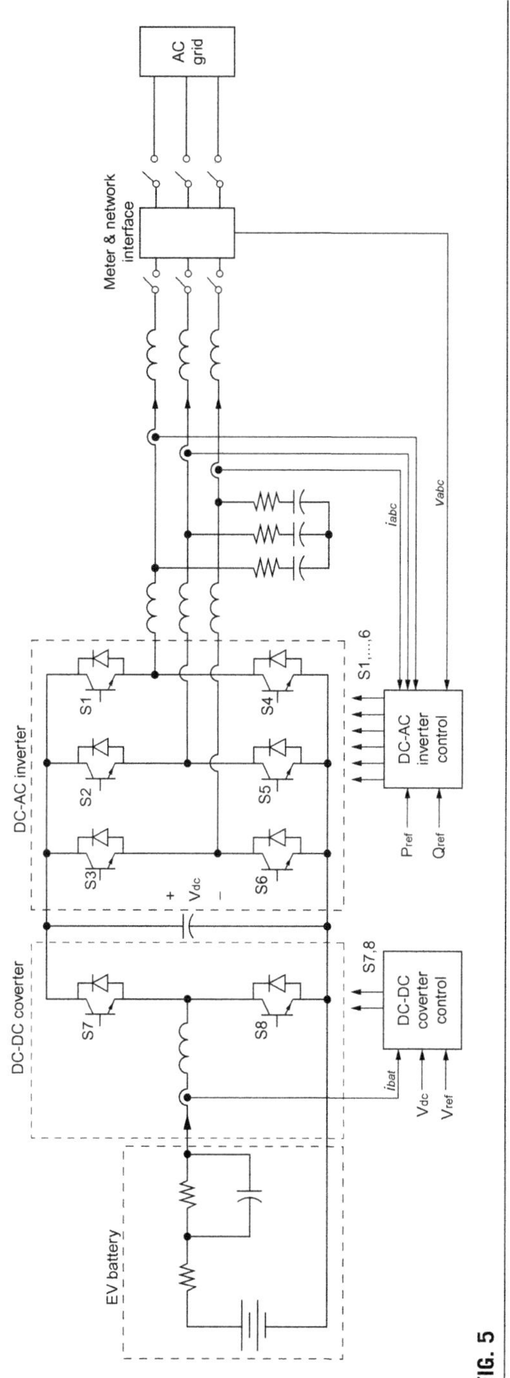

FIG. 5

Overall block diagram of the bidirectional charge and discharge control system.

8.5.1 ARCHITECTURE OF EV INTEGRATION

To integrate EVs into the grid, power system managers are required to monitor and utilize controllable devices, and to update grid infrastructure to manage EV charging and discharging. Fig. 6 shows the hierarchical control framework that is composed of power grid components, and the control signal communication system [12]. The top-level devices supervise activities of devices beneath them. Also, the communication infrastructure can be applied to fit new components into the existing power system structure. As shown in Fig. 6, the control center manages the energy flow by supervising the condition of the bulk transmission system, and issuing commands to dispatch loads or generation resources at each bus. Then each feeder relay, as a controller of a distribution system, receives the higher level command and interprets the control signal for its loads at the distribution system buses.

For charging/discharging EVs, an aggregator takes charge of a group of EVs within a certain region (ie, charging stations or parking lots). Thus, the aggregator provides an interface between a group of EVs and their higher control level. The aggregator acquires the information on its EVs and reports to the distribution system controller. Furthermore, the aggregator assigns the instructions to its individual EVs. The aggregator serves as the intermediary that exempts the higher control level from direct interaction with the massive number of EVs, which is impractical. EVs at the lowest control level follow instructions to adjust their charging/discharging rates and cycles. The hierarchical control framework can be flexible, handling any local problem instead of depending on top-level corrective control.

8.5.2 ISSUES OF MULTIPLE EV INTEGRATION

There are still several key challenges for electric vehicles, such as overall cost, short driving distances, long charging times, and infrastructures that still need building charging stations. To meet cost goals, minimize dependence on foreign material resource, and to ensure safe/efficient large-volume scalability, other key requirements for EVs include:

- Secured materials supply: required technologies should be based on materials without availability barriers and risks when deployed at a large scale.
- Safety: required technologies should be subjected to all applicable safety and environmental standards.
- Recycling: required technologies should be capable of being fully recycled.
- No negative impact on grid reliability: required charging/discharging technologies and infrastructures should be deployed without negatively affecting the reliability of the electricity grid and local distribution networks.

For instance, a distribution transformer is designed to supply power to a specific number of customers based on power requirements and consumption. A distribution transformer can supply power to multiple electric vehicles. However, if the power demand of EVs is higher than the distribution transformer rating, and

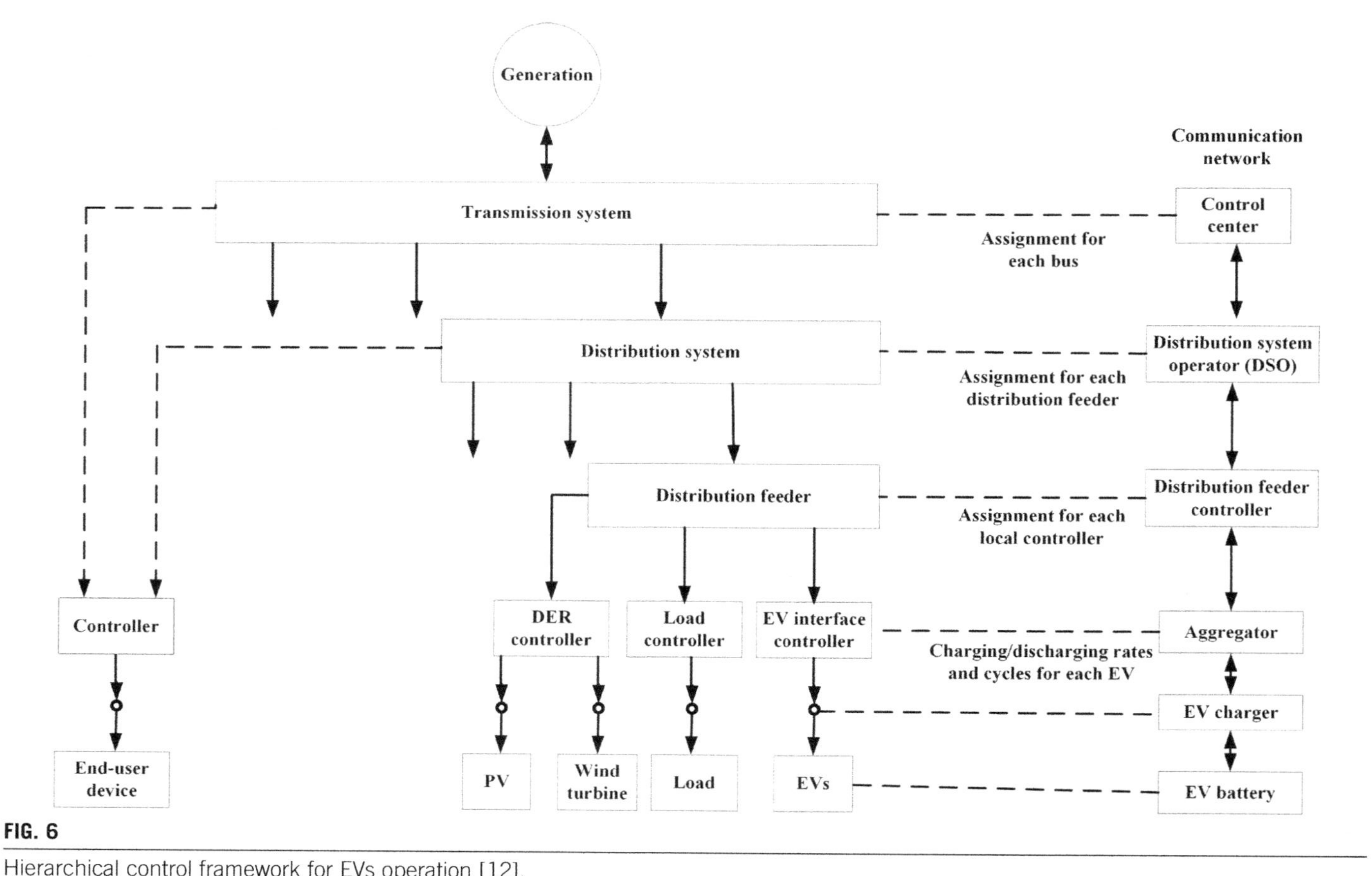

FIG. 6
Hierarchical control framework for EVs operation [12].

there is no communication and appropriate power allocation method, the distribution transformer voltage will collapse. Thus, it will cause problems for other loads supplied by the same transformer. To solve this issue, two-way communication between EVs and the distribution transformer and a power allocation algorithm are required to ensure that the total power demand will not exceed the safe operating limit of the distribution transformer.

Communication methods may not work and sometimes may experience delays. Thus effective control methods without communication will be helpful and promising. Power electronic devices (ie, converters and inverters) operate rapidly. Hence, a fast control method is desired to deal with power management. This can be realized by implementing hierarichal control with a high-level (system plan), real-time allocation algorithm, and a power electronics level control.

8.5.3 POWER DISPATCH BASED ON STATE OF CHARGE

Based on the SOC of EV batteries, the EV controller can be designed to charge/discharge with different rates. For instance, EVs with low SOC can dispatch more power. The energy charged/discharged to/from each EV battery can be determined according to the electricity market, in order to maximize the expected benefits. For example, as shown in Fig. 7, the stored energy in EV batteries can be sent back to the utility grid during peak periods when the electricity market price is high, provided the SOC is relatively high. The battery SOC can be categorized into three groups: urgent charge, regular charge, and mild charge. Urgent charge means that the EV user needs to charge his/her vehicle as soon as possible. Regular charge means that the EV user is not in a hurry, and won't need much power. Mild charge means that the EV user can wait, and needs a small amount of power. Thus different charging/discharging rates should be applied to find out the power dispatched to each vehicle.

8.5.4 LOAD MANAGEMENT BY MANAGING POWER FROM EVs

Renewable-based distributed energy resource (DER) units have an intermittent nature, and customers, such as residential, commercial, and industrial, have a fluctuated load profile. An appropriate solution is to use ESSs to absorb the surplus energy in periods when the generated power is higher than the consumption, and deliver it back in opposite situations. EVs can be considered as controllable loads. Noncontrollable loads are loads whose power cannot be changed. EVs can manage their power to assist the distribution transformer in load management.

- If the required power demand for both noncontrollable loads and EVs is higher than the power capability of the distribution transformer, EVs can reduce their charging power to maintain the system's operational limits. Thus, the power requirement of the noncontrollable loads is fulfilled, and the distribution transformer is fully utilized.

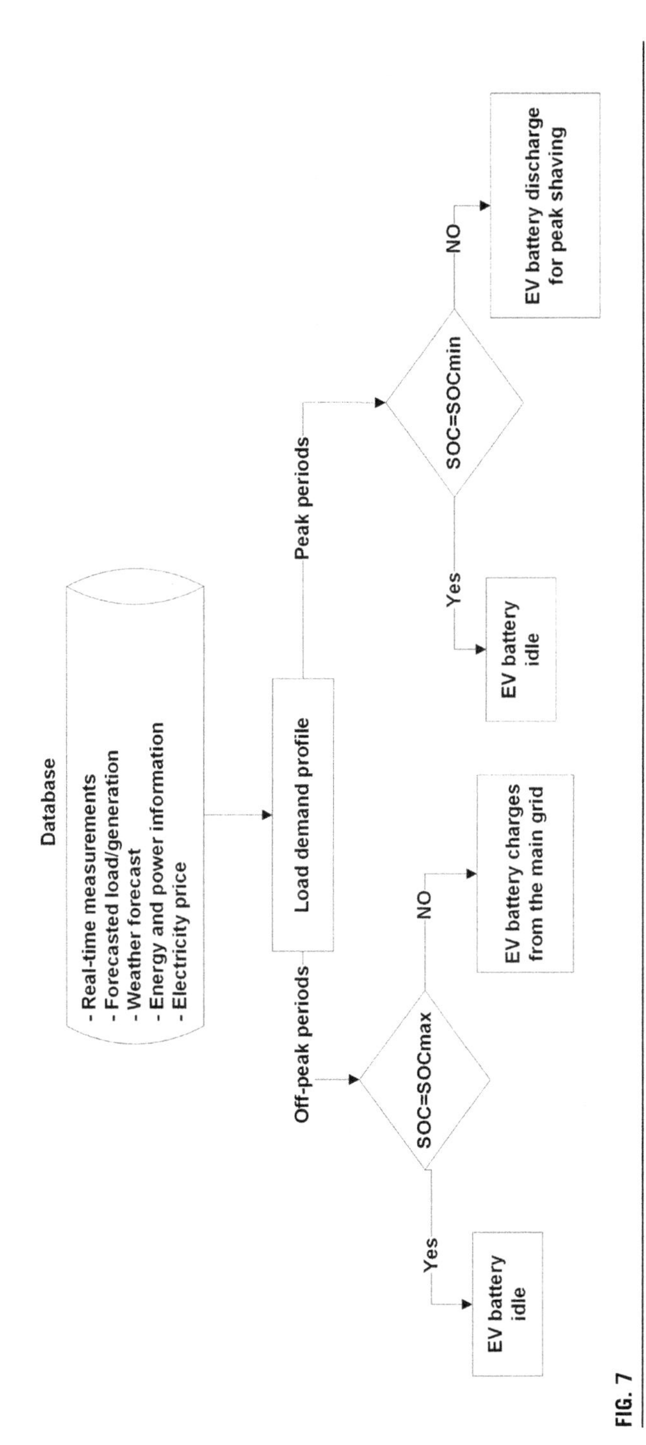

FIG. 7

Power dispatch based on the state of charge (SOC).

- If the power is coming from other renewable resources, the power requirement of the noncontrollable loads should be met first. Then the EVs charging/discharging can be controlled based on the rest of the power.
- If the noncontrollable loads are higher than the power capability of the distribution transformer, the EVs should discharge rather than charge for vehicles that have enough power inside their battery pack.

REFERENCES

[1] Wu D, Aliprantis DC, Gkritza K. Electric energy and power consumption by light-duty plug-in electric vehicles. IEEE Trans Power Syst 2011;26(2):738–46.

[2] Akhavan-Rezai E, Shaaban MF, El-Saadany EF, Karray F. Online intelligent demand management of plug-in electric vehicles in future smart parking lots. IEEE Syst J 2015;PP(99):1–12.

[3] [Online]. Available: http://www.sae.org/smartgrid/chargingspeeds.pdf [Access: February 10, 2016].

[4] Shaaban M. Accommodating a high penetration of plug-in electric vehicles in distribution networks [PhD thesis]. Department of Electrical and Computer Eng., University of Waterloo, 2014 [Online]. Available: https://uwspace.uwaterloo.ca/bitstream/handle/10012/8434/Shaaban_Mostafa.pdf?sequence=1 (Access: February 10, 2016).

[5] Su W, Eichi H, Zeng W, Chow M-Y. A survey on the electrification of transportation in a smart grid environment. IEEE Trans Ind Inf 2012;8(1):1–10.

[6] Liu R, Dow L, Liu E. A survey of PEV impacts on electric utilities. In: IEEE PES innovative smart grid technologies, 17–19 Jan; 2011. p. 1–8.

[7] [Online]. Available: http://www.udel.edu/V2G/docs/LetendDenLil-LoadOrResource06.pdf (Access: February 11, 2016).

[8] Ansean D, Gonzalez M, Garcia VM, Viera JC, Anton JC, Blanco C. Evaluation of life PO_4 batteries for electric vehicle applications. IEEE Trans Ind Appl 2015;51(2):1855–63.

[9] Society of Automotive Engineers (SAE), J1772: Electric vehicle and plug in hybrid electric vehicle conductive charge coupler. [Online]. Available: http://standards.sae.org/j1772_201001/ (Access: February 11, 2016).

[10] Jin K, Yang M, Ruan X, Xu M. Three-level bidirectional converter for fuel-cell/battery hybrid power system. IEEE Trans Ind Electron 2010;57(6):1976–86.

[11] Arancibia A, Strunz K, Mancilla-David F. A unified single- and three-phase control for grid connected electric vehicles. IEEE Trans Smart Grid 2013;4(4):1780–90.

[12] Gao S, Chau KT, Chan CC, Liu C, Wu D. Optimal control framework and scheme for integrating plug-in hybrid electric vehicles into grid. J Asian Electr Veh 2011;9:1473–81.

CHAPTER

Energy storage integration within interconnected micro energy grids

A.M. Othman*,†, H.A. Gabbar*

University of Ontario Institute of Technology (UOIT), Oshawa, ON, Canada Zagazig University, Zagazig, Egypt†*

9.1 INTRODUCTION

Modern power systems are going through dramatic changes in operation requirements, due to the development of renewable energy, energy storage, and distributed generation. The technical challenges related to the operation and control of the micro energy grid (MEG) are immense. Now, MEG system component modeling with control architecture is one of the principal requirements for smooth operation.

Integration of energy storage into an MEG has a significant impact on power flow and operating conditions at the utility equipment and the customer ends. Depending on the type of energy storage and grid connection type, these could positively impact the voltage quality criteria.

Energy storage systems—like battery storage, flywheel, super capacitor, and super conducting magnetic energy storage—are employed as an important part of modern MEGs. They could provide benefits such as improving sustainability, power quality, and reliability of the system. In the case of a disturbance, the energy storage technology could play a very important role in maintaining system reliability.

MEGs can become more effective with the integration of distributed generation and energy storage. Various control strategies are applied if there is a disturbance to the MEG or utility system. Microgrid controllers (MGC) include capacitor control, regulator controls (including substation load changer), distributed generation control, the energy storage control system, switch status control, recloser control, voltage control, and frequency control.

9.2 MEG ARCHITECTURE

Modern MEG architecture assumes an aggregation of distributed generation, storage system, and load operating as a single system providing both electric power and heat. To ensure controlled operation and required flexibility, the majority of the microsources must have a power electronic-based interface.

Smart Energy Grid Engineering.

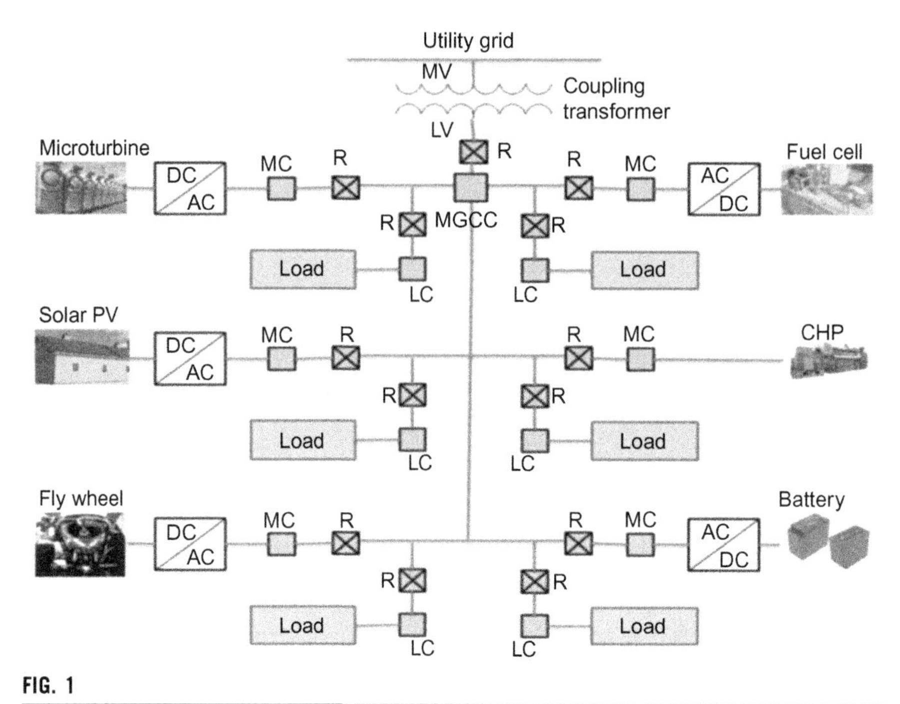

FIG. 1

Micro energy grid control architecture.

The typical MEG architecture is shown in Fig. 1. The key points of the MEG architecture include the interface, load sharing during islanding, microsource protection, power flow control, grid protection, and stability control.

Microsource controller (MC), load controller (LC), and protection are three main functions of the architecture. To emphasize robust control, the MEG control system is based on hierarchical control architecture. The microgrid central controller (MGCC) is used for the upper level of MG operation, through some crucial technical and economical functions. Each microsource, storage device, and electrical load is locally controlled by the MC at the second hierarchical control level. Based on technical and economical requirements, the MGCC provides a set point to the MC and the LC. The MC controls the MS active and reactive power production, whereas the LC acts to control each load (or a group of loads) based on an interruptibility concept.

9.3 ENERGY STORAGE SYSTEM

Recent trends in energy storage and power electronics technologies show a major shift in their contributions to modernize power systems. Viable storage technologies include battery units, super capacitor units, flywheel units, and superconducting energy storage systems (ESSs). These technologies are now considered effective tools for improving system stability, increasing power transfer, and improving the power quality in power systems. An ESS can increase system reliability and dynamic stability, improve power quality, and enhance the transmission capacity of the transmission grid in a high-powered application. For a high-powered application, the use of short-term (cycles to seconds) energy storage integrated with a flexible alternating current transmission system (FACTS) controller could offer the following distinct advantages.

It provides damping components to the system, while keeping constant voltage following a disturbance, and provides additional damping in situations where the dynamic reactive power activated by other conventional FACTS controllers with the same ratings would be inadequate. Alternatively, it could provide the same amount of oscillation damping at a lesser cost by repeatedly interchanging small amounts of active power to the system. It provides energy to maintain the speed of locally connected induction motors during a power system disturbance. This will eliminate a voltage collapse in the areas where there is a heavy demand of induction motors that could otherwise stall.

9.3.1 ENERGY STORAGE TYPES

Energy storage technologies utilize many forms of energy—like chemical and potential energy— that are stored and can be used later in electric form. Most energy storage technologies contain flywheel systems, battery systems, thermal ESSs, pumped-hydro systems, ultracapacitor systems, superconducting magnetic ESSs, compressed air systems, hydrogen systems, and others. Fig. 2 indicates the typical power ratings of various ESSs; the storage capacity are shown against discharge time to clarify the suitability for power quality aspects and energy management criteria. Some energy storage sources (based on their functions) are described in the next section.

ESS by flywheel

A flywheel uses a mechanical form of energy storage harnessed from the kinetic energy of a fast spinning disk, which is capable of considerable energy storage. Flywheels that is coupled to an electric machine could be used to store energy within power systems. To retrieve the stored energy the process is reverse with the motor acting as a generator powered by the braking of the rotating disc. A flywheel could be used in microgrid power quality applications, as it can charge and discharge quickly and frequently.

ESS by battery

Rechargeable batteries, where energy is stored electrochemically, are one of the most cost-effective energy storage technologies. For large-scale energy storage, there are a number of advanced battery technologies to consider such as super capacitor, nickel, lithium, lead-acid, flow, metal-air batteries, and so on. In power systems, a battery could be used for both power quality and load leveling applications.

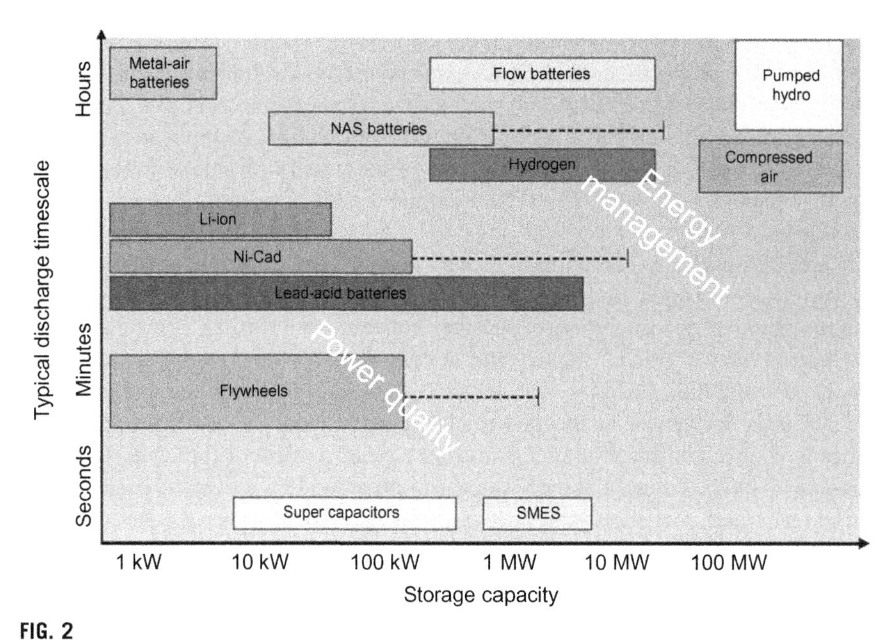

FIG. 2

Schematic for characteristics of energy storage technologies.

Superconducting magnetic energy storage

Superconducting magnetic energy storage (SMES) is a storage unit that stores energy in a magnetic field form while DC current flows through a cooled superconducting coil. The superconducting coil is cryogenically cooled beyond its super-conducting temperature (−269°C). Resistance of the material disappears (very limited) at this temperature and this allows extremely high efficiencies of up to 97% to be achieved, as well as enabling storage. SMES is still undergoing development and research for use in different power quality applications. It could be used to improve power quality as it is able to provide short bursts of energy (in less than a second).

Pumped hydro storage

Pumped storage hydroelectricity is a power generation unit where potential energy is stored in the form of water. When electricity demand is low, water is pumped into the higher reservoir using excess generation capacity. During higher demand, water flows through turbines back into the lower reservoir and generates electricity. Pumped hydro storage is useful and effective for load leveling because it can be developed at high capacity (100–1000 MW) and discharged over long periods of time (4–10 h).

Compressed air energy storage

In the case of compressed air ESSs, off-peak electricity is used to compress air into underground geological formations. During high electricity demand, the compressed air is burned with natural gas and used to run a turbine and generate electricity. With pressurized air, the turbine generates electricity using significantly less natural gas. Compressed air energy storage is also suitable for load leveling because it can be developed in capacities of a few hundred MWs and can be discharged over long (4–24 h) periods of time.

Super capacitor energy storage

Super capacitors are electric devices consisting of two oppositely charged metal plates separated by dielectric materials. The energy density of super capacitors is hundreds of times greater than conventional electrolytic capacitors. Its energy storage capacity can be maximized by either maximizing the capacitance or the voltage stored in the capacitor. The voltage stored is limited by the voltage withstand strength of the dielectric material. Capacitance depends on the area of the plates, their permittivity, and the distance between the plates. Super/ultra-capacitors are most suitable for peak power and low-energy situations. This can provide power availability during voltage sag and momentary interruptions.

9.4 MEG QUALITY IMPROVEMENT WITH ENERGY STORAGE

Although these energy storage technologies do not represent main energy sources, they could provide extra benefits such as improved stability, reliability, and power quality of the supply. The reality is that renewable energy (such as wind and solar) is intermittent, unreliable, and requires additional support to make it a smooth operation. Modern energy storage technologies play an important role, including dynamic voltage stability, transmission enhancement, power oscillation damping, tie-line controller, short-term spinning reserve, load estimating, under-frequency load shedding reduction, circuit break reclosing, subsynchronous oscillations damping, and power quality enhancement.

The selection of energy storage devices depends on their applications, which are determined by the length of discharge. Based on length of discharge, the ESS could be classified in three classes: seconds to minutes, minutes to hours, and hours. For example, in the case of short-term disturbance like impulsive transient frequency fluctuation, the appropriate ESS to maintain power quality could be super capacitor, SMES, or flywheel.

For rapid response and high efficiency, flywheel is suitable for frequency regulations. Super capacitor and SMES have the ability to respond extremely fast when used for power quality applications like transient voltage stability.

Alternatively, energy storage could be used for bride power application during some unavoidable issue like forecast uncertainty and unit commitment errors.

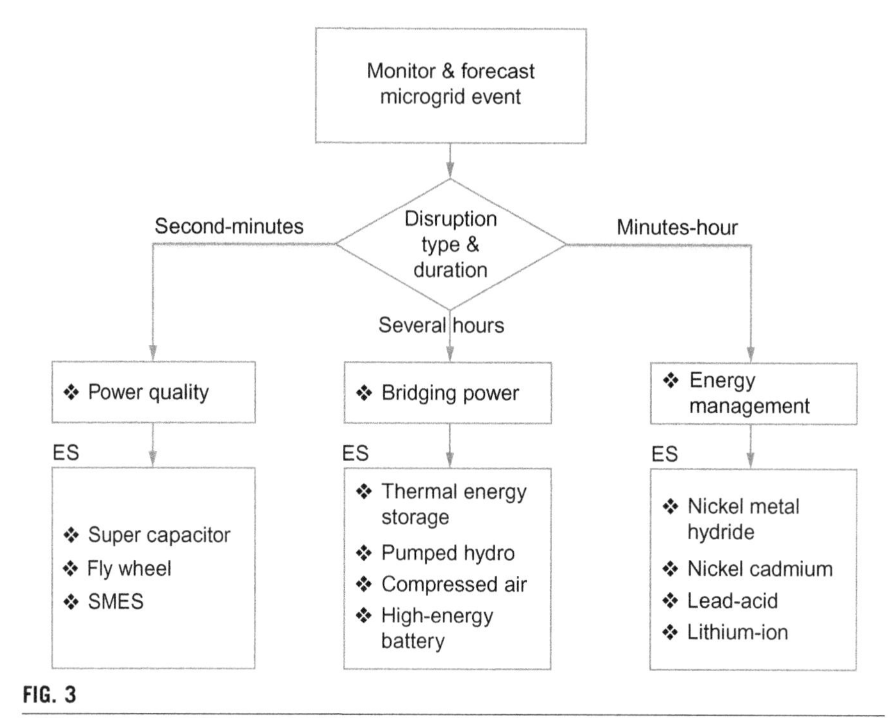

FIG. 3

Application of energy storage technologies for micro energy grids.

Modern battery technology is used for bridge power applications, which include nickel-metal hydride, nickel-cadmium, lead-acid, and lithium-ion. This application requires rapid response (in second to minute) and discharge times should be about 1 h. Sometimes MEGs require longer timescale energy storage to enhance sustainability that has discharge rating with several hours or more. This is another effective storage energy application recognized as energy management. Energy storage technologies for the application are thermal energy storage, pumped-hydro, compressed air, and high-energy batteries.

It is evident that energy storage technologies play a very important role in MEGs, enhancing the grid's sustainability and improving the power quality of the system. Fig. 3 shows different energy storage applications for various MEG scenarios.

9.5 FRAMEWORK OF MEG-ESS-UTILITY GRID INTEGRATION

Management of energy supply and demand can be realized by a combination of distributed energy resources (DERs) incorporated with MEGs, controlled by intelligent control. Cost-effective energy storage enables the connection mode between microgrids and the local power grids. MEGs can operate in two modes to provide critical customers and facilities with power if and when the utility grid gets interrupted.

Energy storage gives the MEG the ability to supply the power to locations and communities that face difficulties as expensive power or practically unattainable power. Based on its energy capacity and lifetime, and suitability to support MEG operations, energy storage plays an important role.

9.5.1 THE CHALLENGES OF FRAMEWORK

The energy demands customers may suffer from clear problem that the supply must instantaneously match demand with almost zero interruptions, that concept needs cost-effective storage elements. The solution will need to meet the energy demands and growth without upgrading the existing infrastructure as to save cost and to lessen the environmental impact.

The first challenge that faces the framework of MEG-ESS-utility grid integration is the need to enhance grid resiliency and reliability. The system may face a problem in maintaining continuity during any disturbance at bus or distribution, and can result in remarkable amounts of demand not serviced. MEG and energy storage will solve this problem by enabling reliable operations as well as ensuring supplying the most critical loads with a secure source in the operation modes, shown in Fig. 4.

The second problem is to enable DERs as stable energy sources. At times, the operation of DERs with the grid faces instability, which limits their impact as an energy source. In addition, the maximum demand can occur in a moment when the supplied power is not at maximum. The solution is the integration of an MEG with energy storage that will enable smooth and stable production, and has the ability of time shift for the supplied power to match the demand profile, shown in Fig. 5.

There are other challenges for maximizing the utilization of existing grid infrastructures and supplying clean and peaking generation. During peak demands, there might be costly and less clean energy generation technologies to supply energy during the peak periods. MEG with energy storage can offer a reliable solution to

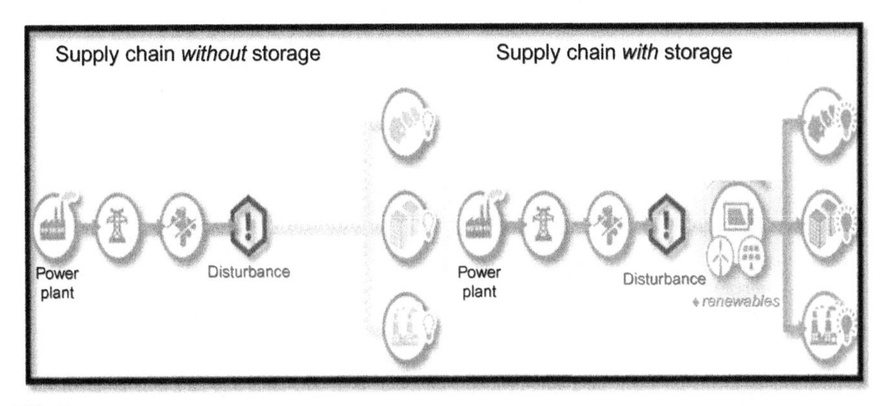

FIG. 4

MEG with storage improves grid resiliency and reliability.

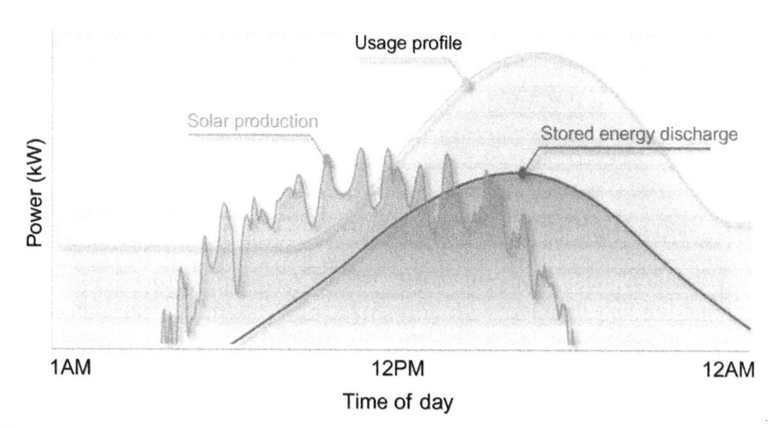

FIG. 5

MEG with storage improves DERs as stable energy sources.

address these challenges. They can supply the peak hours with clean concept supplying to reduce carbon emissions that is shown in Fig. 6.

9.5.2 BENEFITS TO UTILITY

With MEG and energy storage, utilities can depend on an efficient and cleaner energy supply without the high costs of upgrading the infrastructure. This can achieve the following goals:

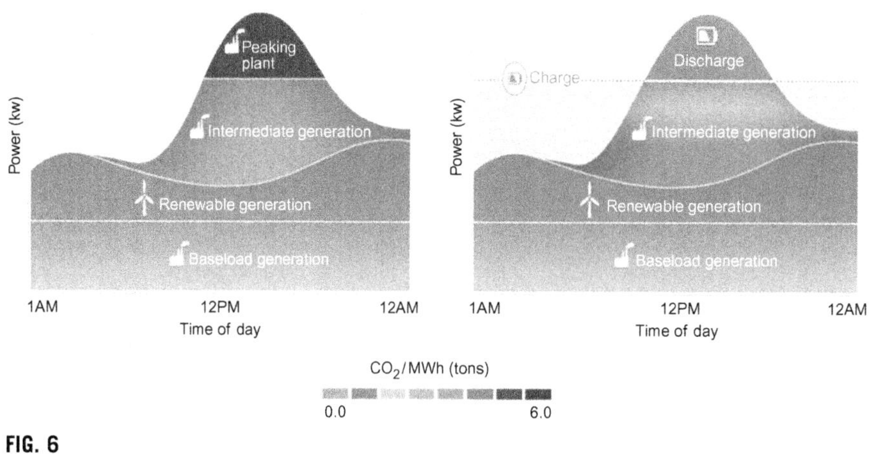

FIG. 6

MEG with storage produces clean and peaking generation.

- *Infrastructure Upgrade Postponement*: Shifting peak load and mitigation congestion will cancel the need for expensive upgrades to transmission and distribution networks, saving money.
- *Energy Balancing*: Storing cheap electricity produced during light loading to be used at peak demands will eliminate the use of unclean generation units, reducing both costs and emissions.

9.5.3 BENEFITS TO CUSTOMERS

The commercial and industrial loads have high power quality and reliability requirements, and they can gain the following benefits from MEG and storage integration:

- Energy management and backup power to support the customers during temporary interruption and outages with no emissions.
- The ability to have firm dispatchable resources.

9.5.4 ROLE DESCRIPTION OF MEG AND ENERGY STORAGE

There are specific roles that are required from the storage system such as to smooth out fluctuations in the grid and to decouple generation and consumption. There are other requirements, which include the following:

At the utility level

- Renewable smoothing/firming
- Grid support services—ancillary support, frequency control, voltage regulation

At the consumer level

- Peak shaving
- Energy management
- Demand analysis and control
- Gains of energy storage

Technology types have differing value contributions, as shown in Figs. 7 and 8.

During MEG and storage integration, there are also specific roles for MEGs that are required, such as:

- Focussed on grid ancillary services
- Traditional designs comprised of rotational generators only
- Strong focus on accepting renewable and coping with intermittency smoothing ramp control of very fast changes
- Firming power at firm level over reasonable time
- Focussed on aspects that will improve the continuity of supply to the consumer, balanced with a reduced level of demand and utilization charges
- Grid backup supply during grid outages
- Peak shaving to reduce the level of energy demands from the network at peak periods when the rate of demand charges is high

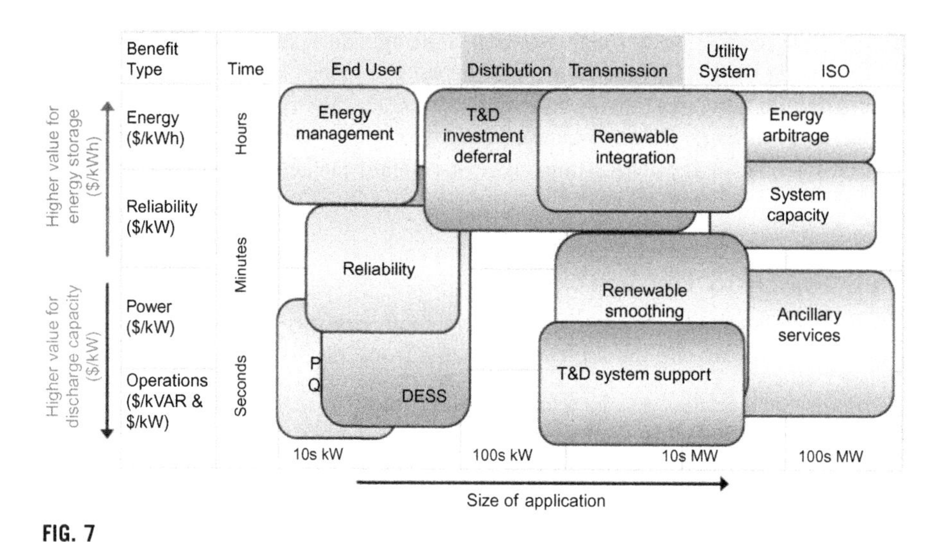

FIG. 7

Storage benefits and applications.

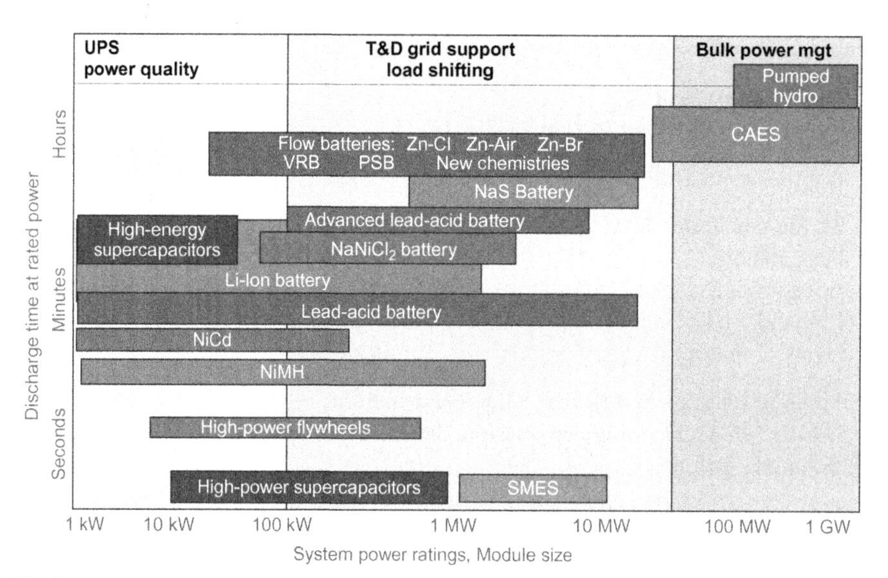

FIG. 8

Storage power and time rates.

- Energy shifting for moving excess energy from one time slot to a later time slot (as movable solar produced midday to support the evening peak)
- Demand response for responding to utility incentives to reduce demands on the grid during specified periods

9.6 CASE STUDY: MEG STABILITY AT FAULT CONDITION WITH ESS

In MEGs, modern storage energy technology could improve stability as well as power quality on various abnormal events. Here, the voltage stability of a DC microgrid is described, using a super capacitor as its energy storage.

During the initial stage of charging, the voltage of the super capacitor rose rapidly, as shown in the first part of Fig. 9. There was an external system fault at 3 s and, at that time, the voltage of the super capacitor stopped rising and began the discharging process. The fault was restored at 4.5 s. The second part of the figure shows the output current of the super capacitor during its charge and discharge process. From Fig. 4 we can see that the initial charging current of the super capacitor is very high—up to 5KA—but gradually decreases and maintains at 50 A. At 3 s the capacitor was reversed, which means the super capacitor was discharging to the external system.

Fig. 10 shows that at 3 s bus voltage had slide variations, but maintained about 400 V. This verifies the effectiveness of the super capacitor control to maintain system stability.

Generally speaking, integrated ESSs with MEGs lead to various power quality criteria, which depends on the type of storage, type of interconnection with the system, total capacity of the storage, and mode of operations. Modern energy storage technology plays an important role in the sustainability and reliability of MEGs, guarding against different electrical hazards. This chapter describes various available energy storage technologies that could be used in different conditions. Suitable

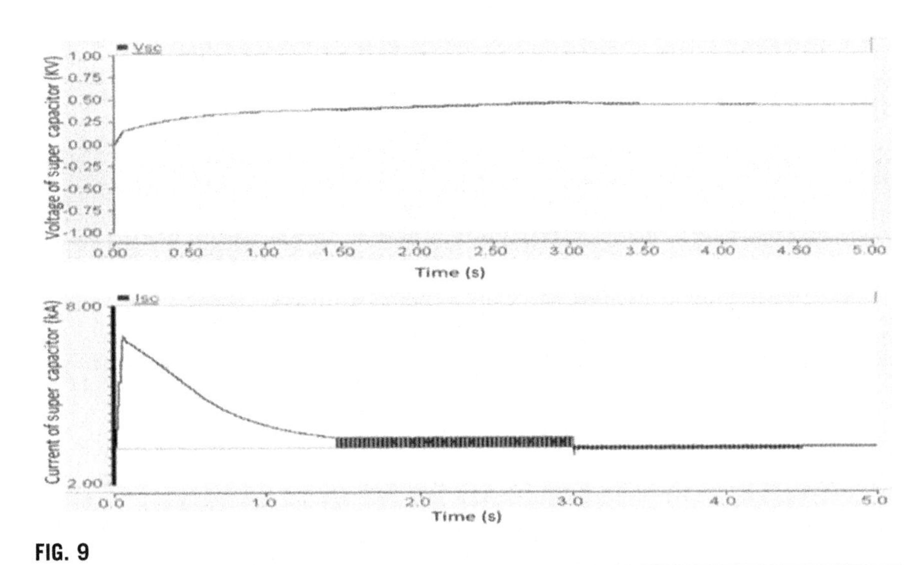

FIG. 9

Terminal voltage and current of super capacitor.

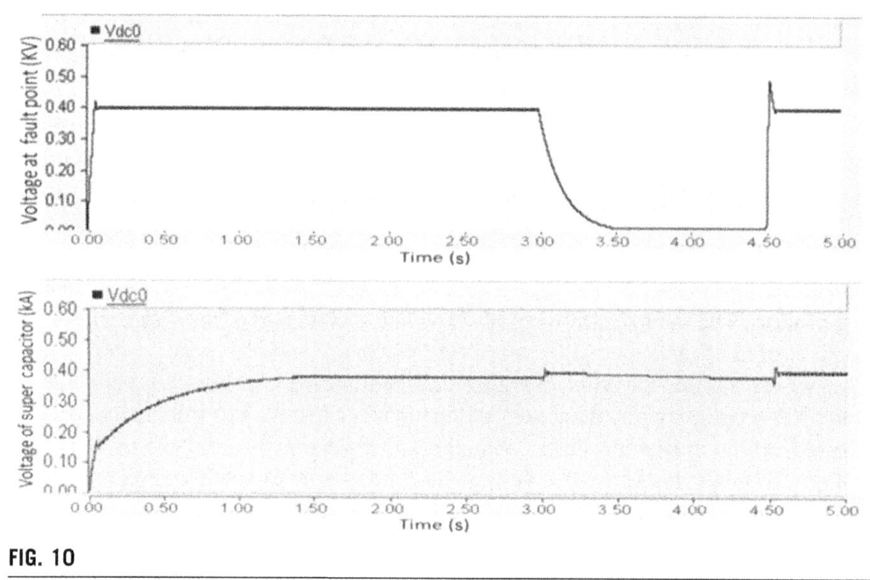

FIG. 10

Voltage of faulted point and DC bus.

ESSs can generally be applied for various abnormal events of MEG. Performance analysis is important for MEG design and operations.

Finally, the case study shows the positive effects of the super capacitor to be used during external fault conditions in order to maintain the stability of the system.

FURTHER READING

[1] Peças Lopes JA, Moreira CL, Madureira AG. Defining control strategies for analysing microgrids islanded operation; 2000, http//www.microgrids.eu/micro2000/presentations/28.pdf.

[2] Collins ER, Jiang J. Voltage sags and the response of a synchronous distributed generator: a case study. 23, 2008. p. 442–8. http://dx.doi.org/10.1109/TPWRD.2007.905562.

[3] Resource Dynamics Corporation, Assessment of distributed generation technology applications, http://www.distributed-generation.com/Library/Maine.pdf.

[4] Kariniotakis GN, Soultanis NL, Tsouchnikas AI, Papathanasiou SA, Hatziargyriou ND. Dynamic modeling of microGrids; 2005, http://dx.doi.org/10.1109/FPS.2005.204227. pp. 0–7.

[5] El-Samahy I, El-Saadany E. The effect of DG on power quality in a deregulated environment, Vol. 3; 2005. p. 2969–76. http://dx.doi.org/10.1109/PES.2005.1489228.

[6] Deng W, Pei W, Qi Z. Impact and improvement of distributed generation on voltage quality in micro-grid; 2008, http://dx.doi.org/10.1109/DRPT.2008.4523687. pp. 1737–1741.

[7] Cobben JFG, Kling WL, Myrzik JMA. Power quality aspects of a future micro grid; 2005, http://dx.doi.org/10.1109/FPS.2005.204282. pp. 0–5.

[8] Meliopoulos APS. Challenges in simulation and design of μGrids. 1; 2002, http://dx.doi.org/10.1109/PESW.2002.985004. pp. 309–314.
[9] Konstantinos Angelopoulos. Integration of distributed generation in low voltage networks: power quality and economics; 2004, http://www.esru.strath.ac.uk/Documents/MSc_2004/angelopoulos.pdf.
[10] Hussain B, Sharkh SM, Hussain S. Impact studies of distributed generation on power quality and protection setup of an existing distribution network; 2010, http://dx.doi.org/10.1109/SPEEDAM.2010.5545061. pp. 1243–1246.
[11] Ribeiro PF, Johnson BK, Crow ML, Arsoy A, Liu Y. Energy storage systems for advanced power applications; 2001, http://dx.doi.org/10.1109/5.975900. pp. 1744–1756.
[12] Fujimoto K, Ota T, Shimizu Y, Ichikawa T, Yukita K, Goto Y, et al. Load frequency control using storage system for a micro grid; 2009, http://dx.doi.org/10.1109/TD-ASIA.2009.5356856. pp. 1–4.
[13] Tang X, Deng W, Qi Z. Research on micro-grid voltage stability control based on super-capacitor energy storage; 2009, http://dx.doi.org/10.1109/ICEMS.2009.5382740. pp. 1–6.
[14] Zhao Q, Yin Z. Battery energy storage research of photovoltaic power generation system in micro-grid; 2010. p. 1–4. http://dx.doi.org/10.1109/CRIS.2010.5617540.
[15] Kwhannet U, Sinsuphun N, Leeton U, Kulworawanichpong T. Impact of energy storage in micro-grid systems with DGs; 2010. http://dx.doi.org/10.1109/POWERCON.2010.5666059. pp. 1–6.
[16] Wang Z, Li G, Li Gang, Yue H. Studies of multi-type composite energy storage for the photovoltaic generation system in a micro-grid; 2011. http://dx.doi.org/10.1109/DRPT.2011.5994000. pp. 791–796.
[17] Xuemei H. Implementing intelligence and distributed execution mechanism for flywheel energy storage system in micro grid; 2011. http://dx.doi.org/10.1109/EMEIT.2011.6022906. pp. 413–416.
[18] R. Lasseter, M. Erickson, Integration of battery-based energy storage element in the certs microgrid, Available in web site: CERTS_Microgrid_Battery_Storage_UWisc-1.pdf.
[19] C. Naish, I. McCubbin, Oliver Edberg, Michael Harfoot, Outlook of Energy Storage, http://www.storiesproject.eu/docs/study_energy_storage_final.pdf.
[20] Denholm P, Ela E, Kirby B, Milligan M. The role of energy storage with renewable electricity generation, http://www.nrel.gov/docs/fy10osti/47187.pdf.
[21] Barker PP. Ultracapacitors for use in power quality and distributed resource applications", 1; 2002. p. 316–20. http://dx.doi.org/10.1109/PESS.2002.1043241.
[22] Arulampalam A, Barnes M, Jenkins N, Ekanayake JB. Power quality and stability improvement of a wind farm using STATCOM supported with hybrid battery energy storage; 2006. http://dx.doi.org/10.1049/ip-gtd:20045269. pp. 701–710.
[23] Jayawarna N, Barnes M, Jones C, Jenkins N. Operating micro grid energy storage control during network faults; 2007. http://dx.doi.org/10.1109/SYSOSE.2007.4304254. pp. 1–7.
[24] R.H. Lasseter, Microgrids and Distributed Generation, http://%20citeseerx.ist.psu.edu.
[25] Meiqin M, Chang L, Ming D. Integration and intelligent control of micro-grids with multi-energy generations: a review; 2008. p.777–80; http://dx.doi.org/10.1109/ICSET.2008.4747112.
[26] Basak P, Saha AK, Chowdhury S, Chowdhury SP. Microgrid: control techniques and modeling. UPEC 2009; 2009. p. 1–5.
[27] Sakimoto K, Miura Y, Ise T. Stabilization of a power system with a distributed generator by a virtual synchronous generator function; 2011, http://dx.doi.org/10.1109/ICPE.2011.5944492. pp. 1498–1505.

[28] Wu X, Zhang Y, Arulampalam A, Jenkins N. Electrical stability of large scale integration of micro generation into low voltage grids; 2000. www.microgrids.eu/micro2000/presentations/35.pdf.
[29] Integration of distributed energy resources the CERTS micro grid concept, http://certs.lbl.gov/pdf/50829.pdf.
[30] White paper on integration of distributed energy resources the micro grid concept appendices, http://certs.lbl.gov/pdf/50829-app.pdf.
[31] Lopes JAP, Moreira CL, Madureira AG. Defining control strategies for microgrids islanded operation; 2006. http://dx.doi.org/10.1109/TPWRS.2006.873018. pp. 916–924.
[32] Jayawarna N, Wu X, Zhang Y, Jenkins N, Barnes M. Stability of a microgrid, In: IET international conference; 2006, p. 316–20.
[33] Bingbing W, Zhongdong Y, Xiangning X. Super-capacitors energy storage system applied in the microgrid. 2010 the 5th IEEE conference on ICIEA; 2010. p. 1002–5. http://dx.doi.org/10.1109/ICIEA.2010.5515758.
[34] Mark Bollman Andrew. An experimental study of frequency droop control in low inertia micro grid. Thesis of MASc, University of Illinois at Urbana-Champaign; 2009.
[35] Kanellos FD, Tsouchnikas AI, Hatziargyriou ND. Micro-grid simulation during grid-connected and islanded modes of operation; 2005, www.ipst.org/TechPapers/2005/IPST05_Paper113.pdf.
[36] A. Schroeder, J. Siegmeier, M. Creusen, Modeling storage and demand management in electricity distribution grids, http://www.diw.de/documents/publikationen/73/diw_01.c.368984.de/dp1110.pdf.

CHAPTER

FACTS-based high-performance AC/DC microgrids

10

A.M. Othman*,[†], H.A. Gabbar*

*University of Ontario Institute of Technology (UOIT), Oshawa, ON, Canada**

Zagazig University, Zagazig, Egypt[†]

10.1 INTRODUCTION

The modern age power grid introduces many limitations and challenges in a world that is heavily dependent on electricity. Many government agencies, utility companies, researchers, and engineers in the electric power industry have envisioned transforming the existing grid into a microgrid. To facilitate domestic consumer's ratings and needs, a microgrid has been developed having distributed energy resources (DER). It is expected to be an intelligent, sustainable, resilient, and reliable energy grid suitable for the modern economy. One of the main concerns for any grid is power quality and efficiency with reactive and active power reliability for the consumer. To manage active and reactive power flexible AC transmission system (FACTS) technologies are used to improve power quality and efficiency. Since FACTS technologies were developed, much research has been done to acquire more accurate and reliable techniques and algorithms to implement FACTS in the power system. Current microgrid schemes implemented with FACTS, an output response monitored with different key performance indicators (KPIs) of the microgrid, developed AC/DC microgrid designs with FACTS technology, including DER consisting of DC batteries, wind turbines, solar photovoltaic (PV) systems, and diesel generators. Modulated power filter compensators (MPFC) for AC, green plug filter compensators (GPFC) for DC are used as FACTS controllers. Optimization and intelligent control are accomplished through self-adapting the controller gains and converter gains in order to achieve the best microgrid energy utilization and stabilizing voltage, and to reduce the inrush of current conditions. KPIs, such as bus voltage stabilizing, feeder loss reduction, power factor enhancement, improvement of power quality, and reduction of the total harmonic distortion (THD) at AC interface buses, are compared with the FACTS criterion. The AC/DC microgrid is modeled in both grid and islanded connected modes.

Smart Energy Grid Engineering.

10.2 GRID-CONNECTED-MODE DER CONTROL

Grid-connected-mode DER control works with the frequency and magnitude of system voltages that are provided by the grid controlled by utility. The main task of the grid-connected mode is to control the active and reactive power exchange within the local networks. For regulation of output powers, it can use a current or voltage-mode control technique. Fig. 1 defines a simplified schematic diagram of a voltage-controlled DER system; a DER system is connected to the grid through a three-phase inductance element, L, and a voltage source converter (VSC). In the voltage-mode control strategy, the output active and reactive powers (P_o, Q_o) are controlled through the phase-angle and magnitude of the VSC on an AC terminal voltage, v_{tabc}, that are close to the host bus voltage, v_{sabc}; two variables in the control strategy can become independent if the resistance in the inductor is bypassed.

Hence it concludes as every phase of voltage is added by the VSC, v_{tabc}, and is calculated by shifting the phase-angle and scaling the amplitude of the substitute phase of v_{sabc}. For the voltage-mode and current-mode control techniques of Figs. 1 and 2, the DER has been assumed to be read as dispatched, that is, its output real and reactive powers can be controlled by the set points P_{oref} and Q_{oref}, which are determined by the main control center. By contrast, the output powers of a nondispatchable DER system are commonly the by-products of an optimal operating condition. For example, a PV system normally operates in the maximum power-point tracking mode, that is, it extracts the maximum possible power from its solar panels.

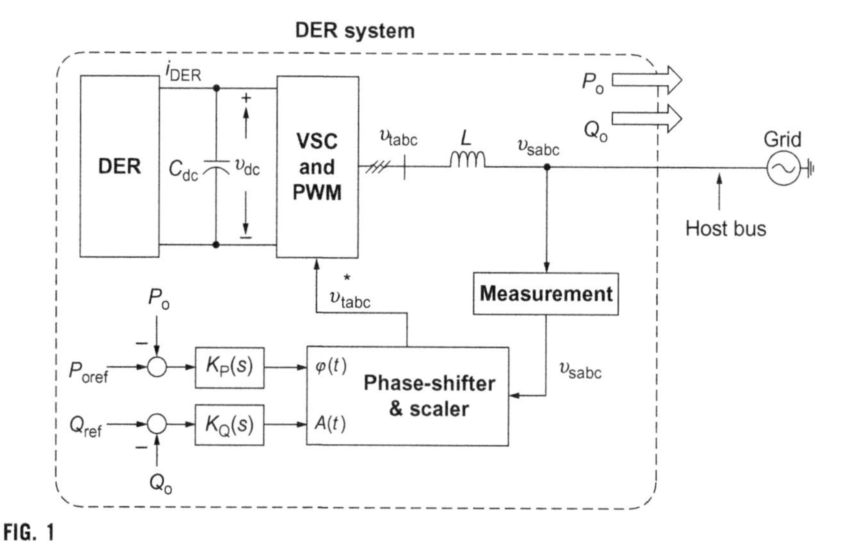

FIG. 1

Schematic diagram of a grid-connected DER system with voltage mode.

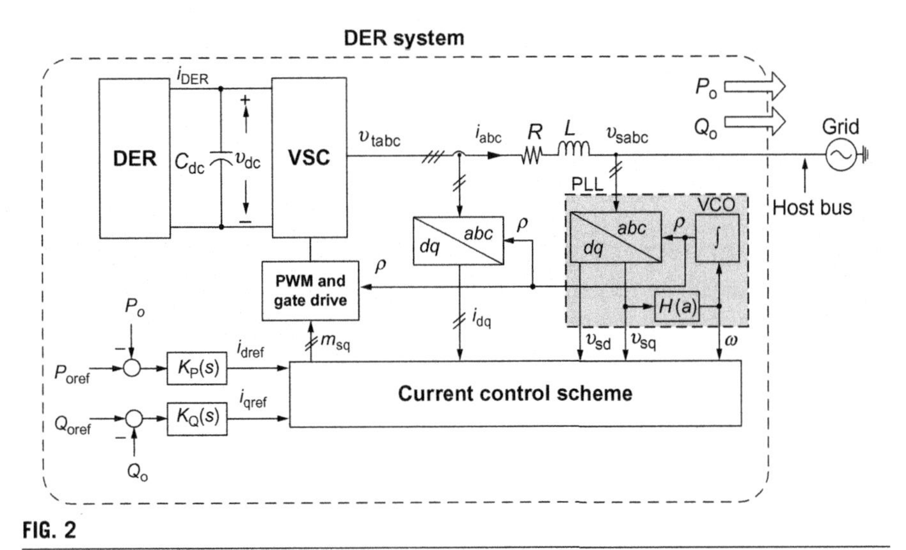

FIG. 2

Schematic diagram of a grid-connected DER system with current mode.

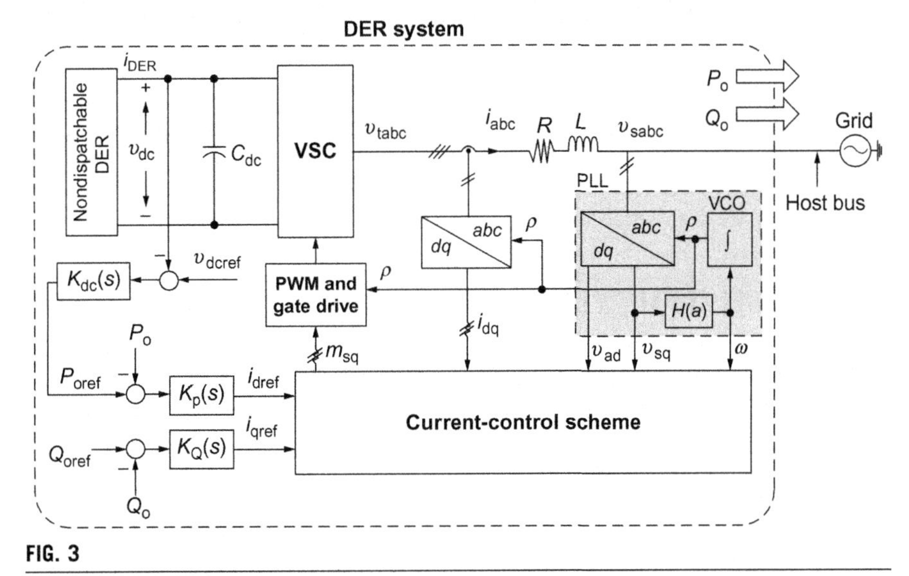

FIG. 3

Schematic diagram of a grid-connected nondispatchable DER system with current.

Fig. 3 indicates a schematic diagram of a nondispatchable DER system, such as a PV system in which the DER has been modeled by a dc voltage source whose voltage, v_{dc}, is related to its current, i_{DER}, through a $v-i$ characteristic. As Fig. 3 illustrates, the kernel of the control system is the real- and reactive-power control scheme (of Fig. 2) by which P_o and Q_o can be controlled independently.

10.3 CONTROL DESIGN OF AN FACTS-BASED MICROGRID

This section gives an overview of the modeling and design of a microgird with FACTS. The objective of this chapter is to investigate modeling, including the equations involved with different generation units, including PV, gas turbines, wind turbines, fuel cells, and batteries.

The proposed microgrid design has several DERs such as PV, fuel cells, batteries, and wind turbines. MPFCs are used on the AC side to improve voltage stability and to reduce the harmonic current and power loss. Also, GPFC are added as a DC-side FACTS device. Multiloop error-driven control strategies are configured so that the converters are properly beneficial to the AC and DC grid with their FACTS. To optimize the controllers and FACTS, an intelligent algorithm, using a genetic algorithm, is used to self-tune the converter and FACTS controller gain.

10.3.1 HYBRID AC/DC MICROGRID DESIGN

The design of the hybrid AC/DC microgrid system is in Fig. 4, where a connected, meshed microgrid with different AC and DC sources and loads are connected to their alternating subgrids. Also, these loads are connected to a utility bus with hybrid local loads. The AC and DC grids are linked to a utility subgrid with a 10- and 5-km feeder, respectively. With an AC subgrid, a step up transformer is linked in order to boost up the AC, and with a DC subgrid a step down transformer is linked. Both AC and DC grids are also linked to an AC/DC, DC/AC converter and inverter to exchange a hybrid connection in the grid. All parameters for microgrid loads, distributed flexible AC transmission system (D-FACTS), distributed generation (DG), and other source specifications are mentioned in Fig. 5.

10.3.2 NOVEL D-FACTS DEVICES

There are two recent and advanced D-FACTS devices, based on alternately switching a switched filter compensator, that are applied to the present design of the hybrid AC/DC microgrid, which are explained in the following sections.

Modulated power filter compensator

The used MPFC scheme, shown in Fig. 6, is a combination of one series-capacitor bank and two shunt-capacitor banks with a tuned arm power filter. The energy discharge path for the two shunt-capacitor banks is formed by the two pulse diode rectification circuit with the resistance (R_f) and inductance (L_f) branch that forms a tuned arm filter at the rectifier's DC side. The two insulated-gate bipolar transistor switches (Sa and Sb) are controlled by two complementary switching pulses (P1 and P2), as shown in Fig. 2, that are generated by the dynamic multiloop error-driven proportional-integral-derivative (PID) controller. The variable MPFC topology can be altered by some complementary switching of the pulse-width modulation (PWM) pulses as follows:

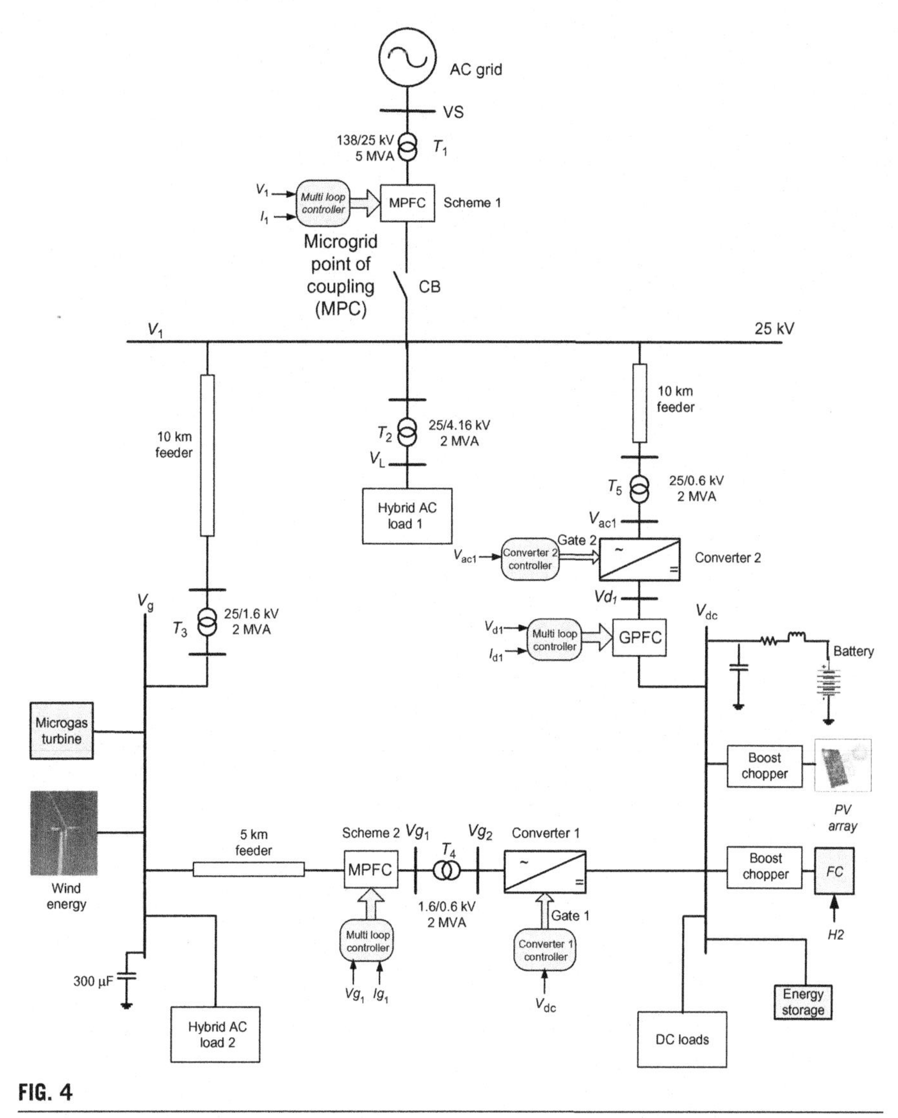

FIG. 4

Hybrid AC/DC microgrid.

Case 1: If P1 is low and P2 is high, the resistor and inductor of the arm filter will be fully shorted, and the combined shunt and series capacitors will provide the appropriate shunt and series capacitive compensated actions to the AC distribution system.
Case 2: If P1 is high and then P2 is low, the resistance and inductance will be in connection to the circuit as a tuned arm filter.

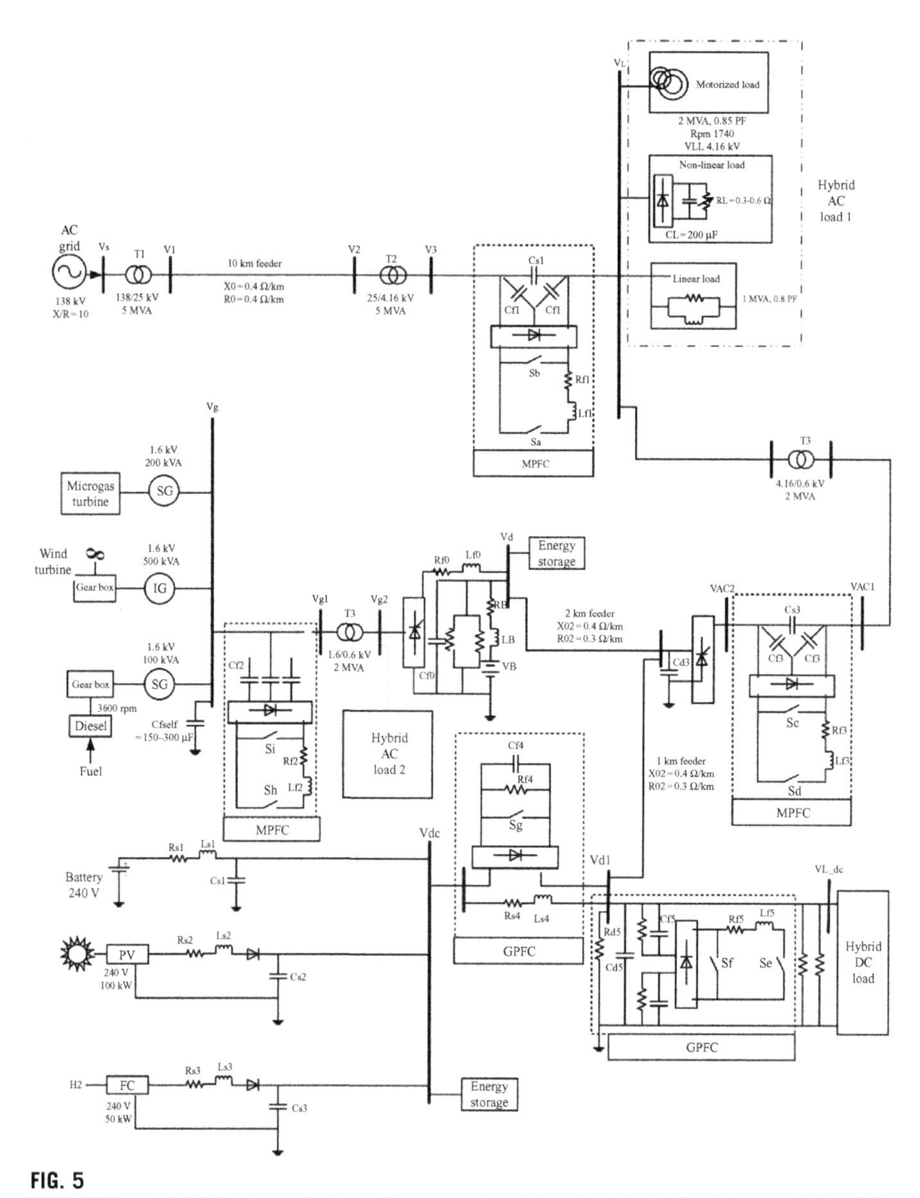

FIG. 5

Full schematic of an AC/DC microgrid.

Green plug filter compensator

To achieve enhancement the performance of the DC microgrid and DC buses voltage stabilization, a GPFC, shown in Fig. 3 comprising a switchable capacitor, is introduced at the DC bus. The GPFC is used to regulate the voltage profile of the DC

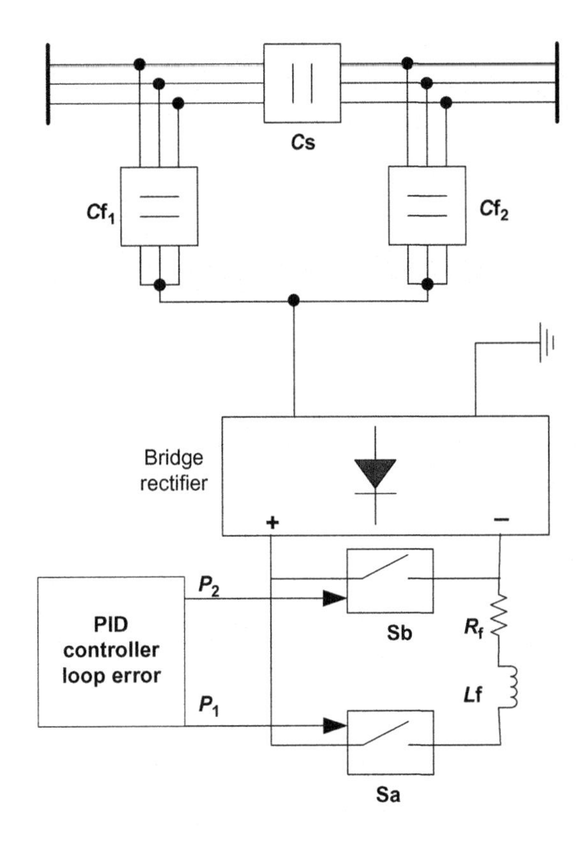

FIG. 6

Scheme of a power filter for modulation/compensation.

bus, and to eliminate, or minimize, the inrush components of current transients and PV and FC nonlinear volt-ampere characteristics.

10.4 CONTROL DESIGN

The proposed control system comprises five controllers listed as follows.

10.4.1 THE MPFC 1 CONTROLLER

In this design form of the control process of the MPFC, shown in Fig. 7, a multiple-loop error-activated dynamic controller is applied to modulate the PWM switching. The global error of the multiple loops (e_1) is the sum of the multiple loop individual errors, including voltage stability, current limiting, and synthesis dynamic power

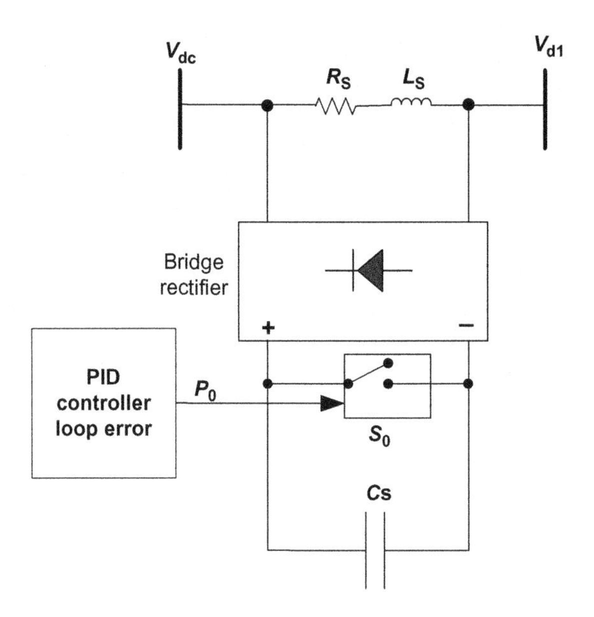

FIG. 7

Scheme of a green plug filter modulation/compensator.

loops. The per-unit 3-D error vector (e_V, e_I, e_P) is controlled by the following formulas:

$$e_{V_1} = \frac{V_{1\text{ref}} - V_1\left(\frac{1}{1+ST_1}\right)}{V_{1\text{base}}} \tag{1}$$

$$e_{I_1} = \frac{I_{1\text{ref}} - I_1\left(\frac{1}{1+ST_1}\right)}{I_{1\text{base}}} \tag{2}$$

$$e_{P_1} = \left(\frac{V_1}{V_{1\text{base}}} \times \frac{I_1}{I_{1\text{base}}}\right) - \left(\frac{V_1}{V_{1\text{base}}} \times \frac{I_1}{I_{1\text{base}}}\right)\left(\frac{1}{1+ST_2}\right) \tag{3}$$

The global multiloop error e_1:

$$e_1 = \gamma_{V_1}|e_{V_1}| + \gamma_{I_1}|e_{I_1}| + \gamma_{P_1}|e_{P_1}| \tag{4}$$

The effective reactance effect of the combination of hybrid assigned capacitors and the modulated tuned-arm filter depends on the duty cycle and the frequency of the PWM output, which, in turn, is a function of the self-tuned PID controller output. The output of the PWM supplier is a generator of pulses with changeable duty cycles and constant frequency. The system control voltage of the MPFC scheme-1 own the following formula related to the time domain for the PID controller:

$$V_{C1}(t) = K_{P1}e_1(t) + K_{I1}\int_0^t e_1(t)dt + K_{d1}\frac{d(e_1(t))}{dt} \tag{5}$$

The modulation of the filter arm is achieved by two complementary pulses (Sa and Sb) generated from the PWM.

10.4.2 THE MPFC 2 CONTROLLER

In this design form of the control process of the MPFC, the modulation of the filter arm is realized by two complementary pulses (Sc and Sd) provided by the PWM, which is modulated using the output of the self-tuned PID controller. The global error (e_2) is the input signal to the self-tuning PID controller, and is a summation of the multiloop errors; $e_{V_{g1}}$, $e_{I_{g1}}$, and $e_{I_{g1}}$ as follows:

$$e_2 = \gamma_{V_{g1}}\left|e_{V_{g1}}\right| + \gamma_{I_{g1}}\left|e_{I_{g1}}\right| + \gamma_{P_{g1}}\left|e_{P_{g1}}\right| \tag{6}$$

$$e_{V_{g1}} = \frac{V_{g1\text{ref}} - V_{g1}\left(\frac{1}{1+ST_3}\right)}{V_{g1\text{base}}} \tag{7}$$

$$e_{I_{g1}} = \frac{I_{g1}}{I_{g1\text{base}}}\left(\frac{1}{1+ST_3}\right)\left(1 - \frac{1}{1+ST_4}\right) \tag{8}$$

$$e_{P_{g1}} = \left(\frac{V_{g1}}{V_{g1\text{base}}} \times \frac{I_{g1}}{I_{g1\text{base}}}\right)\left(1 - \frac{1}{1+ST_4}\right) \tag{9}$$

The modulation signal for the PWM of the MPFC scheme-2 is described by the following formula in the time domain for the PID controller:

$$V_{C2}(t) = K_{P2}e_2(t) + K_{I2}\int_0^t e_2(t)dt + K_{d2}\frac{d(e_2(t))}{dt} \tag{10}$$

10.4.3 THE GPFC CONTROLLER

The design of a GPFC control form depends on the decoupling of the current and voltage components of the DC bus. The GPFC is connected on the terminal of the DC bus to keep a regulated, smooth DC voltage. In this scheme, the multiloop global error is e_3:

$$e_3 = \gamma_{V_{d1}}\left|e_{V_{d1}}\right| + \gamma_{I_{d1}}\left|e_{I_{d1}}\right| + \gamma_{P_{d1}}\left|e_{P_{d1}}\right| \tag{11}$$

$$e_{V_{d1}} = \frac{V_{d1\text{ref}} - V_{d1}\left(\frac{1}{1+ST_5}\right)}{V_{d1\text{base}}} \tag{12}$$

$$e_{I_{d1}} = \frac{I_{d1}}{I_{d1\text{base}}}\left(\frac{1}{1+ST_5}\right)\left(1-\frac{1}{1+ST_6}\right) \quad (13)$$

$$e_{P_{d1}} = \left(\frac{V_{d1}}{V_{d1\text{base}}} \times \frac{I_{d1}}{I_{d1\text{base}}}\right)\left(1-\frac{1}{1+ST_6}\right) \quad (14)$$

The control structure of the PWM signal of the GPFC scheme depends on the following formula in the time domain for the PID controller:

$$V_{C3}(t) = K_{P3}e_3(t) + K_{I3}\int_0^t e_3(t)dt + K_{d3}\frac{d(e_3(t))}{dt} \quad (15)$$

10.4.4 AC/DC CONVERTER 1 CONTROLLER

This controller is used to control the operation of the AC/DC converter by changing the pulse duty cycle so that it is used to switch on/off the converter to stabilize the output voltage on the DC side of the converter.

10.4.5 AC/DC CONVERTER 2 CONTROLLER

The stabilization of the voltage at the AC side of the converter 2 is achieved by this controller.

10.5 CASE STUDY: SIMULATION RESULTS AND DISCUSSIONS

These simulation results are for a hybrid AC/DC microgrid. Different scenarios may be performed with several parameters and configurations. The configuration can consist of an islanded and grid-connected mode.

In this microgrid structure, for the DC section of the grid, the GPFC is used to allow for a stable DC bus voltage, to make compensatory actions on the reactive power, and to provide the exchange transactions for the power between the DC section of the grid, the utility grid, and the AC section of the grid. When the condition of the output power of the DC sources is greater than the DC loads, the converter 1 will operate as an inverter and transfer the power from the DC section to the AC section of the grid. When the condition of the total power generation is less than the total load at the DC side, the converter 1 will transfer the power from the AC section of the grid to the DC section. Also, for the AC section of the grid, the MPFC-2 is to supply stable AC section bus voltages, to allow compensation actions for reactive power, and to transfer the power between the AC section, utility grid, and the DC section of the grid. When the condition of the output power of the AC source is greater than the AC loads, the converter transfers the power from the AC to the DC network. When the condition of the total power generation is less than the total load at the AC section bus, the converter 1 transfers the power from the DC section to the AC section side. When the condition of the total power generation is greater than the total load of the hybrid AC/DC grid, it will transfer power to the utility grid. Otherwise, the hybrid grid will take the power from the utility network.

The power balance formula of a hybrid DC/AC microgrid is as follows:

$$P_{g_{AC}} + P_{g_{DC}} \pm P_S = P_{L_{AC}} + P_{L_{DC}} + P_{Losses} \tag{16}$$

where $P_{g_{AC}}$ the generated real power of the AC sources (microgas turbine and wind turbine); $P_{g_{DC}}$ the generated power of the DC sources (PV, fuel cells, battery); P_S the injected power from/to the utility grid.

10.5.1 GRID-CONNECTED MODE (FACTS EFFECT)

The resilient operation with high performance criteria of this microgrid is emphasized using the optimal values of the PID controller's gains that are shown in Table 1.

As shown in Figs. 8–12, the used D-FACTS have the ability to compensate for the reactive power at the AC buses and DC buses in order to stabilize the voltage profile,

Table 1 Optimal Values of the PID Controller

	Optimal Values of PID Controller Gains		
	K_p	K_i	K_d
MPFC 1	12	2.4	0.5
MPFC 2	15	5	0.2
GPFC	11	1.7	0.2
Converter 1	14	1.2	0.8
Converter 2	19	7	0.3

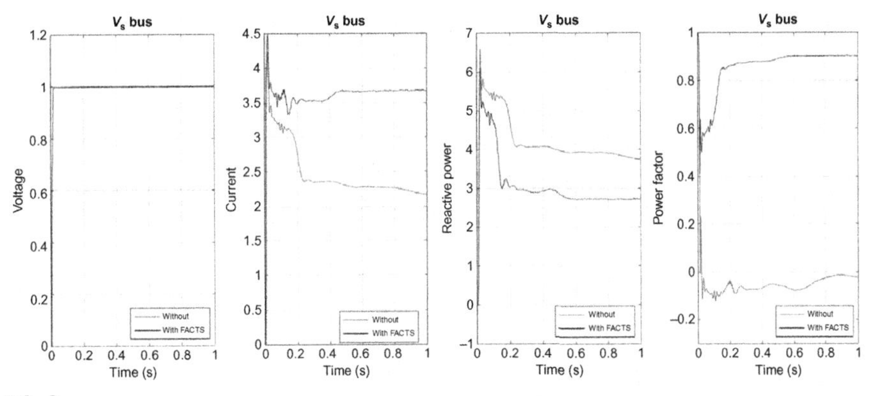

FIG. 8

Voltage, current, reactive power, and power factor at the AC bus (V_S).

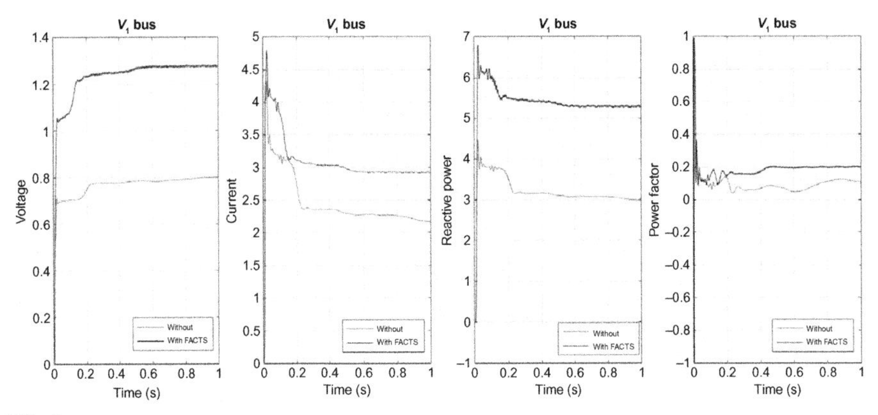

FIG. 9

Voltage, current, reactive power, and power factor at the AC bus (V_1).

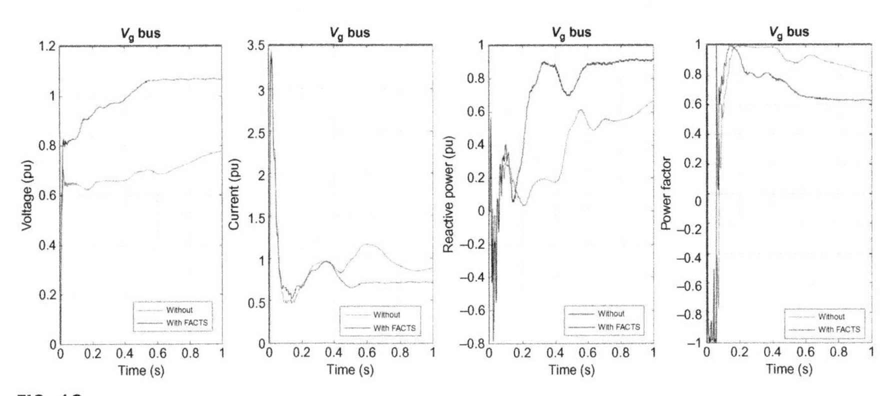

FIG. 10

Voltage, current, reactive power, and power factor at the AC bus (V_g).

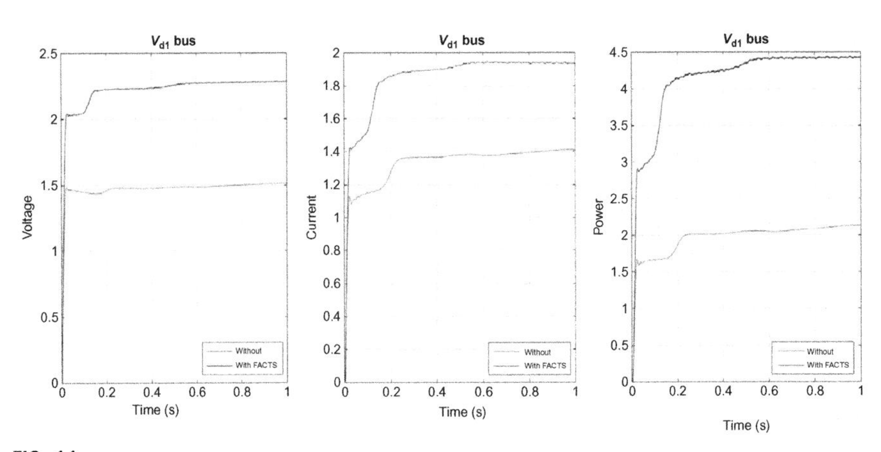

FIG. 11

Voltage, current, and power at the DC bus (V_{d1}).

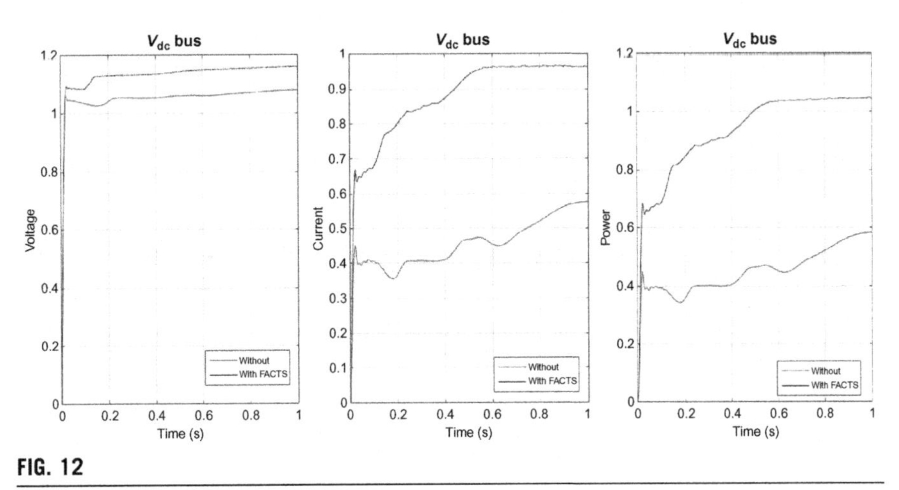

FIG. 12

Voltage, current, and power at the DC bus (V_{dc}).

and to enhance the power factor. Fig. 8 indicates the voltage waveform, current waveform, reactive power waveform, and the power factor waveform via the utility grid bus V_s. Also, the power factor is improved to a 0.9 lag case from a leading case of 0.07 PF, this confirms that the reactive power has been exchanged from the microgrid to the utility grid after using the FACTS. Also, by using the D-FACTS, the voltage profile is stabilized at 1 per-unit, and the power factor is improved at the AC bus, V_1, as shown in Fig. 9. The reactive power compensation by D-FACTS is very clear at the AC subgrid bus, V_g, and the voltage profile is improved and stabilized at 1 per unit, as shown in Fig. 10. Also, the DC section of the grid has a bus voltage that is enhanced and stabilized using the D-FACTS, as shown in Figs. 11 and 12. The THD at all the AC buses are calculated and summarized in Table 2. In this table, the reduction of the harmonic orders of the microgrid voltage and current with D-FACTS installation in comparison with the harmonic orders without D-FACTS installation are shown, in addition to the THD values belongs to the waveforms of the voltage and current are improved to be within limits of the voltage.

Table 2 The %THD of Voltage and Current at All AC Buses

	V_S		V_1		V_L		V_g	
	THD_v	THD_i	THD_v	THD_i	THD_v	THD_i	THD_v	THD_i
Without-FACTS	0.59	6.25	30.5	26	24.3	38	25	20
With-FACTS	0.2	5.65	5.72	5.57	5.6	5.72	5.51	5.3

FURTHER READING

[1] Lasseter RH, Piagi P. Microgrid: a conceptual solution, In: Power electronics specialists conference, 2004. PESC 04. 2004 IEEE 35th annual, 20–25 June 2004, vol. 6; 2004. p. 4285–90.

[2] Smallwood CL. Distributed generation in autonomous and non-autonomous micro grids, In: IEEE Rural Electric Power Conference; 2002. D1–D1_6.

[3] Katiraei F, Iravani MR, Lehn PW. Micro-grid autonomous operation during and subsequent to islanding process. IEEE Trans Power Deliver 2005;20(1):248–57.

[4] Molderink A, Bakker V, Bosman MGC, Hurink JL, Smit GJM. Management and control of domestic smart grid technology. IEEE Trans Smart Grid 2010;1(2):109–19.

[5] Marshall G. Modeling of a micro grid system. Bachelor Thesis, University of Newcastle, Australia; October 2004.

[6] da Silva AMLL, Sales WS, da Fonseca Manso LA, Billinton R. Long-term probabilistic evaluation of operating reserve requirements with renewable sources. IEEE Trans Power Syst 2010;25(1):106–16.

[7] Barker PP, Johnson BK. Power system modeling requirements for rotating machine interfaced distributed resources. In: Power engineering society summer meeting, vol. 1. p. 161–6.

[8] Blazwicz S, Kleinschmidt D. Distributed generation: system interface [Online]. Available http://www.encorp.com/dwnld/pdf/whitepaper/wp_ADL_2.pdf; May 1999.

[9] Ackermann T, Knyazkin V. Interaction between distributed generation and the distribution network: operation aspects. In: Transmission and distribution conference and exhibition 2002: Asia Pacific IEEE/PES, vol. 2. p. 1357–62.

[10] Amstutz A, Luigi K. EGO sensor based robust output control of EGR in diesel engines. IEEE Trans Control Syst Technol 1995;3(5):39–48.

[11] Hatziargyriou ND, Anastasiadis AG, Tsikalakis AG, Vasiljevska J. Quantification of economic, environmental and operational benefits due to significant penetration of Microgrids in a typical LV and MV Greek network. Eur Trans Electr Power 2011;21:1217–37.

[12] DOE microgrid workshop report, Office of Electricity Delivery and Energy Reliability Smart Grid R&D Program, San Diego, California, August 30–31; 2011.

[13] Abu-Sharkh S, Arnold RJ, Kohler J, Li R, Markvart T, Ross J, et al. Can microgrids make a major contribution to UK energy supply? Renew Sust Energy Rev 2006;10:78–127.

[14] Zhang Z, Huang X, Jiang J, Wu B. A load-sharing control scheme for a microgrid with a fixed frequency inverter. Electr Power Syst Res 2010;80:311–7.

[15] Mathur RM, Varma RK. Thyristor-based FACTS controllers for electrical transmission systems. New York-Piscataway, NJ: Wiley-IEEE Press; 2002. p. 495. ISBN 978-0-471-20643-9.

[16] Ambriz-Perez H. Flexible AC transmission systems modelling in optimal power flows using Newton's method. Glasgow: Department of Electronics and Electrical Engineering, University of Glasgow; 1998.

[17] Zhang J, Wen JY, Cheng SJ, Ma J. A novel SVC allocation method for power system voltage stability enhancement by normal forms of diffeomorphism. IEEE Trans Power Syst 2007;22(4):1819–22.

[18] Taylor CW. Improving grid behavior. IEEE Spectr 1999;36(6):40–5.

[19] Mathur RM. Canadian Electrical Association, Committee on Static Compensation. Static compensators for reactive power control. Context Publications: Winnipeg; 1984.
[20] Shin J, Kim B, Park J, Lee KY. A new optimal routing algorithm for loss minimization and voltage stability improvement in radial power systems. IEEE Trans Power Syst 2007;22(2):648–51.
[21] Bonnard G. The problems posed by electrical power supply to industrial installations. Proc IEE Part B 1985;132:335–40.
[22] Mithulananthan N. Hopf bifurcation control and indices for power system with interacting generator and FACTS controllers. PhD Thesis, University of Waterloo, Ontario, Canada; 2002.
[23] FACTS Technology for Open Access. CIGRE JWG 14/37/38/39-24, Final draft report; August 2000, 14 August 2000/EGC.
[24] Li N, Xu Y, Chen H. FACTS-based power flow control in interconnected power systems. IEEE Trans Power Syst 2000;15(1):257–62.
[25] Mihalic R. Power flow control with controllable reactive series elements. IEE Proc Gener Transm Distrib 1998;145(5):493–8.

CHAPTER

Internet of things (IoT) for smart energy systems

11

W.J. Miller[*,a]
*MaCT USA, Washington, DC, United States**

Private industries and governments are accelerating the adoption of, and promoting innovation for smart grid applications that can benefit smart cities. The investment in this technology is made in order to improve critical infrastructures related to power generation and distribution for smart energy systems, which would also enable customer cost savings through their efficient use of consuming devices by scheduling the appropriate time to use devices and when renewable energy, for example, is available which improve the efficiency of the power grid.

The following compares the current prospective of a centralized transactive energy approach for demand response (DR) of collecting data from smart meters and other consuming devices, which allows utilities to make decisions regarding the balance of the power grid. This approach is compared to a decentralized transactive energy approach developed in the Netherlands, known as "PowerMatcher," a open source software by the Flexiblepower Alliance Network (FAN) is used to match energy supply and demand through direct participation by consumers and prosumers in the energy market.

The transition initially started with the installation of new digital "smart meters." There were challenges in deploying digital meters and providing DR capabilities. In the United States, the federal government has provided initial funding for the digital metering transition. However, the utilities that were awarded funding were required to provide matching funds. In many cases, this cost would be passed on to consumers as a rate hike. The Public Utilities Commission (PUC) opposed the approval of rate hikes, so this has delayed the transition to metering. The smart meter equipment faced additional delays due to concerns of interoperability, which also contributed to reluctance by utilities to upgrade their existing metering equipment.

In the last few years, consumers have been told that they must switch to digital metering or would face additional charges. Many areas in the United States that received smart meters are not able to exchange data in real-time, due to the lack of wired or wireless connections. Consequently, a meter reader is required to drive

[a]ISO/IEC/IEEE P21451-1-4 (Sensei-IoT*)

Smart Energy Grid Engineering.

by the homes and buildings, read the meters with, for example, a short-range wireless link which is performed once a month. In areas where there is a direct or wireless connection, the smart meters are remotely accessible and can be read at hourly intervals. This allows utilities to be able to get a more accurate reading of average power consumption. Instantaneous values are averaged over the hour, though a momentary increase in power use will increase the hourly average.

Though instantaneous values can be obtained, the DR system could perform adequately if there is a need to shade load during peak energy demands. This has only been achievable by utilities that have access to consumer devices in homes or buildings that they could remotely control. In the past, energy control has been managed with minimal control of end devices, and by asking large consumers (such as industrial plants) to reduce their usage. However, now our peak demands have changed.

Populations around the world are growing and increasing their demand for electrical energy. With the advent of electric vehicles that require battery rechargers and installation of renewable generation such as solar panels in homes and buildings, energy consumption and generation has put additional strain on the power grid.

Energy demands are increasing as home ownership increase. In addition, solar and wind generation for homes and buildings presents a new challenge for balancing energy on the power grid because the sun and wind generation does not supply constant power. Though there has been some improvement, we are faced with exponential growth of energy usage.

The current energy data analytic model is based on a centralized approach that requires high-speed communications to all end points. The increased use of video streaming, file sharing, and other Internet services can stress communications and increase latency for real-time energy data collection. Currently, there is an increase in wired and wireless communications transitioning from 3G to 4G LTE, and advanced 5G networks, which will offer higher performance in the next 5 years.

The telecom industry recognizes that it will take time to changeover these communications systems, so fiber optic links have generally been used. Wireless services are needed in areas where fiber service is not available. This fiber expansion would be at considerable cost to consumers because wireless carriers have data plan limits that will have higher data usage with the Internet of things (IoT). Major telecommunication carriers are still waiting for the justification to build out their machine-to-machine/Internet of things (M2M/IoT) capabilities. Hindering this process are interoperability and security concerns, because protocols exist in verticals silos and do not provide data sharing and sufficient security to protect the owners privacy.

Today, DR protocols—such as MODBUS and DNP3—are widely used by utilities, manufacturing, industrial plants, and renewable generation such as solar and wind systems. The industry's direction is to adopt the IEC standards for use in the smart grid. IEC 61850 smart grid standards are slowly starting to be adopted by the power industry.

11.1 CENTRALIZED TRANSACTIVE ENERGY

Transactive energy requires utility systems with smart metering to integrate with internal billing systems, as well as generation and distribution operations. Energy transactions have existed for years between utilities that share common connections. The complexity of different systems used in the power grid has become more challenging with the addition of renewable energy sources to the power grid.

11.2 DECENTRALIZED TRANSACTIVE ENERGY

The transactive energy approach has been addressed differently in the Netherlands. They use a decentralized approach that protects the consumer's information and is based on "PowerMatcher" and the Energy Flexibility-Platform Interface (EF-PI) to interface with different sensor systems. The PowerMatcher framework in Fig. 1 offers a distributed decentralized approach where the ability to enter into a price exchange with utilities can be achieved. It evaluates energy consumption from connected devices as well as renewable generation. Europe has experienced the same energy imbalances as the United States, including the addition of electric car rechargers and renewable generation systems.

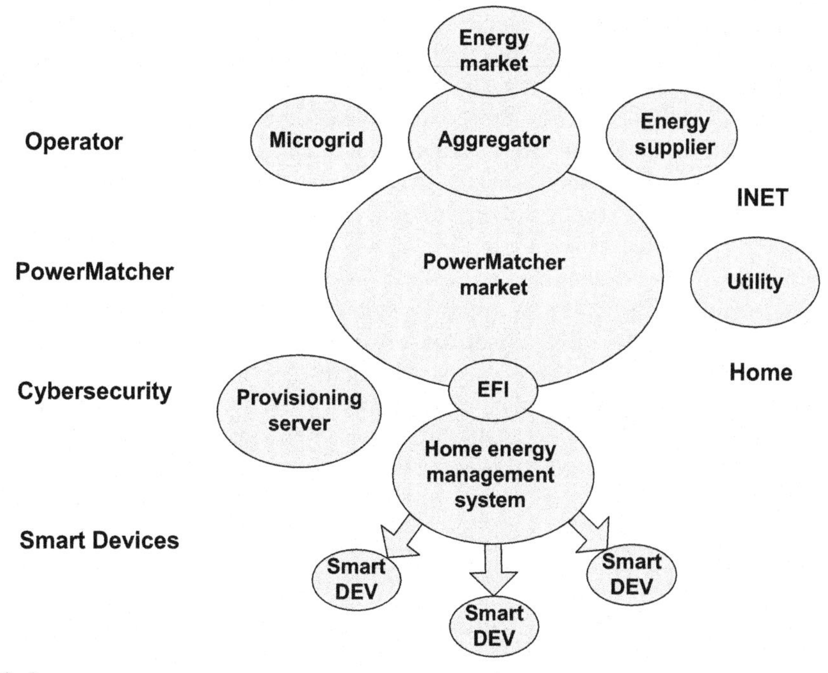

FIG. 1

PowerMatcher framework.

11.3 INSTANT SENSOR MESSAGING FOR TRANSACTIVE ENERGY

Today, smart devices such as cell phones and tablets utilize instant messaging (IM) in social networks. Social networks allow billions of people to collaborate and share information. This capability also exists with web conferencing, which provides chat, audio, and video services. The protocol use is known as eXtensible Messaging and Presence Protocol (XMPP), which was developed by the Jabber Foundation—now known as the XMPP Standards Foundation (XSF) and a Internet Engineering Task Force (IETF) standard.

XMPP utilizes a metadata approach using eXtensible Messaging Language (XML), which is a description language by the World Wide Web Consortium (W3C) that can be used effectively for many purposes including sensor networks. XMPP can be used to secure links between end points and can scalable to hundreds of millions of devices through federation of service brokers. Security is built-in with the use of Transport Layer Security (TLS) encryption, which has been accepted by the Smart Grid Interoperability Panel (SGIP) as a standard for the smart grid.

11.4 SMART TRANSDUCER STANDARD

IEEE SA has facilitated development of a standard for smart transducers. This standard, known as IEEE 1451, has been widely adopted by companies in various industries. It utilizes XML in metadata profiles known as a Transducer Electronic Data Sheets (TEDS) to describe a sensor or actuator device. This attribute is a very important element in defining the characteristics of an endpoint device.

In 2012, the International Standards Organization (ISO), International Electrotechnical Commission (IEC), and the Institute of Electrical and Electronics Engineers (IEEE) formed a joint working group of mutual interest for unique identification of devices and smart transducers (ie, sensors and actuators) monitoring together, utilizing IEEE 1451 as a starting point for new IoT capabilities that are needed, XMPP was chosen as an interface for what is now known as ISO/IEC/IEEE P21451-1-4 (Sensei-IoT*). This standard incorporates new IoT XEPs (XMPP Extensions) to provide metadata isolation and access to sensor data. ISO/IEC/IEEE P21451-1 has adopted ISO/IEC FDIS 29161, which provides unique identification for the IoT using a Jabber ID (JID) and is the identifier used for XMPP messages.

11.5 DECENTRALIZED TRANSACTIVE ENERGY

PowerMatcher is a decentralized transactive energy system that was developed in the Netherlands and will use XMPP as a common secure transport for energy transactions. This allows multiple end points of PowerMatcher nodes to communicate and participate in price exchanges with an aggregators and utilities. The end device provides a demand curve which is used in forward projected price exchanges on the energy market. The various nodes are registered with a provisioning server with XEP-0347 IoT Discovery and XEP-0324 IoT Provisioning, which allows the device owner to have control over who has access to their devices. It eliminates the concern that an intermediary would gain access to their information. The data security concern has been delaying the adoption of the smart grid and, subsequently, for users to gain the benefits of transactive energy (Fig. 2).

Performance is another key issue that will be addressed by P21451-1-4. Using XMPP, it will provide bidirectional streaming of metadata. PowerMatcher will gain the benefit of provisioning, authentication, and encryption using P21451-1-4. In addition, sensors can be registered so that information can be shared (for other purposes) with the owner's permission. This element of shared trust is critical in protecting privacy and data confidentiality for smart energy systems. Provisioning the devices provides the ability to make security decisions.

PowerMatcher's capabilities with EF-PI takes advantage of consuming devices in homes and buildings to obtain flexibility. Because renewable energy (such as solar or wind) is not consistent, its variability must be considered. PowerMatcher evaluates these variable conditions of consumption and renewable generation when it makes energy transactions. The capability injects flexibility, which translates to cost savings while achieving energy balance between energy supply and demand.

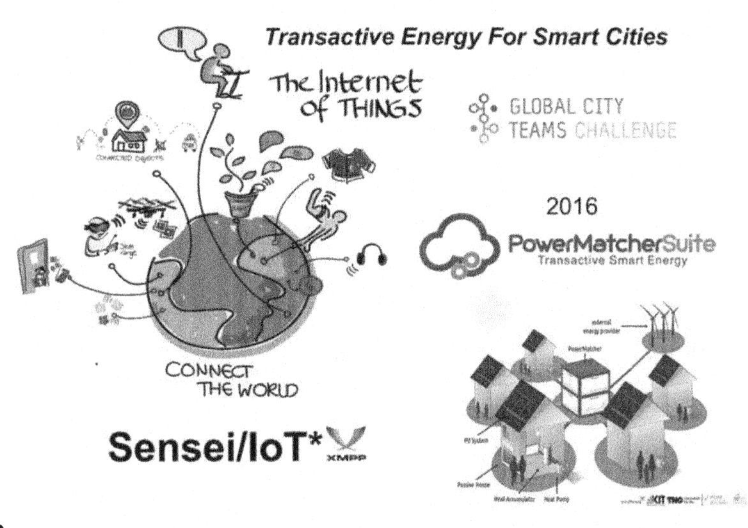

FIG. 2

Transactive energy for smart cities.

11.6 COMPARISON OF IoT PROTOCOLS

The following diagram compares the IoT protocols that have been used and are being considered by the SGIP to connect to IEC 61850-8-2 Open Field Message Bus and provides the information to an IEC 61970-301 Common Information Model.

- *Request/response* is the capacity of a client to request something from a device and have a response returned (momentarily or somewhat delayed).
- *Publish/subscribe* allows publishers to publish items to a message broker, which in turn forwards the item to registered subscribers.
- *Multicast* is the ability of multiple parties to communicate together, that is, messages are broadcast to all members of certain groups.
- *Events or push notification* is the capacity of a device to send data to an interested party without a previous request for each delivery.
- *Bypass firewall* means that messages can be sent and received even if the sender and final receiver lie behind different firewalls.
- *Federation* means that different islands can be interconnected using a federation of message brokers, and that messages can be forwarded between brokers, enabling devices on one island to communicate with devices on the other.
- *Authentication* refers to user authentication, that is, validation of user credentials in networks. This security mechanism is where messages are only allowed to be sent between approved parties, known as "friendships."
- *Network identity* means participants in communication are aware of each other's authenticated network identity when they communicate.
- *Encryption* signifies that encrypted communication is possible.
- *End-to-end encryption* refers to encrypting the communication in such a way that not even message brokers can eavesdrop on messages being transmitted.
- *Compression* is the capacity to compress transmitted data to reduce the number of bytes transmitted over the network.
- *Streaming* is the ability to send unending streams of data bytes over a channel.
- *Reliable messaging* means that the protocol has a mechanism to assure the delivery of messages and that the sender has the option to be informed of the successful receipt of the messages.
- *Message queues* are popular in back-end protocols, but are generally not supported in the available IoT protocols. It allows the ability for multiple publishers to queue items in a queue. These items are then consumed on a first in, first out basis by multiple consumers (or workers). Items are sent using reliable messaging to ensure that each item is processed exactly once.

Table 1 provides a comparison of features of REST/HTTP, CoAP, MQTT, and XMPP. CoAP is a W3C standard. MQTT is currently a standard by OASIS. XMPP now has the new standard-based IoT XEPs (XMPP Extensions), for instant sensor messaging to provide a common transport and a bridge to harmonize protocols used in the IoT.

Table 1 Supported Features for Different Protocols

Feature	HTTP	CoAP	MQTT	XMPP
Request/response	✓	✓	✗	✓
Publish/subscribe	✗	✗	✓	✓
Multicast	✗	✓	✗	✓
Events or push	✓	✓	✓	✓
Bypasses firewall	✗	✓	✓	✓
Federation	✗	✗	✗	✓✓✓
Authentication	✓	✓	✓	✓
Network identity	✓	✓	✗	✓
Authorization	✗	✗	✗	✓✓✓✓
Encryption	✓	✓	✓	✓
End-to-end encryption	✗	✗	✗	✓✓✓✓
Compression	✓	✗	✗	✓
Streaming	✓	✗	✓	✓
Reliable messaging	✗	✗	✓✓✓	2016
Message queues	✗	✗	✗	2016

Further enhancements plan on addressing the integration of a Virtual TEDS for manufacturers, which will define the characteristics of their devices in conformance to the 21451 standard (Table 2). It was also identified that sensor devices have not indicated their health status, so Health TEDS are being added to provide status information for device health conditions. This is particularly important for critical systems, such as the space industry. In summary, the IoT community recognizes that Instant Sensor Messaging offers benefits that have been difficult to achieve with other protocols.

Table 2 IoT XEPs in P21451-1-4

NEW IoT EXTENSIONS

XEP-0322 EXI Compression
XEP-0323-IoT-SensorData
XEP-0324-IoT-Provisioning
XEP-0325-IoT-Control
XEP-0347-IoT-Discovery

The new IoT XEPs are open source and can be found at http://www.xmpp.org. The use of the IoT XEPs is defined in ISO/IEC/IEEE P21451-1-4 and adopted in 2016.

11.7 CYBERSECURITY

Cyber physical security is imperative for transactive energy systems. There are different approaches between those in the United States and in the EU. The benefits have been slow in the United States due to the infrastructure's cost and the system's complexity. The EU has taken a different approach, with the introduction of PowerMatcher, to balance the power grid, where the focus is on privacy and to provide cost benefits for the end-users. It is not necessary to collect all sensor information which may expose cyber vulnerabilities to critical infrastructure and also rise privacy conerns. Though we will see continued growth of the IoT its use in smart energy systems will offer further challenges. There is work ahead to harmonize information including the identification of "Things," finding common metadata definition, assuring interoperability, and addressing cyber security concerns. It may be worthwhile to consider that there is increased risk with greater complexity. The potential risk of cyber-attack exponentially increases, and it may be more difficult to avert damages to the power grid and to protect the privacy of end-users.

11.8 CONCLUSION

Smart energy systems will take time to acquire the cognitive intelligence needed to monitor for energy disturbances. There are challenges to interconnecting various energy systems. IoT protocols are being positioned for use in smart energy systems. XMPP can provide a common transport to harmonize sensor information among different protocols. It has demonstrated its IoT scalability in use by billions of users around the world.

FURTHER READING

[1] Smart meters and smart meter systems: a metering industry perspective. http://www.eei.org/issuesandpolicy/grid-enhancements/documents/smartmeters.pdf.
[2] ISO/IEC/IEEE P21451-1-4 (Sensei-IoT*) Standard. http://www.sensei-iot.rog.
[3] PowerMatcher Suite. http://flexiblepower.github.io/.
[4] NIST big data interoperability framework, vol. 4. Security & Privacy. http://www.nist.gov/nistpubs/SpecialPublications/NIST.SP.1500-4.pdf.

CHAPTER

Design and simulation issues for secure power networks as resilient smart grid infrastructure

12

O. Mohammed, T. Youssef, M.H. Cintuglu, A. Elsayed
Florida International University, Miami, FL, United States

12.1 INTRODUCTION

The increased penetration levels of renewables and distributed energy resources (DERs) result in increased challenges in maintaining reliable control and operation of the grid. Integrating a wide variety of systems governed by different regulations, and owned by different entities, to the grid increases the level of uncertainty, not only on the demand side, but also in terms of generating resource availability. This complicates the process of achieving a generation/demand balance. Renewable energy sources vary by nature and require intelligent forecasting and prediction systems to determine how and when this energy can be used. Most of these distributed resources will be installed on the distribution network, which, in its current state, already lacks the proper communication and control network necessary to control applicable resources. Moreover, the large number and widespread use of these resources make controlling them from a central location difficult.

To overcome these problems, a deep integration between intelligent measurement nodes, communication systems, information technology, artificial intelligence, power electronics, and physical power system components must be made to manage smart grid resources. On the one hand, this type of integration can dramatically improve grid performance and efficiency, but on the other, it can introduce new types of vulnerability to the grid. The risk of vulnerability escalates when the level of integration between the physical and cyber components of the power system increases. A variety of distributed system architectures could be considered ideal for a given application when taken from a purely power-system perspective. However, determining the best choice for the distributed system architecture from a cyber-component aspect is significantly more challenging, given the communication system topology and the interoperability required for control and management capabilities, as well as the resiliency issues embedded in the cyber component of the system. The design of the cyber component must correlate with the requirements and

Smart Energy Grid Engineering.

sensitivity of the physical component, taking into account such matters as the sensitivity of protection devices for communication delay.

The security threats to the grid, due to deep integration with cyber components, are significant and widespread, taking on various forms ranging from compromising smart meters to attacking wide-area monitoring systems (WAMS) and generation control systems. The transmission system and substations represent the backbone of the grid. Attacking the WAMS and substation automation systems could lead to severe damage and blackout.

Considering this type of potential issue in the original design will lead to a more optimum design for cyber and physical components, ensuring continuity of service and system resiliency under various types of events and/or attacks. The design and optimization of such complex systems require coordination between cyber and physical components in order to obtain the best performance, while minimizing vulnerability risk. The challenge is to not only design new, secure cyber physical systems, but to also transition them from the current systems to the new design. Most of the installed components that utilize older protocols and lack security measures will last for decades. It will not be easy to replace these components due to economic factors. Securing this old equipment and maintaining interoperability with new systems are required.

First, for the co-design of such a complex system, the interaction between cyber and physical components needs to be identified. The power system control uses different types of measurements and feedback signals. The impact of compromising these types of signals on the power-system stability should be identified in order to define the level and type of security required for each type of signal. An integrated modeling and simulation framework, that presents an integrated model for cyber and physical components, is required to study this interaction, and the impact of different types of vulnerabilities and signals that compromise the power system stability. Smart grid systems could be affected by integrated modeling types of vulnerabilities from different sources. The first source of vulnerabilities comes from a lack of security measures and data integrity checks in old protocols, control, tools, and software tools. For example, most of the protocols used in WAMS and substation automation, such as IEEE C37.118, were designed for efficiency and do not have any security measures. Even for new protocols that specify some security measures, such as the IEC 62351 protocol, the control and operation requirements possess some restriction on applying these measures. For example, the encryption is not supported for generic object-oriented substation events (GOOSE) messaging, which operates at Layer 2 to meet the performance and 4 ms maximum delay restriction. The second sources of vulnerabilities could result from misconfigured systems and components. Misconfigured equipment such as default accounts, open ports, etc. can leave a back door for an attacker. The last source of vulnerability is the software and implementation bugs. Even if the system utilizes strong security standards with encryption and authentication mechanisms, undiscovered software bugs can lead to dangerous security threats. For example, the famous Heartbleed Bug that affected a large percentage of secured web servers using OpenSSL servers was related to the implementation of this bug. In this particular case, the problem was not related to the secure sockets layer (SSL) standard or the encryption algorithm. The bug came from the software implementation of the OpenSSL server. The affected servers had a buffer overflow software bug, which allowed the attacker to obtain security keys and certificates.

Since the systems vulnerability could result from system protocols, implementation of particular hardware equipment on the integrated modeling framework should be equipped with hardware in the loop, and software in the loop capabilities to test the actual components and firmware-related vulnerabilities.

12.2 SMART GRID DESIGN CHALLENGES

Several technical challenges need to be identified and addressed during the design of the smart grid architecture. These challenges are spread over many multidisciplinary areas including communication, control, and power systems. These challenges can be classified as follows.

12.2.1 INTEGRATION OF DERs

The smart grid is characterized by a large penetration of distributed and renewable energy resources. High penetrations of renewable energy increase the uncertainty of the generation of resource availability. Moreover, most of these resources will be installed on the distribution side. The current grid control model was designed to control a small number of generation stations from the centralized control systems. Applying the current control model for DERs, taking into account a large geographic area and the amount of data needed to be transferred to control centers, possesses large technical and economic challenges. The technical challenges are associated with the design of the communication system that covers large geographic areas and transfers huge amounts of raw measurements and control signals. The processing power needed to process this information in real-time, and take the necessary action in real-time, is another challenge. The implementation of such control and communication networks will be costly.

Moreover, the centralized control model suffers from the reliability and single point of failure problems. Failure of the centralized server, or communication channel, may lead to a severe system problem. To solve this issue, the control model should be changed from a centralized control to a decentralized control model. Instead of a centralized control center, which collects and processes all information locally, the system will be divided into a number of subsystems with a local intelligent controller. The local controller will process the data locally, and perform the necessary control action in their area. Neighboring area controllers will be able to exchange the information in order to coordinate the operation in the local area and support the overall system stability. Only high-level data and control commands will need to be transferred between intelligent controllers and control centers. By processing the data locally and performing local control actions, the amount of data transfer and communication bandwidth will be reduced. The required processing power for control centers will be less, and the system reliability will be improved by avoiding single point of failure. However, the optimum level of decentralized control needs to be identified. A decentralized control model can vary, from dividing the system into subareas with local controllers, as shown in Fig. 1A, to completely distributing a multiagent controller, as shown in Fig. 1B. In the first topology, the system is divided into multiple control areas with a centralized control for each area. The local area controllers are coordinating with central control. If there is a problem in any area, the other area will be able to continue operating normally.

In the second topology, the control system is based on completely distributed multiagent control. Each agent is programmed to manage local resources and coordinate with neighboring agents to reach global system goals.

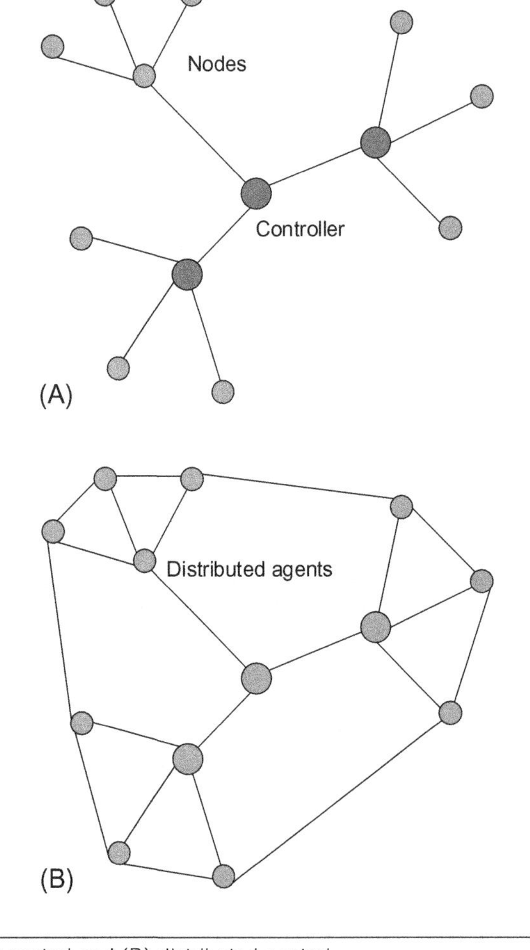

FIG. 1

(A) Decentralized control and (B) distributed control.

12.2.2 INTEROPERABILITY CHALLENGES

Integrating a wide variety of systems governed by different regulations, and owned by different entities, to the grid possesses an interoperability challenge. Several protocols and standards were developed for smart grid operation, such as DNP3 for process automation and supervisory control and data acquisition (SCADA) systems, IEC61850 for substation automation, C37.118 for phasor measurement, and OPC UA for M2M communication. In many application cases, it is required to map the data from one protocol to another. There is a lack of common data bus, or interface, that can be used by application developers to develop smart grid applications, such as energy management systems (EMS), or demand side management systems. Moreover, most of the installed components that utilize older protocols and lack security measures will last for decades. It will not be easy to replace these components due to economic factors. Developing new protocols and communication standards to meet the new requirement for smart grids should take into consideration the interoperability with older equipment and protocols.

12.2.3 COMMUNICATION CHALLENGES

The communication network for smart grid applications should consider the special requirements for real-time control. Smart grid control relies on different types of signals, and each type of signal has a special requirement regarding the bandwidth, delay, and availability. Data availability means the ability to access the right data at the right time. The communication infrastructure design should be coordinated with the control requirement. Some data are sensitive to delay, such as protection and feedback signals in low inertia systems, such microgrids with renewable energy. Other signals are more tolerant to delay, such as power consumption and smart meter data.

Communication middleware is a critical component in smart grid control. The smart grid implementation involves data exchange from local and remote nodes. These nodes represent devices manufactured by different vendors and owned by different entities. The communication middleware provides an abstraction layer that simplifies and manages the communication between different nodes without being concerned with the complexity of the hardware layers or network details. The communication middleware should also provide a wide range of quality of service profiles to meet the different needs of controllers and data types. For example, the measurement data could be discarded in case of a delay when a new sample becomes available, while the circuit-breaker states should be persistent to ensure its proper operation.

Moreover, the middleware should provide a standard application programming interface (API) to different applications and controllers. Using a standard API interface reduces the efforts needed to develop new devices and applications for the smart grid. Several types of communication middleware are available and are used in different industrial and control applications.

The modern grid is very susceptible to future expansions, thus the used middleware and communication infrastructure should be capable of handling these new expansions. An expansion process should be done without the need to redesign or modify the implemented protocols. Furthermore, the communication middleware must provide a standard communication interface to ensure interoperability among different vendors and devices.

To ensure system operation and reliability, the communication network should be protected by a proper encryption and authentication mechanism. The middleware should provide the security features embedded in the implementation in order to secure the data exchange and prevent the altering of data or violating customer privacy. Since most smart grid controllers and intelligent electronic devices (IEDs) use low power processors with limited memory and hardware resources, the middleware implementation must support a small footprint for limited resource devices.

The communication middleware can be classified into message and data centric middleware. Most old protocols use the message centric approach.

Message centric middleware approach

The data exchange in the message centric approach is based on defining a set of messages and data formats to support the expected data types and use scenarios. These messages are predefined and embedded in nodes. The unit of the information is the message; the message can contain different types of information. The communication middleware has no information about message content or data types, thus the message parsing, data filtering, and integrity check is done at the application level. Each node must be responsible for assuring the correctness of the data types it receives as any mismatch can cause malfunctioning of the application. Each node must locally track the state of the data of interest. This approach has several drawbacks. Implementing the message parsing and integrity checks on the application level puts more responsibility on the control application developer, which makes the development more complex and time consuming. Filtering the data of interest at the application layer causes poor network utilization, wasting bandwidth, and adding extra processing overhead on the application processor.

Using a set of predefined messages puts some limitations on the system expandability when expansion requires defining new data types or operation scenarios. Since the message handling is done at the application level, any change in the message formats or data types requires major changes on the application implementation. Increasing the complexity of the control application by using the message centric approach can increase the chance of software bugs, and decrease the overall system reliability.

Data centric middleware approach

In this approach, the application is developed to deal with interested data types only without any concern with the message structure or network details. The message is built by the middleware to update the change in the system state. The message structure is derived directly from the system data model. Since the message is created by the middleware in the data centric infrastructure, it will be aware of the message contents and data types. The data centric infrastructure does all message parsing, data filtering, and integrity checks at the middleware level to ensure the delivery of the correct data types and system states to all nodes.

This approach offers more capabilities over the traditional message centric approach. Moving the message processing responsibility from the application layer to the middleware layer not only simplifies application development, but also increases system reliability by reducing the numbers of error results from the different implementations of message parsing. Implementing data filtering at the middleware layer could result in more optimum utilization of the network bandwidth. The infrastructure awareness of data types makes it possible to assign different quality of service (QoS), priorities, and security levels based on data types, instead of the message in the message centric approach. Since the middleware is responsible for all message processing tasks, and the applications are only concerned with the data object, adding new data types will not require modification of existing applications. This feature is essential for an expandable system, such as a smart grid. Fig. 2 depicts both middleware approaches, with message centric on the left and data centric on the right.

Communication bandwidth challenges

In smart grid control, usually several applications are needed to access the same data at the same time. In the unicast communication, the data source sends a copy of the data for each receiver node, as shown in Fig. 3A. For example, EMS and demand side management systems could request to receive the price and consumption data measured by a smart meter. In this case, two copies of the same data will be sent over the network. If a smart appliance requests the same data, a third copy will be sent to the new appliance. The bandwidth used to send the data will increase linearly as the number of the nodes requesting the data increases. This method of communication could be suitable for local high-speed networks, and it is a simple configuration, but it is not the ideal method when considering transmitting data over a WAN or a low-speed communication line in the case of wide area measurements and phasor measurement unit (PMU) data. For the data requested by multiple readers, it is better to use a multicast communication scheme. In the multicast, the data source sends only one copy of the data for remote receiving, as shown in Fig. 3B. The bandwidth is independent of the number of receiver nodes. Only one copy will be sent over the WAN communication line. At the receiving end, the router will forward a copy from the data for each subscriber.

Modeling and simulation tools

To study the complete system behavior of modern smart grids with deep integration between cyber and physical components, a new set of tools for modeling and analysis of the complex cyber physical system is required. The currently available simulation and testbeds focus only on the physical or cyber part. The expected capabilities of simulation and physical testbeds differ significantly. Simulation is typically used to evaluate overall system performance at scale. To do so, the simulation needs to abstract away unnecessary details in order to obtain "the big picture." Physical testbeds, on the other hand, offer

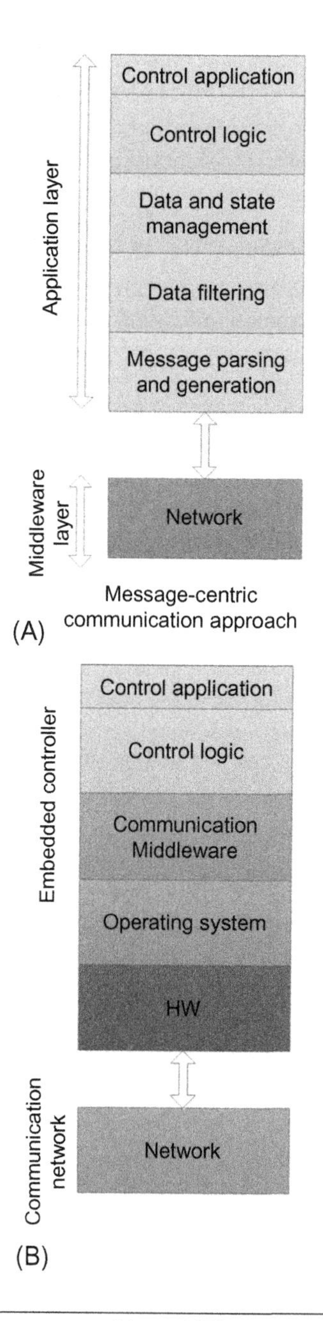

FIG. 2

Middleware approaches (A) message centric and (B) data centric.

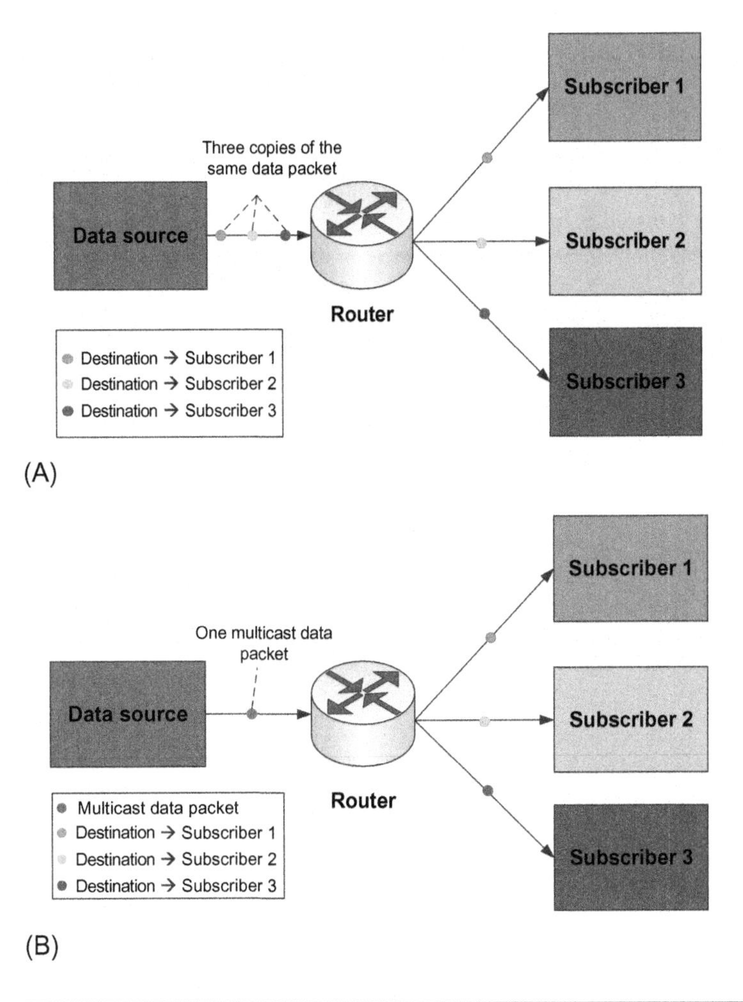

FIG. 3

(A) Unicast communication and (B) multicast communication.

the important capability of being able to operate on a real system that produces detailed responses. Physical testbeds provide the real operation of the microgrids, which can be used to evaluate the actual behaviors of the embedded systems on power system operation and stability. Because of resource limitations, however, physical testbeds cannot represent all elements of the entire large-scale, cyber-physical system. Conversely, simulation offers greater flexibility and scalability, but cannot provide operational realism. To allow high-fidelity, high-performance, large-scale experiments a hybrid software/hardware testbed environment is required. Fig. 4 shows a block diagram for the hybrid software/hardware smart grid testbed block diagram. The testbed consists of scaled power systems with four synchronous generators, a reconfigurable transmission network, and programmable loads.

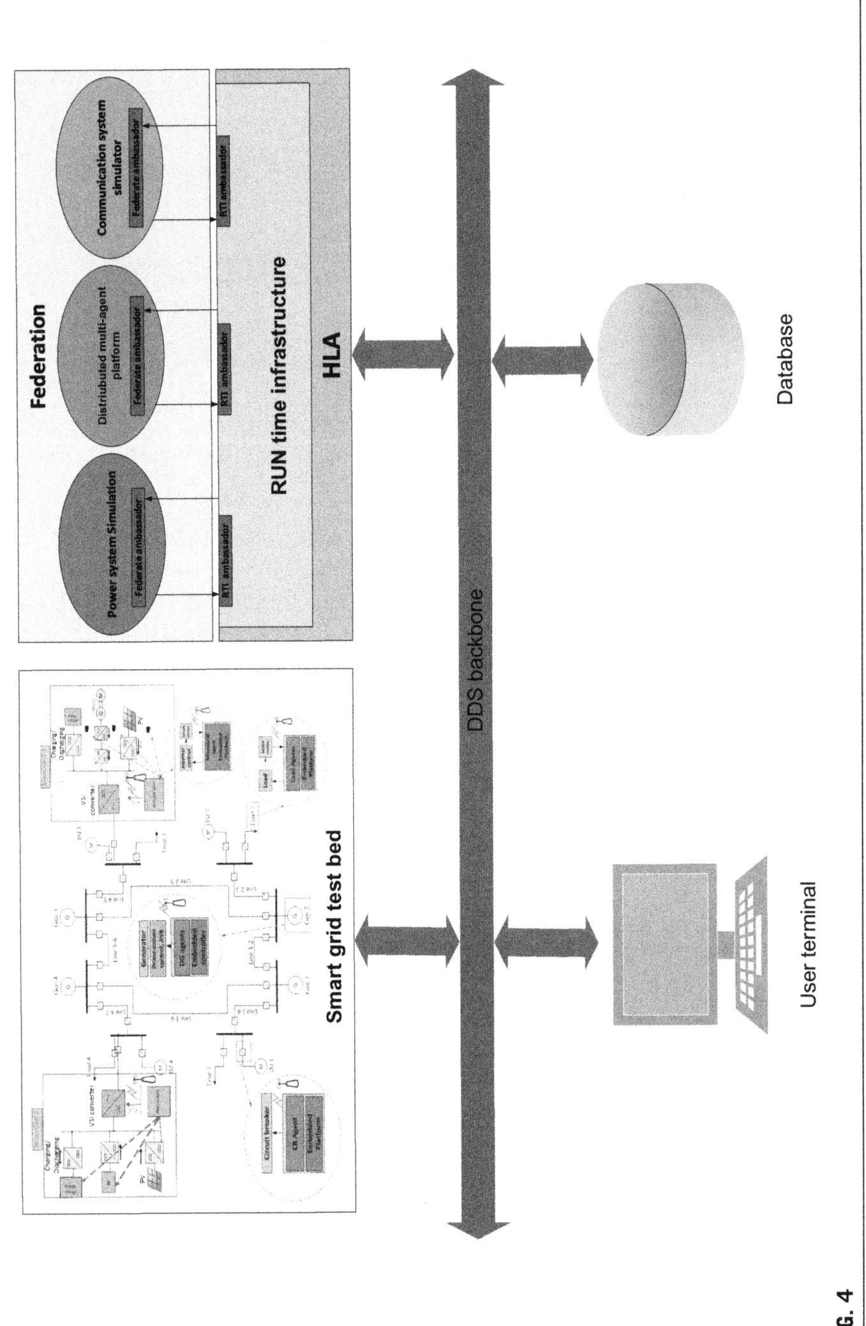

FIG. 4

Hybrid software/hardware testbed block diagram.

Renewable energy resources are emulated by wind turbine and PV system emulators. Also, the system has three microgrids with all power electronic converters emulating an electric vehicle (EV) charging park, a hybrid microgrid, and a DC microgrid. Phasor measurement units are utilized to measure the voltage and phase angle at different bus bars. Phasor measurement protocols C37.118 are used to collect the PMU data. An IEC61850 substation automation protocol is implemented for protection relays. Smart meters measure power consumption and transmit it using ZigBee protocols. Other protocols, such as RS-485 and MODBUS, are used to control the generator's programmable power supplies. To provide an interoperability layer, a data distribution service (DDS) middleware is used at the high-speed, real-time data bus. The DDS is a data centric communication middleware standard set by the Object Management Group.

The DDS uses a publisher/subscriber communication scheme, which enables peer-to-peer communication without a message broker or servers. These peer-to-peer communication capabilities enable fully distributed control and increase system reliability by avoiding single point of failure. Fig. 5 shows a comparison of a publisher/subscriber communication scheme and a client server communication.

In client/server communication, the server serves as a message broker. If the server fails, the communication between all nodes will fail. In DDS, the publisher/subscriber communication data is shared in the global data space without a message broker. Each node can communicate directly with other nodes, which eliminate the single point of failure. Eliminating the message broker also reduces latency and increases the update rate. For simulation and analyses, a multiscale and multiresolution co-simulation technique is needed to model the integrated cyber-physical system in response to various cyber and physical events. The co-simulation system seamlessly integrates various communication and power system simulators using the high-level architecture (HLA). The HLA provides the standard for federated architecture that incorporates different simulation models and simulation packages. Federated simulation, in a fully integrated cyber and physical component, can be

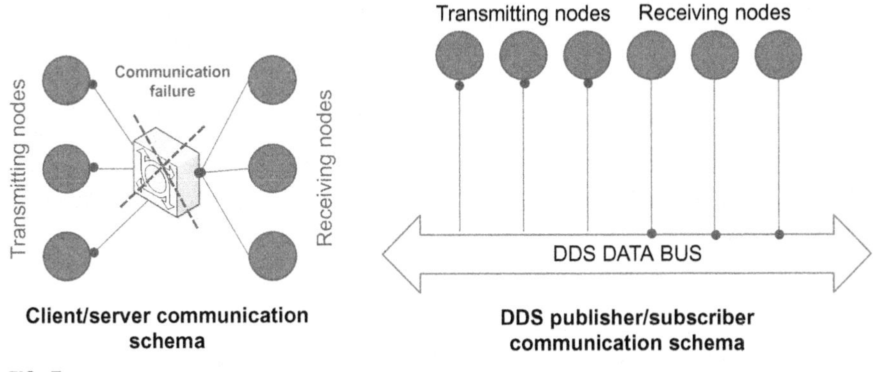

FIG. 5

Client/server and publisher/subscriber comparison.

modeled to capture the full system dynamics. The simulation model can send control/command and receive/feedback from the physical testbed components in real-time. In this way, the physical testbed capability can be extended within the simulation environment, while capturing more realistic data from real hardware. The actual implementation of standard protocols and real devices makes it possible to test and discover different vulnerabilities that may exist in the protocols or device design.

12.3 SMART GRID CONTROL INFRASTRUCTURE

The rapidly expanding infrastructure, catastrophic failures, and deteriorating attacks on power systems have led to an increased interest on the resilience and reliability of the electrical grid. Resilient grid infrastructure allows the power system to better sustain and more quickly recover. These kinds of practices include self-healing outage management, fault protection, system restoration, and general readiness. Two-way communication between different power system domains and hierarchical control levels are imperative to improve grid flexibility and control for such kind of circumstances. With the transmission of a domain layer, wide area, time-synchronized, real-time monitoring helps in understanding the overall system by obtaining normal, alert, and emergency conditions that enable wide area, closed-loop control of large-scale power system networks, and provide compensation for voltage and angle stability with time-synchronized feedback control.

Since distribution networks influence the complete power system's behavior, due to intermittent and high penetration of renewables, aggregated distributed resources (DERs) and energy storage devices will undoubtedly provide numerous benefits to the grid, such as frequency regulation and reactive power compensation. Online information exchange between transmission and distribution domains are necessary to enable wide area closed-loop control. Nowadays, distribution networks are in an on-going change to become similar to transmission networks due to bi-directional energy flow, resulting from a high penetration of DERs. Thus the power system is becoming more active and complicated as compared to the existing structure. While the system operation becomes active, the aging network infrastructure is rehabilitated by replacing or adding new components into the system, such as PMUs and IEDs. Microgrids are small-scale decentralized electricity networks featuring a variety of DERs: inverter based, synchronous, and induction generator based. As the penetration level of DERs increases, concerns regarding microgrids, during the grid-connected and isolated-mode, rise. The main challenges of microgrid operation can be roughly classified as bi-directional power flow management, adaptive protection of microgrids, dynamic reconfiguration upon prevailing conditions, open electricity market conditions, and interoperability and cyber-physical infrastructure of information and the communication network.

The control hierarchy in a power system has three major levels: primary, secondary, and tertiary. Primary control is the immediate response of the turbine governors to arrest deviations in the power system frequency; secondary control is the centrally coordinated actions taken to reinstate frequency to its nominal value; tertiary control involves the dispatch actions of generation units to move a new operating condition while the system is in balance. During emergency conditions, such as islanding and generator shutdown, the primary control action is locally taken by generators by enabling droop-controller and load-shedding schemes. However, the secondary control comes into question due to a lack of generation reserve. Although a bulk load-shedding scheme appears to be the easiest solution, in reality, the determination of nondeferrable loads is quite complex.

Therefore, instead of load shedding, an aggregated, decentralized control of active distribution networks can be used to determine the amount of deferrable load and stored energy reserve for each distribution feeder. In addition to the conventional primary control scheme, advanced secondary control can be achieved by decentralizing the active distribution aggregators. The unsolved highlighted problem requires sophisticated wide area monitoring, and a central and decentralized communication infrastructure for both transmission and distribution network domains. An aggregated control of active distribution networks for wide area power system controls would help in extreme conditions with secondary control of unintentional islanding situations. The goal of such aggregated control is to obtain self-sustained islands until the complete power system recovery is established.

Human interactions are involved mainly with the tertiary control level through smart meters of the customer's loads, or microgrid operators with an IED interface. However, during emergency conditions, the corporation of this bulk grid becomes quite complex. Introducing stochastic and uncertain process into the grid results in the necessity for more sophisticated, intelligent, and autonomous controls of the electric grid. The design and optimization of such complex systems require coordination between the cyber and physical components. Vast numbers of distributed resources should be able to operate, incorporating a large number of control functions. Multiagent-based frameworks are tools to manage complex operations inside the local grid, and its connection to the whole system. Agents are compact software packages embedded within power system devices. They are capable of operating autonomously in the decision-making process. At the same time, they are instantaneously broadcasting their operating status, informing neighboring IEDs. Intelligent power sharing, optimization of unit commitment, power quality, and self-healing protection schemes are the main concerns in agent-based decision making.

Microgrids provide many benefits to power systems by integrating intermittent and conventional DERs. The existing distribution protection coordination schemes are becoming inefficient due to continuously varying conditions, such as bi-directional power flow, frequent connection/disconnection of DERs, and islanding operation of microgrids. A remarkable research effort is being put into the design of special tailor-made protection schemes that can adapt to grid-connected and standalone operation modes of the microgrids. These kinds of protection schemes are widely favored by system operators for their fast adaption to prevailing network conditions and ease of maintenance. However, adaptive protection by itself cannot fully address the microgrid security and reliability challenges. A complete solution can be accomplished when adaptive protection and reconfiguration are provided together. Adaptive protection and dynamic reconfiguration approaches can be roughly classified as centralized and distributed methods. In the centralized approach, the optimization tools are basically applied to minimize the coordination times of the relays, while maintaining the protection coordination. The dynamic reconfiguration problem is considered an optimization problem as well, which maintains the objective as the sum of power delivered to the loads. The major drawbacks of these centralized methods are high communication capability with a powerful central controller requirement, and being fragile to single point failure, which can easily jeopardize the system with a complete collapse. Due to its more resilient nature, distributed methods draw more attention as compared to centralized methods, for both adaptive protection and dynamic reconfiguration. Agent-based distributed approaches have the ability to self-check and to accordingly react to prevailing environment conditions. The emerging international standard, IEC 61850, provides an efficient communication and interoperability framework while power system operations are in transition from centralized control to some form of decentralized functionality. Based on the logical node definition, IEC 61850 resembles multiagent systems with a horizontal and vertical messaging structure.

Interoperability is one of the major challenges in establishing a complete smart grid infrastructure, due to large numbers of data from different vendors and communication protocols. Utilities and independent system operators seek the most proper way to reach required information easily and securely for different application layers, such as metering, protection, automation, and market segments. For this reason, the NIST Framework and Roadmap for Smart Grid Interoperability Standards Release defines three major goals for establishing interoperability standards and protocols for the smart grid. According to a recently published release, it is believed that the identification of smart grid standards need to be accelerated, a robust smart grid interoperability panel should be established, and a conformity-testing and certification infrastructure needs to be created. The IEEE Std. 2030 establishes three integrated architectural perspectives: power systems, communications technology, and information technology. The guidelines also define the design criteria, the reference model applications with communication connections, and the data flows. Aligned with the future smart grid visions, it is imperative to develop comprehensively equipped testbeds to test interoperability issues. Interoperability is defined as the ability of multiple devices, or systems, to cooperate together by exchanging online information. The scope of this shared information can be very large, covering a wide range of power system practices and services. Power system applications, synchrophasor (or PMUs), IEDs, and programmable logic controllers (PLCs) are main system operator devices. It is imperative to provide accessible information from aforementioned devices and platforms. When it comes to interoperability in the smart grid, IEC 61850 is the most promising standard for future grids. IEC 61850 is the new international standard of communications, which enables the integration of all substation functions in itself, such as protection, control, measurement, and monitoring. However, IEC 61850 expands the area of influence in many parts of the power system due to a wide acceptance in the industry. Communication systems for hydroelectric power plants and DERs have been recently applied to other domains as IEC 61850 extension standards. The ability of self-describing devices and object-oriented, peer-to-peer data exchange is the most significant indicator of the superiority of IEC 61850 over other common standards. Use of names for all data, virtualized models, a standardized configuration language, and lower cabling and transducer installation cost are some of the numerous key features and benefits. Logical nodes (abstract data objects) are the main elements of the IEC 61850 object-oriented virtual model, which consists of standardized data and data attributes. IEC 61850 defines the abstract communication service interface (ACSI), which creates objects and services independent of any protocols. This enables a hierarchical class model, where all information can be accessed from a communication network using services that operate in this classes. The abstract interface allows data objects to be mapped with any other protocol, such as with a manufacturing messaging specification (MMS), or a sampled measured value (SMV) on an Ethernet data frame. The virtual model aims to express a physical (logical) device and a number of logical nodes. The IEC 61850 standardized 91 logical nodes into 13 logical groups. Each logical node contains data elements (DATA), which are standard and related to logical node function. Most of the data objects are composed of common data classes (CDC), which involve basic data objects, status, control, and measurement. Each data element consists of a number of data attributes with a data attribute type (DATA Type), which belongs to functional constraints (FC).

PMUs are deployed to provide a wide-area situational awareness by displaying the disruptions that arise over a large geographic area. Aimed transmission points can be observed by time-synchronized collected data from PMUs with a phasor data concentrator (PDC). A PDC can be a hardware- or software-based communication station, where PMUs provide time-synchronized measurements. The collected information is sorted according to time-stamp values. This enables a comparable real-time monitoring infrastructure with a high-precision sampling. The collected data can be stored in a database system for future analysis. Transmission network operators seek new ways to use the PMU measurements for power system operation and control, as well as for wide area monitoring. For instance, PMU data can be utilized for generator excitation and power system stabilizer input, islanding of a substation process, or as a flexible AC transmission device. Therefore, the time-synchronized data should be mapped to other related domains in order to be readily and accurately utilized as online feedback control.

The IEEE Std. C37.118.1 and C37.118.2 define synchrophasor measurements and data transfer for power systems, respectively. A PMU supports high reporting rates, which can go up to 60 frames per second. Although nowadays many substation vendors are moving away from serial to Ethernet-based IEC 61850, a vast number of old communication protocols for legacy power system automation devices have been installed on the field. Most of these devices are vendor specified interfaces with low bandwidth, and are not able to expend the network. Since it is not wise to get rid of these at this time, these devices should be operated and incorporated with new protocols and devices. Modbus is the most popular open industrial protocol widely used, due to its simplicity and ease to use. Numerous PLCs deployed in the industry use the Modbus protocol. It uses master-slave communication, where the master sends a message that includes a targeted slave-device address. All the devices in the network receive the message, however only the device with the correct address responds. The most common Modbus versions are ASCII, RTU, and TCP/IP. Object linking and embedding for process control (OPC) was originally released for the purpose of abstracting various PLC protocols into an interoperable interface for a secure and reliable data exchange. The advent of the smart grid interoperability efforts led to the development of open connectivity unified architecture (OPC UA), while keeping all the functionality of a classic OPC, but switching from Microsoft-COM/DCOM technology to state-of-the-art Web service technology. OPC UA is not directly compatible with Classic OPC, since they use different technology. OPC UA uses a framework based on client and server architecture, in which the server provides real-time process data to clients. Moreover, it can be implemented with Java, or .NET platforms, thereby eliminating the need to use Microsoft Windows-based platforms. This provides a perfect opportunity to model multiagent-based systems (MAS) on Unix/Linux running systems.

In open market conditions, utilities are no longer monopolized, hence DER stakeholders are private entities that can compete with regional utilities. The electricity trade can be handled locally in a region by several DER investors or by cross-border trading by utility wholesale markets and neighboring regional systems. In distribution networks, feeder-based electricity trade is more reasonable, avoiding transmission routes and loss costs. A vast number of DERs should be able to operate incorporating a large number of complicated operational functions. New generation grids require fast intelligent decision-making algorithms and a communication infrastructure, since power system operators will be inefficient at dealing with highly active and changing operations in the future grid. Multiagent-based decision-making algorithms, incorporating with real-time communicative devices, offer a reliable solution for unit commitment in competitive market conditions. Agent-based marketing models have applications on a variety of scales from large bulk generation units to small, home-based entities. There has been a growing interest in agent-based market scenarios that maximize the benefits of generation units by analyzing the effects and evaluation of bidding strategies.

12.3.1 MICROGRID CONTROL

Microgrids offer a significant solution for enabling a resilient grid infrastructure, since they have the ability to continue operating in case of a utility outage. A microgrid can be viewed as a cluster of DERs, which have a connection to the main utility grid. Microgrids could be composed of conventional or renewable DERs. Microgrids mostly depend on intermittent renewable resources. This may introduce important stability problems especially during islanded operations. Minor stability problems can cause cascading outages, which would result in a large-scale blackout if proper action was not taken in time. An effective way to prevent such risks, and to ensure robustness and resiliency, is to operate the grid in smaller regions by splitting the network when necessary.

The islanding process can be classified as intentional or unintentional. The unintentional islanding process is most likely due to the disconnection of the microgrid when a fault or utility outage occurs in the system. A microgrid should continue to operate after it is islanded by maintaining the voltage and frequency in the islanded area with the appropriate control techniques. A facility island can be assumed as one single source in a microgrid, which is connected through only one circuit breaker to the main utility grid. A secondary island can be considered as multiple DERs with customers connected to the secondary side of a distribution transformer. Microgrid functionality can be considered in four modes: grid-connected, transition-to-island, island, and reconnection.

Grid-connected operation is the mode in which the microgrid is coupled to the main utility grid in order to deliver/receive power to/from the grid. In the grid-connected mode, the frequency and voltage of the microgrid can be supported by the stronger power system. Therefore, a complex regulation control is not necessary. The transition process from the grid-connected to the islanded operation mode should be smooth, that is, not violating power quality indicators, such as voltage and frequency levels. Microgrids should be able to provide the real and reactive power requirements of the loads in the islanded area while regulating the voltage and frequency within limits. A genetic microgrid infrastructure is shown in Fig. 6.

Control of islanded microgrids

Control of the voltage and frequency subsequent to the islanding operation of a microgrid is a major challenge for proper operation. In islanded microgrids, conventional DERs have a slow response to load changes compared to inverter-based DERs due to their high inertia. Inverter-based DERs, which have power electronics interfaces, have a faster response to load changes than conventional DERs. Most of the DERs cannot connect directly to the grid since renewable energy generation is mostly based on DC sources, such as solar panels. Therefore, a power electronic interface is required to enable the AC grid side connection. DERs and combined battery storage systems contribute to microgrids by enabling transient power sharing.

Frequency is directly related to the active power balance in the system. Similarly, voltage is regulated by a reactive power balance. The unintentional islanding process can cause severe problems due to unexpected frequency and voltage stability issues if either active, or reactive, power imbalances exist between the microgrid and the main utility grid. Load shedding is one of the easiest solutions for disconnecting the required amount of a load inside the microgrid in order to achieve supply and demand balance. However, many systems involve critical loads for which a shortage of electricity is prohibited. In this study, it is assumed that the generation capacity is enough to meet the load demand in the islanded area. Frequency and voltage should be kept within limits for the reconnection mode (between 59.3 and 60.5 Hz). Although the

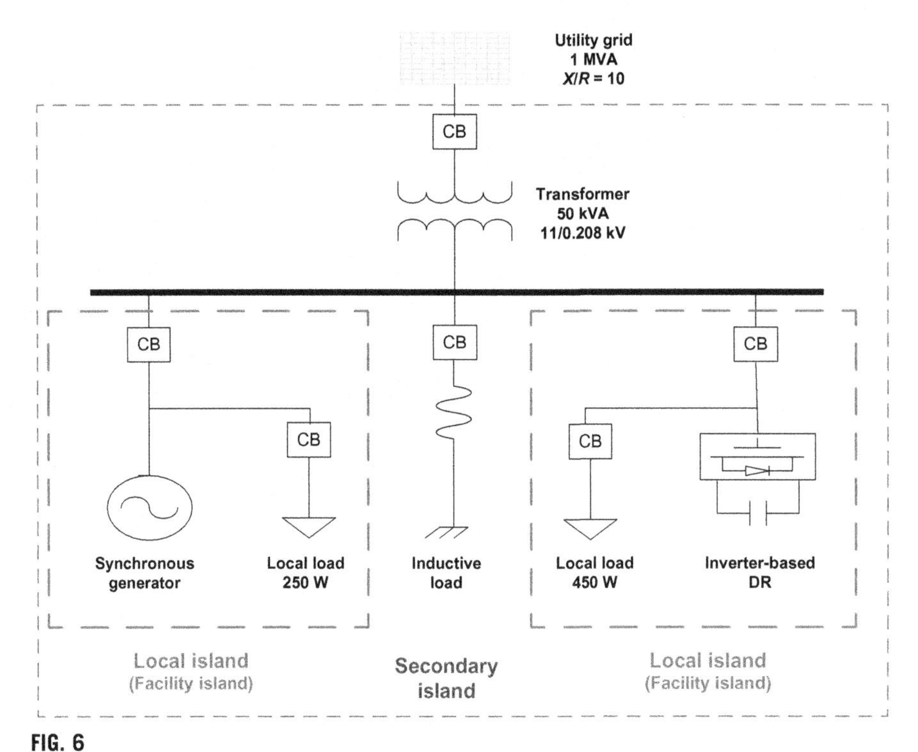

FIG. 6

Microgrid architecture.

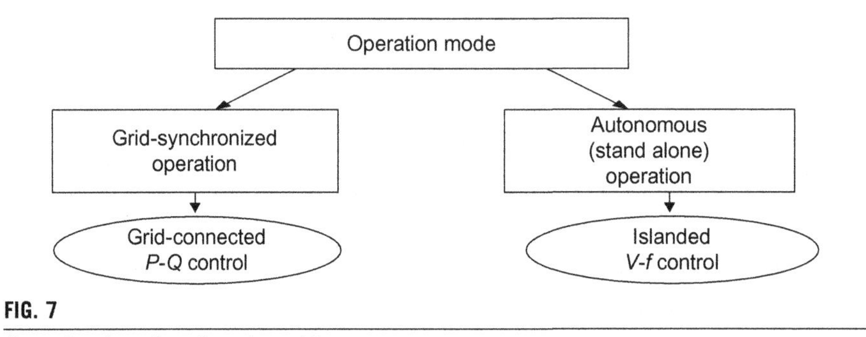

FIG. 7

Operational modes of a microgrid.

ultimate goal is to regulate frequency and voltage, the control of the synchronous generator and inverter-based DER are different. Once unintentional islanding occurs, the main grid support to the microgrid is no longer available. According to the scheme, the DER should be able to switch operation modes according to predetermined islanding conditions to regulate frequency and voltage in the islanded area. Operation modes of islanded microgrids are shown in Fig. 7.

If the form of the island is local (facility), where only a single generation unit exists, the DER should be switched to the isochronous mode of operation in order to supply all the electrical power of the host facility. Alternatively, if the form of the island is secondary, where multiple generation units will exist, a proper load sharing mode should be applied. If the mismatch in power demand is less than what a generation unit can handle, one can be selected as an isochronous machine to regulate system frequency and voltage. Otherwise, droop control can be applied to the generation units for the desired load sharing operation.

Operation modes of DERs

The major challenge is to decide the appropriate operation mode of each individual DER while taking into consideration intermittent, renewable, and limited fuel resources.

Synchronous generator grid parallel operation

Synchronous generators can operate in *P-V* (constant active power and terminal voltage) or *P-Q* (constant active and reactive power) modes while in a grid-connected operation. The synchronous generator follows the grid frequency and delivers the required power to the grid by applying the required amount of torque calculated by the generator governor control. In *P-V* mode, the synchronous generator excitation system arranges the current flow from the excitation field windings to regulate the voltage by generating the required reactive power. In this paper, studies are carried out using the *P-V* control mode. The synchronous generator is equipped with an automatic voltage regulator (AVR) in grid-connected and islanded operations.

Synchronous generator islanded operation

Subsequent to the islanding process, if a synchronous generator is operating in *P-V* or *P-Q* mode, a power mismatch would appear between the supply and demand. This would cause a frequency drop or increase directly related to the power mismatch ratio. For a stable operation, supply and demand should exactly match inside the islanded area. The voltage is regulated by the AVR, if reactive power demand is within the synchronous generator limits. In order to prevent system frequency collapse, the synchronous generator should be switched to the isochronous operation mode. In this mode, the generator simply follows the speed or frequency reference, which is defined for the microgrid operating condition. By increasing the applied torque, according to the speed error reference, the active power supplied is adjusted. If the microgrid is delivering power to the utility side, the speed reference is negative, and the governor decreases the applied torque to regulate the generator frequency. *V-f* control of a synchronous generator is shown in Fig. 8.

The synchronous generator motion and excitation equations in the $dq0$ domain are given as:

$$\frac{d\delta}{dt} = \omega_0 \Delta\omega \tag{1}$$

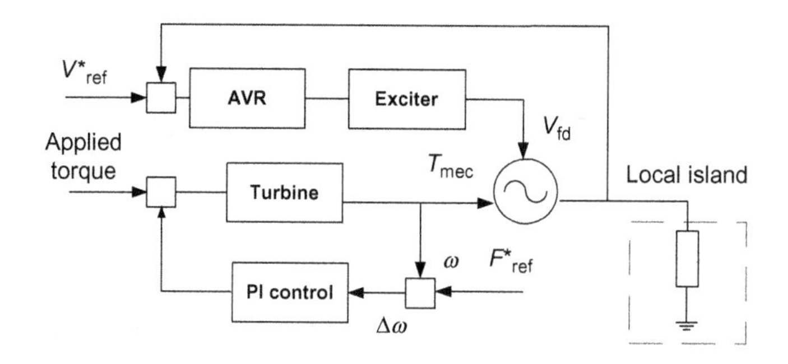

FIG. 8

Synchronous generator standalone (*V-f*) operation.

$$\frac{d\omega}{dt} = \frac{1}{2H}(T_{\mathrm{m}} - T_{\mathrm{e}} - K_{\mathrm{D}}\Delta\omega) \tag{2}$$

where δ is the rotor (load) angle, ω_0 is the nominal synchronous speed, ω is the rotor speed, H is the inertia constant, K_{D} is the damping factor, T_{m}, and T_{e} are the mechanical and electrical torques, respectively. Electrical output torque, terminal active and reactive power:

$$T_{\mathrm{e}} = \frac{V_d i_d + V_q i_q}{\omega} \tag{3}$$

$$P = \omega_0 T_{\mathrm{e}} = V_d i_d + V_q i_q \tag{4}$$

$$Q = V_q i_d - V_d i_q \tag{5}$$

where i_d, i_q are the currents on the d-axis and q-axis. V_d and V_q are the d-axis and q-axis output voltages.

Inverter-based DR grid parallel operation

Based on the defined control, an inverter-based DER can operate whether in grid-connected or islanded mode. If the inverter is operating in a grid-connected mode, voltage and frequency will be regulated by the grid. Therefore, the inverter is responsible for delivering reactive and/or active power according to the current control strategy. In instantaneous power theory, on the synchronous reference frame, the state-space representation of the injected current and system voltage relations are given as Eq. (6).

$$\frac{d}{dt}\begin{bmatrix} i_a \\ i_b \\ i_c \end{bmatrix} = \frac{R_{\mathrm{s}}}{L_{\mathrm{s}}} V_{\mathrm{g}} \begin{bmatrix} i_a \\ i_b \\ i_c \end{bmatrix} + \frac{1}{L_{\mathrm{s}}}\left(\begin{bmatrix} V_{ga} \\ V_{gb} \\ V_{gc} \end{bmatrix} - \begin{bmatrix} V_a \\ V_b \\ V_c \end{bmatrix}\right) \tag{6}$$

L_{s} and R_{s} represent the resistance and inductance of the filter. V_{g} is the grid voltage. When Park's transformation is applied, the state equation can be expressed in a $qd0$ frame as:

$$[i_{qd0}] = T[i_{abc}] \tag{7}$$

$$\theta = \omega t + \theta_0 \tag{8}$$

$$T(\theta) = \frac{2}{3}\begin{bmatrix} \cos(\theta) & \cos\left(\theta - \frac{2\pi}{3}\right) & \cos\left(\theta + \frac{2\pi}{3}\right) \\ \sin(\theta) & \sin\left(\theta - \frac{2\pi}{3}\right) & \sin\left(\theta + \frac{2\pi}{3}\right) \\ 1/2 & 1/2 & 1/2 \end{bmatrix} \tag{9}$$

where ω is the angular frequency, θ is the electrical angle, and θ_0 represents its initial value.

$$\frac{d}{dt}\begin{bmatrix} i_d \\ i_q \end{bmatrix} = \begin{bmatrix} -\frac{R_s}{L_s} & \omega \\ -\omega & -\frac{R_s}{L_s} \end{bmatrix}\begin{bmatrix} i_d \\ i_q \end{bmatrix} + \frac{1}{L_s}\left(\begin{bmatrix} V_{gd} \\ V_{gq} \end{bmatrix} - \begin{bmatrix} V_d \\ V_q \end{bmatrix}\right) \tag{10}$$

The power injected into the grid for a three-phase system results in the following:

$$S = P + jQ = 3V_{qd}I_{qd}^{*} = 3\left(\frac{V_q - jV_d}{\sqrt{2}}\right)\left(\frac{I_q + jI_d}{\sqrt{2}}\right) \tag{11}$$

$$\begin{aligned} P &= \frac{3}{2}(V_q i_q + V_d i_d) \\ Q &= \frac{3}{2}(V_q i_d - V_d i_q) \end{aligned} \tag{12}$$

Inverter-based DER islanded operation

When the islanded condition is detected, but the inverter still operates in grid-connected mode, the phase locked loop (PLL) will fail to operate since the utility grid reference will no longer exist. This situation can cause severe conditions, and may interrupt the microgrid services. In order to improve resiliency and maintain continuous service, the inverter controller must be automatically reconfigured as a voltage control upon islanding detection. The reconfiguration process can be triggered by the trip signal from the islanding detection algorithm, which is basically a comparison of the DER and utility phase angle difference.

Fig. 9 shows the block diagram for the proposed reconfigurable converter control algorithm. If the grid signal is no longer available, the inverter will not be able to track the grid frequency reference. Thus the control will be switched to generate its own internal frequency reference, and to operate in voltage-control mode on the terminal point by the adjusted modulation index. The system's voltage and frequency will be regulated by supplying the required amount of active and reactive power to the local load.

In the three-phase VSI (voltage source inverter) SPWM (sinusoidal pulse width modulation) technique, the maximum amplitude of the fundamental phase voltage in the linear region ($m_a \leq 1$) is $V_{dc}/2$, where the modulation index is m_a. The maximum amplitude of the fundamental AC output line voltage is:

$$V_{ab} = m_a \frac{\sqrt{3}}{2} V_{dc} \tag{13}$$

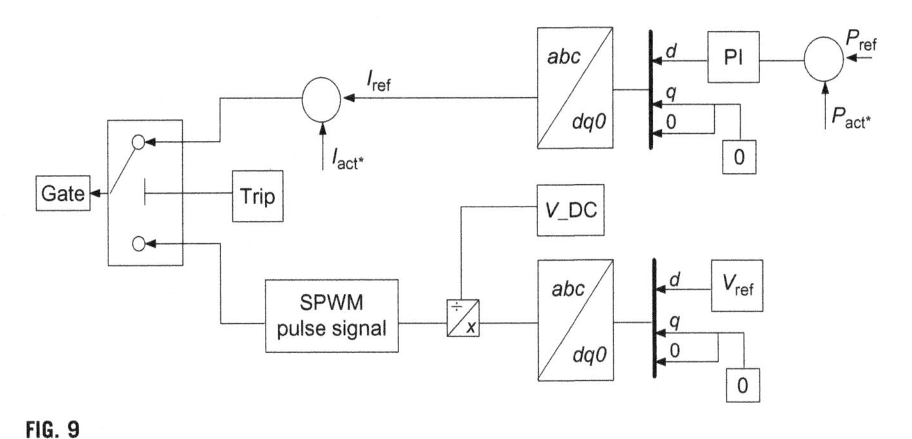

FIG. 9

Reconfigurable converter control.

Inverter-based DER and synchronous generator secondary island operation

In a secondary island situation, multiple DERs are connected to the same system feeder or bus. In this case, one of the generation units should take the responsibility to regulate frequency and voltage in the islanded area. Since the inverter has a faster response and the ability to track system changes, in this study, the inverter-based DER is utilized to operate in the isochronous mode. The required amount of active and reactive power will be injected into the inverter-based DER, in order to cover the power mismatch between the supply and demand. In addition to the basic current control strategy, one more control loop is implemented to regulate the frequency and voltage.

$$\begin{aligned} i_d^* &= (V_{\text{ref}} - V_{\text{meas}})\left(K_{\text{pv}} + K_{\text{iv}}/s\right) \\ i_d^* &= (f_{\text{ref}} - f_{\text{meas}})\left(K_{\text{pf}} + K_{\text{if}}/s\right) \end{aligned} \tag{14}$$

Study system model in simulation

The system shown in Fig. 6 is modeled in the simulation environment to investigate possible scenarios. The studied system consists of one conventional synchronous generator model, which can be considered as a diesel or gas turbine-driven generator, and one inverter-based DER model with combined storage. Each DER is modeled for flexible operation, whether in grid-connected or islanded mode. This control structure is achieved by switching the operation modes of the generators from the (P-Q) mode to the (V-f) mode. The synchronous generator is equipped with excitation, an AVR, and governor systems. Both DERs are coupled to the main utility through circuit breakers. Each generation unit has a local load of 250 and 450 W, respectively, directly coupled to a point of common coupling (PCC) of each DER. The utility grid is connected to the microgrid through a 50 kVA 11/0.208 kV transformer.

Grid-connected operation mode

In this case, the synchronous generator and the inverter-based DER are operating in the (*P-Q*) grid-connected mode. The frequency and voltage are regulated by the main grid. Fig. 10 shows the operating conditions prior to the islanded operation. Since the synchronous generator is equipped with an AVR, the voltage on the generator bus is regulated to a constant reference point through an excitation system by injecting the required amount of reactive power.

Creating a secondary island

At the 5th second, a secondary islanding case was created by opening the circuit breaker at the grid's PCC. The resulting island consists of a synchronous generator, an inverter-based DER, and local loads. An instant power mismatch occurred between the generation and system load. If an unintentional islanding protection method is not enabled, due to this mismatch between generation and demand, the

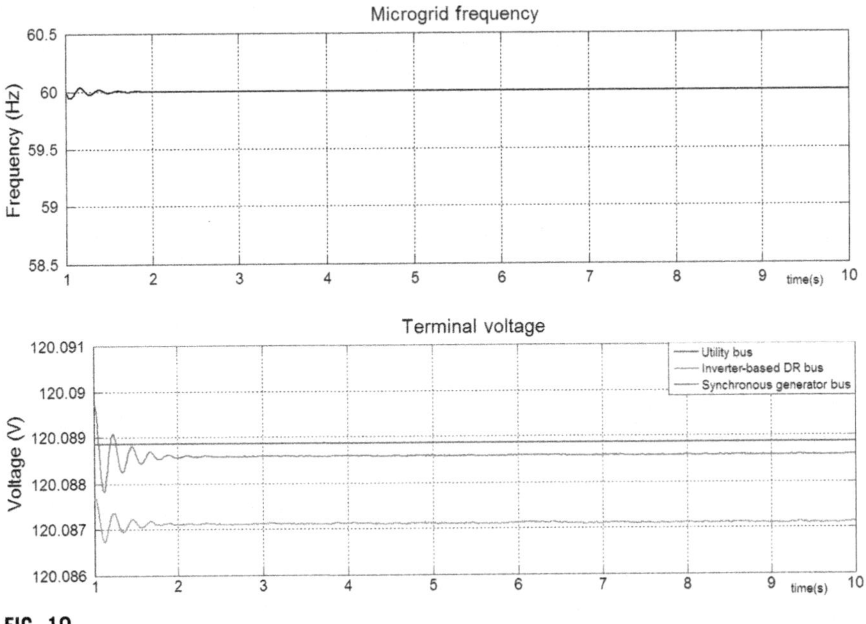

FIG. 10

Grid-connected operation.

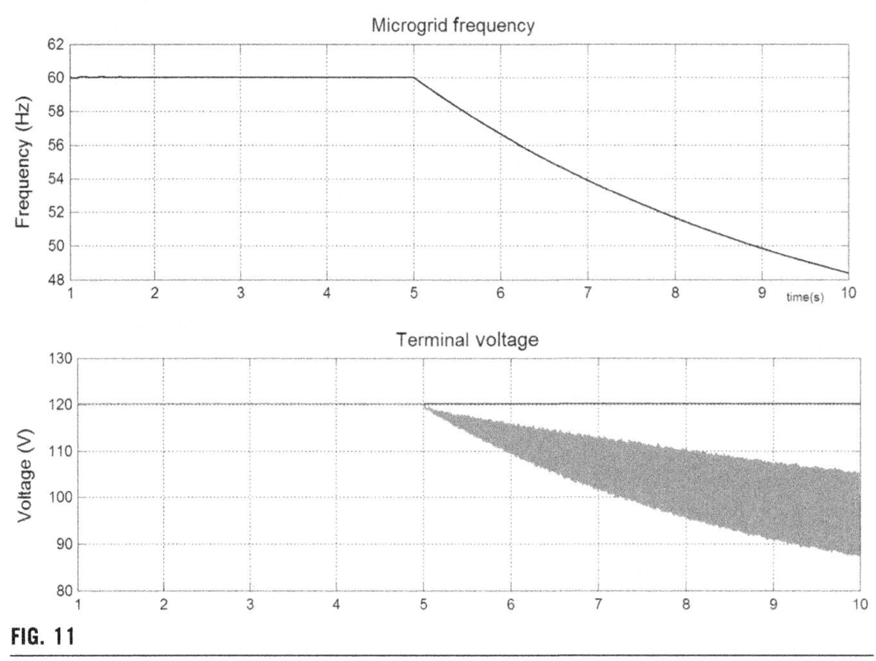

FIG. 11

System collapse.

system will be subject to a frequency and voltage collapse. Fig. 11 shows the system's frequency and voltage collapse, as expected.

The frequency measured from the system bus starts to drop gradually, depending on the system inertia. This results in a tripping of the generator due to under frequency, phase angle difference, or rate of change of frequency (ROCOF) relays. The frequency drift of the island, with respect to the power mismatch, is given by Eq. (15). The system inertia and mismatch ratio are the main effects, which accelerates the frequency change rate during the islanding process.

$$\frac{df}{dt} = \frac{\Delta P}{2HG} \mathrm{f} \tag{15}$$

where ΔP is the power difference between the load and generation [W]; H is the moment of inertia generator/system [W s/VA]; and G is the rated generation capacity of the generator [VA].

The same scenario is repeated by adding the proposed islanding control [enabling the automatic switching of the inverter-based DER operation mode from (P-Q) to (V-f)]. This is enabled by the islanding detection relay modeled in the simulation environment. Under frequency, phase angle difference, and ROCOF methods are used to enable islanding detection. The synchronous generator remains in (P-Q) mode. Fig. 12 shows the measured frequency, voltage, and active power for each unit.

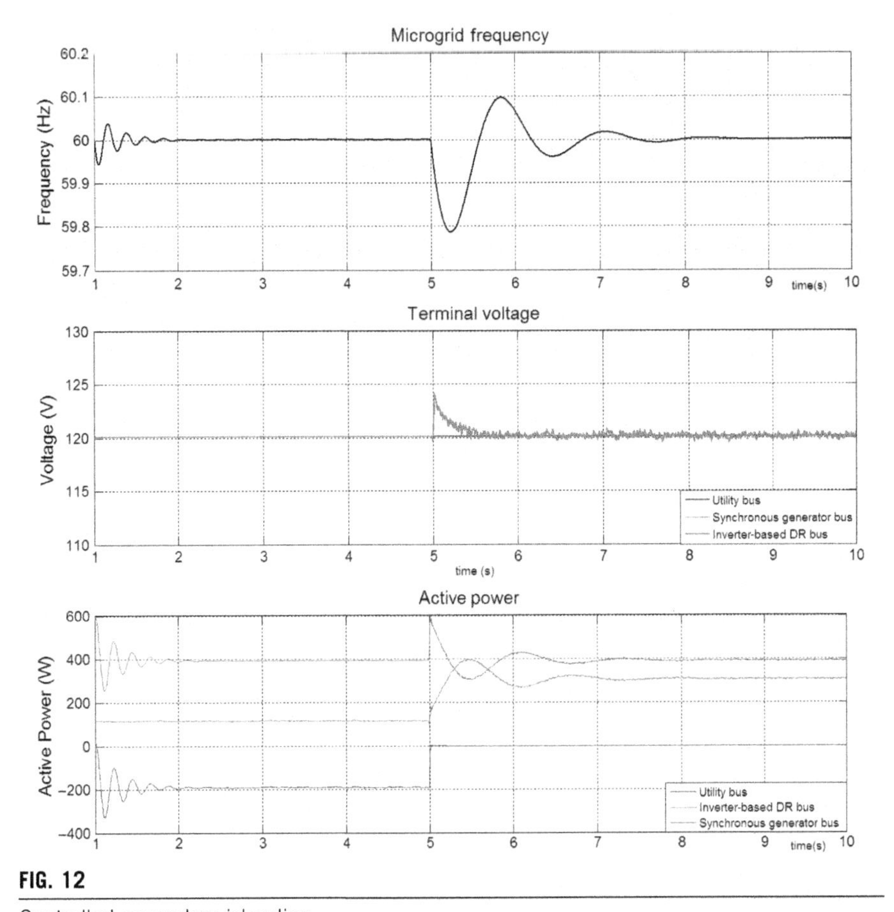

FIG. 12

Controlled secondary islanding.

It is noticed that the system frequency experiences an oscillation during the transition which eventually settles to the reference value. The terminal voltage experiences temporary spikes during transition as well, and it settles within limits due to the reactive power control inside the island. The delivered active power from the synchronous generator remains the same. The power drawn from the grid becomes zero due to disconnection. The inverter-based DER starts to generate the required amount of active power to keep the frequency within limits.

Creating local (facility) islands

In this case study, while the system is operating in secondary island mode, two separate local islands are formed by opening two circuit breakers on the PCC of the synchronous generator, and the inverter-based DER. The synchronous generator is switched to speed (frequency) control mode. The required amount of power supplied is adjusted by applying torque to the generator shaft, which is regulated by the speed governor according to the frequency error input. In the inverter-based DER local island control, neither the grid nor the synchronous generator frequency signal is available for the PLL to track. Therefore, the inverter-based DER operation mode is switched from the (*V-f*) mode to voltage control. During this transition, the generator bus frequency experiences a sharp spike since the generated power is more than required. By employing speed control regulation, the generated power is forced down to match the local load requirement. Unlike the synchronous generator, the generated power by the inverter-based DER is less than the local load demand. The power drawn from the grid remains zero. The terminal voltage of the synchronous generator is adjusted by the generated reactive power and by the AVR control. The predefined modulation index determines the terminal voltage of the inverter-based DER, as is illustrated in Eq. (13). Fig. 13 shows the measured frequency, voltage, and active power.

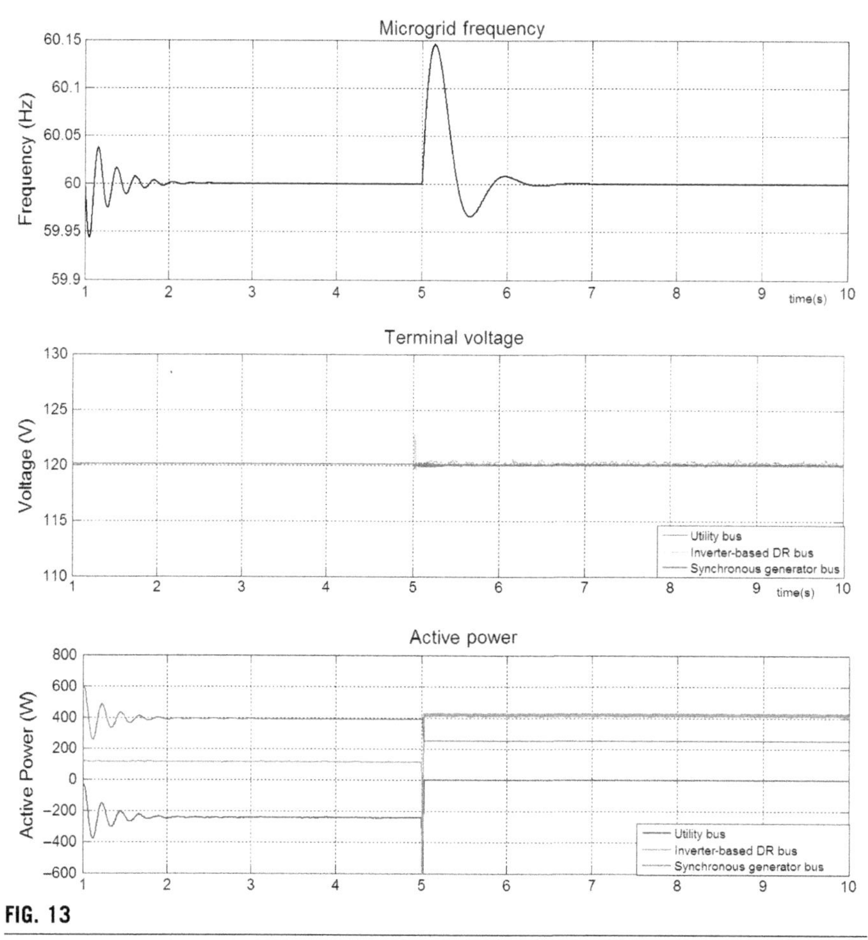

FIG. 13

Controlled local islanding.

FIG. 14

Microgrid experimental setup.

Study system model in real-time experiment

The same model is implemented on a small-scale power system at Florida International University. Fig. 14 shows a hybrid microgrid structure used for real-time experiments [1].

Grid-connected operation mode

In this experiment, a synchronous generator and an inverter-based DER synchronize with the grid to supply some portion of local load demand. The remaining power is drawn from the grid as in the simulation scenario. Fig. 15 shows the system frequency and active power measurement on the synchronous generator bus. The system frequency is 60 Hz. The delivered active power from the synchronous generator is around 400 W, and power drawn from the grid is around 250 W. Fig. 16 shows that the active power injected into the grid by the inverter is around 110 W, and terminal voltage is constant at 120 V.

Creating a secondary island

At the 10.5th second, the secondary island is created by opening a circuit breaker between the external grid and the microgrid. Islanding is detected when frequency drops to a defined value, which triggers the switching inverter-based DER to operate in an isochronous mode. The inverter-based DER is assigned to operate in an isochronous mode to regulate system stability by increasing the amount of active and reactive power so that it regulates system frequency and voltage.

The secondary island experiences deep frequency decreases when utility support is not available. Fig. 17A shows the recovery of a system frequency change by

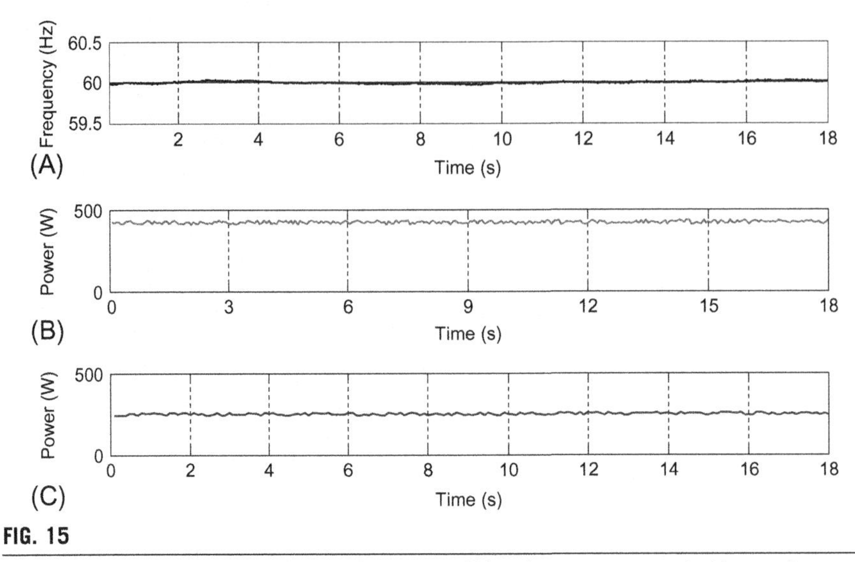

FIG. 15

Grid-connected operation. (A) System frequency, (B) active power generated by synchronous generator, and (C) active power received from utility grid.

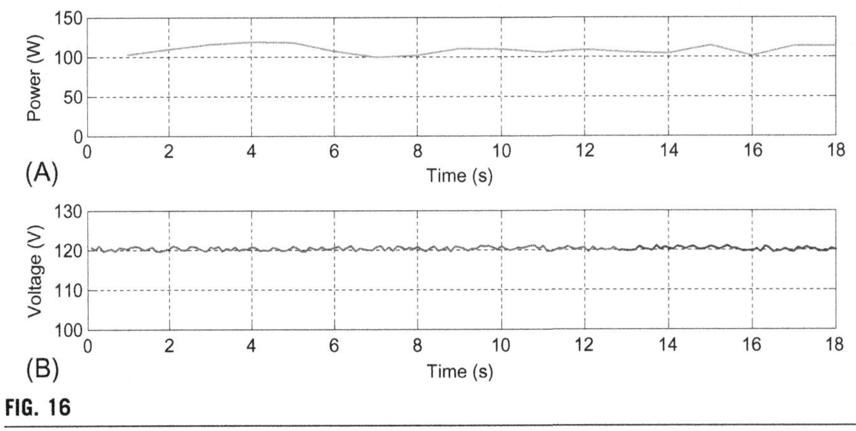

FIG. 16

Grid-connected operation. (A) Active power generated by inverter-based DER and (B) system voltage.

settling on a constant acceptable limit. Frequency decreases to 57 Hz during the detection and recovery process. Frequency is regulated gradually up to 59.5 Hz, which is within the accepted limits according to IEEE 1547.4. Fig. 17B shows that the active power exchanges with the grid dropped to zero. This shows that the utility support was no longer available. The active power generated by the synchronous generator remains almost constant at 420 W, as shown in Fig. 17C.

The inverter-based DER takes responsibility for recovery. Fig. 18A shows the power system values on the inverter-based DER bus. The active power injection increases from 150 to 350 W. Fig. 18B shows the voltage drop and settling around 112 V.

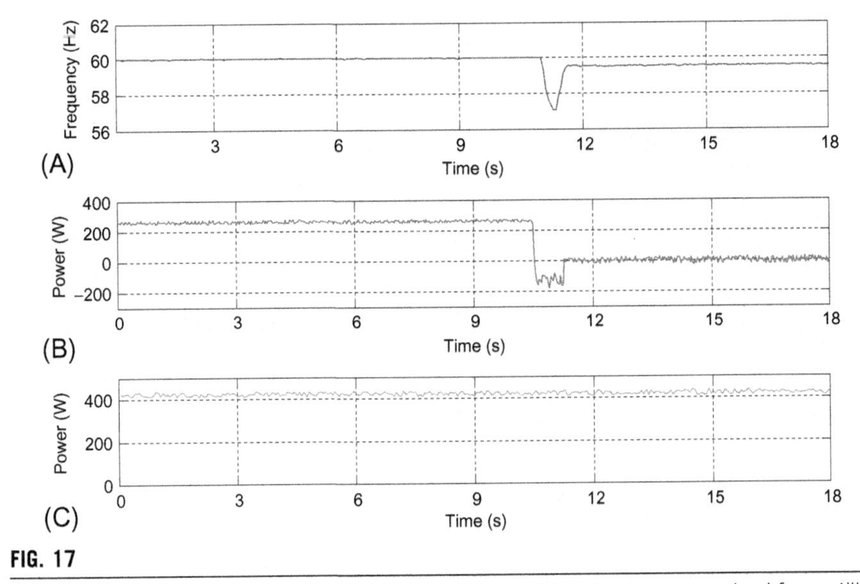

FIG. 17

Secondary islanding operation. (A) System frequency, (B) active power received from utility grid, and (C) active power generated by synchronous generator.

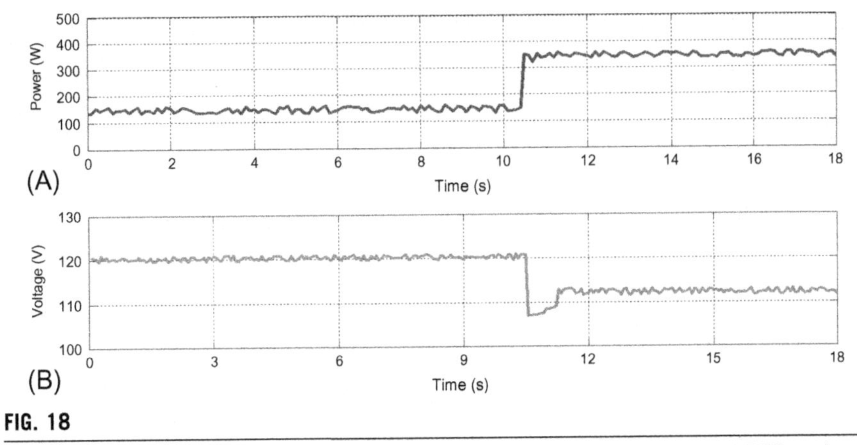

FIG. 18

Secondary islanding operation. (A) Active power generated by inverter-based DER and (B) system voltage.

Creating local (facility) islands

Local islands are created for a synchronous generator and an inverter-based DER. There is no active power injection from the external grid. The synchronous generator gets separated from the inverter-based DER. Once the local islanding is detected, then the synchronous generator is switched to the frequency regulating mode. Each generation unit starts to supply its own local load. The local load for a synchronous generator, in this case, is around 250 W. The synchronous generator decreases the generation from 400 to 250 W, as shown in Fig. 19A. This load rejection is associated

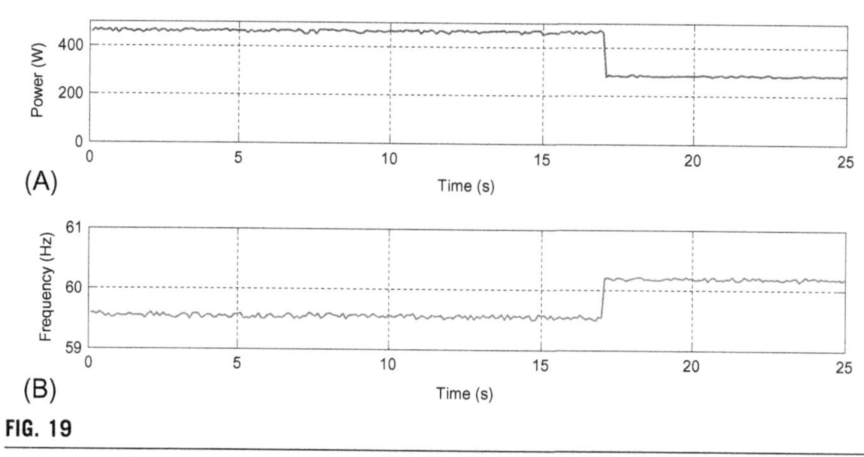

FIG. 19

Facility islanding operation of synchronous generator. (A) Active power generated by synchronous generator and (B) system frequency.

with a frequency increase. Fig. 19B shows that the generator frequency is increased at the moment of islanding, from 59.6 to 60.2 Hz, and settles at this value for the remaining operation period. The inverter-based DER local island is created in the same sense. Fig. 20 shows the performance of the new created island. It can be seen in Fig. 20A that the generated active power from the inverter increases from 250 to 450 W to meet the local load requirement. The fast response of the inverter is obvious

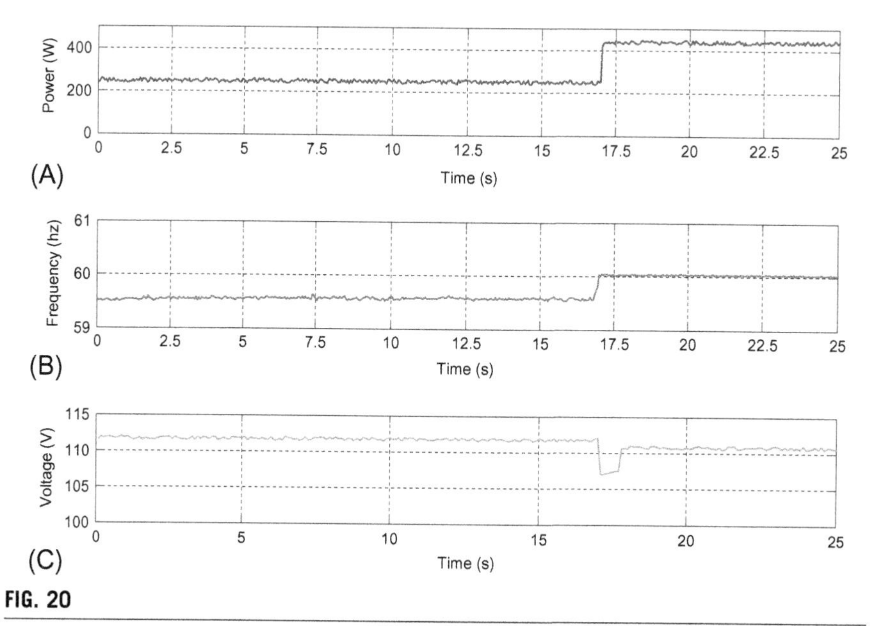

FIG. 20

Facility islanding operation of inverter-based DER. (A) Active power generated by inverter-based DER, (B) system frequency, and (C) system voltage.

due to its very low inertia. Fig. 20B shows that the frequency is regulated to the nominal value of 60 Hz. During the transition period, the voltage drops to 107 V and is recovered again to be settled around 111 V. In this case, the inverter succeeded to regulate the frequency and meet the load demand by the local load.

12.3.2 MICROGRID AUTOMATION AND PROTECTION

As microgrid penetration increases, the concerns regarding the protection and remedial action schemes (RAS) for distribution networks are becoming more important. Conventional control and management schemes for the microgrids, where only voltage magnitudes are measured and utilized, would undermine the new grid dynamics. Furthermore, conventional methods may potentially lead to severe complications in operations. This has resulted in a growing interest in utilizing synchrophasors in microgrid protection and RAS applications.

A microgrid, which is the main component of the future smart grids, requires sophisticated measurement and control capabilities. Conventional RTU-based automation approaches suffer from a slow measurement update and a lack of time-stamp values. Protection event analysis, islanding detection, resynchronization, and online remote–generation unit dispatch studies are realistically demonstrated.

A PMU, which is commonly referred to as synchrophasor, is one of the most important measurement devices of power systems. PMUs enable control centers to collect synchronized measurements from the network. In addition to conventional protection schemes, PMUs generate time-synchronized event reports to understand an event trigger, to track corrective actions in order to prevent its recurrence, and to provide lessons learned from that particular application. This becomes imperative for system security analysis, and for revealing the reasons of blackouts or power outages.

Recent catastrophic events have focussed attentions on a resilient and reliable electric grid, which would require proper risk assessment and emergency preparedness. Essentially, the power system is able to withstand disturbances and contingencies, depending on the design and operating conditions. However, an unusual combination of circumstances and events may cause a portion of the interconnected system to separate completely, and form one or more electrical islands. Although the separation process is expected to be controlled by enabling a minimum load-generation imbalance, most of the cascading power system collapses result in uncontrolled islands, where system recovery may not be assured. Rapid emergency restoration actions should be directed toward primary and secondary controls in order to ensure continuous operability.

RAS are designed to remediate disturbances by changing the network configuration according to predetermined rules, including generator tripping, load shedding, and shunt or series capacitor insertion. RAS methods are mainly classified as: (i) event based, (ii) parameter based, (iii) response based, and (iv) combined type. Event-based schemes directly initiate the actions following a fault detection with an open-loop type of control. Parameter-based and response-based methods are indirect methods mainly seeking a significant disturbance confirmed by various measurements. Event-based schemes are more appropriate for microgrid applications, especially for low-inertia systems involving a high number of power converters.

Microgrid hardware setup

A microgrid is the core structure of the complicated active distribution network concept, which enables multiple DERs to run collectively by a central control entity. The communication of a microgrid with the central control entity is enabled through various network infrastructures, which consist of a wired network, such as high-speed Ethernet serial communications, and wireless networks with global positioning system (GPS) synchronized timeservers.

This part is performed using a developed microgrid testbed. The remainder of this section provides a brief explanation about the developed platform. For the microgrid setup, it is necessary to follow widely used common industrial protocols, which are already implemented in field devices, in order to enable interoperability of the new and existing assets. The PLC and feeder protection relays are examples of main electrical field devices. A microgrid can involve conventional DERs, such as combined heat and power, back-up generator sets, and small hydro; as well as renewable DERs, such as wind-turbines and solar panels. Conventional DERs, such as hydro and generator sets, are mainly a synchronous generator type. Renewable units, such as wind and solar, are inverter-based DERs. Since the generated power from renewable resources is mainly DC, they are connected to the AC side with the inverter interface. The generic hybrid microgrid structure under study is shown in Fig. 21.

The hybrid microgrid involves a synchronous generation unit and an inverter-based DER. In addition to the local loads, which are connected to the PCC of the generation unit, the system has a global load on the main bus. The microgrid has a connection to the utility grid which enables both grid-connected and islanded operation. The operational voltage level of the microgrid is 208 V RMS line-to-line, which is the utility voltage level.

Synchronous generator and inverter-based DERs

A 13.8 kVA, 60 Hz, 208 V, and 1800 rpm AC synchronous generator represents a conventional synchronous generation unit. The machine is equipped with a half-wave, phase-controlled, thyristor-type AVR. The excitation current is derived directly from the generator terminals. An induction motor, driven by a frequency driver, is used to emulate the generator governor model. The torque, or speed reference, of the frequency driver is controlled by the PLC. The inverter-based DER is interfaced with an AC microgrid via an AC/DC to DC/AC converter that allows bi-directional power flow between the AC and DC parts of the microgrid. In this study, only the AC part of the inverter-based DER is analyzed, in which the DC bus is modeled to be a constant source. A 6 kW programmable DC power supply is used to emulate typical renewable energy resources.

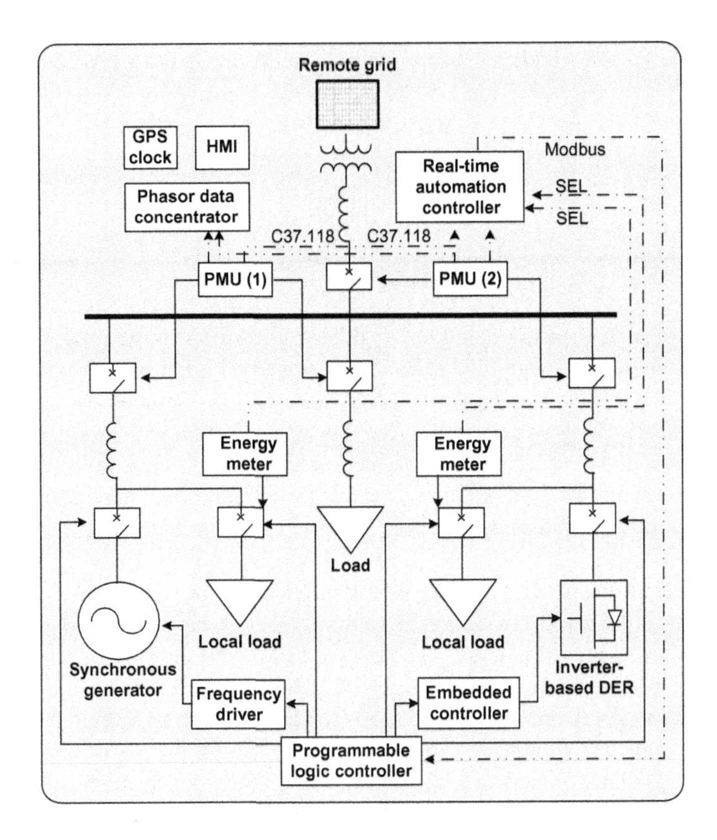

FIG. 21

Synchrophasor deployment in microgrid.

Transmission line, load, and bus models

The transmission lines are modeled as medium length π-type lines, which are composed of series inductors and two parallel capacitors. The load module is composed of ten levels of parallel resistive loads from 0 to 3 kW in steps of 300 W at nominal voltage. $[2 \times 72\,\Omega + 4 \times 144\,\Omega]$ resistance models, which can be controlled by the PLC, are utilized. For emulating the circuit breaker, 530 V, 25 A solid state relays are used. The trip/close signals can be controlled by a 3–32 V DC, supplied by the PLC.

Synchrophasors

Synchrophasor applications explicitly deal with the system disturbances of dynamic operation of complicated systems. High-sampled data acquisition can reach up to 60 messages per second. In this study, the reporting rate is 30 frames per second. The phasor representation of a sinusoidal AC signal is given in Eq. (16), where the phasor magnitude is X_m, f_o is the nominal system frequency, and ψ is the phase angle relative to the synchronized coordinated universal time (UTC). Frequency drift is one of the trustable indicators of system disturbance. The frequency and rate of change of frequency are given in Eqs. (17), (18).

$$x(t) = X_{\mathrm{m}} \cos [\psi(t)] \tag{16}$$

$$f_{\mathrm{o}}(t) = \frac{1}{2\pi} \frac{d\psi(t)}{dt} \tag{17}$$

$$\mathrm{ROCOF}(t) = \frac{df(t)}{dt} \tag{18}$$

The synchrophasor measurements are evaluated according to the total vector error expression, where the theoretical values of a sinusoid signal vary from the values obtained from a PMU. The standard IEEE Std. C37.118 includes the data transfer protocol for PMUs for power systems. The messaging format with required reporting rates are defined. The locally available measurements are frequency, ROCOF, primary voltages, currents, real and active power flows, and other analog/digital quantities defined by the system operators.

Phasor data concentrator (PDC) and the GPS clock

A PDC can be considered as a station in a communication network where PMUs populate time-aligned measurements. The deployed PMUs in the system send acquired data with time-stamp information to the PDC. The collected data from a number of PMUs are sorted and correlated according to the time-stamp value. This enables comparable real-time monitoring of the system with high-precision sampling. The collected data is also stored in a large database system for accurate post-mortem applications, such as fault-event monitoring, loss of mains, and blackout analysis.

Synchrophasor measurements time referencing critical applications, such as fault-event analysis and protection relays, require highly reliable satellite clocks. Inaccurate time-stamps can cause misdiagnosing of the network and degrade the controllability of the distribution network. The Inter-Range Instrumentation Group (IRIG-B) time code is widely accepted for time distribution in substations. In this study, we use a GPS clock with IRIG-B outputs.

Real-time automation controller (RTAC)

Interoperability is one of the greatest challenges facing an active distribution network infrastructure, as many new and existing devices need to securely and effectively communicate to each other. RTAC is an industrial automation platform that enables the collection and combination of a number of measurements from the system, with the ability to support many communication protocols. This platform enables the central controller to communicate with other devices that exist in the network, such as IEDs, PMUs, and PLCs. In this study, we use some of the commonly used industrial communication protocols such as Modbus, SEL as well as the C37.118 synchrophasor protocol. As shown in Fig. 21, RTAC has an interface with PMUs, PLC, and the energy meter through C37.118, Modbus, and SEL (Schweitzer Engineering Laboratories), respectively. Fig. 22 shows the synchrophasor experimental setup used in this study.

FIG. 22

Synchrophasor experimental setup.

Programmable logic controller

Active distribution networks require much more field actuator controls due to DERs, compared to the passive grid structure. Generator process control and field sequential relay control are some of the applications of PLCs in the field. The SCADA system is capable of providing the interface of the PLCs through built-in communication ports with options for RS485 and TCP/IP. In this study, the RTAC and PLC interface is established with the very common Modbus protocol.

Frequency driver (governor) and embedded controller

The frequency drive is used to drive a three-phase AC induction motor, where the rotor is coupled to the synchronous generator prime mover in order to emulate the governor model. Torque and frequency references are controlled through the PLC. The digital signal processor (DSP)-based embedded dSpace control platform is used to control the inverter-based DER model. The internal active/reactive power generation of an insulated-gate bipolar transistor (IGBT)-based inverter is programmed by the Matlab/Simulink platform. The communication of this field actuator and PLC control is established through a built-in RS232 serial port.

Data acquisition and monitoring

A complete human machine interface and SCADA system is necessary to establish network management by distribution network operators. For this study, an SCADA system is built to monitor and control the generic microgrid model, which is proposed

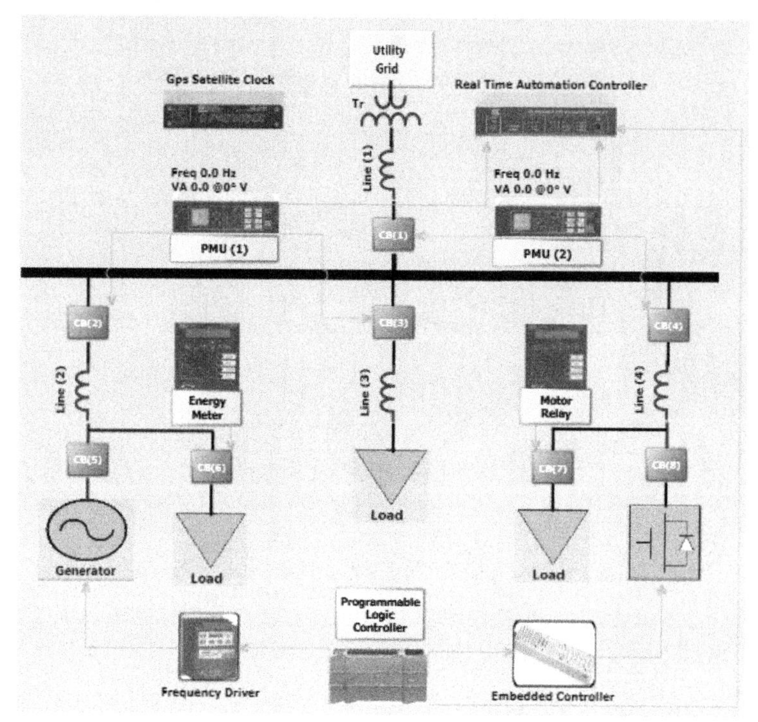

FIG. 23

SCADA main page.

in Fig. 23. The system is running on a webserver, and is available from the Internet with valid access. Fig. 23 shows the main page of the implemented SCADA.

Protection and RAS experiments

The microgrid can operate in grid-connected or islanded mode. The RAS are usually applied to microgrids after a fault instance, and, accordingly, to the microgrid transitions to an islanded (isolated) operation mode. In the grid-connected mode, frequency and voltage regulation is handled by the host grid. However, in an islanded operation, a microgrid must be able to regulate internal frequency and voltage with a proper control. Droop control is the commonly accepted operation for power sharing among DERs in a microgrid. In the droop control scheme, the frequency can deviate from the nominal value, based on loading conditions. Selecting one of the DER units to enable the secondary control to restore the frequency to the nominal value is a common practice in islanded operations.

A single phase fault is created by realistically examining the protection scheme with a tripping event. Fig. 24A shows the dynamic change of voltage phasors before the fault instant. Initially, the system runs normally. Once the breaker is closed, phase c is grounded. Over-current protection logic causes the tripping of the

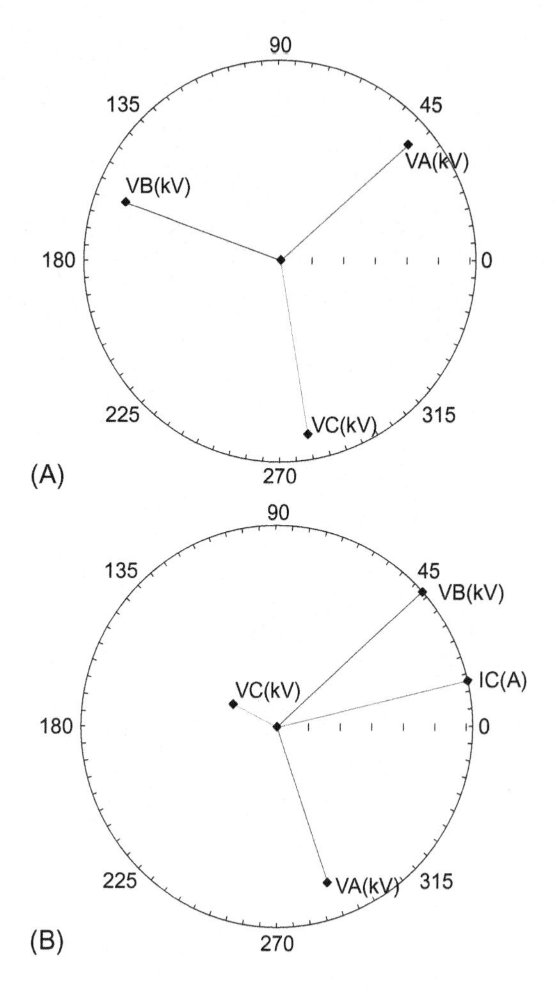

FIG. 24

(A) Voltage phasors before fault and (B) phasors during fault.

synchrophasor circuit breaker, flagging a trip event. Hence, the synchrophasor captured the event and protected the circuit. The current on phase C jumped to 25 A, and the voltage on phase C dropped to zero. Fig. 24B shows the voltage and current phasors during the fault. Fig. 25 shows the magnitude change of phase voltages. The grounded phase voltage c becomes zero after the fault. Fig. 26 shows the short-circuit current that appeared during the fault.

Upon detecting the fault, the microgrid gets disconnected from the host grid. Prior to separation, the microgrid was importing power from the host grid. During the islanding situation, an immediate microgrid frequency dip was detected due to the power imbalance. For an accurate phase angle comparison, the phasor measurements should be correlated by means of a time-stamp with a high data frame rate.

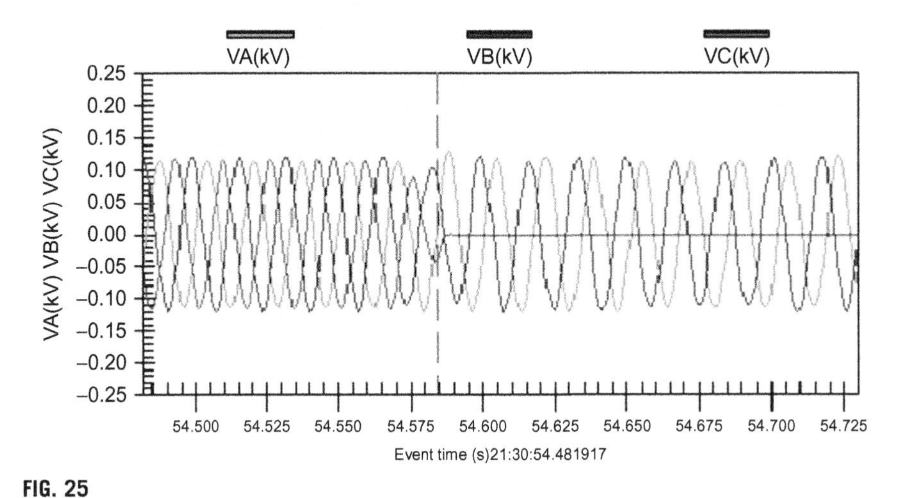

FIG. 25

Magnitude change of phase voltages.

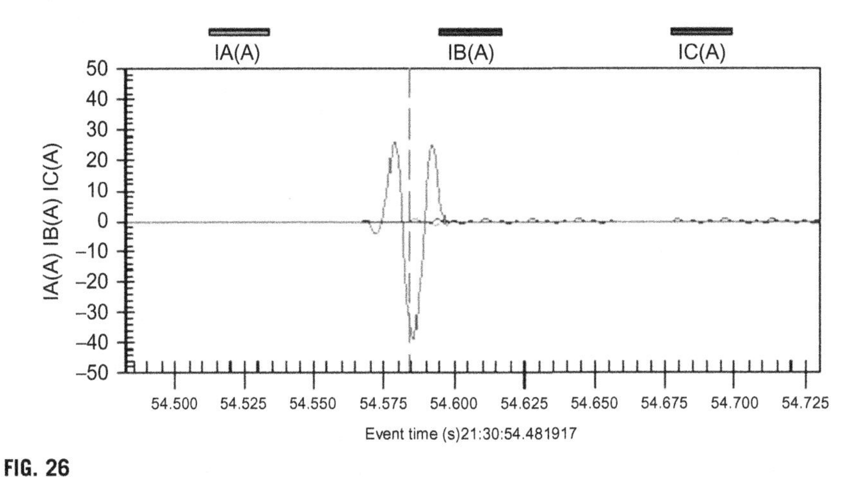

FIG. 26

Short-circuit current appearing during the fault.

Thus, the synchrophasor measurements can be effectively used to determine whether the microgrid is islanded. ROCOF, phase angle difference, and over/under frequency are the most accurate indicators for islanding detection. The frequency drift in the islanded area is related to a power mismatch and system inertia. Since microgrids have very low inertia, during the load changes the frequency drift has larger values. Fig. 27 shows the islanding logic used in this study.

The active power imbalance introduces a frequency deviation in the islanded microgrid (Eq. 20), where H_{tot} is the total inertia, f_n is the nominal frequency, and f_s is the system frequency.

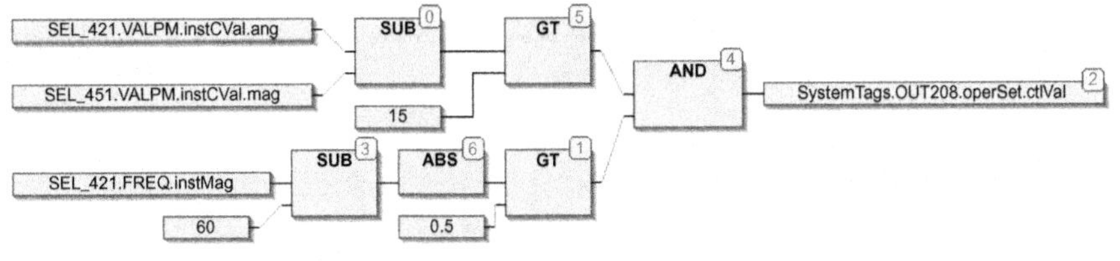

FIG. 27

Islanding detection logic.

$$\Delta P(t) = P_{\text{gen}}(t) - P_{\text{load}}(t) - P_{\text{import}}(t) = 0 \tag{19}$$

$$\Delta P(t) = \left(P_{\text{gen}}(t) - P_{\text{load}}(t)\right) = \frac{2H_{\text{tot}}}{f_{\text{n}}}\frac{df_{\text{s}}}{dt} \tag{20}$$

A consecutive islanding detection algorithm is used to enable islanding detection, which initially senses the under/over frequency setting; then the frequency gradient is compared to the set value. When islanding is detected, the DERs switch operation to droop control to enable accurate power sharing. The droop-based primary control deviates the frequency from the nominal value according to the system loading conditions.

Primary control is the immediate response of the turbine governors to arrest deviations in power system frequency; secondary control is the centrally coordinated actions to reinstate frequency to its nominal value. In a conventional power system operation approach, if an uncontrolled islanding is formed due to emergency conditions, the primary control responds rapidly according to droop adjustments of the generators inside the island. Upon separation, it may be necessary to shed some of the predetermined loads in the islanded area in order to balance generation and load.

Automatic generation control (AGC)-based secondary control is used to restore system frequency to nominal value. A common way to enable AGC in power systems is to implement a proportional-integral (PI) controller. An area control error (ACE) in a power system is given in Eq. (21), where B is the frequency bias factor, ΔP_{T} is the deviation of active power balance in area, and ΔP_{AGC} is the control command to be sent to the governor. $\beta 1$ and $\beta 2$ are the PI control coefficients.

$$\begin{aligned} \text{ACE} &= \Delta P_T + B\Delta f \\ \Delta P_{\text{AGC}} &= -\beta_1 \text{ACE} - \beta_2 \int \text{ACE}dt \end{aligned} \tag{21}$$

Fig. 28 illustrates the islanding transition of the microgrid and the corresponding primary and secondary controls. During the synchronized operation, the islanded power system was importing power from the remote power system. When the islanding situation takes place at 130th second, the imported power becomes zero. As per Eq. (20), the power imbalance results in a frequency drift in the islanded area, therefore the phase angle difference between two areas increases. Fig. 28 shows the

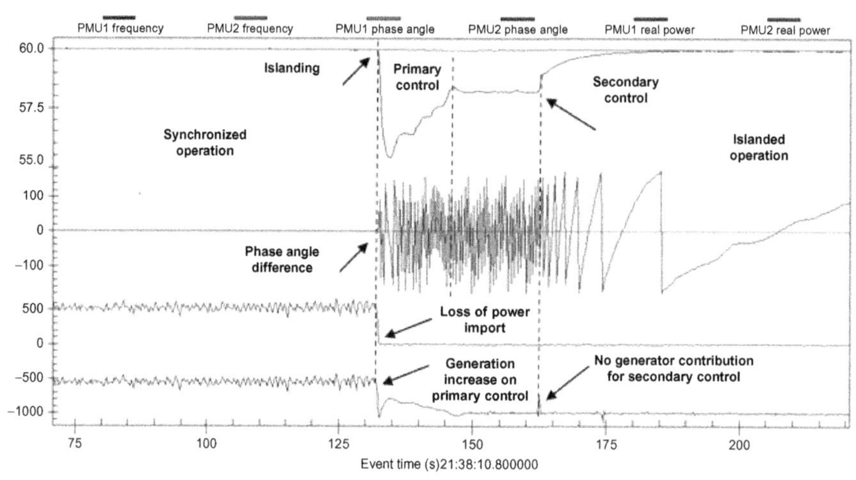

FIG. 28

Islanding instant, primary control, and secondary control.

oscillation of the phase angle of the system generators during primary control. Between the 130th and 147th second, the primary control is established. The generation in the islanded area increases, and the frequency settles in a stable region. At the 163rd second, the secondary control is initiated to regulate the system frequency to a nominal value by power injection from the storage system of the microgrid. At the 181th second, the system frequency reaches an optimal value. The inverter control scheme is shown in Fig. 29.

Islanded microgrids have larger margins of operation, such that the allowable operational limits are 59.3–60.5 Hz and 5% voltage deviation from the maximum

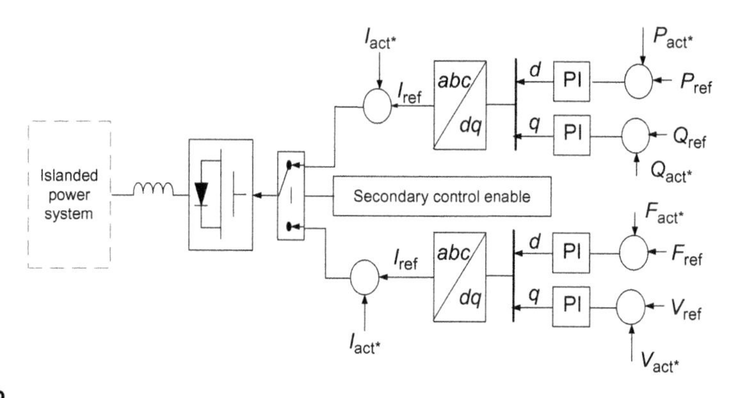

FIG. 29

Inverter-based DER control.

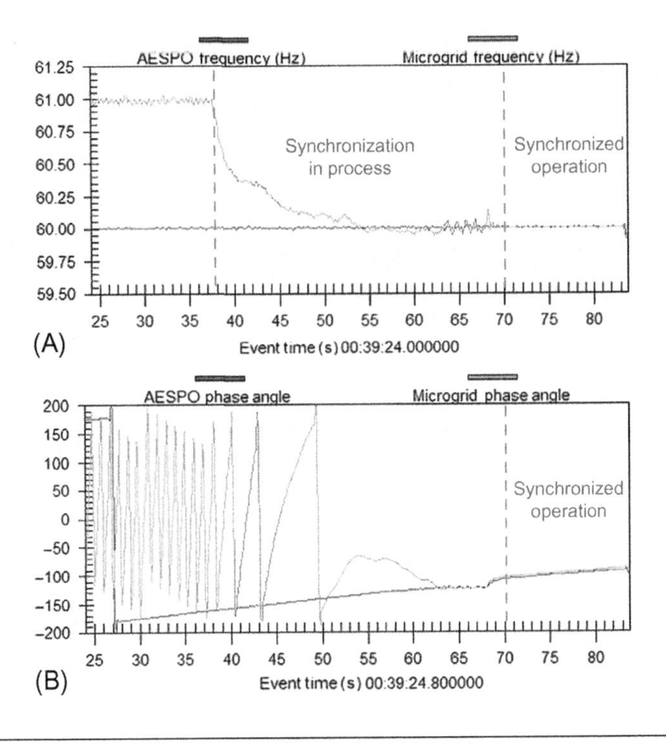

FIG. 30

Resynchronization of the microgrid. (A) System frequency during resynchronization and (B) phase angles during resynchronization.

permissible limits. When it comes to synchronization, the main grid frequency is very close to 60 Hz nominal values, and the variation is low.

$$
\begin{aligned}
&T\left(\left|f_{\text{host}} - f_{\text{microgrid}}\right| \geq f_{\text{threshold}}\right) \geq T_{\text{threshold}} \\
&T\left(\left|\phi_{\text{host}} - \phi_{\text{microgrid}}\right| \geq \phi_{\text{threshold}}\right) \geq T_{\text{threshold}}
\end{aligned}
\tag{22}
$$

Fig. 30A and B shows the frequency and phase angle difference of the area electric power system operator (AESPO) and microgrid. The figures cover 30 s of the synchronization process. Initially, the microgrid is operating at 61 Hz. From the 35th to the 65th second, the generator output frequency decreased by manually applying torque to the generator shaft from the governor. At the 70th second, the AESPO and microgrid frequency match, thus the synchronizer switch is closed.

12.3.3 MICROGRID DATA INFORMATION AND CYBER-PHYSICAL INFRASTRUCTURE

In contrast to the centralized control in the existing grid, the emerging smart grid concept compels utilities to adopt decentralized methods, as a result of the highly dynamic behavior of the active distribution networks. Decentralized control approaches intend to provide autonomy for different control layers by enabling an event-driven peer-to-peer communication structure, where central control schemes mainly rely on master-slave interactions. In power system applications, the implementation of decentralized control is established using multiagent frameworks, which are composed of interacting multiple agents to achieve a global, or local, objective function. An agent requires interaction with its environment through sensors and actuators. A sensor acquires the data from the outside world, and the actuator responds according to the agent's decision. Embedded decision-making algorithms facilitate the benefit maximization of the agents' autonomy.

Multiagent-based schemes are widely applied to power system controls, including self-healing, resilient grid automation, and power system protection. Multiagent-based microgrid control draws considerably more attention than any other smart grid applications. For actual implementation of decentralized control schemes in power systems, it is imperative to link multiagent objects to distributed industrial control systems such as IED and PLC. The required interface is established through the combination of the information data and protocols.

Interoperability is one of the major challenges in accomplishing a complete smart grid infrastructure due to the large amount of processed data from different vendor and communication protocols. Utilities and independent system operators seek the most proper way to reach the required information easily and securely for different application layers, such as metering, protection, automation, and market segments. For this reason, The NIST Framework and Roadmap for Smart Grid Interoperability Standards Release defines three major goals for establishing interoperability standards and protocols for the smart grid. The IEEE Std. 2030 establishes three integrated architectural perspectives: power systems, communications technology, and information technology. The guidelines also define the design criteria, and reference model applications with communication connections and data flows.

IEC 61850 is the new international standard for communications, which enables the integration of all substation functions within itself, such as protection, control, measurement, and monitoring. IEC 61850 expands the area of influence in many parts of the power systems due to its wide industry acceptance. Communication systems for hydroelectric power plants and DERs have been recently applied to other domains as IEC 61850 extension standards. However, the smart grid concept covers extensive control, automation, and protection applications, such that a single standard may not meet all the required forms of monitoring and information exchange demands.

Considering the emerging active distribution networks, new energy market policies are necessary, such as the implementation of real-time auction models, and scheduling the dispatch of DERs. Hierarchical control of microgrids requires interaction with utilities for dynamic adjustment of the primary, secondary, and tertiary control levels. Taking into account the aforementioned requirements of the future grid, advanced multiagent frameworks are necessary with a flexible ability to create tailor-made decentralized control schemes, while following the legacy protocols.

The Foundation for Intelligent Physical Agents (FIPA) is an organization which intends to evolve inter-operable agent communications with semantically meaningful messages, such as how messages are transferred and presented as objects. Java agent development framework (JADE) is a software framework that develops agents compliant with FIPA standards, with flexible agent behavior methods.

Cyber-physical framework

A cyber-physical framework consists of physical and cyber components. Actual physical components are the sensors, actuators, generation units, circuit breakers, and distribution lines. Cyber components are the data informational representation of the actual physical models with standardized protocols. This section briefly explains the hardware and the data information model of the proposed framework.

IEC 61850 framework

When it comes to interoperability in a smart grid, IEC 61850 is the most promising standard for future grids. Self-describing devices and object-oriented, peer-to-peer data exchange capabilities are the most significant superiorities of IEC 61850 over other common standards. The use of names for all the data, virtualized models, standardized configuration language, lower cabling, and transducer installation costs are some of the numerous key features and benefits [2]. Logical nodes (abstract data objects) are the main elements of the IEC 61850 object-oriented virtual model, which consists of standardized data and data attributes. IEC 61850 defines the ACSI, which creates objects and services independent of any protocols. This enables a hierarchical class model, in which all class information, services that operate on these classes, and associated parameters, can be accessed from a communication network. The abstract interface allows the data objects to be mapped to any other protocol, such as with an MMS protocol and SMV protocol on an Ethernet data frame.

The virtual model aims to express a physical (logical) device and a number of logical nodes. IEC 61850 standardized 91 logical nodes into 13 logical groups. Each logical node contains DATA, which are standard and related to logical node functions. Most of the data objects are composed of CDC, which involve basic data objects, status, control, and measurement. Each data element consists of a number of data attributes with a DATA Type, which belongs to FC. Fig. 31 shows a sample anatomy of an object name for a breaker position value. A physical device is defined by a network address.

GOOSE is a multicast model based on a publisher-subscriber mechanism within the IEC 61850 framework, which ensures fast messaging with a 4 ms period of time. GOOSE messages are periodically sent from the publisher IEDs to subscribers within a retransmission time period. Should an event occur related to the GOOSE control,

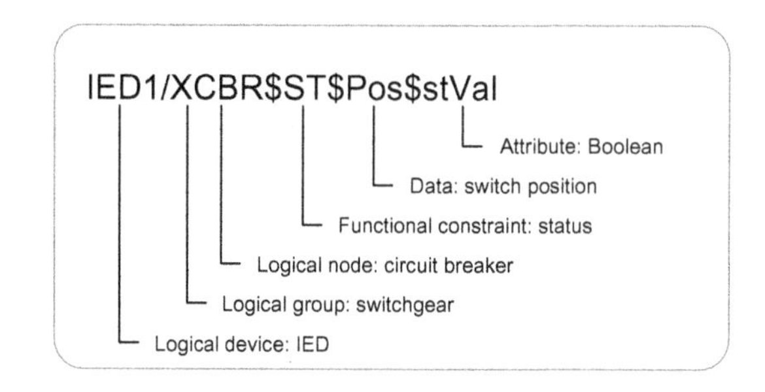

FIG. 31

Object name of a circuit breaker position value.

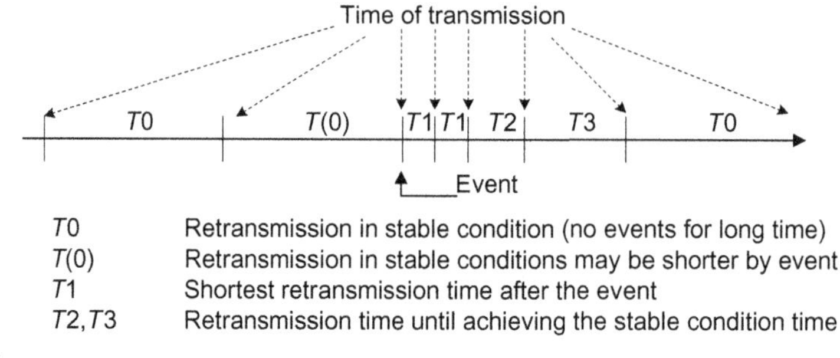

FIG. 32

GOOSE messaging.

a new message is generated momentarily, then the message is continuously retransmitted with variable time periods ($T_1, T_2, \ldots, T_n$) until it reaches the T_0 value again, as shown in Fig. 32.

This retransmitting scheme ensures the appropriate level of reliability. The fast messaging capability of the GOOSE model is widely used in modern power system protection applications, bringing forth a new era of advanced high speed peer-to-peer communication (Fig. 33).

OPC UA

OPC UA modeling is based on nodes, and references between nodes. A node can have different sets of attributes connected through references. A node class is composed of objects, variables, and methods. A variable contains the value, and clients can read, write, and subscribe to the changes of the value. A method is similar to a function called by the client that returns a result. The OPC UA address space is structured with objects containing only the node attributes.

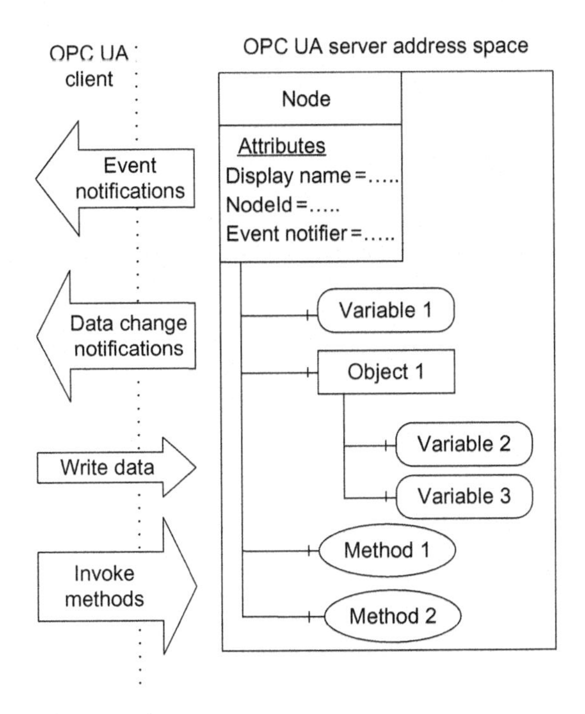

FIG. 33

OPC UA node class.

FIPA specifications and JADE platform

An agent is an interacting object with its own thread of control that operates autonomously. The ideal agent is also expected to have semantic interoperability based on internal decision making. The standardization of agent-based technologies is an ongoing field of research, and few standards have yet to be realized to their fullest potential. FIPA specifications help to allow an easy interoperability between the agent systems with the agent communication language (ACL) and transport level protocols. ACL represents a communicative act, or message intended to perform some action, with precisely defined syntax and semantics. Fig. 34 shows the components of a sample ACL message.

The beginning message structure of an ACL message expresses communicative acts, such as inform, request, refuse, and so on. Sender and receiver parameters designate the name of the sender and intended recipient agents, respectively. Content involves the object of the action and the parameters passed through the message. Message parameters define the expression of the agent responding to the received messages, and which parameter is sent through the message.

The JADE platform is based on FIPA specifications, which enables developers to create complex agent-based systems with a high degree of interoperability using ACL messages. A JADE agent, at its simplest, is a Java class that extends the core agent class allowing it to inherit behaviors for registration, configuration, and general

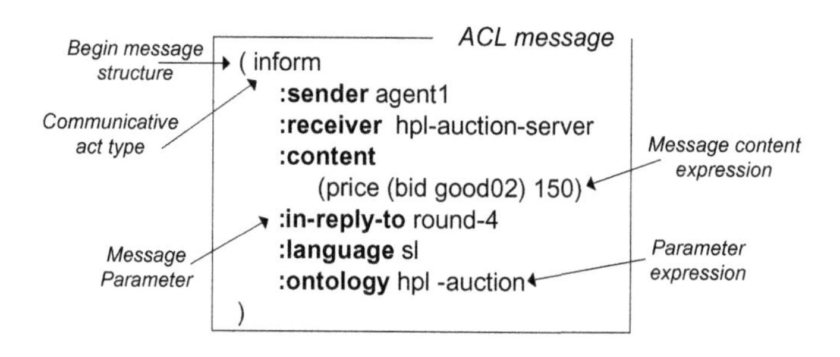

FIG. 34

ACL message components.

management of agents. Send/receive messages can be implemented by calling basic methods using standard communication protocols and registering in several domains. External software can be integrated with the use of behavior abstraction, which enables a link with the OPC UA nodes along with the JADE agent messages.

IEDs and hardware components

The intelligent multiagent framework is implemented in a reconfigurable small-scale power system available at Florida International University's smart grid testbed [3]. The platform consists of conventional and nonconventional generation units, transmission and load models, field sensors, and actuators. Further information about the testbed can be obtained from previous references. The IEDs are located on system buses to enable monitoring, control, and protection. The agent platform was implemented on a single personal computer. However, since the information is accessible through the network, the computation can be easily distributed. An off-the-shelf OPC UA server was implemented to acquire IEC 61850 logical node measurements. An OPC UA client was embedded in the Java platform to enable JADE to access mapped IEC 61850 measurements. The overview of the laboratory setup is shown in Fig. 35.

Sample use case: DER ancillary service

A DER-based autonomous ancillary service use case has been demonstrated in real-time for validation of the proposed multiagent framework, using the combination of IEC 61850 and FIPA standards. The IEEE guide 1547.3 defines DER interoperability issues by means of monitoring, information exchange, and control. Some use cases are demonstrated as business operations of the DERs and stakeholder entities with direct communication interactions.

This case study was adopted from the IEEE 1547.3 guide to provide a prototype demonstration of the developed framework. The DER units can be utilized to provide ancillary services, such as load regulation and reactive power support in distribution feeders. Especially at peak hours, excessive energy demands may result in the overload of the distribution lines due to the drawing of excessive currents. This would result in thermal overheating and voltage drops beyond permissible limits on different parts of the feeder. A scheduled operation of the DERs

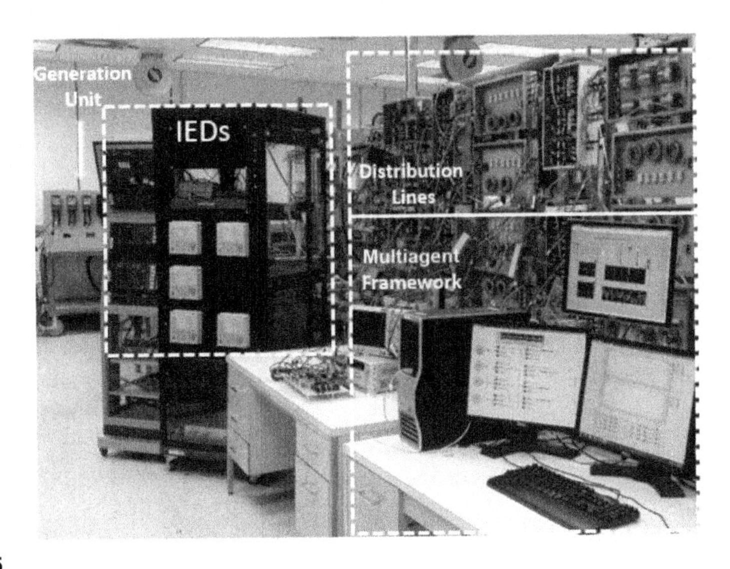

FIG. 35

Agent platform and laboratory setup.

would provide a solution for relieving such overloading problems by contributing either active or reactive power support.

DER operators and the AESPO interact with each other though messages. These two entities can be considered intelligent agents with the following duties and attributes: (i) the AESPO is the responsible entity for safe and reliable operation of the distribution power system. The complete utility grid model is the property of the AESPO. (ii) the DER Operator is the main responsible entity for DER generation units. Monitoring, protection, and control of the units are handled by DER operators.

AESPO and DER operators are the two entities that carry out the stated ancillary service mechanisms. The decentralized collaboration of AESPO and DER operators is based on an autonomous event-driven information exchange model, which requires implementation of agents featuring actuators, sensors, embedded intelligent decision-making algorithms, and communication channels. Fig. 36 illustrates the laboratory deployment of the multiagent framework, IEDs, PLCs, line models, loads, DERs, AVRs, and governors.

The cooperation agreement between the entities is established through sent/received messages from each party, in a certain form with an understandable content. This common, or shared-structured, vocabulary is referred to as the system ontology. Ontologies are constructed with a consistent relationship within an application domain. Abstraction is a way to express real-world objects with their characteristics and attributes, as well as their interaction with other entities. Prior to actual computer code implementation, unified modeling language (UML) tools are utilized to provide a meaningful abstract modeling of the emulated case. UML is the terminology used to document the process and information modeling of entity interactions. In the

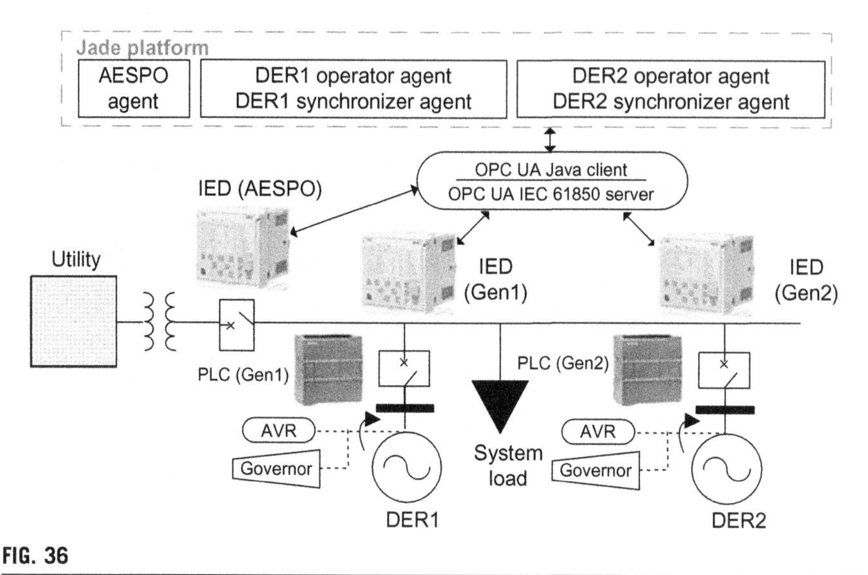

FIG. 36

IED deployment and agents.

emulated case, the UML tool is utilized to visualize the process of defining functions, and illustrating how they influence others. Fig. 37 illustrates the ancillary service use case UML flowchart defined in this study.

The AESPO agent and DER operator agents are defined in the JADE platform. The AESPO agent is intended to continuously check the critical current flow value from the beginning point of the feeder through the IEC 61850 three-phase current measurement (CMMXU) logical node.

When the current flow from the feeder reaches its critical value, the high-alarm node (LDO.CMMXU.HiAlm.stVal) of the function block becomes high. The AESPO agent is monitoring this value through the OPC UA client. According to the embedded intelligent decision-making algorithm, an ancillary service support request message is published to the DER operator agents, which are registered to the directory service (yellow pages). The yellow page is a service mechanism, in which an agent can find other agents providing it with the services it requires in order to achieve its goals. The directory facilitator (DF) is the agent that provides the yellow page service to the agent platform. The AESPO agent periodically looks up available operators from the DF agent. A random availability function is defined for each DER operator to define whether to issue an "agree" or "reject" message in return. Fig. 38 shows the correspondence between the AESPO agent and two DERs.

If the DER operator agrees to provide an ancillary service, it autonomously enables the DER synchronizer agent. The DER synchronizer agent is the IEC 61850 synchronism check (RSYN) logical node of the IED, and it is not defined in the JADE platform. The DER synchronizer agent continuously checks the condition across the circuit breaker from the bus to the line parts of the power system, and

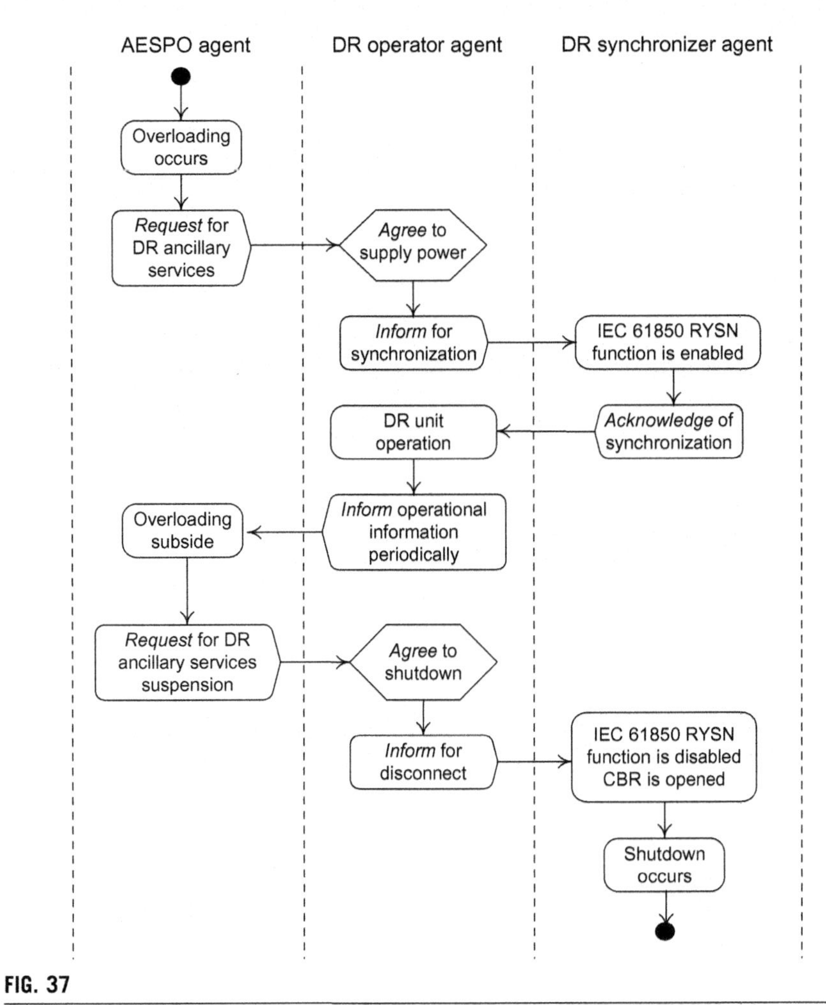

FIG. 37

Ancillary service use case UML flowchart.

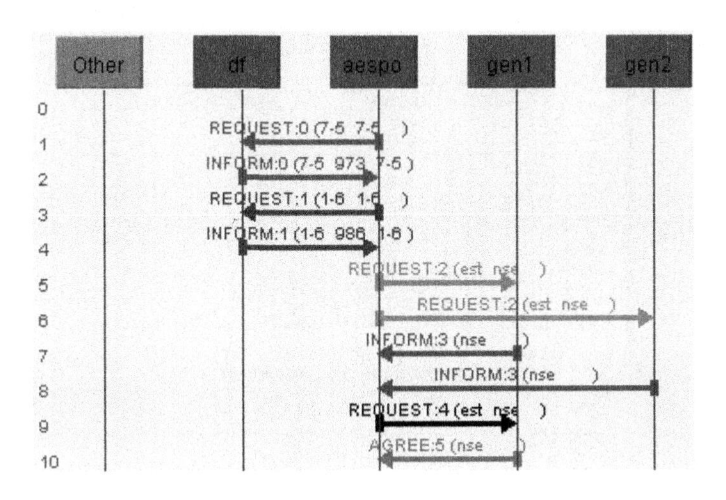

FIG. 38

Correspondence between JADE agents.

gives permission to close the circuit breaker when the synchronization conditions are satisfied. The determination of the closing signal is defined according to frequency and phase angle difference. The monitored frequency and phase angle difference value is continuously read by the PLC in order to adjust the governor speed.

Bus voltage and line voltage measurements are U_BUS and U_LINE, respectively. The synchronization status can be obtained from the function block by synchronization in progress (SYNC_INPRO) or in synchronism (SYNC_OK). Live line, dead bus (LLDB), live line, live bus (LLLB), dead line, live bus (DLLB), dead line, dead bus (DLDB) outputs designate the health of the line and bus. Blocking, bypassing, and external closing request inputs are also defined.

Fig. 39A and B shows the frequency and phase angle difference of the DER and AESPO. The figures cover 30 s of the synchronization process. From the 35th to the 65th second, the generator output frequency decreased by manually decreasing the applied torque to the generator shaft from the governor. At the 70th second, the utility and generator frequency match, thus the synchronizer switch is closed. At the 76th second, the applied torque to the generator shaft was increased to deliver more power to the system. Fig. 39B shows the phase angle difference between the AESPO and DER voltage. As synchronization occurs at the 70th second, the phase angle difference decreases to a value almost equal to zero. This clearly shows that the generator is synchronized to the utility.

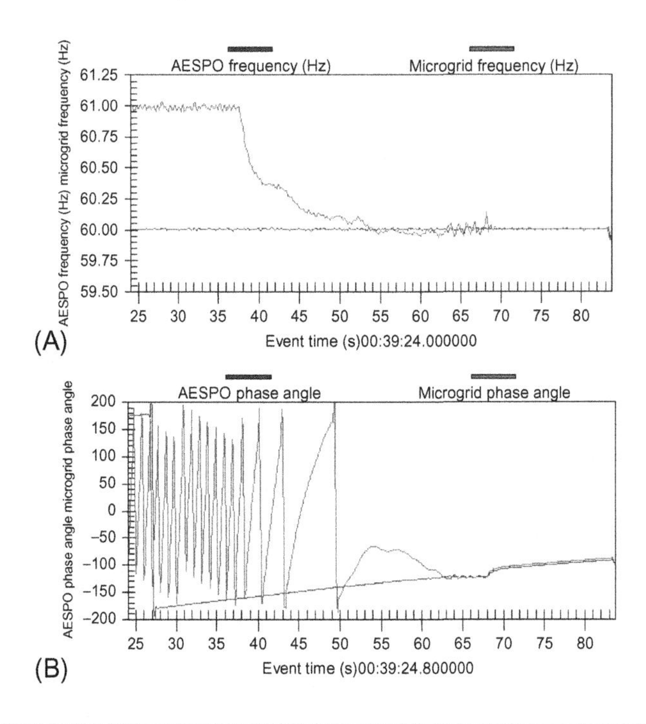

FIG. 39

(A) Frequency change and (B) phase angle difference.

12.3.4 MICROGRID MARKET OPERATIONS

Conventional electricity distribution networks encounter a high penetration of commercial basis renewable energy and distributed resources. Smart grid and active distribution networks are terms that express an operation with a high-renewable penetration to power network. The term smart grid was introduced as an intelligent energy supply system, and comprises the networking and control of intelligent generators, storage facilities, and loads with the aid of information and communication technologies. This is an era of major transition for existing electricity networks, transitioning from passive to active distribution networks with distributed resources, resulting in bi-directional electricity flow. This situation brings many challenges upon operation due to the bi-directional power flow and the intermittent source of renewable energy generation. However, DERs bring a significant impact on power system reliability, as long as they are operated efficiently. Active distribution networks offer solutions for the demand side of management, since they have the ability to partially control the local generation capacity incorporated within storage devices and dispatchable generation units.

The DERs can also be utilized to reduce peak demand costs by enabling them to operate at only peak hours, reducing transmission network congestion and stress during peak hours. Direct grid-connected renewable DERs can be considered as a negative load source with an intermittent generation profile. This uncertain generation profile should be controlled by the assessment of optimal unit scheduling. Economical unit commitment, scheduling of DERs, and power sharing between multiple units in a microgrid are some of the most important concerns in the operation of active distribution networks in open market conditions. Microgrids are small-scale electricity networks that comprise diverse distributed generation (DG) units, which can operate in a grid-connected or islanded mode. Microgrid economics is highly dependent upon local conditions, such as the availability of fuel, gas, wind, and solar irradiation.

In open market conditions, utilities are no longer monopolized; hence DER stakeholders are private entities that can compete with regional utilities. The electricity trade can be handled locally in a region by several DER investors or by cross-border trading by utility wholesale market and neighboring regional systems. In distribution networks, feeder-based electricity trade is more reasonable for avoiding transmission route and loss costs.

A vast number of DERs should be able to operate, incorporating a large number of complicated operational functions. New generation grids require fast intelligent decision-making algorithms and communication infrastructure, since the power system operators will be inefficient in dealing with the highly active and changing operation of the future grid. Multiagent-based decision-making algorithms, incorporating with real-time communicative devices, offer a reliable solution for unit commitment in competitive market conditions.

Agent-based marketing models have applications on a variety of scales from large bulk generation units to small, home-based entities. There has been a growing

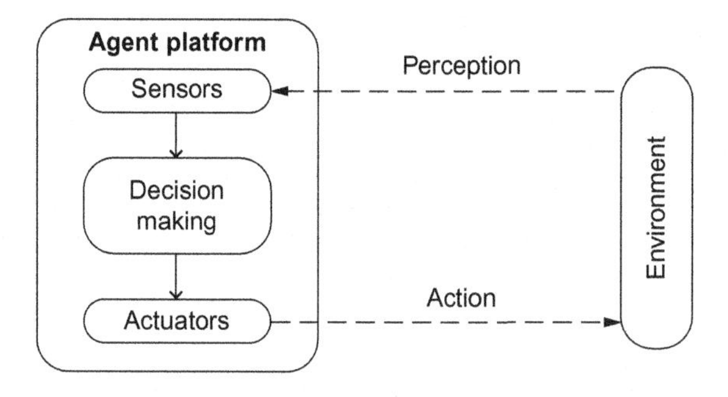

FIG. 40

Basic agent behavior.

interest for agent-based market scenarios for maximizing the benefits of generation units by analyzing the effects and evaluation of bidding strategies. Fig. 40 illustrated the basic agent behavior to be modeled.

For many years, applications of multiagent systems have been deployed in simulation and prototype laboratory applications. Few efforts have been put forth to implement multiagent systems in real-time utility-based or industrial applications in power system engineering. Aligned with maturing technology, new methods should be discovered to enable utility and industrial multiagent system applications. Hence, multiagent systems should be able to operate with existing industrial components and future power system assets. PMUs, DERs, and IEDs are main assets of future electric power systems. The aging network structure should be rehabilitated by replacing, or adding, new components into the system. Currently, the cost of the PMU and IED is high for particular applications, such as distribution network deployment; however, it is expected that costs will decrease with maturing technology and new market players. Such new components will enable us to maintain a sustainable, reliable, and efficient operation of future generation grids.

Multiagent systems

An agent is an object that can be considered as perceiving its environment through sensors, and acting upon the environment through actuators. The difference between a single agent and multiagent environment is straightforward. If any of the agent's value is influenced by another agent's behavior, that environment is considered to be a multiagent environment. Intelligent agents are mainly assigned to maximize their benefits in a specific environment by interacting with other agents.

Decentralized intelligent agents are capable of autonomous information exchange management and decision making in order to improve the system performance and to provide the maximum benefit to the stakeholder. For electricity procurement, multiagent-based algorithms can be utilized to purchase the cheapest electricity by exploring independent sellers and negotiating with them using real-time market prices. Similarly, for generation units, it can be utilized to find the highest priced buyer in order to maximize the benefit. The cost of electricity changes according to the state of charge (SOC) in the storage device of the renewable DER, and the fuel cost of the conventional-type DER.

Interoperability is the ability of two or more devices to exchange information and work together in a system. This is achieved by using a published object, data definitions, standard commands, and protocols. Interoperability of distributed resources is primarily concerned with establishing the requirements for interconnection, protection, and power system operation functions. Intelligent agents and other programs make use of shared ontology to facilitate interoperation. The data exchanges between distributed resource controllers and local electric power systems with common industrial, or newly released, protocols can be used to achieve the proper data exchange. Protocols are set by rules that determine the behavior of functional units in achieving a meaningful communication. They are implemented in one or more layers of the open systems interconnection model.

The agent platform is the software for development of the agent, which is implemented for system operation. In this study, we selected the JADE system, which complies with FIPA standards, and provides the implementation of the ACL, which is implemented in the Java programming language.

Game-theoretic auction model

The goal of restructuring the electricity market is to provide benefits to society by enabling a competitive environment. Volatility of fuel/gas prices or weather conditions shape the DER stakeholder's attitude toward operation time and duration of the power generation. The market structure has three entities: sellers, buyers, and aggregators. An effective way to decide which seller will qualify to supply the required demand for the system load is to implement trade contracts. There are two main types of contracts available for the short term: the day-ahead market and hour-ahead mechanism. In the day-ahead market, the auction is organized to balance the required amount of power for the next day; similarly, in the hour-ahead mechanism, the auction interval takes place on an hourly basis. The availability of generation units, bids, and trades are assigned each hour. The DER can be deployed at a fixed price in monopoly markets. However, for perfect competition, a wider policy should be enabled.

An auction is a public sale in which property or merchandise is sold to the highest bidder by determining the resource allocation according to bids received from participants. The auction can be used to sell the commodity of electricity or to buy it. A procurement auction, or reverse auction, is the process that is ruled by an entity that announces that it is interested in purchasing a certain amount of product; it then solicits competitive bids in order to acquire it at the lowest cost. Several countries have used reverse auctions to enable competitive renewable energy deployment in their systems. Auctions can be open or seal-bid type, which means whether other participants know what other participants are bidding or not. One-sided auctions only allow a buyer or seller to submit a bid, whereas, a two-sided auction allows both sides to submit bids. The bidding process can occur once or continuously until it reaches an equilibrium point. Designing a proper auction model is tough work. There are many questions to be answered before designing the appropriate auction model. Some of them can be stated as follows: number of objects to be auctioned, reserve price, acknowledgment of reserve price, how bids are to be collected, and selecting a qualified bidder.

The most common types of auctions in practice are the English, Dutch, and Vickrey auction models. A reverse auction is where the prices go down, in contrast to typical auction types. Normally, in a reverse auction, multiple sellers bid to sell goods to a single buyer. However, more than one seller could be picked as a qualified bidder when a single seller cannot meet the required demand. Although there are a variety of implemented types, it has been realized that the reverse auction is a useful tool in power system market operations.

The contracts can be done bilaterally or by using the pool approach. Many game-theoretic auction models are available. The two main types are individual and cooperative games. In individual game, the agents behave autonomously with a noncooperative approach. Hence, the game is a competition between coalitions of players that fix a problem influencing all players in the game. For example, in an actual power system, during an islanded operation of a microgrid, the frequency of the island should be stable; hence, supply and demand should match. Cooperative DERs should supply the required demand in order to maintain the stability of the system, regardless of the cost of the electricity. However, in a grid-connected mode of operation, if the price of the DERs is higher than the utility, then none of the DERs will join the game, since the frequency will be regulated by the stronger utility grid. This example can be considered as an individual game. In this study, multiple-grid connected DERs compete to sell electricity for a lumped load in a microgrid.

Auction framework

The IEEE guide 1547.3 addresses the guidelines for monitoring information exchange and control of the DERs interconnected to electric power systems. The stakeholders for the DER interaction can be assumed to be the AESPO, DER aggregator, DER maintainer, and DER operator.

AESPO is the entity responsible for safe and reliable operation of the power system. It is necessary to interact with DER units in the system by means of exchanging messages regarding its status. Utility grid interface and distribution line models are the properties of AEPSO.

DER Aggregator is a centralized controller that manages multiple DERs for main and ancillary services. The lumped load model is managed by the DER aggregator by organizing hour-ahead reverse auctions.

DER Maintainer is responsible for the safe operation of the DER by enabling a maintenance schedule. The availability of the generation unit can be defined by the DER maintainer.

DER Operator is the main entity responsible for the local control of generation units. Monitoring and measurement of the power system parameters and controlling the outputs, such as governor, is handled by the DER operator. The DER operator manages the sensors and actuators.

Interoperability of the stakeholders with a multiagent approach is highly dependent on the information exchange capability of the field devices, such as sensors and actuators. An interface is the point of contact that a software component has with its interacting partners. Some interfaces are defined by protocols, where they have a generic set of services. Protocols are a set of rules for the communication of industrial devices. Electricity network field devices are manufactured by different vendors. Each vendor's equipment has its own communication protocols for interacting with the outside world. This problem is one of the main challenges for the implementation of multiagent-based, decision-making platforms within the existing grid infrastructure.

Implemented agents on the Jade platform used to operate the complete system are shown in Fig. 41. Sensor agents are responsible for system parameter measurements and sending the information to other agents, or upper control areas if necessary. Actuator agents are responsible for operating the generation units according to the power reference received from the bidder agents. Bidder agents inform the actuator agent duration and power generation amount, if they are qualified bidders, after the auction process is organized by the aggregator agent. All available bidder agents register themselves to the yellow page, so that the aggregator agent will able to communicate with them. If the DER is available for the next hour, a bidding agent will decide whether to join an auction or not according to the base price information received from the cost function or SOC agents.

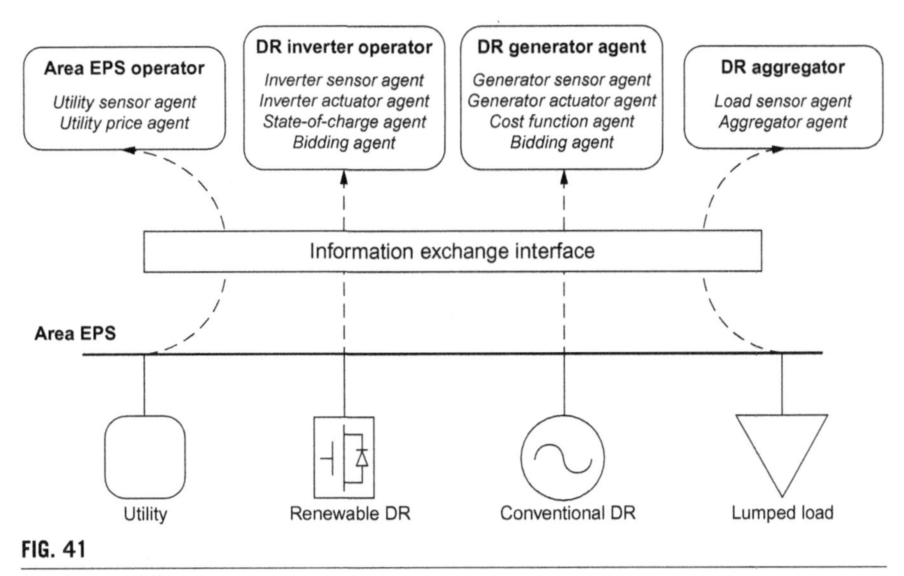

FIG. 41

Auction model.

Table 1 Inverter Bidding Agent Modes

State of Charge	Mood of Agent	Lowest Bidding
0–30%	Relax	0.15–0.13 $/kWh
30–60%	Moderate	0.13–0.11 $/kWh
60–100%	Greedy	0.11–0.08 $/kWh

The cost function agent is specifically defined for the fuel/gas-dependent generation unit type, which is a synchronous generator in this study. The fuel/gas price is assumed to be volatile; hence, generation cost is changing. Each conventional generator has a quadratic cost function, which affects the cost of electricity according to volatile fuel costs at different hours, where i is the unit number, P is the electrical output, and a, b, and c, are random fuel coefficients.

$$F_c(P) = \sum_{i=1}^{n} a_i + b_i P_i + c_i P_i^2 \tag{23}$$

Renewable resources are assumed to be coupled with storage devices. According to the SOC of the battery, the mood of agent changes from being greedy, to moderate, to relaxed. According to the mood of agent, the base price changes; therefore, it affects the bidding agent's duration of stay in the auction as a qualified bidder. Table 1 shows the base prices that the inverter bidder agent can go down, according to mood. The utility cost agent basically broadcasts the utility electricity cost for each hour. During the peak hours, it is assumed that the cost of electricity is much higher than at regular hours of the day. In this study, utility peak, and regular, hour prices are defined, respectively, as 0.33 and 0.14 $/kWh.

The aggregator manages communication and information exchange between participants to try to minimize the cost of the electricity to be purchased by the lumped load. The aggregator agent is the main entity that operates the auction process, asks for the yellow pages, and asks for the DERs to participate in the auction, Hence, it will be possible to get the required amount of power at the lowest cost. DERs submit bids to the DER aggregator, declaring the power available to sell, and the price. Each DER has a generation capacity maximum, marginal cost, and availability.

A reverse auction model has been implemented and realistically operated. Algorithm 1 briefly expresses the simplified pseudo-code of the proposed reverse auction generic architecture. The process in Algorithm 1 is continuous, unless the program is terminated. All available generators in the microgrid should register to the yellow pages in order to participate in the bidding process. The required amount of the load is assumed to be estimated for each following hour. Therefore, an aggregator agent has this information ready to inform all agents registered to the yellow page. The main purpose of the aggregator agent is to purchase the required amount of energy with the least cost. The aggregator agent collects the submitted bids, and checks whether the submitted supply matches the demand.

Algorithm 1 AGENT NEGOTIATIONS

Begin
Initialize: All available generators register to yellow page;
Loop
  Aggregator agent: **inform** utility base price, power demand for next hour;
  Aggregator agent: **request** bids;
  if *(DER base price <utility base price)* **then**
  Bidder agents: **inform** submit bids;
else
    Bidder agent: **drop** not participating;
   end
until *(demand ≥total bid submitted supply)* **do**
    Aggregator agent: **inform** new base price:
    Aggregator agent: **request** new bids;
      Bidder agents: **inform** submit bids;
    end until
     Aggregator agent: **inform** market clearing price:
    if *first qualified bidder covers all demand* **then**
   Aggregator agent: **agree** send power reference to qualified bidder;
else
    Aggregator agent: agree send power references to first and the other qualified bidders;
end

If the supply is higher than the demand, the aggregator sends a new bidding request to the bidding agents by lowering the price. If the bidding agents are still interested in the new submission, they will submit the new price; otherwise, they will drop and quit the auction for that particular hour. The process continues until only one qualified bidder remains in the auction. Accordingly, the market-clearing price is established, which is the submitted price of the last, second qualified bidder, as in the Vickrey auction type. One more process is defined to check whether the qualified bidder's submission is able to cover all the demand. Although the first qualified bidder is supposed to cover the main portion of the total demand, if there is still more power required, the second qualified bidder would cover the remaining amount. Once the auction process is over, the aggregator informs the final power references to the qualified bidders. Accordingly, bidder agents inform actuator agents to increase their generation for the next hour. The same process is repeated each hour.

If all generation unit base prices in the microgrid are higher than the utility base price, and none of the generators are interested in participating in the auction, the aggregator agent will buy all required power from the utility. If the total supply is less than the demand, and some of the bidders are participating in the auction, the aggregator agent will buy the submitted amount of power from the DERs, and the remaining power will be purchased from the utility.

While managing the change of power output variety of the generation units, the power system quality should be maintained. Frequency and voltage stability of the system are main concerns for continuous operation. A microgrid can be operated in both grid-connected and islanded mode. While utility connection is available, voltage and frequency are regulated by the grid. The synchronous generator and the inverter-based renewable generation units are operated by the power generation reference.

In this study, only the grid-connected operation mode is studied. The active power balance is directly related to the system frequency; hence it should be regulated by the utility, if proper load sharing cannot be established by the DERs. The dynamic model for frequency is given as follows, where H is the system inertia constant in seconds, ω is the electrical angular velocity or frequency, and P_{m} and P_{e} are mechanical and electrical powers, respectively.

$$2H\frac{\partial \omega}{\partial t} = P_m - P_e \tag{24}$$

Synchronous generators have inertia, however inverter-based DERs not. Although the assigned power reference to each generation unit matches the demand and supply in theory, in real-time implementation, losses and other system behaviors would result in an improper power balance. This mismatch would result in a frequency drift inside the system. Therefore, in the utility-connected mode, the remaining power will be drawn from the grid. Power line loss is an important factor in which large-scale power generation units, and long transmission lines exist. However, in short feeders, since DERs are spread all over the feeder, the power line loss can be neglected. In this study, line losses will be drawn from the utility.

Hardware setup for real-time implementation

The system under study is shown in Fig. 42. PMUs have lately become a reliable measurement and monitoring choice for electric power systems, since they provide system variable measurements, synchronized with microsecond precision. This has been enabled by the availability of global positioning systems (GPS). Local frequency, voltage, current, and power measurements are the main available measurements provided by the PMUs. Thanks to precise sampled data, network statuses in different buses become comparable.

Two PMUs and a revenue meter are deployed on the system as sensor devices. The PMUs are capable of measuring two network buses at the same time since they have two analog measurement channels. Four generation buses are monitored by the PMUs, and the lumped load model is monitored by the revenue meter. In order to understand and visualize the overall view of the complete distribution network, exact time synchronization for all measurements are necessary. A satellite clock is available for the PMUs to get a time synchronized measurement. Apart from readings in the agent platform, a PDC has been implemented to measure the system parameter through the PMUs. The rate of data update from the PMUs have been adjusted to five messages per second, which is quite enough for this application. Monitored parameters from the PMUs are real-power measurements from the utility, generator, and

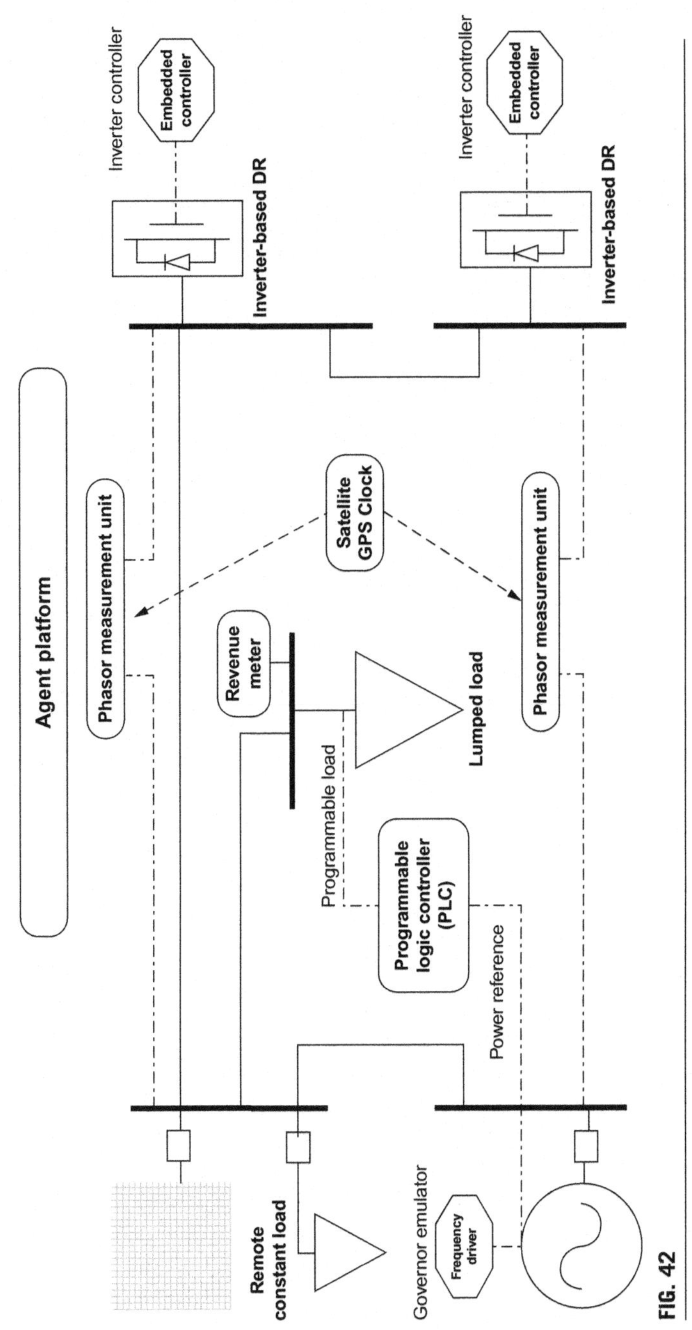

FIG. 42

Hardware setup and agent platform.

the inverter 1 and inverter 2 buses. Voltage variation at the utility PCC is also monitored in real-time.

The PLC is used to enable the load switching function according to the predetermined load profile. The AC synchronous generator model, used for this experiment, is a 13.8 kVA, 60 Hz, 208 V, and 1800 rpm synchronous machine. The AVRmodel is a half-wave, phase-controlled, thyristor-type regulator. The excitation current is derived directly from the generator terminals. The synchronous generator turbine shaft is directly coupled to an induction motor to emulate the governor model. The frequency driver provides the required amount of torque to the induction motor; hence the synchronous generator operates.

Inverter-based DER interfaced with an AC microgrid via a bi-directional AC/DC to DC/AC converter to allow a bi-directional power flow between the AC and DC sources. A 5 kW programmable DC power supply is used to emulate the typical renewable energy resource. This can be considered to be a solar or wind resource. The complete inverter system control was enabled by the DSP-based platform dSpace control desk.

A variety of industrial communication protocols have been implemented to acquire field data, and send information to the actuators. The agent platform is located at a central PC, with a PCI express card mounted to enable the use of multiple serial-communication ports. Each agent is assigned to communicate through individual serial or Ethernet ports. Utility, generator, inverter 1, and the inverter 2 sensor agents access the bus parameter readings by using the IEEE Std. C37.118 synchrophasor data transfer protocol. The load sensor agent interface through the revenue meter interface is enabled by the SEL Fast Message protocol. Modbus is a de facto standard protocol, which is used by many power systems and industrial electronic device vendors. Agent platform and PLC communication is established by the Modbus TCP protocol through Ethernet ports. Modbus RTU is used to communicate with the generator actuator agent through a variable frequency drive, which emulates a synchronous generator governor model. In order to control the inverter 1 and inverter 2 actuator agents, raw data exchange between PC and the dSpace Control Desk is established through the RS232 serial port.

Results and discussions

The game is played for approximately 1 h in a real-time experiment; where each hour of the 24 h of the day is assumed to last 2 min. Table 2 shows the random output of the played game in the microgrid. The predefined daily load profile shows a regular day's power demand in a household which has two specific peak hours: at morning and evening times. As seen from Table 2, the system experiences a morning peak between 0700 and 0900, and an evening peak between 1800 and 1900. The utility peak, and regular, hour prices are defined, respectively, as 0.33 and 0.14 $/kWh. Where the generator is always available, availability of the inverter-based DERs

Table 2 One Hour Random Game Output

Auction for Next Hour				Availability of Units			Base Price			Current Charge		Power Reserved for Present Hour		Power Can Be Bid for Next Hour		Present Irradiation		Participation in Auction			Qualified Bidders			
Hour	Hour for Auction	Utility Price ($/kwh)	Load for Hour (W)	G1 (T/F)	INV1 (T/F)	INV2 (T/F)	G1 Price ($/kwh)	INV1 Price ($/kwh)	INV2 Price ($/kwh)	Inv1 (kWh)	Inv2 (kWh)	Inv1 (kWh)	Inv2 (kWh)	Inv1 (kWh)	Inv2 (kWh)	Inv1 (kWh)	Inv2 (kWh)	Gen1	Inv1	Inv2	Gen1	Inv1	Inv2	Market-Clearing Price
8	9	0.33	1200	T	T	T	0.23	0.12	0.12	400	400	0	0	400	400	500	200	T	T	T	400	400	400	0.23
9	10	0.33	1200	T	T	T	0.22	0.08	0.1	900	600	400	400	500	200	100	300	T	T	T	500	500	200	0.22
10	11	0.14	900	T	T	T	0.21	0.13	0.1	600	500	500	200	100	300	100	300	F	T	T	N/A	100	300	0.13
11	12	0.14	600	T	T	T	0.2	0.13	0.1	200	600	100	300	100	300	400	500	F	T	T	N/A	100	300	0.13
12	13	0.14	600	T	T	T	0.2	0.11	0.1	500	600	100	300	400	300	200	100	F	T	T	N/A	300	300	0.11
13	14	0.14	600	T	T	T	0.22	0.1	0.12	600	400	300	300	300	100	300	500	F	T	T	N/A	300	100	0.13
14	15	0.14	600	T	T	T	0.21	0.1	0.1	600	600	300	100	300	500	200	300	F	T	T	N/A	300	300	0.1
15	16	0.14	600	T	T	T	0.2	0.11	0.1	500	600	300	300	200	300	500	100	F	T	T	N/A	200	300	0.13
16	17	0.33	600	T	T	T	0.24	0.1	0.12	700	400	200	300	500	100	400	500	T	T	T	N/A	500	100	0.23
17	18	0.33	900	T	T	T	0.2	0.08	0.1	900	600	500	100	400	500	200	400	T	T	T	N/A	400	500	0.19
18	19	0.33	1200	T	T	T	0.19	0.1	0.1	600	600	400	500	200	100	0	0	T	T	T	600	200	100	0.32
19	20	0.33	1200	T	F	F	0.23	N/A	N/A	200	100	200	100	0	0	0	0	T	F	F	600	N/A	N/A	0.32
20	21	0.14	900	T	F	F	0.22	N/A	N/A	0	0	0	0	0	0	0	0	F	F	F	N/A	N/A	N/A	0.14
21	22	0.14	600	T	F	F	0.24	N/A	N/A	0	0	0	0	0	0	0	0	F	F	F	N/A	N/A	N/A	0.14
22	23	0.14	600	T	F	F	0.21	N/A	N/A	0	0	0	0	0	0	0	0	F	F	F	N/A	N/A	N/A	0.14
23	24	0.14	600	T	F	F	0.23	N/A	N/A	0	0	0	0	0	0	0	0	F	F	F	N/A	N/A	N/A	0.14
24	1	0.14	600	T	F	F	0.24	N/A	N/A	0	0	0	0	0	0	0	0	F	F	F	N/A	N/A	N/A	0.14
1	2	0.14	600	T	F	F	0.24	N/A	N/A	0	0	0	0	0	0	0	0	F	F	F	N/A	N/A	N/A	0.14
2	3	0.14	600	T	F	F	0.24	N/A	N/A	0	0	0	0	0	0	0	0	F	F	F	N/A	N/A	N/A	0.14
3	4	0.14	600	T	F	F	0.23	N/A	N/A	0	0	0	0	0	0	0	0	F	F	F	N/A	N/A	N/A	0.14
4	5	0.14	600	T	F	F	0.23	N/A	N/A	0	0	0	0	0	0	0	0	F	F	F	N/A	N/A	N/A	0.14
5	6	0.33	600	T	F	F	0.24	N/A	N/A	0	0	0	0	0	0	0	0	T	F	F	600	N/A	N/A	0.32
6	7	0.33	900	T	F	F	0.22	N/A	N/A	0	0	0	0	0	0	0	0	T	F	F	600	N/A	N/A	0.32
7	8	0.33	1200	T	F	F	0.23	N/A	N/A	0	0	0	0	0	0	0	0	T	F	F	600	N/A	N/A	0.32

N/A, not available.

depends on the solar irradiation and stored energy; hence they are not available in the evening or at night due to the absence of irradiance.

Generator 1 has a volatile operation cost, which is defined by a random cost function. The irradiation profile is randomly generated in the software, which can be assumed due to different weather and shadow conditions. The current charge value shows how much power is stored in the renewable resource coupled battery system, as well as the present output for that particular hour. If the unit is not able to discharge the generated power by failing to qualify for the auction, its irradiated power is stored. Inverter 1 and inverter 2 batteries are limited to 600 and 900 kWh, respectively. According to the SOC of the battery, the mood of agent changes, hence the base price changes linearly. The utility price is assumed to be much higher during peak hours as compared to regular hours. This situation enables renewable resources and the generator to compete in supplying power to a lumped load. It can be seen that the competition is higher during peak hours, such as at the 16th hour. When renewable DERs are online at peak hours, the market-clearing price is less than the utility base price, which means that the aggregator agent purchases the electricity at a cheaper cost. In order to determine the next hour of qualified bidders, Fig. 43 illustrates the aggregator agent auction progress for the 16th hour. The 17th hour required power is anticipated to be 600 W, and the utility base price is 0.33 $/kWh.

The auction starts one cent lower than the utility base price. The aggregator agent solicits bids from registered DERs on the yellow page. All generation bidder agents participate in the auction since their base price is lower than the utility base price. The base prices for the generator, inverter 1, and inverter 2 are 0.24, 0.10, and 0.12 $/kWh, respectively. If the submitted bid amount for power is higher than the demand, the aggregator agent runs another round by lowering the base price by one cent in each turn. At the end of nine rounds, the generator bidding agent drops. According to Table 2, the power bid for the 17th hour for inverter 1 and inverter 2 are 500 and 100 W, respectively. Since the total submission meets the demand, the auction process ends and the market-clearing price is established. Since two generation units qualify at the same time, the market-clearing price is defined as a last bid. Upon receiving the qualification information from the aggregator agent, the bidder agents send the new power reference to the actuator agents.

Fig. 44 shows a snapshot of the correspondence between an aggregator agent and other bidder agents during the auction process for the 17th hour. Fig. 45 shows the output power measured in watts by the PMU at the inverter 1 bus. All measurements are collected from the PDC in order to have an exact time reference between measurements. The horizontal numbers on the top of the figure presents the hour of the day. It can be seen that the generation pattern successfully follows the output profile stated in Table 2. Fig. 46 shows the output power measured in watts by the PMU at the inverter 2 bus. It is seen that, in Figs. 45 and 46, renewable DERs are operating between the 8th and 19th hours when solar irradiance exists. Fig. 47 shows the power measured at the generator bus. The generator is mainly operating at peak hours when

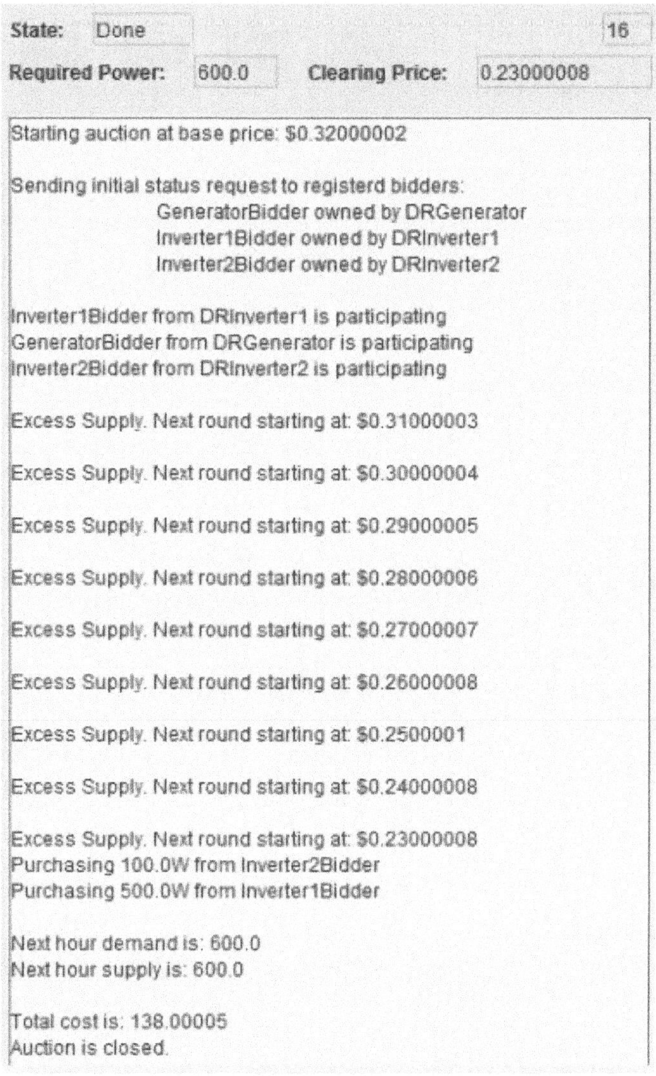

State: Done 16
Required Power: 600.0 Clearing Price: 0.23000008

```
Starting auction at base price: $0.32000002

Sending initial status request to registerd bidders:
        GeneratorBidder owned by DRGenerator
        Inverter1Bidder owned by DRInverter1
        Inverter2Bidder owned by DRInverter2

Inverter1Bidder from DRInverter1 is participating
GeneratorBidder from DRGenerator is participating
Inverter2Bidder from DRInverter2 is participating

Excess Supply. Next round starting at: $0.31000003

Excess Supply. Next round starting at: $0.30000004

Excess Supply. Next round starting at: $0.29000005

Excess Supply. Next round starting at: $0.28000006

Excess Supply. Next round starting at: $0.27000007

Excess Supply. Next round starting at: $0.26000008

Excess Supply. Next round starting at: $0.2500001

Excess Supply. Next round starting at: $0.24000008

Excess Supply. Next round starting at: $0.23000008
Purchasing 100.0W from Inverter2Bidder
Purchasing 500.0W from Inverter1Bidder

Next hour demand is: 600.0
Next hour supply is: 600.0

Total cost is: 138.00005
Auction is closed.
```

FIG. 43

Auction process at the 16th hour.

renewable DERs are not available due to a lack of solar irradiation. Fig. 48 shows the power measured at the utility PCC. In addition to the power remainder, when none of the DERs are operating, the constant remote load and transmission losses are fed from the utility. Fig. 49 shows the voltage profile at the utility PCC. Small variations are observed according to the power drawn from the utility.

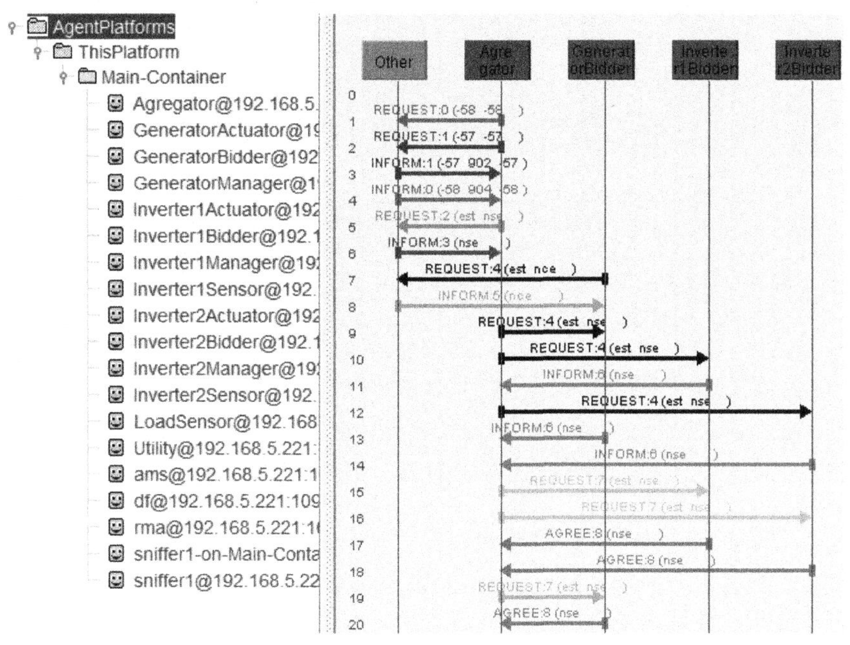

FIG. 44

Sniffer agent: aggregator interaction with other bidder agents.

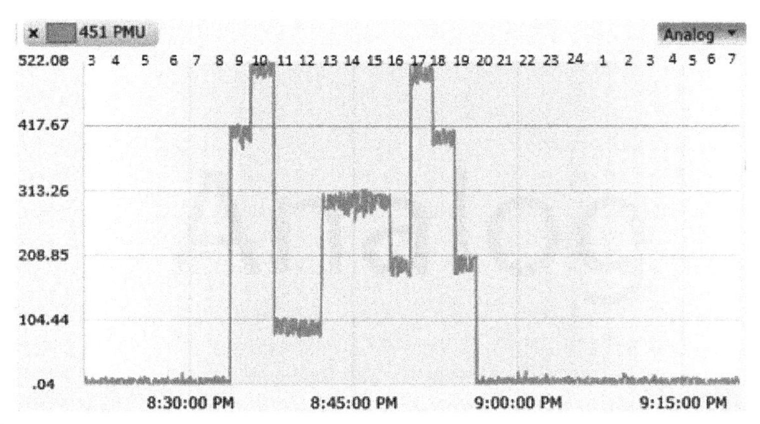

FIG. 45

Inverter 1 power output.

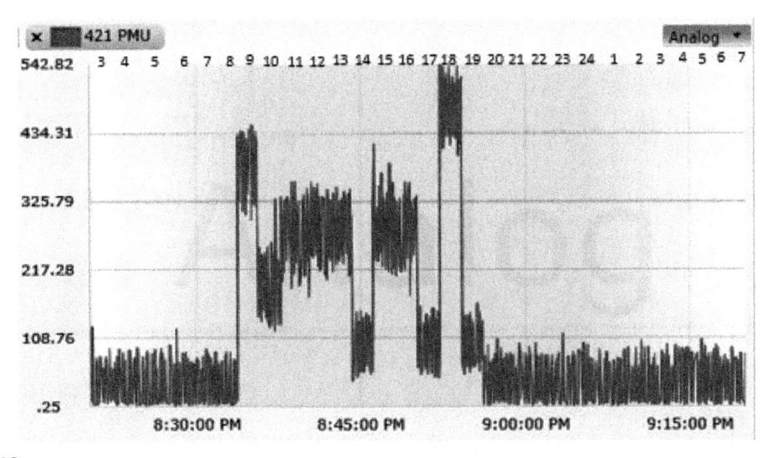

FIG. 46

Inverter 2 power output.

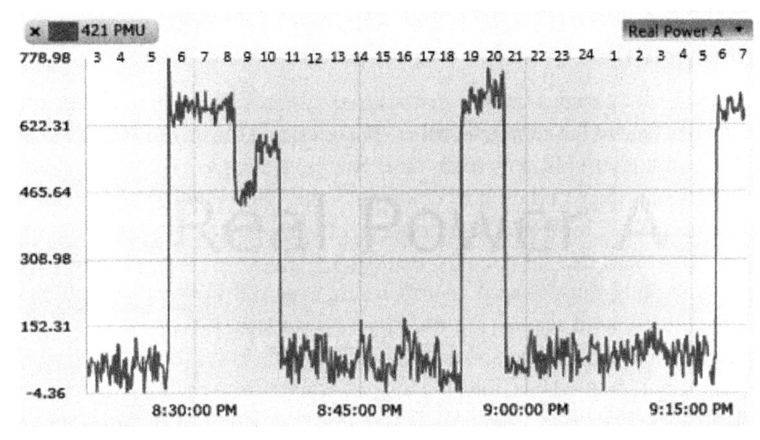

FIG. 47

Generator power output.

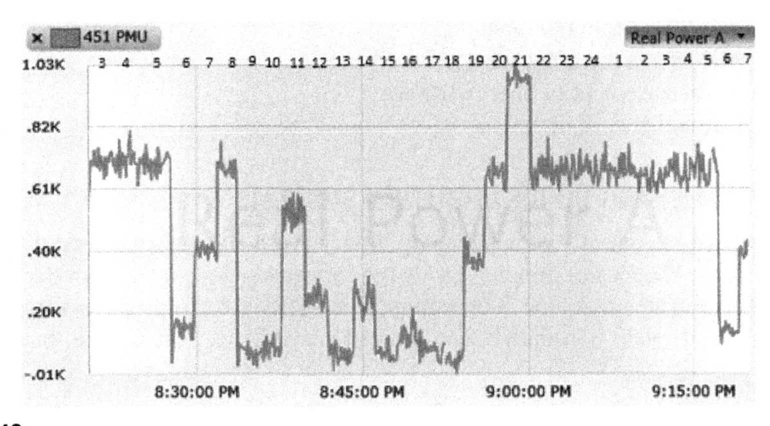

FIG. 48

Power drawn from utility.

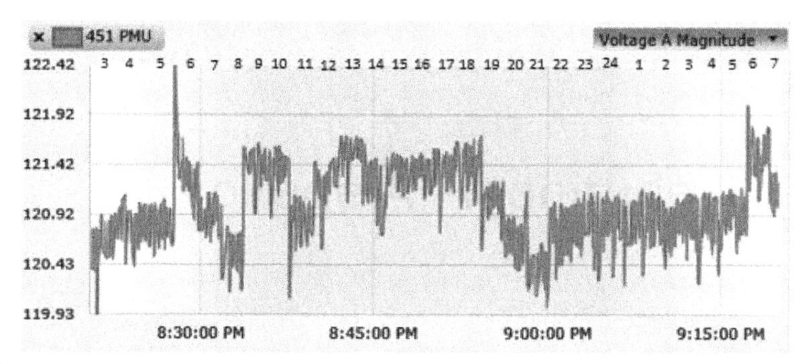

FIG. 49

Voltage profile at utility point of common coupling.

12.4 MODERN DISTRIBUTION ARCHITECTURES IN SMART GRID

During the last two decades, power distribution systems have encountered unprecedented changes. Due to the change in daily life styles, and the incredible advancements in various technology fields, new loads have emerged, such as electronic-based loads. These changes introduce certain new challenges and issues, including but not limited to: low power quality, high losses, uncertainty, high voltage, frequency fluctuations, and so on. Moreover, this new nature mandates high-power supply, continuity, and reliability. Electricity shutdowns do not represent a simple problem, partially preventing some activities, as it was in the old days. Nowadays, it represents a significant impediment for the simplest personal, and professional, activities, and introduces tremendous financial losses. Thus, introducing new solutions to optimize and support the distribution of the system is essential. This chapter provides a discussion on a set of modern solutions that has emerged recently under the umbrella of smart grid technologies. These solutions are typically involved in research community discussions and literature. These solutions are:

1. utilizing DC distribution systems,
2. flywheel as green energy storage,
3. optimal management of energy storage,
4. distribution network optimization.

12.4.1 UTILIZING DC DISTRIBUTION SYSTEMS

One idea that has recently come up to the surface is utilizing the DC-based distribution systems. The idea of utilizing DC is not completely new; it was introduced for the first time at the turn of the 20th century, when a fierce battle over how electricity would be generated, transmitted, and utilized was initiated. This battle, famously known as the "war of currents," was waged by G. Westinghouse and N. Tesla, supporting AC, on one side, and T. Edison, the leading proponent of DC, on the opposite side. Obviously, the debate ended with the predominant implementation of AC distribution in the vast majority of our power systems. This was due to reasons that made a lot of sense at the time. One of these reasons was the invention of the transformer, which offered a great, and simple, means for stepping up the voltage, which consequently widened the area covered by a distribution system, while changing DC voltage levels was an impediment. Moreover, the invention of the polyphase AC machine helped people find an alternative to DC machines, which had remained the only option for some time back then. However, DC systems did not completely disappear from the distribution scene. For instance, there is an old system used by Pacific Gas and Electric (PG&E) in San Francisco that feeds variable speed DC-motored elevators in several historic buildings.

The advances achieved in power electronics, which made DC voltage regulation a simple task, in addition to the increasing penetration of DC loads and sources, encouraged researchers to reconsider DC distribution for at least a portion of today's power systems for increasing overall efficiency by eliminating conversion stages. Dealing with DC power systems became significantly easier due to the stunning advent of semiconductor technology, and the continuous developments of power electronic converters. In addition, there are several other valid reasons for rethinking DC deployment. These reasons are related to the loads, sources, and storage elements.

Loads

Many of today's consumer loads are DC supplied. Electronic-based office and home appliances, such as computers, laptops, tablets, phones, printers, TVs, microwave ovens, and lighting, consume electricity in DC form. Newer, more efficient, lighting technologies, such as compact fluorescent fixtures and solid-state lighting, involve a DC stage, and hence, it is more efficient to utilize them in a DC distribution system. DC power is used in variable speed drives for pumps, HVAC systems, fans, elevators, mills, and traction systems. In addition, for industrial applications, the steel industry has been employing more DC electric arc furnaces because they consume less energy than their corresponding AC ones, and cause less light flicker. The electrochemical industry is almost pure DC application. Supplying these loads through the predominant AC distribution systems adds conversion stages, and consequently, adds inefficiencies to the delivery chain. According to some studies, nearly 30% of the generated AC power passes through a power electronic converter before it is utilized. The amount of lost energy varies, but generally it lies within the range of 10–25%. In another study, the authors mentioned that the power conversion efficiency could be increased by about 8% if a DC bus system were used, and further savings of around 25% could be achieved as a result of removing one rectifier and one power factor correction stage.

Renewable energy sources

Current power systems encounter changes on the source side; these changes are milestones in the history of power systems. Motivated by environmental and economic conditions, there has been a global trend toward the utilization of more renewable energy sources (RES). Some of the RES are natively DC, such as Photovoltaic (PV) and fuel cells (FC), or AC with variable frequency, such as wind, which is integrated into the AC grid through a DC link. Hence, wind farms can be integrated into a DC distribution system with more efficiency. Microturbines are also considered to be a DG resource. Microturbines generating high-frequency AC are also easier to connect to a DC system than to an AC system, where inversion and generating a synchronized sinusoidal AC current is required.

Energy storage

One of the great benefits of DC microgrids is their inherited capability of facilitating static storage integration. Most storage elements are purely DC, such as batteries and ultra-capacitors. Moreover, flywheels, even though they are mechanical energy storage systems (ESS), are mostly coupled to a permanent magnet synchronous machine (PMSM) that is integrated into the distribution system through a DC link. A study carried out by the NTT facility in Japan, comparing an AC uninterruptible power supply (UPS) with a DC one, from an availability perspective, showed that the reliability of the DC supply was higher.

Data centers

Data centers store and transfer huge amounts of digital information, such as internet, cellular communications, and credit card transactions. The main feature that must be maintained in a data center power system is high reliability. Therefore, data centers are typically equipped with UPSs, which require multiple conversion stages in order to connect the batteries to a DC bus. These conversion stages create losses that can be avoided if the power is distributed in DC form, leading to increased overall system efficiency. Consequently, energy cost, which contributes to around 20% of the total operating cost of a data center, is decreased. Therefore, DC distribution is a more economical and efficient option for data centers.

Firstly, in 2006, the idea of utilizing DC-based power distribution systems in data centers was reported in a comprehensive study. The conclusion was based on comparing the efficiencies of five distribution architectures (two AC- and three DC-based). In 2008, a more recent, and accurate, study prepared by Lawrence Berkeley National Laboratory (LBNL) revealed that converting the typical AC distribution systems in data centers to DC-based systems could achieve up to 28% energy savings. LBNL prepared a research roadmap toward a high-performance data center, they emphasized the importance of the conversion of the main power in-feed to DC as a step in improving the reliability and efficiency of data center power supplies. In another study, the authors implicitly promoted the idea of utilizing DC distribution for data centers; they proposed a grounding scheme and made recommendations for a 380 V DC distribution system for data centers.

Even though most of the existing data centers use AC distribution, some of them use DC. Duke Energy data center in Charlotte, North Carolina, is employing a 380 V DC distribution system. Duke Energy and the Electric Power Research Institute (EPRI) prepared a study showing that the system uses 15% less energy than a typical AC system with double conversion UPS. The data center at the University of California, San Diego, is a 2.8 MW DC-based data center, which is powered through a large fuel cell stack. The data center was brought into service in August 2010. "Green," which is one of the top ICT (information and communication technology) service providers in Switzerland, announced the opening of their 1 MW DC-based data center in May 2012. Hewlett-Packard has provided IT equipment supporting DC input, commercial availability of DC enabled IT equipment is a stunning and encouraging step toward a wide deployment of DC data centers. Although DC distribution is not utilized in Google data centers, they have managed to save $30/year per server by optimizing the power path by eliminating two AC/DC conversion stages and bringing the batteries on the server rack.

Electric vehicles

The global call for reduced CO_2 emissions, the investments that pioneering automotive companies have been making in the advancement of plug-in all electric, and hybrid EV technologies, and the problems inherently associated with fuel availability and price stability will inevitably lead to a significant increase in the number of EVs in the near future. The problem of coordinating the charging process of a large number of EVs has lately acquired the attention of many researchers, and is still under study.

It has not yet been determined whether EVs will be charged casually at home, like any other home appliance; at a fast charging station, similar to a gas-fueling station for conventional vehicles; at a place where a discharged EV battery is replaced with a completely charged one; or at a smart charging park where EVs are coordinated centrally at a smart garage that enable vehicle-to-vehicle (V2V) and vehicle-to-grid (V2G) services. Each of these different techniques has supporters and opponents for reasons that are outside the scope of this chapter. However, the last model relates to DC distribution, since some of the researchers who work on the concept of smart charging parks believe that they should operate as DC microgrids, with a common DC bus at which the EV batteries, and any DG units, should be integrated.

It is worth mentioning that a vehicle-to-grid (V2G) concept has been recently introduced, where aggregated batteries of grid-connected vehicles serve as a bulk energy storage that is able to support the grid operation. Imagine a parking lot where 20 TESLA model S cars are parked, with a 70 kWh battery bank in each car, there is a bulk storage of 0.7 MWh (assuming 50% available capacity) available for different ancillary services for the grid. This point seems to be a strong supporting point to allow V2G to dominate. However, this strategy is facing stronger challenges, which is lacking public acceptance. V2G techniques imply violation of the recommended charging pattern and an increased number of charging/discharging cycles. Consequently, this causes accelerated wear and tear of the batteries, reducing their lifetime and performance. Taking into account the current battery technology, it seems that this strategy will—for a while—remain not practically applicable, unless incredible incentives are offered for the car owners.

12.4.2 FLYWHEEL AS GREEN ENERGY STORAGE

Driven by the latest developments in different engineering realms, such as superconducting bearings, frictionless, vacuum encased machinery, and power electronic switching, flywheels gained much interest as a reliable energy storage element. Flywheel energy storage systems (FESS), or sometimes known as electromechanical batteries, have been used lately in several applications, such as for data centers, aerospace, shipboard power systems, UPS, electrification of rural areas, fast charging of EVs, and improving renewable energy integration. The FESS operates in three operating modes: charge, stand-by, and discharge. Since flywheels are classified as short-term energy storage, the transition among these three modes should be performed rapidly, unlike other types of energy storage elements. In order for this process to be done seamlessly, a fast acting, flexible and reliable driving system is required. Moreover, the control loop should be designed accurately to avoid slow action, high-overshoot, or steady state errors.

Generally, flywheels can be classified according to their speed into low speed and high speed. High-speed systems feature a much lower weight and smaller size. However, they entail sophisticated technologies for reducing friction, and their power output is limited by its cost and difficulty in cooling. A variety of machines have been discussed in the literature for use in low-speed flywheel applications including induction machines, and doubly fed induction machines (DFIM). In this system, the stator was connected to the grid through a step up transformer, while the secondary machine (rotor) was connected to a cycloconverter via slip rings. Thus the existence of slip rings is acceptable in low-speed flywheels while it is not in high speed ones. Consequently, a brushless DC, homopolar inductor, PMSM, axial flux permanent magnet, and synchronous reluctance machines are preferred in high-speed flywheels.

FESS operates in three operating conditions: charge, stand-by, and discharge. In the charging mode, the power grid injects energy into the flywheel through a bi-directional converter. When the flywheel reaches the maximum stored kinetic energy limit, the FESS moves into the standby mode in which the charging current is kept small in order to maintain its charge and spin at the rated speed. If a power outage occurs, the FESS switches into the discharge mode, the PMSM acts as a generator to provide energy to loads through a bi-directional power converter. When the power grid recovers from the failure, the FESS reenters the charge mode and is ready to handle the next power quality event.

This chapter briefly addresses the control and performance of DC machine-based flywheels in DC microgrids. DC machines feature a rugged construction and reliable operation. Furthermore, they can be interfaced to the DC distribution network through DC/DC converters, which are simpler and more efficient than their AC/DC counterparts. The detailed modeling of a DC/DC converter-based driving system for a FESS, including the parasitic resistances for all elements, was carried out. In order to validate the derived model, it was compared to another model, which

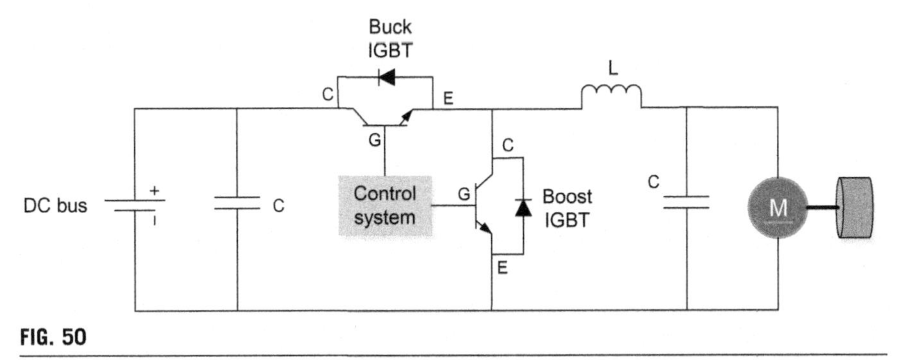

FIG. 50

Schematic diagram for flywheel driving system.

was estimated using the MATLAB/Simulink environment. Two control loops were designed independently for each mode of operation. Improved voltage/current controllers were introduced to achieve steady operation during the charging and discharging modes.

DC-based FESS modeling

The flywheel driving system is based on a DC/DC bi-directional, buck-boost converter. Fig. 50 shows a schematic diagram for the system. As shown in the figure, the utilized topology features two IGBT switches; during each mode of operation, one switch is operated while the other one is disabled. Given that the DC bus voltage is higher than the machine terminal voltage over the entire operation range, the converter acts as a buck converter in the charging mode while acting as a boost during discharge. Each mode was modeled separately with its two switching states then an averaged state space model was obtained.

Discharging state (boost mode)

In the boost mode, the source will be the machine, ie, the machine will be operated as a generator. The capacity of the DC network is much larger than the FESS and the voltage of the DC bus is controlled through a large stable source (eg, a large voltage source converter (VSC) or DC generator). Thus, the DC network can be represented by a DC voltage source with a small resistance. This small impedance in the converter's output can lead to an impedance mismatch, and consequently, this mismatch can cause instability in the whole system due to the violation of the Nyquist stability criterion. In the boost mode, the control system controls the boost IGBT within a duty cycle (D) based on a reference current sent from the main distribution network controller, while it disables the buck IGBT.

1. ON state

When the IGBT is in its ON state, the converter equivalent circuit tends to be as shown in Fig. 51A. In order to obtain the most accurate model, a model of the DC machine represented by an R-L-E branch is added to the converter model. The state variables are the inductor current, i_L, the machine current, i_m, the input

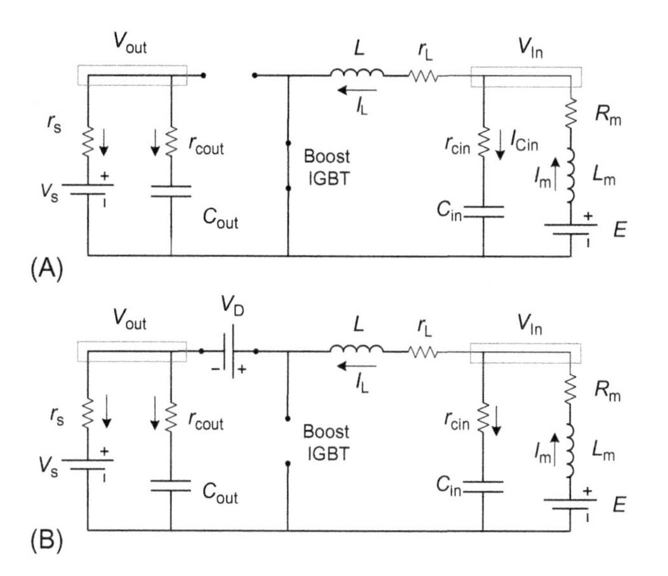

FIG. 51

Equivalent circuit for a converter in boost mode during (A) ON state and (B) OFF state.

capacitor voltage, v_{cin}, and the output capacitor voltage, v_{cout}. By applying KVL and KCL, the dynamic equations of the system are derived as the following:

$$\frac{di_L(t)}{dt} = \frac{V_{in}(t) - i_L(t) \cdot r_L}{L} \tag{25}$$

where V_{in} is the input voltage, i_L, r_L, and L is the inductor current, resistance, and inductance, respectively. Then by applying KCL at the input node $(i_{cin}(t) = i_m(t) - i_L(t))$ to find V_{in} as a function of the state variables.

$$V_{in}(t) = v_{cin}(t) + i_m(t) \cdot r_{cin} - i_L(t) \cdot r_{cin} \tag{26}$$

where r_{cin} is the input capacitor resistance. By substituting from Eq. (26) into Eq. (25) yields:

$$\frac{di_L(t)}{dt} = -\frac{1}{L} \cdot [r_{cin} + r_L] \cdot i_L(t) + \frac{r_{cin}}{L} i_m(t) + \frac{1}{L} \cdot V_{cin}(t) \tag{27}$$

The derivative of the machine current i_m can be evaluated as:

$$L_m \frac{di_m(t)}{dt} = E - V_{in}(t) - r_m \cdot i_m(t) \tag{28}$$

where L_m and r_m are the machine equivalent inductance and resistance, respectively. Similarly, substituting from Eq. (26) into Eq. (28), and rearranging the equation terms, yields:

$$\frac{di_m(t)}{dt} = \frac{r_{cin}}{L_m} \cdot i_L(t) - \frac{1}{L_m} \cdot (r_{cin} + r_m) \cdot i_m(t) - \frac{1}{L_m} V_{cin}(t) + \frac{1}{L_m} E \tag{29}$$

The derivative of the input capacitor voltage can be expressed as:

$$\frac{dv_{\text{cin}}(t)}{dt} = -\frac{1}{C_{\text{in}}} \cdot i_{\text{L}}(t) + \frac{1}{C_{\text{in}}} \cdot i_{\text{m}}(t) \tag{30}$$

where the capacitance of the input capacitor is C_{in}. In order to get V_{out} as a function of the state variables, apply KCL at the V_{out} node:

$$V_{\text{out}} = V_{\text{cout}} \cdot \frac{r_{\text{s}}}{(r_s + r_{\text{cout}})} + V_{\text{s}} \cdot \frac{r_{\text{cout}}}{(r_{\text{s}} + r_{\text{cout}})} \tag{31}$$

where V_{s}, and r_{s} are the voltage, and the equivalent resistance of the voltage source representing the DC bus, r_{cout}, is the resistance of the output capacitor. The derivative of the output capacitor voltage can be given by:

$$\frac{dv_{\text{cout}}(t)}{dt} = -\frac{1}{C_{\text{out}} \cdot (r_s + r_{\text{cout}})} \cdot V_{\text{cout}} + \frac{1}{C_{\text{out}} \cdot (r_{\text{s}} + r_{\text{cout}})} \cdot V_{\text{s}} \tag{32}$$

where C_{out} is the capacitance of the output capacitor. Using Eqs. (27), (29), (30), and (34) to formulate the state space model as in Eq. (33) and writing it in the matrix form result in Eq. (34). It should be noted here that the output equation is not used because the controlled variables are state variables.

$$\frac{dx(t)}{dt} = A_{\text{on}}x(t) + B_{\text{on}} \cdot u(t) \tag{33}$$

$$\frac{d}{dt} + \begin{bmatrix} 0 & 0 & 0 \\ 0 & \frac{1}{L_{\text{m}}} & 0 \\ 0 & 0 & 0 \\ \frac{1}{C_{\text{out}} \cdot (r_{\text{s}} + r_{\text{cout}})} & 0 & 0 \end{bmatrix} \cdot \begin{bmatrix} V_{\text{s}} \\ E \\ V_{\text{D}} \end{bmatrix} \tag{34}$$

$$\begin{bmatrix} i_{\text{L}}(t) \\ i_{\text{m}}(t) \\ v_{\text{cin}}(t) \\ v_{\text{cout}}(t) \end{bmatrix} = \begin{bmatrix} -\frac{1}{L} \cdot \left[r_{\text{cin}} + r_L + \frac{(r_s \cdot r_{\text{cout}})}{(r_s + r_{\text{cout}})} \right] & \frac{r_{\text{cin}}}{L} & \frac{1}{L} & -\frac{r_{\text{s}}}{L \cdot (r_{\text{s}} + r_{\text{cout}})} \\ \frac{r_{\text{cin}}}{L_{\text{m}}} & -\frac{1}{L_{\text{m}}} \cdot (r_{\text{cin}} + r_m) & -\frac{1}{L_m} & 0 \\ -\frac{1}{C_{\text{in}}} & \frac{1}{C_{\text{in}}} & 0 & 0 \\ \frac{r_s}{C_{\text{out}} \cdot (r_s + r_{\text{cout}})} & 0 & 0 & -\frac{1}{C_{\text{out}} \cdot (r_{\text{s}} + r_{\text{cout}})} \end{bmatrix}$$

$$\cdot \begin{bmatrix} i_{\text{L}}(t) \\ i_{\text{m}}(t) \\ v_{\text{cin}}(t) \\ v_{\text{cout}}(t) \end{bmatrix} \frac{d}{dt} + \begin{bmatrix} -\frac{r_{\text{cout}}}{L \cdot (r_{\text{s}} + r_{\text{cout}})} & 0 & -\frac{1}{L} \\ 0 & \frac{1}{L_{\text{m}}} & 0 \\ 0 & 0 & 0 \\ \frac{1}{C_{\text{out}} \cdot (r_{\text{s}} + r_{\text{cout}})} & 0 & 0 \end{bmatrix} \cdot \begin{bmatrix} V_{\text{s}} \\ E \\ V_{\text{D}} \end{bmatrix} \tag{35}$$

2. OFF State

Fig. 51B shows the equivalent circuit of the converter during the off state. In this case the antiparallel diode with the buck IGBT will be forward biased. Using a similar approach, the state space model for the converter in the OFF state can be written as shown in Eq. (36). An averaged state space model is obtained by multiplying each state by its corresponding duty cycle.

$$A = A_{on} \cdot D + A_{off} \cdot (1 - D) \tag{36}$$

$$B = B_{on} \cdot D + B_{off} \cdot (1 - D) \tag{37}$$

After averaging, the system is perturbed around a steady state operation point, then linearized by neglecting the second order terms. By applying Laplace transformation, the steady state operating point can be obtained by Eq. (38).

$$X = -A^{-1} \cdot B \cdot U \tag{38}$$

Given that U is the input vector, the solutions of the state variables are given by Eq. (39), and then the transfer function of the duty cycle for each state variable can be obtained. In this case, the control loop of the boost converter mode will be controlling the injected current to the DC bus to fulfill the pulsed-load required energy. Therefore, the important transfer function is the duty cycle-inductor current. This transfer function will be used for the accurate design and tuning of the PI controller.

$$\hat{x} = (SI - A)^{-1} \cdot [A_{on} - A_{off}] \cdot X - (B_{on} - B_{off}) \cdot U] \cdot \hat{d} \tag{39}$$

Charging state (buck mode)

$$\frac{d}{dt}\begin{bmatrix} i_L(t) \\ i_m(t) \\ v_{cin}(t) \\ v_{cout}(t) \end{bmatrix} = \begin{bmatrix} -\frac{1}{L}\cdot\left[\frac{(r_s \cdot r_{cin})}{(r_s + r_{cin})} + r_{cout} + r_L\right] & \frac{r_{cout}}{L} & \frac{r_s}{L \cdot (r_s + r_{cin})} & -\frac{1}{L} \\ \frac{r_{cout}}{L_m} & -\frac{1}{L_m}\cdot(r_{cout} + r_m) & 0 & \frac{1}{L_m} \\ \frac{r_s}{C_{in} \cdot (r_s + r_{cin})} & 0 & -\frac{1}{C_{in} \cdot (r_s + r_{cin})} & 0 \\ \frac{1}{C_{out}} & -\frac{1}{C_{out}} & 0 & 0 \end{bmatrix} \cdot \begin{bmatrix} i_L(t) \\ i_m(t) \\ v_{cin}(t) \\ v_{cout}(t) \end{bmatrix} + \begin{bmatrix} \frac{r_{cin}}{L \cdot (r_s + r_{cin})} & 0 & 0 \\ 0 & -\frac{1}{L_m} & 0 \\ -\frac{1}{C_{in} \cdot (r_s + r_{cin})} & 0 & 0 \end{bmatrix} \cdot \begin{bmatrix} V_s \\ E \\ V_D \end{bmatrix} \tag{40}$$

$$\frac{d}{dt}\begin{bmatrix} i_{\mathrm{L}}(t) \\ i_{\mathrm{m}}(t) \\ v_{\mathrm{cin}}(t) \\ v_{\mathrm{cout}}(t) \end{bmatrix} = \begin{bmatrix} -\frac{1}{L}\cdot[r_{\mathrm{cout}}+r_{\mathrm{L}}] & \frac{r_{\mathrm{cout}}}{L} & 0 & -\frac{1}{L} \\ \frac{r_{\mathrm{cout}}}{L_{\mathrm{m}}} & -\frac{1}{L_{\mathrm{m}}}\cdot(r_{\mathrm{cout}}+r_{\mathrm{m}}) & 0 & \frac{1}{L_{\mathrm{m}}} \\ 0 & 0 & \frac{1}{C_{\mathrm{in}}\cdot(r_{\mathrm{s}}+r_{\mathrm{cin}})} & 0 \\ \frac{1}{C_{\mathrm{out}}} & -\frac{1}{C_{\mathrm{out}}} & 0 & 0 \end{bmatrix}$$

$$\cdot\begin{bmatrix} i_{\mathrm{L}}(t) \\ i_{\mathrm{m}}(t) \\ v_{\mathrm{cin}}(t) \\ v_{\mathrm{cout}}(t) \end{bmatrix} + \begin{bmatrix} 0 & 0 & -\frac{1}{L} \\ 0 & -\frac{1}{L_{\mathrm{m}}} & 0 \\ \frac{1}{C_{\mathrm{in}}\cdot(r_{\mathrm{s}}+r_{\mathrm{cin}})} & 0 & 0 \end{bmatrix} \cdot \begin{bmatrix} V_{\mathrm{s}} \\ E \\ V_{\mathrm{D}} \end{bmatrix} \tag{41}$$

The converter will be operated in this mode during the off pulse (the pulsed load is not energized), so there is excess energy in the system to charge the flywheel. The power flow direction will be reversed from the DC network to the FESS. In this mode, the DC machine is working as a motor, in contrast to the boost mode where the machine was working as a generator. Fig. 52A and B show the converter equivalent circuits in both IGBT states. By using a similar procedure to the one used for the boost mode, the dynamic equations of the buck converter are determined during ON and OFF states, and are given in Eqs. (40) and (41), respectively.

The transfer functions of the duty cycle to each state variable are determined, however, the controller in the buck mode will be controlling the machine terminal voltage

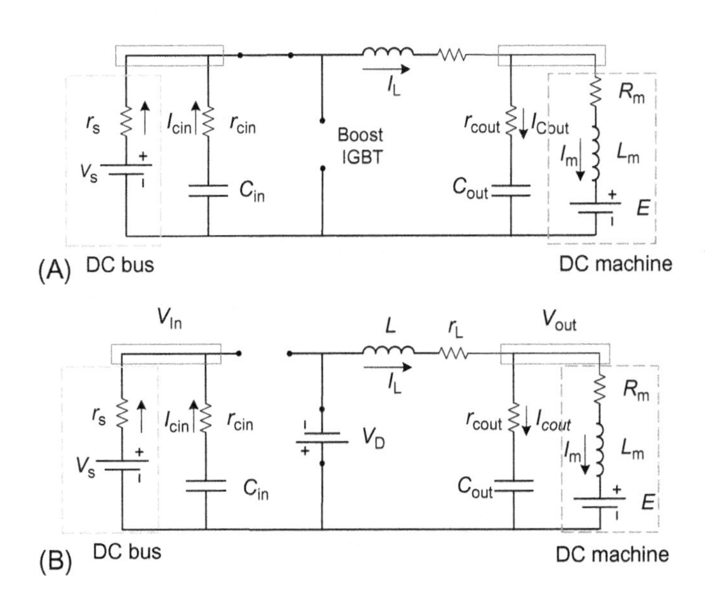

FIG. 52

Equivalent circuit for the converter in buck mode during (A) ON state and (B) OFF state.

in order to control the charging rate. Thus, the transfer function of the duty cycle to the output capacitor voltage is used to design the PI control loop for the buck mode.

FESS controller design

Two PI control loops were designed, one for each operation mode. Switching between both controllers is performed through the main supervisory controller of the distribution network based on its needs. The main purpose for this is to utilize the available energy in the FESS, to fill the energy gap caused by the pulsed load, and to prevent any voltage fluctuations. The controller of the buck mode is designed to control the charging rate through the control of the machine voltage. It is known that the DC machine can withdraw high-currents during the start until the motor rotation builds up the counter electromotive force (EMF). For this reason, a constant voltage reference cannot be used; rather, a soft starting approach has to be utilized. The controller generates an internal ramping up voltage reference. One should keep in mind that the duration of the pulsed load is unknown, consequently the SOC of the FESS at the end of a given discharging cycle is unknown. Since the SOC is directly dependent on the machine speed, the machine speed and voltage are not known at the end of the pulsed load period as well. To assure a stable operation over a wide range of pulses, it is mandatory that the controller picks the last voltage value before switching to the buck mode (charging). Then the controller generates the ramping voltage reference starting from this value and ending at the full speed of the machine (full SOC). This approach allows the utilization of a very wide speed range of the machine, which can be an advantage over the DFIMs. Since the operation speed range of the DFIM is limited by the allowed slip, which is typically within $\pm 15\%$ to $\pm 30\%$.

A block diagram showing the controller design is given in Fig. 53. There are two independent PI loops for the buck/boost modes controlling the charging and discharging of the flywheel, respectively. The transfer function for the buck mode controller is C_{buck} and C_{boost} for the boost mode controller. It should be emphasized here that when the FESS is operating in a certain mode, the other IGBT should be turned off. For example, if the machine is charging, the buck mode controller is activated and the buck IGBT is working (please refer to Fig. 50). Hence, the boost IGBT should be

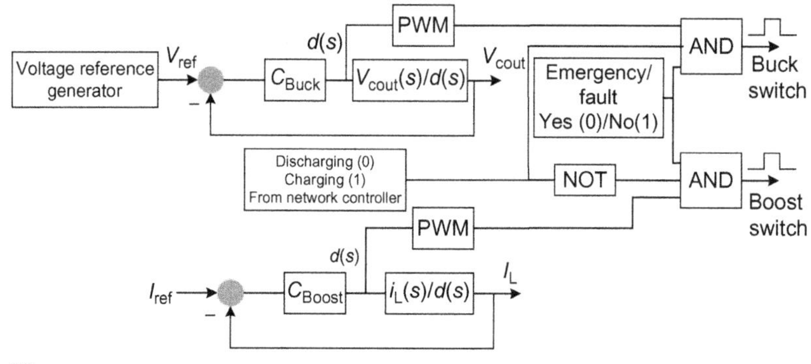

FIG. 53

Block diagram for the FESS controller.

kept off and vice versa. A simple interlocking logic circuit is provided for this function. Moreover, this design is prepared to be suitable for hardware and practical implementation. An input is provided to disable the converter and turn off both switches in case of an emergency or fault. One of the merits of this controller is its simplicity, by avoiding heavy computations it can be easily implemented on any microcontroller or field-programmable gate array chip.

As mentioned previously, the converter in the boost mode works in current control to control the injected current into the DC bus. The reference current (I_{ref}) is calculated and passed from an outer control loop, which is typically resolved by a slower controller.

Case study and discussion

A model for a DC distribution network with a pulsed load and the proposed FESS is built in Matlab/Simulink environment. The parameters of the machine and converter models are listed in Tables 3 and 4, respectively. The flywheel inertia (J) is taken as 0.75 kg m^2. The simulation results are shown in Fig. 54, the pulsed load profile is as shown in Fig. 54B. This pulse load can be either an electromagnetic gun, or an integrated fight through power (IFTP) system. Also, a large number of EVs requesting fast charging, or large cranes, can represent a pulsed load. In this case, the duty cycle of the pulsed load is variable (ie, a nonuniform pulsed load profile) and its amplitude is 30 A.

Table 3 Machine Parameters

Arm. voltage	100 V
Arm. current	19 A
Field voltage	100 V
Field current	0.6 A
r_m	0.44 Ω
L_m	12.9 mH
r_f	207 Ω
L_f	115 H
Speed	1750 rpm

Table 4 Converter Parameters

DC bus voltage	325
r_s	0.01 Ω
C_{out}	1200 μF
r_{cout}	0.008 Ω
L	12.7 mH
r_L	0.125 Ω
C_{in}	1200 μF
r_{cin}	0.008 Ω
V_D	1.75 V

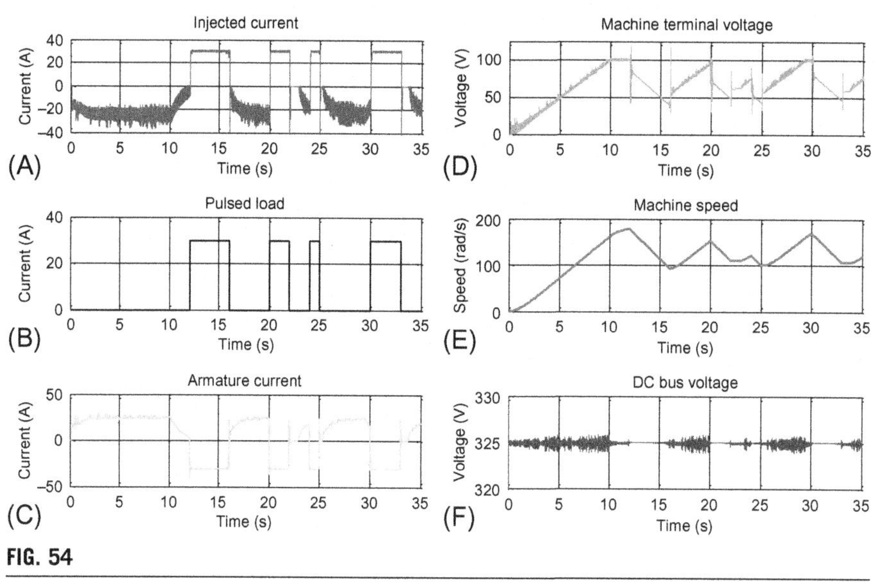

FIG. 54

FESS operation under pulsed load.

The injected current to/from the DC bus is shown in Fig. 54A. It can be seen that during the pulse, the injected current perfectly follows the pulsed shape to fulfill the energy gap and meet the pulse load requirements. This is due to the fast, yet stable response of the controller. The injected current is steady around the 30 A level, which is the amplitude of the pulse.

When the pulsed load is off, the machine charges, which is denoted by a negative current in Fig. 54A. The machine terminal voltage and speed are shown in Fig. 54D and F, respectively. It can be seen that the machine is accelerating smoothly during charging, and then it decelerates during the load, indicating a loss of charge. Because of the nonuniformity of the pulse load profile, at some points, the span between any two consequent pulses is very short. Thus, the next pulse occurs while the FESS is not completely charged. The FESS can flexibly react to this situation and switch seamlessly from charging to discharging without any interruptions. Minor spikes are detected in the machine voltage during charging, this is because the higher current was withdrawn during the build-up of the counter EMF. This is reflected as fluctuations in the DC bus voltage. However, these fluctuations are less than 0.5%, which is acceptable by the standards. On the other hand, there are no fluctuations during the pulse, which indicates a complete mitigation of the pulsed load. The armature current is shown in Fig. 54C, which is a reverse of the current shown in Fig. 54A.

12.4.3 OPTIMAL MANAGEMENT OF ENERGY STORAGE

The structure and resiliency of the emerging smart grid will rely heavily on storage facilities in cooperation with renewable energy to provide uninterrupted service to the customer. The usage of battery arrays continues to grow due to their capability in restoring system voltage and frequency following an outage. Moreover, promoting smart grid concepts, along with the high penetration of RES and DG in modern grid architecture, drives the deployment of new ESS. Currently, lead acid batteries are being widely deployed for usage in grid energy storage due to their versatility and low cost, but are burdened by a number of factors, which has resulted in a decline in their performance without a proper management system. As lead acid ages, the series resistance will begin to increase, as it is susceptible to many factors that impact its lifespan, most notably, the number of cycles that it has charged and discharged, effects of temperature, and the types or levels of discharge currents it has been exposed to. These factors lead to electrochemical changes inside a battery, which result in decreased usable capacity and inefficient charging. A number of studies have been presented in the literature on battery management system (BMS) architectures, but most have only looked at case studies under smaller configurations. In [4], a management system solution is presented to highlight the importance of including the battery state of health (SOH) in measurements for grid applications. Focussing on a range of discharge rates, and the depth of discharge, for two parallel-configured, lead-acid batteries, the life consumption rate is modeled for each of the two batteries independently. Battery stack configurations are regulated to produce a total SOC through the means of the current integration method, or in some simplified cases, voltage-based measurements. In a study, an advanced method of depicting the SOC of a lithium-ion utility array was tested in grid-connected and islanded modes during microgrid operations. Three modes of operation are proposed to indirectly measure the SOC of the battery array. This system was able to demonstrate the effectiveness of its control strategy, but was still unable to access individual battery modules.

Though individual modules have been difficult to access in series configurations, research has been conducted with SOC balancing for parallel connections. Chaoui and Sicard [5] proposed a supervisory energy management algorithm designed to regulate the charge flow to a bank of three batteries. A constant current is used to charge each battery independently with respect to the load profile and individual SOC measurements. Balancing of the array was accomplished simply through switching, and was unable to adjust the charging current. Different methods of cell equalization for lithium ion batteries, including the flying capacitor charge shuttling method, as well as shared single and multiple transformer methods are discussed extensively in the literature. Qian et al. [6] proposed a charge equalization technique similar to a transformer-based one. This technique employs a topology, which is developed specifically for lithium ion cells. In this topology, the charging voltage is pulsed through a control signal and passed through a transformer. The current from the cell stack induces currents in each of the secondary coil windings, and the secondary (connected to each battery) with the least reactance will have the highest induced current. This ensures each cell has a charge current proportional to its SOC. Lawder et al. [7] proposed a comprehensive review of the existing BMSs for grid-scale applications. A notional model for implementing a BMS into a battery ESS is presented. In this model, a number of objectives are identified, including the source peak power demand, cell balancing, and thermal control, but it still lacks the capability of extracting individual cells from a stack, or even parallel configuration. In this case, SOH and SOC methods are identified for lithium ion and redox flow batteries only. Although lead acid is identified as a medium of technological maturity and low-cost, it is not presented for use in this survey. Two simplified battery management schemes are presented by Wen and Qiu. The latter emphasized the importance of not neglecting the differences in a battery's internal resistance during charging.

Tran and Khambadkone [8] presented an EMS that was developed with respect to extending the lifetime and efficiency of an ESS. Using the Peukert lifetime energy throughput model, energy efficiency of the ESS is improved. An optimization algorithm is presented by Grillo et al. [9] where a discrete-time model of an electrochemical storage device is developed to introduce a battery system, but is limited to simply wind turbine and sodium nickel chloride battery combinations.

Pulsed charging has introduced a revolutionary control over battery charging behaviors, accelerating charging rates and providing battery charge balancing. Pulsed charging is not only used as a method to regulate a charging current. In the case of a damaged battery, charging current pulses can be used as an attempt to characterize or improve the battery SOH by neutralizing the internal electrolyte. Introducing this capability to each battery in a stack configuration would not only provide controlled current charging, but also provides a tool to potentially revive batteries in the event of a failure.

In this brief discussion, there is an imperative need for individual and independent control of each unit (battery or stack) in the bank to achieve efficient operation. Another major requirement is to prevent failures on a single battery from impacting the operation of the whole energy storage bank. The system presented can be suitable for applications ranging from a microgrid to a utility scale.

In this chapter, energy storage management systems (ESMS) are presented as monitors for individual batteries inside a series configuration that identify independent voltages, current contributions, and SOC levels for each energy storage element. A method to balance SOC and to diagnose SOH is established, where an ESMS fully decouples the battery from the system and applies it to a charging and diagnostics bus, while still maintaining a connection to the load. The system bypasses the decoupled battery to guarantee the continuity of supply and normal operation of the whole stack. The proposed system offers the capability of charging more than one battery at the same time with different charging levels by applying pulsed charging currents with different duty cycles and frequencies. The voltage fluctuations accompanied by the coupling/decoupling of batteries are mitigated by a boost DC/DC converter to maintain bus voltages, preventing the propagation of problems from the ESMS to the utility side.

The introduced systems do not only provide the optimal management (charging/discharging) for the energy storage arrays, but also offers a variety of features and capabilities, which can be summarized by the following:

- It is capable of controlling each individual unit within a series/parallel array, ie, each single unit can be treated, controlled, and monitored separately from the others.

- The capability to charge some units within a battery array, while other units continue to serve the load. By applying a pulsed charging profile at different frequencies and duty cycles, the BMS can regulate the charging energy to each unit. Hence, SOC balancing can be accomplished without the need for power electronic converters.
- It is able to electrically disconnect a unit and allow the operator to perform the required maintenance, or replacement, without affecting the performance of the whole array.
- Considering the appropriate design, and the selection of the relays and other components, the system can be expanded for controlling other types of energy storage.
- The system can be designed and implemented at low cost.

Based on the presented literature survey, most of these energy storage management systems focus either on cell equalization, SOC and SOH estimation, or pulsed charging. Moreover, some of these schemes involve transformers or large complicated power electronic devices. To the best of the author's knowledge, no system has all the aforementioned capabilities and features like the proposed one.

ESMS design

This section discusses the philosophy of the ESMS design. It is worthy to mention that this design is generic, and the developed ESMS can be connected across an individual battery, or stack of batteries, in large-scale applications. In this context, the design process and implementation is explained for 12 V battery, however, it should be kept into consideration that the system can be scaled up to the utility scale by adequately considering the current and voltage ratings. The overall topology for connecting ESMS units for N_{batt} batteries is shown in Fig. 55. The 14.7 V battery charging and diagnostic bus is connected in parallel to all BMS links. The terminals of the battery bank are connected to a DC/DC boost converter to stabilize the Bus DC voltage. This converter is unidirectional as the charging of the batteries is accomplished through another bus.

The schematic for a single BMS unit is shown in Fig. 56. The battery is placed in-between a network of relays in order to provide for complete coupling and decoupling. The various components of the system are described as the following:

DC bus connectivity: In order to achieve full isolation, two normally closed (NC) relays connect the positive and negative terminals of the battery to the DC bus. A normally open (NO) relay is connecting the positive terminal of the ESMS to its negative side to offer a battery bypass circuit, decoupling it from the array while still providing an alternative path to maintain the continuity of supply. An interlock is provided between the three relays to avoid simultaneous connection. This would fully isolate the battery in the case of performing maintenance, or coupling of the charging circuit.
Current measurement: Current measurement is provided directly at the battery terminals. The LA 25-NP current transducer is installed in a series on the

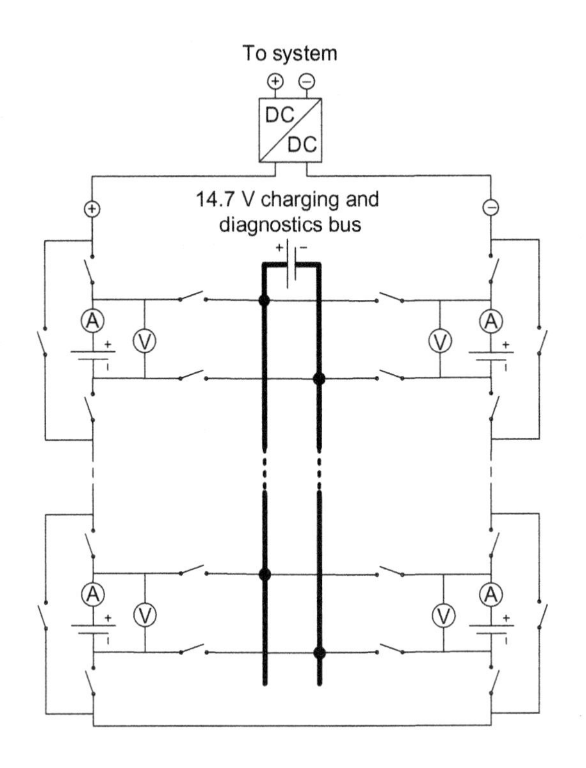

FIG. 55

Battery management unit array for N_{batt}.

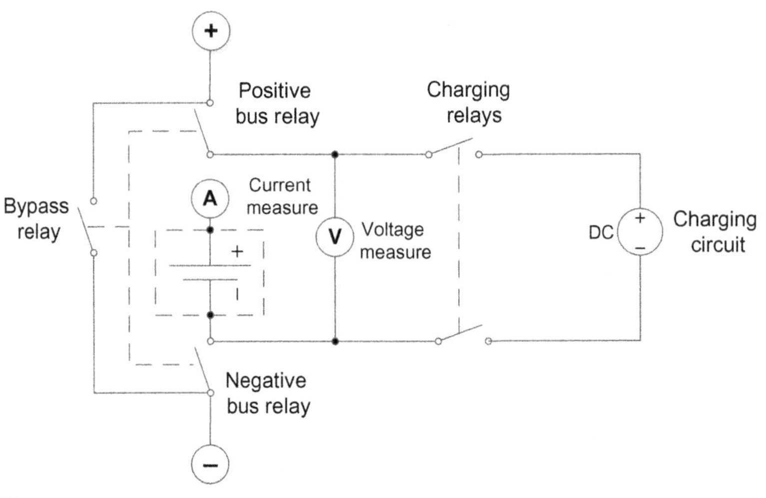

FIG. 56

Proposed individual energy storage management unit schematic.

current path from the positive terminal. The LA 25-NP can measure current up to 36 A. The battery used for testing is rated at 110 Ah, thus the nominal C20 current is rated at 5.5 A. A constant current of up to 7 × the nominal (~C3 rate) can be passed and measured accurately though this transducer.
Voltage measurement: Due to limitations of most data acquisition equipment available on the market, a voltage range of ±10 VDC must be adhered to. Since stack voltages will range from 10.5 VDC to 14.7 VDC, this had to be stepped down with a voltage transducer, LV25-P. With respect to keeping the ESMS universal for usage with most of the energy storage chemistries and cell configurations, it was decided to set the maximum voltage to handle up to an eight-cell lithium ion series battery configuration (≤29.6 VDC).
Charging and diagnostic bus connectivity: This bus is connected at the same terminals of the switches to couple/decouple the DC bus. Fig. 55 identified the bus operating voltage at the typical charging voltage for a lead acid battery (14.7 VDC), but it also has the versatility to operate at a very wide range of voltages to accommodate different types of batteries. This is one of the major flexibilities added by the proposed system. Two NO relays offer connection, or total isolation, from this bus, depending on the operating scenarios. This bus can provide the charging current for multiple batteries in parallel, or the isolation of a single battery where diagnostics can be performed. Battery diagnostic signals can be sent directly to the battery to evaluate its performance or individual SOH. This useful feature can allow an operator to initiate test procedures and identify a consistently failing battery, while the system is in operation. This mode can be of great usefulness for online maintenance.

ESMS implementation

Fig. 57 shows the implemented laboratory prototype for the BMS unit implemented on a 10 × 16 cm printed circuit board, which reflects the compactness and simplicity of the proposed design. The components on the board are: (1) bypass relay; (2 and 3)

FIG. 57

Implemented unit of energy storage management system.

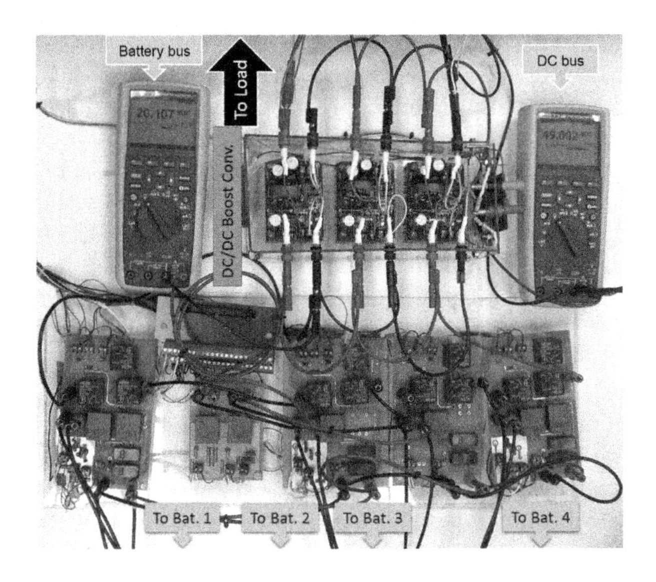

FIG. 58

Hardware setup with four ESMS units.

positive and negative DC bus relays, respectively; (4) Freewheeling diode for a data acquisition device protection; (5) Battery terminals; (6) current transducer; (7) voltage transducer; (8) DC bus terminals; (9) charging bus relays; and (10) charging bus terminals. Unified color coding is followed for all terminals, red is for positive, and black is for negative. Fig. 58 shows the laboratory implementation of Fig. 55, where four units are connected in series and interfaced to a DC bus using a DC/DC boost converter.

ESMS modes of operation

The ESMS discussed here can be operated in three modes. Each mode requires switching the switches in certain states. The transition from one mode to another is done seamlessly. The modes are described as the following and illustrated in Fig. 59.

1. Normal Operation (Discharging): In order to achieve full isolation, two NC relays connect the positive and negative terminals of the battery to the DC bus. A NO relay bypasses the ES to provide an alternative path to maintain the continuity of supply. An interlock is provided between the three relays to avoid simultaneous connection. In this mode, the positive and negative bus relays are closed, and the bypass relay is open as shown in Fig. 59B.
2. Charging Mode: One of the important options featured by the ESMS is the capability to charge one of the energy storage elements, while the rest of the stack is operating normally. During this mode, the positive and negative bus relays are open; and the bypass relay is closed to offer an alternate path for the current so that the system continues operation. After an adjustable short delay, the energy storage element is connected to the charging circuit through the charging relays.

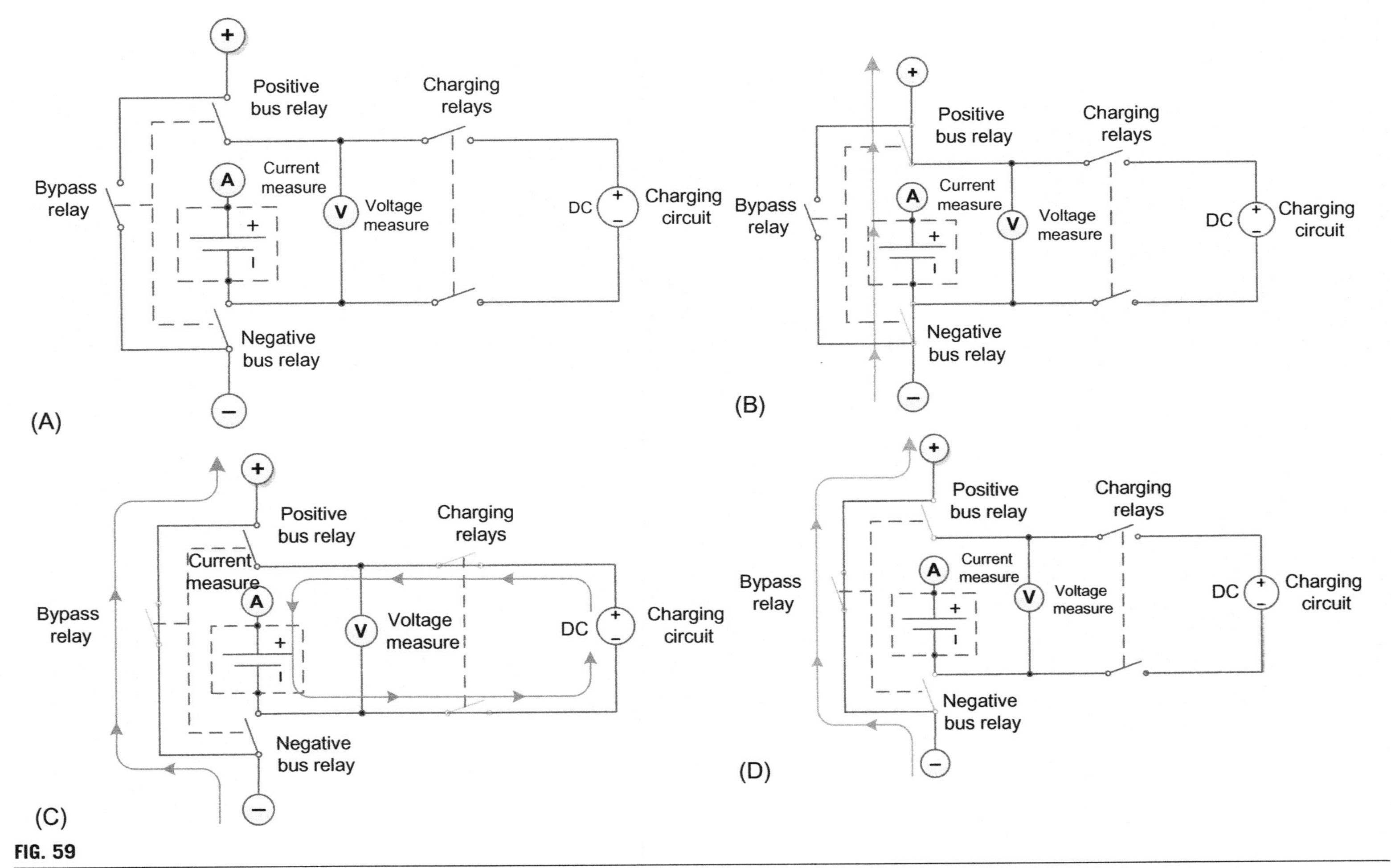

FIG. 59

ESMS operation modes with current flow. (A) A schematic diagram for the ESMS, (B) discharge mode, (C) charge mode, and (D) ideal mode.

This functionality allows the application of a rolling algorithm to recover the battery or restore a portion of its charge until the next pulse (or load peak) occurs. Fig. 59C shows the current path in this mode.

3. Ideal mode: This mode can be used for maintenance, or ES replacement. In this mode, the topology is as depicted in Fig. 59D, the positive and negative bus relays are open to isolate the battery from the DC bus. The charging relays are open as well to isolate the battery from the charging circuit, while the bypass relay is closed to provide an alternative path for the current.

12.4.4 DISTRIBUTION NETWORK OPTIMIZATION

As explained earlier in this chapter, modern distribution power systems encounter significant changes, and become more complicated and interconnected. Therefore, an optimization tool becomes essential to help system planners and decision makers to determine the best architecture to serve their objectives. The optimum architecture refers to the optimum state of the switches, as well as the type, location, and size of the energy storage to satisfy the imposed constraints by standards or operator objectives. These objectives can either be loss minimization, the increase of economic benefits or profits, reliability improvement, and/or enhancing power quality. This problem is formulated as a multiobjective optimization problem. Solving such a problem requires the determination of a set of points that all fit a predetermined definition of an optimum solution. This set of solutions is called the Pareto front. The concept of defining optimal solutions is called Pareto optimality. In this research work, it is proposed to use the nondominated sorting genetic algorithm II (NSGA II) to generate a nondominated Pareto front.

NSGA was a popular nondomination-based genetic algorithm for multiobjective optimization. However, it is criticized for its computational complexity. Then, a modified version, NSGA II was developed, which had a better sorting algorithm and less computational complexity. First, let us breakdown the name. N is for nondominated because it finds the nondominated solution, the nondominated solution is the one achieving the minimum objectives in a minimization problem, or maximum in a maximization one. Assuming that the objective is to enhance power quality through minimizing the voltage and frequency variations, the fitness function of this problem will be (min ΔV, and min Δf). Without loss of generality, more objectives can be considered, including maximizing a stability index, or minimizing the equipment weight and size. Relating to our specific problem, the nondominated solution is the architecture/configuration that achieves the minimum voltage and frequency fluctuations. S stands for sorting, because it sorts the population based on the prespecified fitness function, and puts the nondominated individual in the front to form the Pareto front, or Pareto efficient solution set. Furthermore, it implements the elitism concept, which stores all nondominated solutions, and allows them to survive to the next generation without changes. This concept enhances convergence properties.

How it works: The population is initialized, similar to that in regular genetic algorithms. Once the population in initialized, the population is sorted, based on nondomination, into each front. The first front being the completely nondominant set in the current population, and the second front being dominated by the individuals in the first front only, and then the front goes so on. Each individual in each front is assigned a rank (fitness) value. Individuals in the first front are given a fitness value of 1, and individuals in the second are assigned a fitness value of 2, and so on. Some examples in the literature showed that it was successful in avoiding the local minima/ maxima, or local traps. In complicated multiobjective-multicriterion problems, such as in the power systems case, the problem exhibits nonlinear behavior, which can give a local minima and global minima. Fig. 60A shows an example for a function

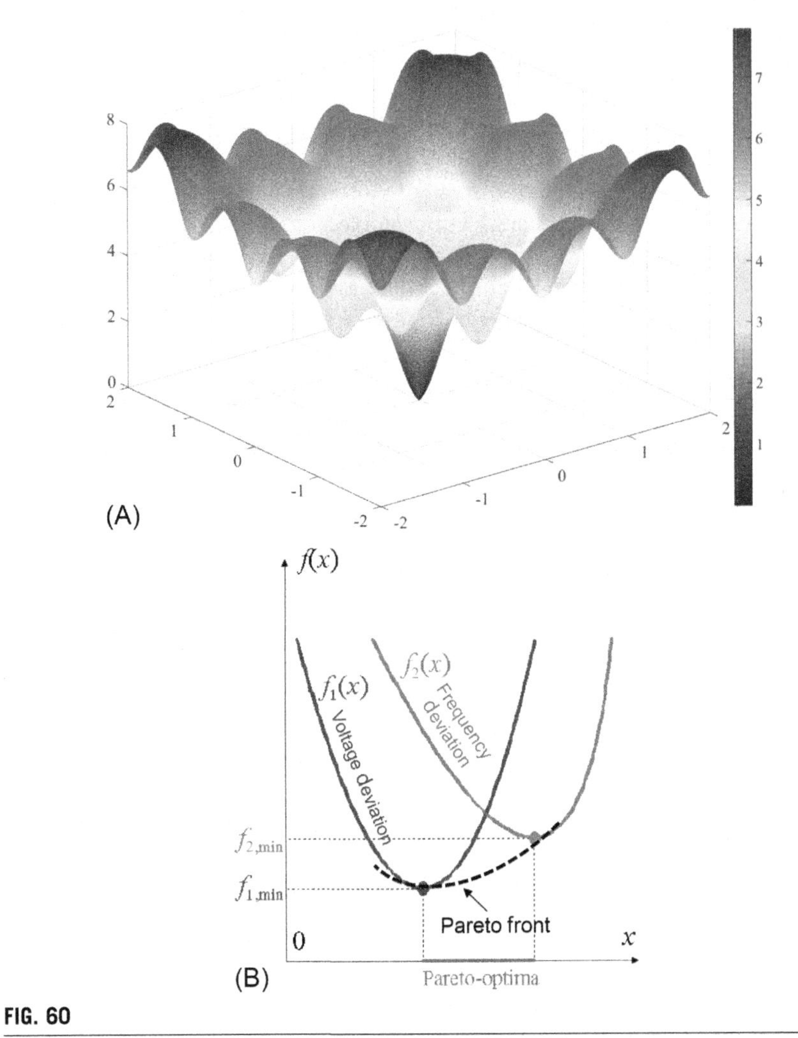

FIG. 60

(A) Global minima and local minima and (B) defining a Pareto front.

with multiple local minima, and only one global. Nondominated NSGA II has the capability to avoid the local minima due to its sorting functionality. For further understanding of the concept of the Pareto Front, please refer to Fig. 60B. The probability of failure (PF) is the set of solutions that meet the objective function criterion. In the context of our specific problem, which is determining the optimal distribution architecture, the objective function is the power flow problem. In order to evaluate each candidate architecture, the power flow problem has to be solved for this architecture.

Power flow in hybrid AC/DC systems

Earlier studies show that optimal selection of energy storage locations, the PCC between the AC and DC sides, and load connection points, have a desirable impact on the power systems, including improving system reliability and reducing transmission losses. In this regard, an accurate solution for the power flow problem is mandatory. It is used to evaluate the available decisions in the design space, and to make sure that a certain candidate solution does not violate the voltage and power quality constraints imposed by the standards. For example, for certain architecture (representing a candidate solution), the optimization algorithm checks this solution through solving the power flow, and determining the voltages at all the buses. If the voltage of one bus or more exceeds the limits, or the power flow through one of the transmission line exceeds its thermal capacity, then this solution is rejected, or marked as infeasible.

Power flow algorithms for AC systems are well developed and extensively studied; however, for the reasons mentioned earlier in this chapter, new power system architectures are evolving. These architectures are incorporating AC and DC networks interfaced with each other through power electronic converters. Solving the power flow for the hybrid AC/DC power system is challenging and, is not straightforward. The most widely used power flow method is the Newton Raphson (NR) method, because it has a very high convergence rate. For example, for the IEEE 30 bus system, the power flow using NR converges in three iterations, while using Gauss Seidel it takes 670 iterations to converge. However, the NR method is well studied for the legacy AC systems, and applying this algorithm to hybrid AC/DC systems requires some manipulations. There are two methods adopted in the literature for solving such a problem: the unified and sequential methods.

- Unified Method: The AC and DC system equations are solved simultaneously, ie, formulate a single Jacobian matrix (matrix of active and reactive power derivatives) for the entire system. A disadvantage of this method is that the entire Jacobian matrix has to be updated at each iteration.
- Sequential Method: The AC system equations are solved, and then their output parameters are passed to the DC system equations to be solved. The AC and DC set of equations are solved sequentially. For several considerations, including computational simplicity, this method will be used in this study.

The main difference is in formulating the Jacobian matrix. The unified approach solves the entire system at one time, ie, it formulates one Jacobian matrix for the entire system with its AC and DC sides. This matrix is massive, and inversing it requires a large computational effort. The other approach is the sequential approach, it solves the AC system first then passes the required parameters to solve the DC, so this why it is called sequential, ie, it solves one side at a time. So, to reduce the computational effort, the NR algorithm with a sequential approach will be used in this study, considering the simplified diagram shown in Fig. 1, where the system is consolidated into two sections AC and DC. The AC and DC sides of the system are interfaced through the VSC. The steps of applying the sequential power flow algorithm to the system are depicted in the flowchart in Fig. 61.

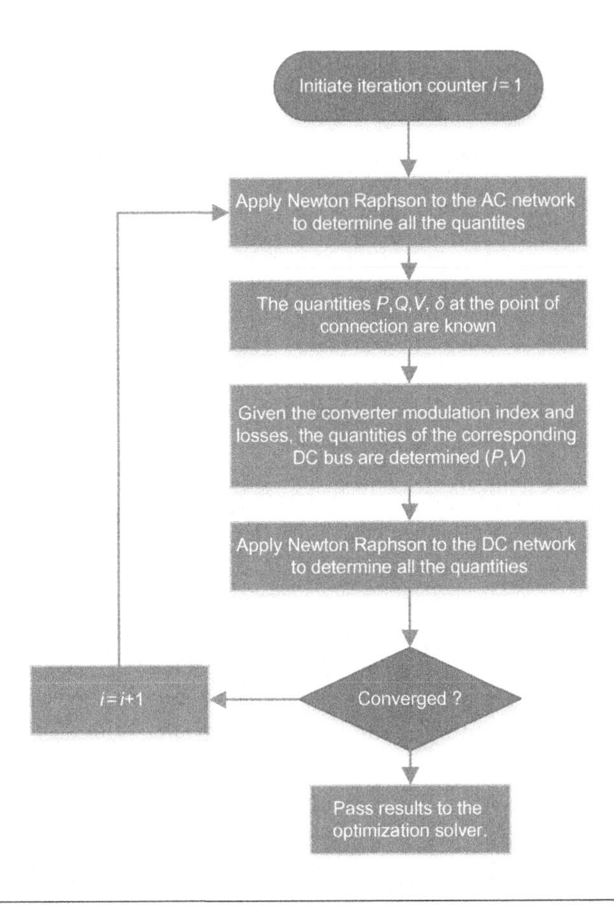

FIG. 61

Flow chart for the sequential power flow approach.

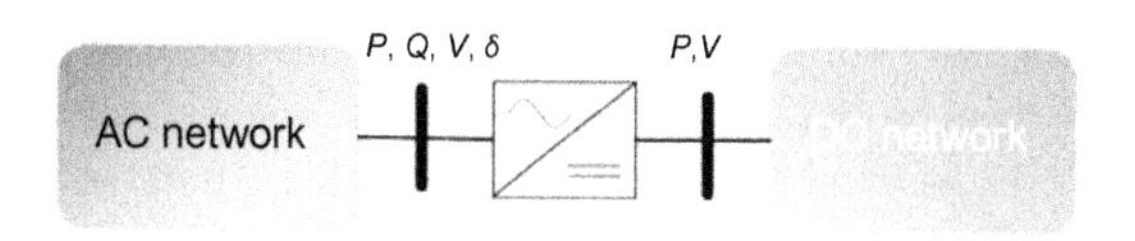

FIG. 62

Simplified schematic for AC/DC power system.

AC power flow

The AC power flow problem is well studied and investigated extensively in the literature, so it will be outlined briefly in this chapter. As depicted in Fig. 62, the system is divided into two main networks, AC and DC, and then each network is broken into zones. Thus, for zone Z the bus admittance matrix, Y_z, is formed. The injected active and reactive power at the bus, i, are given through the following equations:

$$P_i = U_i \sum_{\substack{j=1 \\ j \neq i}}^{M} U_j \left(G_{ij} \cos\left(\Theta_i - \Theta_j\right) + B_{ij} \sin\left(\Theta_j - \Theta_j\right) \right) \tag{42}$$

$$Q_i = U_i \sum_{\substack{j=1 \\ j \neq i}}^{M} U_j \left(G_{ij} \cos\left(\Theta_i - \Theta_j\right) - B_{ij} \sin\left(\Theta_j - \Theta_j\right) \right) \tag{43}$$

Then the NR method is employed to solve the nonlinear AC power flow equations. It is worth mentioning here that the simplified version of the NR method, which is the fast-decoupled method, cannot be used here. The main reason for this is because it neglects some of the terms, which affects the accuracy of the load flow results. This claim coincides with the conclusions of a large number of publications. The following equations calculate the power mismatch vector.

$$0 = \left(S_i^{\text{sch}}\right)* - \left(V_i^{[k]}\right)* \sum_{j=1}^{n} Y_{ij} V_j^{[k]} + \varepsilon \tag{44}$$

$$\Delta V^{[k+1]} = \left[J\left(V^{[k]}\right)\right]^{-1} \cdot \Delta S^{[k]} \tag{45}$$

where $\Delta S^{[k]}$ is the power mismatch at the kth iteration, $[J(V^{[k]})]^{-1}$ is the inverse of the Jacobian matrix at the kth iteration. $\Delta V^{[k+1]}$ is the voltage incremental vector at the next iteration, which is $k+1$. S_i^{sch} is the scheduled apparent power. The power mismatch vector $\Delta S^{[k]}$ can be written as:

$$\Delta P_i^{[k]} = P_i^{\text{gen}} - P_i^{\text{dem}} - P_i^{[k]} \tag{46}$$

$$\Delta Q_i^{[k]} = Q_i^{\text{gen}} - Q_i^{\text{dem}} - Q_i^{[k]} \tag{47}$$

where P_i^{gen}, P_i^{dem}, Q_i^{gen}, and Q_i^{dem} are the generated, demanded active, and reactive, power. These quantities are either scheduled or obtained from the previous iterations. $P_i^{[k]}$, $Q_i^{[k]}$ are the active and reactive power of the current iteration.

DC load flow

By taking the same notation of the AC systems, the bus admittance matrix $Y_{z,\mathrm{DC}}$ for zone z is formed. However, the bus admittance matrix will be much simpler, as all its elements are real. The admittance of the line connecting bus i and j is:

$$y_{ij} = {}^{1}/_{Rdc,ij} \tag{48}$$

The current injected at bus i on the DC side can be given as:

$$I_{dci} = \sum_{\substack{j=1 \\ j \neq i}}^{n} Y_{dcij}\left(U_{dci} - U_{dcj}\right) \tag{49}$$

By assuming bipolar configuration, the transferred power over a transmission line is given as:

$$P_{dci} = 2U_{dci} \cdot I_{dci} \tag{50}$$

Combining both equations, yields:

$$P_{dci} = 2U_{dci} \cdot \sum_{\substack{j=1 \\ j \neq i}}^{n} Y_{dcij}\left(U_{dci} - U_{dcj}\right) \tag{51}$$

The Jacobian matrix can be formed as the following:

$$\left(U_{dc\_j}\frac{\partial P_{dc\_i}}{\partial U_{dc\_j}}\right)^{(j)} = -2U_{dc\_i}^{(j)} Y_{dc\_ij} U_{dc\_j}^{(j)} \tag{52}$$

$$\left(U_{dc\_i}\frac{\partial P_{dc\_i}}{\partial U_{dc\_i}}\right)^{(j)} = P_{dc\_i}^{(j)} + 2U_{dc\_i}^{(j)} U_{dc\_i}^{(j)} \sum_{\substack{j=1 \\ j \neq i}}^{n} Y_{dc\_ij} \tag{53}$$

Similar to the AC systems, one of the converters will be responsible for regulating the DC bus voltage; this converter is called the DC slack bus. The equations and terms corresponding to the slack bus are removed since its voltage is known prior to the DC network power flow. Since all the AC quantities are known from the previous iteration, including $P_{ac}^{[k]}$, the power flow from the DC side is given as:

$$P_{dc,\,i}^{[k]} = P_{ac,\,i}^{[k]} \pm P_{Loss\_Convi} \tag{54}$$

where P_{Loss_Convi} is the converter losses. The sign in the previous equation is dependent of the power flow direction. It should be noticed that the *Pac* is known from the AC power flow. In the next step, the DC power flow can be solved using the NR method.

$$\left(U_{\rm dc}\frac{\partial P_{\rm dc}}{\partial U_{\rm dc}}\right)^{(j)}\cdot\frac{\Delta U_{dc}{}^{(j)}}{U_{dc}}=\Delta P_{\rm dc}^{(j)} \tag{55}$$

$$\Delta P_{\rm dc\_i}^{(j)}=\begin{cases}P_{\rm dc\_i}^{(k)}-P_{{\rm dc}_i}\left(U_{\rm dc}^{(j)}\right)\forall i\leq k\\ -P_{{\rm dc}_i}\left(U_{\rm dc}^{(j)}\right)\forall k\leq i\leq n\end{cases} \tag{56}$$

where $\Delta P_{dc}^{(j)}$ is the power mismatch vector for iteration j.

REFERENCES

[1] Cintuglu MH, et al. Frequency and voltage control of microgrids upon unintentional cascading islanding. In: SoutheastCon 2015. New York: IEEE; 2015.
[2] Mackiewicz RE. Overview of IEC 61850 and benefits. In: Power systems conference and exposition, PSCE'06. 2006 IEEE PES; 2006.
[3] Cintuglu MH, Martin H, Mohammed OA. An intelligent multi agent framework for active distribution networks based on IEC 61850 and FIPA standards. In: 18th international conference on intelligent system application to power systems (ISAP), IEEE 2015; 2015.
[4] Babazadeh H, Asghari B, Sharma R. A new control scheme in a multi-battery management system for expanding microgrids. In: Innovative smart grid technologies conference (ISGT), 2014 IEEE PES; 2014. p. 1–5.
[5] Chaoui H, Sicard P. Hierarchical energy management scheme for multiple battery-based smart grids. In: 2014 IEEE 23rd international symposium on industrial electronics (ISIE); 2014. p. 2020–3.
[6] Qian H, Zhang J, Lai J-S, Yu W. A high-efficiency grid-tie battery energy storage system. IEEE Trans Power Electron 2011;26(3):886–96.
[7] Qiu W, Qiu Z. Design for symmetrical management of storage battery expert system based on single battery. In: Proceedings of the 2006 IEEE international conference on mechatronics and automation, 25–28 June 2006; 2006. p. 1141–6.
[8] Tran D, Khambadkone AM. Energy management for lifetime extension of energy storage system in micro-grid applications. IEEE Trans Smart Grid 2013;4(3):1289–96.
[9] Grillo S, Marinelli M, Massucco S, Silvestro F. Optimal management strategy of a battery-based storage system to improve renewable energy integration in distribution networks. IEEE Trans Smart Grid 2012;3(2):950–8.

FURTHER READING

[1] Martínez J-F, Rodríguez-Molina J, Castillejo P, De Diego R. Middleware architectures for the smart grid: survey and challenges in the foreseeable future. Energies 2013;6:3593–621.
[2] Ozansoy CR, et al. The real-time publisher/subscriber communication model for distributed substation systems. IEEE Trans Power Deliv 2007;22(3):1411–23.
[3] Zaballos A, Vallejo A, Selga JM. Heterogeneous communication architecture for the smart grid. IEEE Netw 2011;25(5):30–7.

[4] Hahn A, Ashok A, Sridhar S, Govindarasu M. Cyber-physical security testbeds: architecture, application, and evaluation for smart grid. IEEE Trans Smart Grid 2013;4 (2):847–55.
[5] Komninos N, Philippou E, Pitsillides A. Survey in smart grid and smart home security: issues, challenges and countermeasures. IEEE Commun Surv Tutor 2014;16 (4):1933–54.
[6] RTI Whitepaper, Data centric middleware. http://www.rti.com/docs/RTI_Data_Centric_Middleware.pdf.
[7] Schmidt DC, Van't Hag H. Addressing the challenges of mission-critical information management in next-generation net-centric pub/sub systems with opensplice DDS. In: IEEE international symposium on parallel and distributed processing (IPDPS), 14–18 April 2008; 2008. p. 1–8.
[8] Von Dollen D. Report to NIST on the smart grid interoperability standards roadmap. Palo Alto, CA: Electric Power Research Institute for NIST; 2009.
[9] Ieee. Vision for smart grid communications: 2030 and beyond. New York: IEEE; 2013.
[10] Cintuglu MH, Elsayed AT, Mohammed OA. Microgrid automation assisted by synchrophasors. In: Innovative smart grid technologies conference (ISGT). New York: IEEE Power and Energy Society; 2015.
[11] Odell J, Nodine M. The foundation for intelligent physical agents. http://www.fipa.org.
[12] Bellifemine F, Poggi A, Rimassa G. JADE-A FIPA-compliant agent framework. In: Proceedings of PAAM, vol. 99; 1999. No. 97-108.
[13] CODE, PRICE. Communication networks and systems in substations—part 5: communication requirements for functions and device models. 2003.
[14] Mahnke W, Leitner SH, Damm M. OPC unified architecture. New York: Springer; 2009.
[15] Cintuglu MH, Martin H, Mohammed O. Real-time implementation of multiagent-based game theory reverse auction model for microgrid market operation. IEEE Trans Smart Grid 2015;6(2):1064–72.
[16] Wooldridge M. An introduction to multiagent systems. New York: Wiley; 2009.
[17] Alibhai Z, Gruver WA, Kotak DB, Sabaz D. Distributed coordination of micro-grids using bilateral contracts. In: IEEE international conference on systems, man, and cybernetics, The Hague, The Netherlands, vol. 2; 2004. p. 1990–5.
[18] Ieee-1547.3-2007. Photovoltaics, dispersed generation. In: IEEE guide for monitoring, information exchange, and control of distributed resources interconnected with electric power systems; 2007.
[19] Fairley P. DC versus AC: the second war of currents has already begun. IEEE Power Energy Mag 2012;10(6):101–4.
[20] Ryu S-H, Ahn J-H, Lee B-K, Cho K-S. Single-switch ZVZCS quasi-resonant CLL isolated DC-DC converter for low-power 32″ LCD TV. In: 2013 IEEE energy conversion congress and exposition (ECCE); 2013. p. 4887–93.
[21] Tsai C-H, Bai Y-W, Lin M-B, Jhang RJR, Chung C-Y. Reduce the standby power consumption of a microwave oven. IEEE Trans Consum Electron 2013;59(1):54–61.
[22] Salomonsson D, Sannino A. Load modelling for steady-state and transient analysis of low-voltage DC systems. IET Electr Power Appl 2007;1(5):690–6.
[23] Amin M, Arafat Y, Lundberg S, Mangold S. Low voltage DC distribution system compared with 230 V AC. In: IEEE electric power and energy conference (EPEC); 2011. p. 340–5.

[24] Techakittiroj K, Wongpaibool V. Co-existance between AC-distribution and DC-distribution: in the view of appliances. In: 2nd international conference on computer and electrical engineering (ICCEE); 2009. p. 421–5.
[25] Ryu M, Kim H, Kim J, Baek J, Jung J. Test bed implementation of 380 V DC distribution system using isolated bidirectional power converters. In: 2013 IEEE energy conversion congress and exposition (ECCE), Denver, Colorado, USA; 2013. p. 2948–54.
[26] Thomas BA. Edison revisited: impact of DC distribution on the cost of LED lighting and distribution generation. In: Twenty-seventh annual IEEE applied power electronics conference and exposition (APEC); 2010. p. 588–93.
[27] Yu W, Lai J-S, Ma H, Zheng C. High-efficiency DC–DC converter with twin bus for dimmable LED lighting. IEEE Trans Power Electron 2011;26(8):2095–100.
[28] Lazaroiu GC, Zaninelli D. A control system for DC arc furnaces for power quality improvements. Electr Power Syst Res 2010;80:1498–505.
[29] Wang P, Goel L, Liu X, Choo FH. Harmonizing AC and DC: a hybrid AC/DC future grid solution. IEEE Power Energy Mag 2013;11(3):76–83.
[30] Maniscalco PS, Scaini V, Veerkamp WE. Specifying DC chopper systems for electrochemical applications. IEEE Trans Ind Appl 2001;37(3):941–8.
[31] Reed GF. DC technologies: solutions to electric power system advancements. IEEE Power Energy Mag 2012;10(6):10–7.
[32] Patterson BT. DC, come home: DC microgrids and the birth of the Enernet. IEEE Power Energy Mag 2012;10(6):60–9.
[33] Wu T, Chang C, Lin L, Yu G, Chang Y. DC-bus voltage control with a three-phase bidirectional inverter for DC distribution systems. IEEE Trans Power Electron 2013;28(4):1890–9.
[34] Dierckxsens C, Srivastava K, Reza M, Cole S, Beerten J, Belmans R. A distributed DC voltage control method for VSC MTDC systems. Electr Power Syst Res 2012;82:54–8.
[35] Aragüés-Peñalba M, Egea-Àlvarez A, Arellano SG, Gomis-Bellmunt O. Droop control for loss minimization in HVDC multi-terminal transmission systems for large offshore wind farms. Electr Power Syst Res 2014;112:48–55.
[36] Al-Diab A, Sourkounis C. Integration of flywheel energy storage system in production lines for voltage drop compensation. In: 37th annual conference of IEEE on industrial electronics (IECON); 2011. p. 3882–7.
[37] Kenny BH, Jansen R, Kascak P, Dever T, Santiago W. Integrated power and attitude control with two flywheels. IEEE Trans Aerosp Electron Syst 2005;41(4):1431–49.
[38] Becker D, Sonnenberg B. DC microgrids in buildings and data centers. In: IEEE 33rd international telecommunications energy conference (INTELEC); 2011. p. 1–7.
[39] Gross P, Godrich K. Total DC integrated data centers. In: Twenty-seventh international telecommunications conference: INTELEC '05; 2005. p. 125–30.
[40] Rajagopalan S, Fortenbery B, Symanski D. Power quality disturbances within DC data centers. In: 32nd international telecommunications energy conference (INTELEC); 2010. p. 1–7.
[41] Ailee G, Tschudi W. Edison redux: 380 Vdc brings reliability and efficiency to sustainable data centers. IEEE Power Energy Mag 2012;10(6):50–9.
[42] Kim D, Yu T, Kim H, Mok H, Park K. 300 V DC feed system for Internet data center. In: IEEE 8th international conference on power electronics and ECCE Asia (ICPE & ECCE; 2011. p. 2352–8.
[43] Pratt A, Kumar P, Aldridge T. Evaluation of 400 V DC distribution in telco and data centers to improve energy efficiency. In: 29th international telecommunications energy conference, INTELEC, September 30–October 4; 2007. p. 32–9.

[44] Rasmussen N. AC vs. DC power distribution for data centers. APC White paper # 63; 2006.
[45] California Energy Commission. DC power distribution cuts data center energy use. Sacramento, CA: CEC; 2008. http://hightech.lbl.gov/documents/data_centers/CEC-TB-40.PDF.
[46] Koomey JG. High-performance data centers—a research roadmap. Berkeley, CA: Lawrence Berkeley National Laboratory; 2007 http://hightech.lbl.gov/documents/datacenters_roadmap_final.pdf.
[47] Zhang L, Jabbari F, Brown T, Samuelsen S. Coordinating plug-in electric vehicle charging with electric grid: valley filling and target load following. J Power Sources 2014;267:584–97.
[48] Mills G, Macgill I. Potential power system and fuel consumption impacts of plug in hybrid vehicle charging using Australian National Electricity Market load profiles and transportation survey data. Electr Power Syst Res 2014;116:1–11.
[49] Liu C, Chau KT, Wu D, Gao S. Opportunities and challenges of vehicle-to-home, vehicle-to-vehicle, and vehicle-to-grid technologies. Proc IEEE 2013;101 (11):2409–27.
[50] Ota Y, Taniguchi H, Baba J, Yokoyama A. Implementation of autonomous distributed V2G to electric vehicle and DC charging system. Electr Power Syst Res 2015;120:117–83.
[51] López MA, Martín S, Aguado JA, De La Torre S. V2G strategies for congestion management in microgrids with high penetration of electric vehicles. Electr Power Syst Res 2013;104:28–34.
[52] Tabari M, Yazdani A. A DC distribution system for power system integration of plug-in hybrid electric vehicles. In: IEEE power and energy society general meeting (PES), 21–25 July; 2013. p. 1–5.
[53] Mohamed A, Salehi V, Ma T, Mohammed O. Real-time energy management algorithm for plug-in hybrid electric vehicle charging parks involving sustainable energy. IEEE Trans Sust Energy 2014;5(2):577–86.
[54] Noriega BE, Pinto RT, Bauer P. Sustainable DC-microgrid control system for electric-vehicle charging stations. In: 2013 15th European conference on Power electronics and applications (EPE), 2–6 September 2013; 2013. p. 1–10.
[55] Roggia L, Rech C, Schuch L, Baggio JE, Hey HL, Pinheiro JR. Design of a sustainable residential microgrid system including PHEV and energy storage device. In: Proceedings of the 2011–14th European conference on power electronics and applications (EPE 2011), 30 August 2011–1 September 2011; 2011. p. 1–9.
[56] Hebner R, Beno J, Walls A. Flywheel batteries come around again. IEEE Spectr 2002;39(4):46–51.
[57] McCluer S, Christin J-F. Comparing data center batteries, flywheels, and ultracapacitors. White paper No. 65, Rev. 2. Schneider Electric-Data Center Science Center; 2011.
[58] Kenny BH, Jansen R, Kascak P, Dever T, Santiago W. Integrated power and attitude control with two flywheels. IEEE Trans Aerosp Electron Syst 2005;41 (4):1431–49.
[59] Kenny BH, Kascak PE, Jansen R, Dever T, Santiago W. Control of a high-speed flywheel system for energy storage in space applications. IEEE Trans Ind Appl 2005;41(4):1029–38.
[60] Toliyat HA, Talebi S, Mcmullen P, Huynh C, Filatov A. Advanced high-speed flywheel energy storage systems for pulsed power applications. In: IEEE electric ship technologies symposium; 2005. p. 379–86.

[61] Mcgroarty J, Schmeller J, Hockney R, Polimeno M. Flywheel energy storage systems for electric start and an all-electric ship. In: IEEE electric ship technologies symposium; 2005. p. 400–6.
[62] Lawrence RG, Craven KL, Nichols GD. Flywheel UPS. IEEE Ind Appl Mag 2003;9 (3):44–50.
[63] Okou R, Sebitosi A, Khan MA, Barendsa P, Pillay P. Design and analysis of an electromechanical battery for rural electrification in sub-Saharan Africa. IEEE Trans Energy Convers 2011;26(4):1198–209.
[64] Dragicevic T, Sucic S, Vasquez JC, Guerrero JM. Flywheel-based distributed bus signalling strategy for the public fast charging station. IEEE Trans Smart Grid 2014;5 (6):2825–35.
[65] Suvire GO, Molina MG, Mercado PE. Improving the integration of wind power generation into AC microgrids using flywheel energy storage. IEEE Trans Smart Grid 2012;3 (4):1945–54.
[66] Kisacikoglu MC, Ozpineci B, Tolbert LM. EV/PHEV bidirectional charger assessment for V2G reactive power operation. IEEE Trans Power Electron 2013;28(12):5717–27.
[67] Su M, Wang H, Sun Y, Yang J, Xiong W, Liu Y. AC/DC matrix converter with an optimized modulation strategy for V2G applications. IEEE Trans Power Electron 2013;28 (12):5736–45.
[68] Wang B, Venkataramanan G. Dynamic voltage restorer utilizing a matrix converter and flywheel energy storage. IEEE Trans Ind Appl 2009;45(1):222–31.
[69] Arghandeh R, Pipattanasomporn M, Rahman S. Flywheel energy storage systems for ride-through applications in a facility microgrid. IEEE Trans Smart Grid 2012;3 (4):1955–62.
[70] Ran L, Xiang D, Kirtley JL. Analysis of electromechanical interactions in a flywheel system with a doubly fed induction machine. IEEE Trans Ind Appl 2011;47 (3):1498–506.
[71] Samineni S, Johnson BK, Hess HL, Law JD. Modeling and analysis of a flywheel energy storage system for Voltage sag correction. IEEE Trans Ind Appl 2006;42 (1):42–52.
[72] Daoud MI, Massoud A, Ahmed S, Abdel-Khalik AS, Elserougi A. Ride-through capability enhancement of VSC-HVDC based wind farms using low speed flywheel energy storage system. In: 2014 twenty-ninth annual IEEE applied power electronics conference and exposition (APEC), 16–20 March 2014; 2014. p. 2706–12.
[73] Daoud MI, Abdel-Khalik AS, Elserougi A, Ahmed S, Massoud AM. DC bus control of an advanced flywheel energy storage kinetic traction system for electrified railway industry. In: 39th annual conference of the IEEE industrial electronics society, IECON 2013, 10–13 November 2013; 2013. p. 6596–601.
[74] Abdel-Khalik AS, Elserougi AA, Massoud AM, Ahmed S. Fault current contribution of medium voltage inverter and doubly-fed induction-machine-based flywheel energy storage system. IEEE Trans Sust Energy 2013;4(1):58–67.
[75] Abdel-Khalik A, Elserougi A, Massoud A, Ahmed S. A power control strategy for flywheel doubly-fed induction machine storage system using artificial neural network. Electr Power Syst Res 2013;96:267–76.
[76] Kairous D, Wamkeue R. DFIG-based fuzzy sliding-mode control of WECS with a flywheel energy storage. Electr Power Syst Res 2012;93:16–23.
[77] Goncalves De Oliveira J, Schettino H, Gama V, Carvalho R, Bernhoff H. Study on a doubly-fed flywheel machine-based driveline with an AC/DC/AC converter. IET Electr Syst Transp 2012;2(2):51–7.

[78] Gowaid IA, Elserougi AA, Abdel-Khalik AS, Massoud AM, Ahmed S. A series flywheel architecture for power levelling and mitigation of DC voltage transients in multi-terminal HVDC grids. IET Gen Transmis Distrib 2014;8(12):1951–9.
[79] Akagi H, Sato H. Control and performance of a doubly-fed induction machine intended for a flywheel energy storage system. IEEE Trans Power Electron 2002;17 (1):109–16.
[80] Gurumurthy SR, Agarwal V, Sharma A. Optimal energy harvesting from a high-speed brushless DC generator-based flywheel energy storage system. IET Electric Power Appl 2013;7(9):693–700.
[81] Tsao P, Senesky M, Sanders SR. An integrated flywheel energy storage system with homopolar inductor motor/generator and high-frequency drive. IEEE Trans Ind Appl 2003;39(6):1710–25.
[82] Zhou L, Qi Z. Modeling and control of a flywheel energy storage system for uninterruptible power supply. In: International conference on sustainable power generation and supply. SUPERGEN '09, 6–7 April 2009; 2009. p. 1–6.
[83] Flynn MM, Mcmullen P, Solis O. Saving energy using flywheels. IEEE Ind Appl Mag 2008;14(6):69–76.
[84] Nguyen TD, Tseng KJ, Zhang S, Nguyen HT. A novel axial flux permanent magnet machine for flywheel energy storage system: Design and analysis. IEEE Trans Ind Electron 2011;58(9):3784–94.
[85] Park J-D, Kalev C, Hofmann HF. Control of high-speed solid-rotor synchronous reluctance motor/generator for flywheel-based uninterruptible power supplies. IEEE Trans Ind Electron 2008;55(8):3038–46.
[86] Middlebrook RD, Cuk S. A general unified approach to modelling switching-converter power stages. In: IEEE power electronics specialists conference record PESC; 1976. p. 18–34.
[87] Radwan AAA, Mohamed YA-RI. Linear active stabilization of converter-dominated DC microgrids. IEEE Trans Smart Grid 2012;3(1):203–16.
[88] Mira MC, Knott A, Thomsen OC, Andersen MAE. Boost converter with combined control loop for a stand-alone photovoltaic battery charge system. In: 2013 IEEE 14th workshop on control and modeling for power electronics (COMPEL), 23–26 June 2013; 2013. p. 1–8.
[89] Roberts BP, Sandberg C. The role of energy storage in development of smart grids. Proc IEEE 2011;99(6):1139–44.
[90] Such MC, Hil C. Battery energy storage and wind energy integrated into the smart grid. In: Innovative smart grid technologies (ISGT), 2012 IEEE PES; 2012. p. 1–4.
[91] Wong YS, Lai LL, Gao S, Chau KT. Stationary and mobile battery energy storage systems for smart grids. In: 2011 4th international conference on electric utility deregulation and restructuring and power technologies (DRPT); 2011. p. 1–6.
[92] Xu L, Miao Z, Fan L. Control of a battery system to improve operation of a microgrid. In: 2012 IEEE power and energy society general meeting; 2012. p. 1–8.
[93] Awad ASA, El-Fouly THM, Salama MMA. Optimal ESS allocation and load shedding for improving distribution system reliability. IEEE Trans Smart Grid 2014;5 (5):2339–49.
[94] Kim J, Lee S, Cho B. Discrimination of battery characteristics using discharging/charging voltage pattern recognition. In: IEEE energy conversion congress and exposition, ECCE 2009; 2009. p. 1799–805.
[95] Zhu D, Yue S, Wang Y, Kim Y, Chang N, Pedram M. Designing a residential hybrid electrical energy storage system based on the energy buffering strategy. In: 2013

international conference on hardware/software codesign and system synthesis (CODES+ISSS); 2013. p. 1–9.

[96] Treptow RS. The lead-acid battery: its voltage in theory and in practice. J Chem Educ 2002;79(3):334–8.

[97] Miao Z, Xu L, Disfani VR, Fan L. An SOC-based battery management system for microgrids. IEEE Trans Smart Grid 2014;5(2):966–73.

[98] Moore SW, Schneider PJ. A review of cell equalization methods for lithiumion and lithium polymer battery systems. In: Presented at the SAE 2001 world congress, Detroit, MI; 2001.

[99] Lawder MT, Suthar B, Northrop PWC, De S, Hoff CM, Leitermann O, et al. Battery energy storage system (BESS) and battery management system (BMS) for grid-scale applications. Proc IEEE 2014;102(6):1014–30.

[100] Wen J, Jiang J. Battery management system for the charge mode of quickly exchanging battery package. In: IEEE vehicle power and propulsion conference, 3–5 September 2008. VPPC '08; 2008. p. 1–4.

[101] Yifeng G, Chengning Z. Study on the fast charging method of lead-acid battery with negative pulse discharge. In: 2011 4th international conference on power electronics systems and applications (PESA); 2011. p. 1–4.

[102] Jwo W-S, Chien W-L. Design and implementation of a charge equalization using positive/negative pulse charger. In: Conference record of the 2007 IEEE industry applications conference, 2007, 42nd IAS annual meeting; 2007. p. 1076–81.

[103] Coleman M, Lee CK, Hurley WG. State of health determination: two pulse load test for a VRLA battery. In: 37th IEEE power electronics specialists conference, 2006. PESC'06; 2006. p. 1–6.

[104] Shi Y, Ferone C, Rao C, Rahn CD. Nondestructive forensic pathology of lead-acid batteries. In: American control conference (ACC); 2002. p. 1350–5.

[105] Cheung TK, Cheng KWE, Chan HL, Ho YL, Chung HS, Tai KP. Maintenance techniques for rechargeable battery using pulse charging. In: 2nd International conference on power electronics systems and applications, ICPESA'06; 2006. p. 205–8.

[106] Cao J, Du W, Wang HF, Bu SQ. Minimization of transmission loss in meshed AC/DC grids with VSC-MTDC networks. IEEE Trans Power Syst 2013;28(3):3047–55.

[107] Chen C, Duan S, Cai T, Liu B, Hu G. Optimal allocation and economic analysis of energy storage system in microgrids. IEEE Trans Power Electron 2011;26(10):2762–73.

[108] Department of Defense Interface Standard for Shipboard Systems. Electric power, direct current, MIL-STD-1399, Section 390; 1987.

[109] Earley MW, Sargent JS, Coache CD, Roux RJ. National electrical code handbook. 12th ed. NFPA; 2011.

[110] Dragicevic T, Lu X, Vasquez JC, Guerrero JM. DC microgrids-part II: a review of power architectures, applications and standardization issues. IEEE Trans Power Electron 2015;99(1). http://dx.doi.org/10.1109/TPEL.2015.2464277.

[111] Elsayed AT, Mohamed AA, Mohammed OA. DC microgrids and distribution systems: an overview. Electr Power Syst Res J 2015;119:407–17.

[112] Baradar M, Ghandhari M. A multi-option unified power flow approach for hybrid AC/DC grids incorporating multi-terminal VSC-HVDC. IEEE Trans Power Syst 2013;28(3):2376–83.

[113] Beerten J, Cole S, Belmans R. Generalized steady-state VSC MTDC model for sequential AC/DC power flow algorithms. IEEE Trans Power Syst 2012;27(2):821–9.

CHAPTER

Applications of energy semantic networks

13

H.A. Gabbar, A. Eldessouky, J. Runge
University of Ontario Institute of Technology (UOIT), Oshawa, ON, Canada

13.1 INTRODUCTION

Energy consumption in infrastructure constitutes the largest share of total worldwide energy consumption (35%). That is one-third of total energy consumption, with the industry sector accounting for 31%, the transportation sector 30%, and other sectors (services, fishing, agriculture, forestry, and nonspecified) 4% [1]. Hence, reducing energy consumption in infrastructures can significantly reduce global energy consumption and improve its economic and environmental impacts. Infrastructure energy conservation is an active and growing research discipline that helps engineers design and modify infrastructures in order to reduce consumption [2]. Major programs have been repurposed to apply energy conservation measures to reducing the energy consumed by heating [3–6]. This is due to the fact that energy consumed for thermal comfort and climate control represents about 60% of the total consumed energy in infrastructures [7]. There are a number of simulation tools such as EnergyPlus that generate heat models and simulations for infrastructures. Others combine both the Energy Conservation Infrastructure Code (ECBC) and the advanced energy efficiency measures, leading to up to 61.75% savings in energy consumption [8,9]. Other effective procedures for infrastructure energy reduction have been identified for the deployment of distributed energy sources. Accordingly, the infrastructure will be able to generate a large share of the energy required, while reducing the energy losses associated with long-distance power lines. Further research efforts address the use of net-zero or zero energy infrastructures to reduce total energy consumption [10–12].

Infrastructures with complex architectures and energy profiles require a methodology for creating potential deployment scenarios involving distributed energy resources (DERs) with effective validation and evaluation capabilities. Such methodologies must provide a clear view of the different alternatives that can be implemented in order to generate and distribute energy within infrastructures. Energy simulation tools, such as HOMER and EnergyPlus, are used to evaluate the different energy conservation scenarios for infrastructures. However, they require the manual

Smart Energy Grid Engineering.

setting of scenario parameters, with a given number of assumptions that affect the accuracy of infrastructure energy modeling practices. Therefore, these tools can only evaluate a specific number of scenarios on a case-by-case basis, and they are not suitable for simulating and evaluating multiple scenarios at once.

In Ref. [13], the authors proposed an algorithm, called GreenCharge, for managing renewable energy and storage in infrastructures. Their intention was to determine the charging and discharging profiles for battery banks used to support energy demand and energy harvesting periods. Other efforts focussed on implementing solar windows in residential infrastructures [14]. In previous work [10–14], no systematic methodologies were able to generate all possible scenarios for energy deployment or conservation.

This proposed methodology allows for the simulation and evaluation of multiple scenarios for the deployment of energy sources in infrastructures. The results of the simulation and evaluation are exported to spreadsheet, so that they can be easily accessed and analyzed using other available tools. The proposed algorithm accepts basic information regarding the infrastructure type and structures, and it generates possible scenarios, using the energy knowledge database related to energy semantic networks (ESNs). An ESN contains environmental and economic information such as local irradiance profiles, wind speeds profiles, feed-in-tariff (FIT) costs, electricity costs, natural gas costs, and so on. According to infrastructure systems, the algorithm generates all possible scenarios for DER structures, and it analyzes the deployment and performance evaluation results for all generated scenarios. Based on the evaluation of different alternative energy scenarios, the selection of an effective energy solution can be achieved to match the desired performance.

ESN structures encode the knowledge of wide-spectrum expertise so that it can be available for the energy designer. For complex buildings with complicated energy demands, establishing the design parameters for energy conservation measures, sources, and loads can be a challenge. In Ref. [11], the authors developed an ESN for building thermal performance that studied the effect of insulation materials, dimensions, loads, and human-interactions schedules on selected key performance indicators (KPIs). They focussed on evaluating different heat energy conservation technologies with respect to the dynamic thermal changes, while tracking the dominant sinks resulting from these changes. Other approaches for reducing costs and emissions involve integrating a micro energy grid (MEG) within the building itself or among its facilities. The term *net-zero* or *zero energy* building is used to identify efforts to minimize the total consumption of the utility energy network [12,13].

13.2 SMART ENERGY GRID INFRASTRUCTURES

Energy infrastructures include energy generation, multigeneration, transmission, and distribution infrastructures, as well as their integration with utilization networks from transportation, industrial, and commercial infrastructures, as shown in Fig. 1.

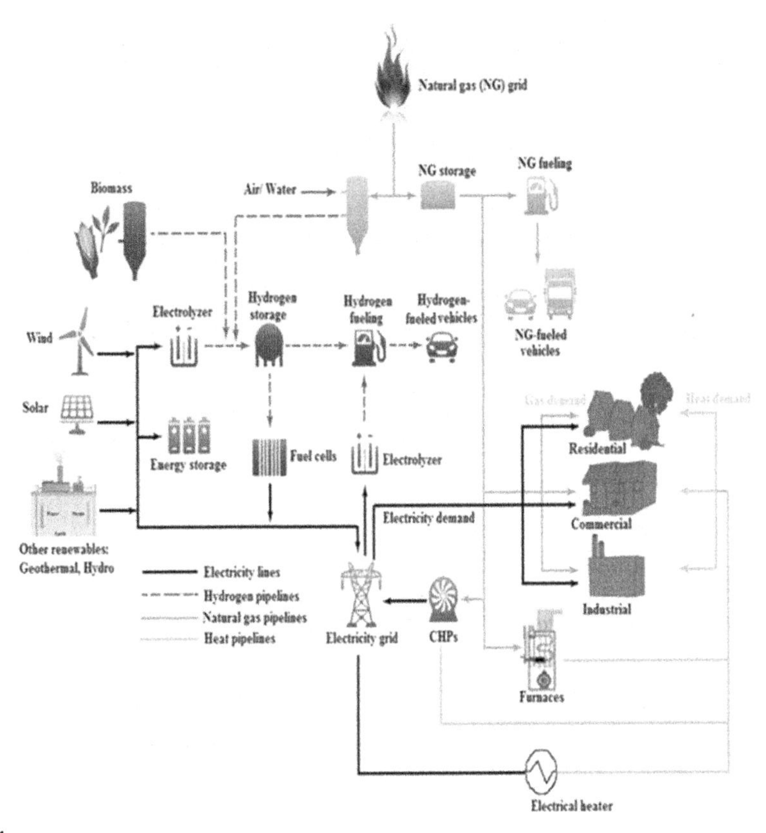

FIG. 1

Energy infrastructures.

13.2.1 ESN FOR SMART ENERGY GRIDS

An ESN is a network that describes the relationships between different energy generation and load classes based on supply, storage, and utilization scenarios. The ESN classifies infrastructure energy types as electric, thermal, or gas. Infrastructure energy classes include thermal zones, loads, storage, and energy sources, and the ESN recognizes all energy requirements associated with these classes. The semantic connections among energy classes represent all possible energy scenarios. The ESN can be thought of as a tool for mapping between the problem domain that describes infrastructure requirements and capabilities and the design domain that describes possible energy scenarios. The ESN model was developed by the authors in Ref. [15].

The ESN can be thought of as a dynamic and adaptive superstructure for modeling and simulating energy supply and production chains [1]. The ESN provides the

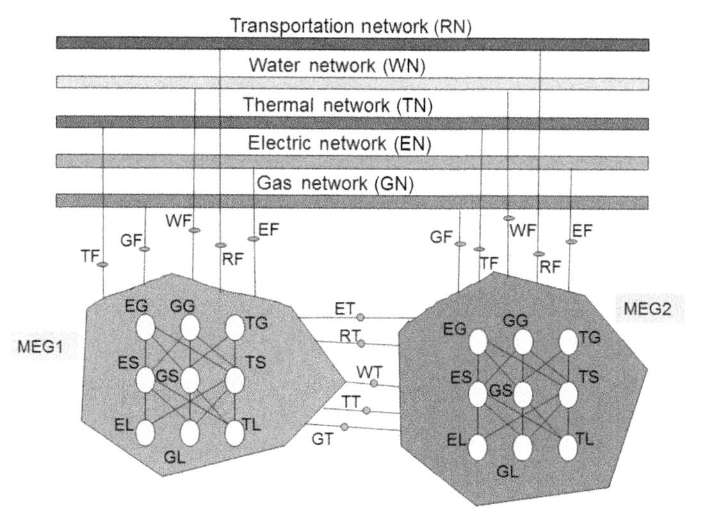

FIG. 2

Energy semantic network (ESN) with interconnected micro energy grids (MEGs).

means for integrating generation (G), storage (S), and loads (L). The ESN is also used to integrate electricity (E), gas (G), and thermal (T) grids, while combining MEGs with transfer lines (T). Fig. 2 shows the proposed ESN for modeling the transmission lines (TL) or distribution lines (DL) of electricity networks (EN), gas networks (GN), thermal networks (TN), water networks (WN), and transportation networks (RN). The ESN includes the integration of TL/DL with MEGs via electricity feeder (EF) lines, gas feeder (GF) lines, thermal feeder (TF) lines, water feeder (WF) lines, and transportation feeder (RF) lines. The ESN includes nodes of energy generation: electricity generation (EG), gas generation (GG), and thermal generation (TG). It also accounts for energy storage: electricity storage (ES), gas storage (GS), and thermal storage (TS), as well as energy loads such as electricity loads (EL), gas loads (GL), and thermal loads (TL). The interconnection between MEGs includes electricity transfer (ET) lines, thermal transfer (TT) lines, gas transfer (GT) lines, water transfer (WT) lines, and transportation transfer (RT) lines. The ESN's static structures are synthesized and dynamically tuned with computational intelligence techniques using real-time data and simulation.

13.2.2 PERFORMANCE INDICATORS

The performance of the target smart energy grid is evaluated at different levels: overall microgrid, energy generation, energy storage, power lines, power compensators, regulators, flexible alternating current transmission system devices, controllers, sensors, and loads. An initial study has investigated KPIs for evaluating microgrids. The ultimate goal of analyzing smart energy grid performance indicators is to identify the best set of microgrid parameters for meeting the following multiobjective optimization criteria: minimum change in the fundamental frequency load bus voltage under steady-state conditions, minimum feeder current with maximum DC generation and maximum AC system grid capacity release, minimum feeder power losses due to reduced fundamental root mean square current magnitude, and minimum dominant harmonic distortion in the host electric grid using the hybrid power filter. Simulation results show optimized performance of the proposed algorithm.

13.3 MODELING AND SIMULATION USING ESN

Infrastructure energy conservation is an active research discipline that influences infrastructure design, structure, and energy deployment strategies related to the extensive energy used in both the residential and commercial sectors. The analysis and simulation of energy conservation scenarios are essential for applying the efficient use of energy. However, modern societies have complex infrastructures with vast and diverse applications. Accordingly, analyzing the infrastructures and their energy requirements gets more complicated. Energy zones, loads, and sources depend on the infrastructure class, type, and size [1]. Hence, the design of an efficient energy system that satisfies the energy requirements of the infrastructure, with minimal faults, requires accumulated knowledge of the infrastructure energy, as well as validation checks of all possible energy scenarios.

The ESN for infrastructures is an essential tool for designing efficient energy infrastructures. The ESN should be open, dynamic, and minimal [2]. *Open* means the ESN's standard structure allows for the incorporation of new components. *Dynamic* refers to the ESN's ability to include new properties of existent components. *Minimal* indicates that the ESN can reduce the computational burden during simulation and evaluation process.

The ESN is used as a tool for mapping between the problem domain and the design domain, as shown in Fig. 2. The inputs to the ESN are the infrastructure type, size, capabilities, number of inhabitants, governmental constraints, environmental constraints, and customer and stockholder constraints. The output of the ESN includes the possible scenarios for the energy design and energy conservation measures that can be implemented. The structure of the ESN is a collection of nodes representing the classes of the problem domain instants and objects and the design domain instants and objects, with relation links encoding the relations between the classes and objects.

13.3.1 MODELING STRATEGIES

1. Construct a dynamic infrastructure ESN: the semantic net is able to modify, add, or delete energy classes and rules.

2. Identify different classes.
   a. Energy load classes (machine, set, room temp, chillers, etc.)
   b. Energy generation classes (PV), wind, combined heat and power (CHP) (boiler, geothermal, furnace, etc.)
   c. Storage classes (battery, thermal storage, gas storage such as fuel cell (FC))
   d. Energy conversion classes (electrical and gas heaters, chillers, gas generator, electrolyzers)
   e. Transmission classes
   f. Energy zone classes
3. Identify the process variables of each class.
   a. Static variables
      i. Power rating
      ii. Physical statistics (weight, size, etc.)
      iii. Initial cost
      iv. Material characteristics (the isolation, the windows, etc.)
      v. Deployment
   b. Dynamic variables
      i. Energy profile
      ii. Running cost profile
4. Develop a methodology for generating the rule base, according to the following.
   a. The methodology generates the possible rules for different scenarios.
   b. The methodology allows the implementation of different constraints implied by the designer (governmental, environmental, costumers, or stockholder constraints).
   c. The generated rule base should be feasible.
   d. The rules should be finite.

13.3.2 ELECTRICITY INFRASTRUCTURE MODELING

The proposed environment for modeling and simulation allows hierarchical and network levels of abstraction and specifications of smart energy grid components, including the electricity grid, as connected to other networks such as thermal and gas networks. Each component is linked with a knowledge base for different model components: business process, regulations, power, energy, asset integrity, physical topology, and geographical or environmental information.

Model libraries are then developed with graphical symbols or icons that represent the class-level knowledge, so that the user can drag and drop the symbols in the drawing area, thus forming or instantiating the corresponding objects. This allows utilities to build a detailed grid model and to dynamically tune this model with different design and operational alternatives. The modeling user interface is developed independent of any simulation engine, using a standard Microsoft library, as per Table 1. Fig. 3 shows the typical infrastructures of SEGs, where distributed generation and storage are combined with information technology infrastructures to integrate phasor measurement units, which are implemented in regional power grids.

Table 1 Transmission Line Infrastructures

#	Type of Transmission Line/Grid/Power Source	Code
1	High-voltage direct current line (HVDC)	TLD
2	High-voltage alternating current line (HVAC)	TLA
3	Low-voltage distribution line	DL
4	Distribution grid/substation	DG
5	Microgrid	MG
6	Microgrid lines	ML
7	Power plant such as nuclear, hydro, coal, etc.	PP
8	Transformer	T
9	Renewable sources such as solar PV, wind farms, etc.	RE
10	Energy storage	ES
11	Fueling/charging unit (gas/PHEV)	CU

13.3.3 ELECTRIC LOAD MODELING

Voltage-dependent, reactive, and active powers are specified using the following equations [16]:

$$P = P_0 \left(\frac{V}{V_0}\right)^{np} \tag{1}$$

$$Q = Q_0 \left(\frac{V}{V_0}\right)^{nq} \tag{2}$$

where P is the active power absorbed by the load, P_0 is the rated active power absorbed by the load (when the voltage across it is equal to V_0), Q is the reactive power absorbed by the load, Q_0 is the rated reactive power absorbed by the load (when the voltage across it is equal to V_0), V is the voltage across the building load, V_0 is the rated voltage across the building load, np is the active power variation coefficient, and nq is the reactive power variation coefficient.

Typical values for np and nq are specified in Table 7.2, page 311 of Ref. [16]. For a typical residential load during the summer, values of $np = 1.2$ and $nq = 2.9$ are preferred. Values of $np = 1.5$ and $nq = 3.2$ should be used for a winter residential load. This type of load model is typical and is commonly used in many simulation tools. A three-phase version is available in the SPS toolbox of MATLAB/Simulink, which inspired the implementation of the single-phase version of the building model.

13.3.4 GAS-POWER MODELING

A detailed steady-state model for a list of candidate generation technologies has been developed and represented using an ESN. The selected technologies are as follows:

1. Natural gas turbines (NGTs)
2. Natural gas fuel cells (NGFCs)

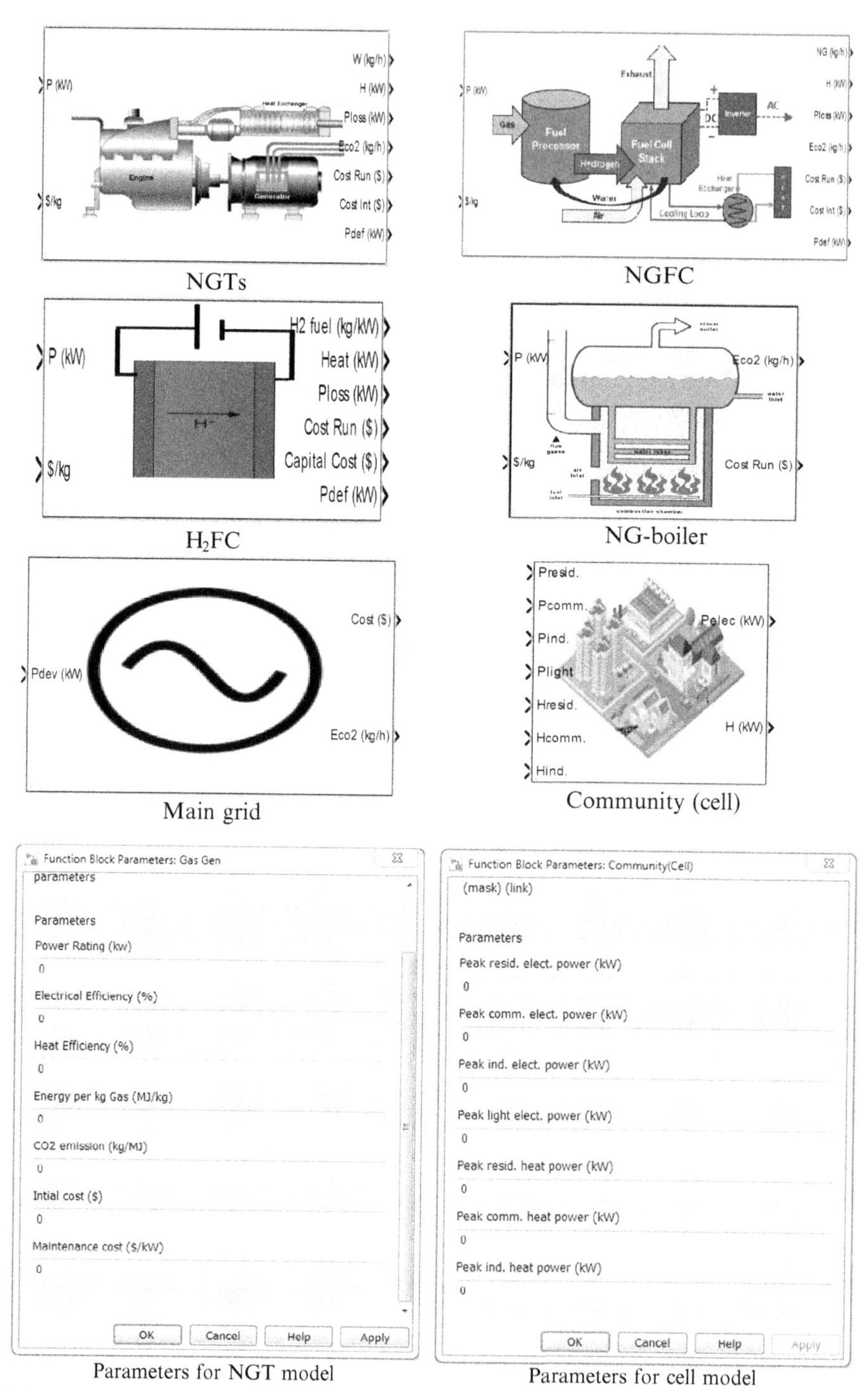

FIG. 3

Sample of models and their parameters.

3. Hydrogen gas fuel cells (H_2FCs)
4. Renewable energy sources (RES) (wind, solar)
5. Main electrical grid
6. Natural gas-based boilers
7. Electrical heaters
8. Thermal storage

A detailed model of heat and electrical load demands has also been developed. The model allows the user to define the type of load (residential, commercial, industrial, and public light), the number of customers for each type, the peak load for each type, and the typical daily load profile. As shown in Eqs. (3)–(6), the models include the evaluation of gas fuel consumption, CO_2 emission, power losses, operational and maintenance (O&M) costs, and capital costs:

$$f_{\text{fue},i}(t) = \frac{P_i(t)}{u_i \eta_{i,P}} \quad \forall t \in T,\ 0 \le P_i(t) \le P_{r,i},\ i \in G \tag{3}$$

$$H_i(t) = P_i(t) \frac{\eta_{i,H}}{\eta_{i,P}} \quad \forall t \in T,\ i \in G \tag{4}$$

$$E_i(t) = K_i u_i f_{\text{fue},i}(t) \quad \forall t \in T,\ i \in G \tag{5}$$

$$P_{\text{loss},i}(t) = \frac{P_i(t)}{\eta_{i,P}} \left(1 - \eta_{i,P} - \eta_{i,H}\right) \quad \forall t \in T,\ i \in G \tag{6}$$

where G is the set of DGs, $f_{\text{fue},i}$ is the fuel consumed by DG i at hour t, P_i is the output power from DG i at hour t, T is the set of hourly periods (ie, 8760 for the entire year), u_i is the energy density of the fuel consumed by DG i in kWh/kg, $\eta_{i,P}$ is the power efficiency of DG i, $P_{r,i}$ is the rated power for DG i, H_i is the heat power generated by DG i at hour t, $\eta_{i,H}$ is the heat efficiency of DG i, E_i is the CO_2 emissions from DG i at hour t, K_i is the carbon footprint for the energy produced by DG i in kg CO_2/kWh, and $P_{\text{loss},i}$ is the power losses for DG i at hour t.

Natural gas turbines

Natural gas is one of the primary energy sources. NG-based generators are leading choices for on-site power generation. NGTs operate by burning natural gas. Hot gasses produced during the combustion process turn the turbine and generate electricity. NGTs are favored for meeting peak loads, as the turbines can quickly achieve full generation capability. An NGT model is represented by Eqs. (3)–(6) using its own parameters (ie, i=NGT).

Natural gas fuel cells

Fuel cells convert the energy present in a fuel to electricity based on an electrochemical process. Thus, fuel cells extract electricity without a combustion process. Fuel cells use hydrogen as a fuel source. However, hydrogen can be produced from natural gas (NGFCs). An NGFC model is represented by Eqs. (3)–(6), using its own parameters (ie, i=NGFC).

Hydrogen fuel cells

In a hydrogen fuel infrastructure, the hydrogen gas fuel feeds directly into a hydrogen fuel cell (H_2FC) to produce electrical power. Thus, the H_2FC is a clean source of power because it produces no CO_2 pollutants (ie, $E_{H_2FC} = 0$ in Eq. 5). An H_2FC model is represented by Eqs. (3), (4), (6), using its own parameters (ie, $i = H_2FC$).

Renewable energy sources

Renewable energy sources (wind turbine; photovoltaic, PV) have become key elements in energy production systems. Because they depend on climate conditions, renewable energy sources are characterized as fluctuating power sources. WTs and PVs are clean sources without CO_2 pollutants, and they don't have fuel costs (operating and maintaining costs are considered in the model).

The WT output power is calculated using Eq. (7):

$$P_w(v) = \begin{cases} 0 & 0 \le v \le v_{ci} \\ P_{rated} \times \dfrac{(v - v_{ci})}{(v_{co} - v_{ci})} & v_{ci} \le v \le v_r \\ P_{rated} & v_r \le v \le v_{co} \\ 0 & v_{co} \le v \end{cases} \tag{7}$$

where v_{ci}, v_r, and v_{co} are the cut-in speed, rated speed, and cut-off speed of the wind turbine, respectively; $P_w(v)$ is the output power; and v is the wind speed.

The output power of PV modules is calculated using Eqs. (8)–(12):

$$T_c = T_A + s\left(\frac{N_{OT} - 20}{0.8}\right) \tag{8}$$

$$I = s[I_{sc} + K_i(T_c - 25)] \tag{9}$$

$$V = V_{oc} - K_v \times T_c \tag{10}$$

$$FF = \frac{V_{MPP} \times I_{MPP}}{V_{oc} \times I_{sc}} \tag{11}$$

$$P_s(s) = N_m \times FF \times V \times I \tag{12}$$

where T_c is the cell temperature (°C), T_A is the ambient temperature (°C), K_v is the voltage temperature coefficient (V/°C), K_i is the current temperature coefficient (A/°C), N_{OT} is the nominal operating temperature of the cell (°C), FF is the fill factor, N_m is the number of modules, I_{sc} is the short circuit current (A), V_{oc} is the open circuit voltage (V), I_{MPP} is the current at the maximum power point (A), V_{MPP} is the voltage at the maximum power point (V), $P_s(s)$ is the output power, and s is the irradiance.

Thermal storage

Thermal storage (TS) stores surplus thermal energy for later use. TS systems are used in buildings and industrial heating/cooling applications. TS systems reduce peak demand, energy consumption, CO_2 emissions, and costs. A TS model is represented by Eqs. (13)–(17):

$$\mathrm{HS_{TS}}(t) \le H_{\mathrm{TS},r} \quad \forall\, t \in T \tag{13}$$

$$H_{\mathrm{TS,in}}(t) \le C_{\mathrm{TS}} \quad \forall\, t \in T \tag{14}$$

$$H_{\mathrm{TS,out}}(t) \le D_{\mathrm{TS}} \quad \forall\, t \in T \tag{15}$$

$$\mathrm{HS_{TS}}(t) = \mathrm{HS_{TS}}(t-1) + H_{\mathrm{TS,in}}\,\eta_{\mathrm{TS}} - \frac{H_{\mathrm{TS,out}}}{\eta_{\mathrm{TS}}} \quad \forall\, t \in T \tag{16}$$

$$\mathrm{HS_{TS}}(24) = \mathrm{HS_{TS}}(0) \tag{17}$$

where $\mathrm{HS_{TS}}$ and $\mathrm{HS}_{\mathrm{TS},r}$ are the heat stored and the rated installed capacity of the TS, respectively; $H_{\mathrm{TS,in}}$ and $H_{\mathrm{TS,out}}$ are the heat sent/received to/from the TS; C_{TS} and D_{TS} are the discharging rates/discharging rates for the thermal storage; and η_{TS} is the turn-around efficiency of the TS.

13.3.5 MODELING AND SIMULATION TECHNOLOGIES

One important benefit of using an ESN is to provide synchronization between real-time grid data and the simulation engine. ESN has two important capabilities: (1) starting simulation scenarios with initial conditions using real-time data from the grid, and (2) fine-tuning simulation models using real-time data. Synchronization is achieved by the infrastructure's intelligent interface program that captures real-time data from the grid and maps it to the state variables in the simulation engine. The simulation engine then tunes simulation models using genetic programming with limited iteration to reduce the error between simulation results and real time. It is also possible to predict the future status of grid loads and demands by using a smaller time step for the simulation, with the current condition of the grid as the initial condition. This provides an accurate estimation of loads and demands, allowing the designer to identify and plan the best recovery operation. And, in order to support decision-making, intelligent algorithms can convert simulation data into qualitative models with symbolic representations that support decision-making. To accommodate the simulation engine, and to support the wider range of commercialization through interoperability, the simulation interface is designed to work with major simulation engines such as DIgSILENT PowerFactory, CYMDIST, ETAP, Paladin Design, EMTP/PSCAD, SimPowerSystems (SPS), and PSS1. Simulation tools for energy systems, such as HOMER and EnergyPlus, are used to evaluate the different energy conservation scenarios for infrastructures. Ideally, the parameter passes between the simulation engine and the simulation interface program in two ways, so that the simulation results are captured for each time step, and the database is updated for tuning simulation models using the specified data analysis techniques.

13.4 ESN IMPLEMENTATIONS

Fig. 4 shows a simplified ESN for infrastructures. The infrastructure is broken down based on its classification (residential, commercial, industrial), which is further broken down into subclasses. Within each subclass, and according to the ESN knowledge base, the infrastructure can be further broken down into zones and available areas for the deployment of DERs. The ESN provides a generic structure for the energy system within the infrastructure. The energy deployment scenarios program (EDSP) unitizes the generic structure to generate detailed scenarios with an evaluation tool using predefined KPIs, such as economic, environmental and reliability,

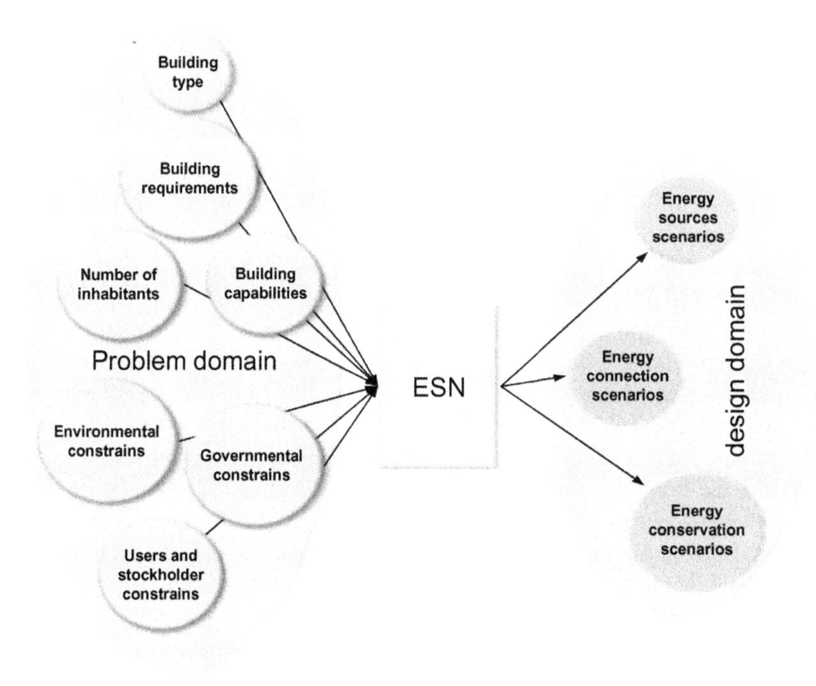

FIG. 4

Mapping the problem domain to the design domain.

and quality KPIs [17]. The ESN allows the implementation of restriction rules to limit generated scenarios to only those that are feasible.

The careful and efficient design of the ESN avoids the generation of tremendous rules at the rule base that can have a large negative affect the simulation and evaluation process. The structure of the ESN should provoke (fire) the portion of the semantic net that related with the design problem. To achieve this objective, the ESN should have a hierarchical design. At the top level, infrastructure classes, source classes, and energy storage classes are presented. At the next level, energy zone classes are introduced. At the bottom level, load classes, energy conversion classes, and energy transmission classes are introduced. Each class is presented by class frame. The frame structure allows for the presentation of multidimensional knowledge (Fig. 5).

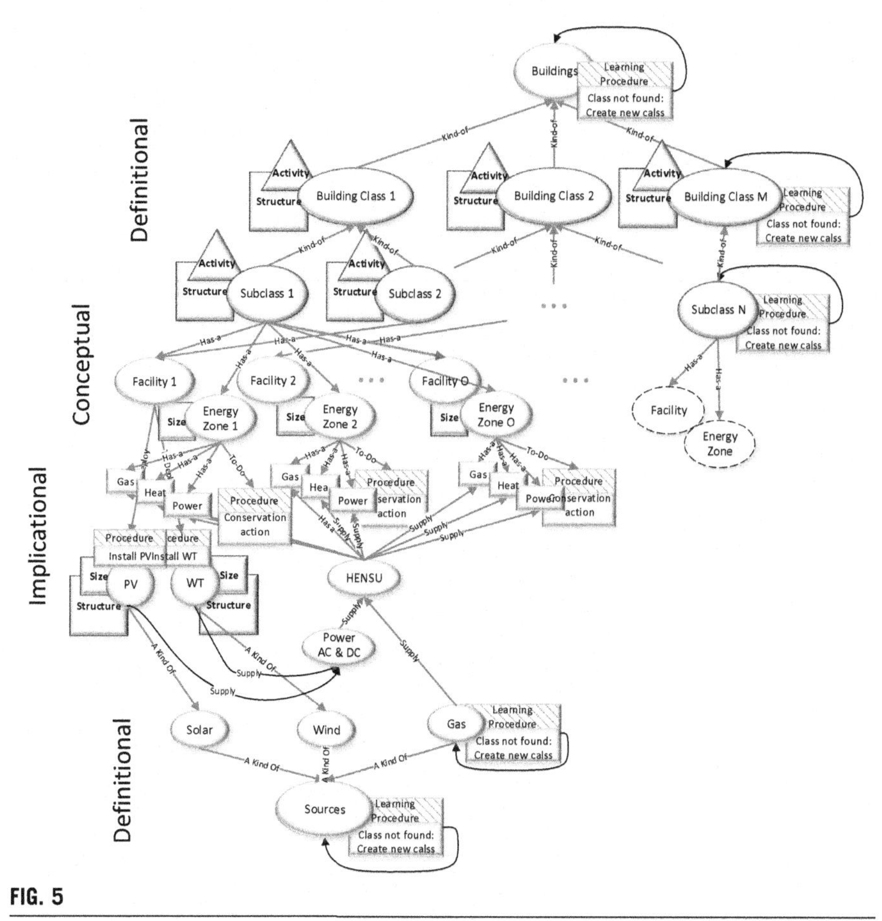

FIG. 5

ESN detailed structure.

13.4.1 CONCEPTUAL REPRESENTATION OF ESN

Fig. 6 demonstrates a tree structure (parent-child structure) [1]. The figure shows the two possible arrangements of the tree structure, with child nodes at the root level connected to parent nodes at the top level. The result is a series of kind-of links or an inverted tree [4] with connections that describe the relations between classes. In the tree, has-a links can also represent characteristics shared by multiple objects (common cause and common effect trees) [5]. A semitree semantic network is used to describe the ESN. The difference between the tree semantic network and semitree semantic network is that the latter allows multiparent connections to interface between two different global classes (eg, infrastructure classes and sources classes) at the bottom level only. A schematic presentation of the application of the ESN for an infrastructure, using a conceptual semitree semantic network, is shown in Fig. 7, where the infrastructure classes and subclasses are presented in the tree semantic network and the sources are presented using an inverted tree.

The mathematical presentation of the ESN can be described by connection matrices. Each matrix represents the connection between the higher-level class and the successor subclass. A general connection matrix is given by the equation:

$$C_i = \begin{bmatrix} P_1^i & \cdots & P_1^i \\ \vdots & P_j^i & \vdots \\ P_{J_i}^i & \cdots & P_{J_i}^i \end{bmatrix} \tag{18}$$

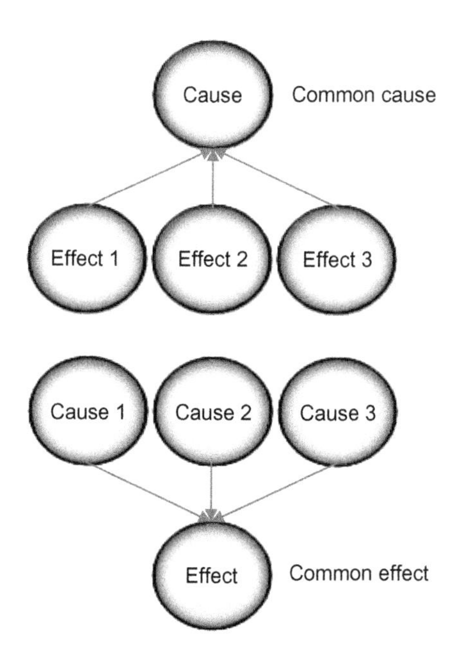

FIG. 6

Illustration of tree structure (common cause and common effect trees).

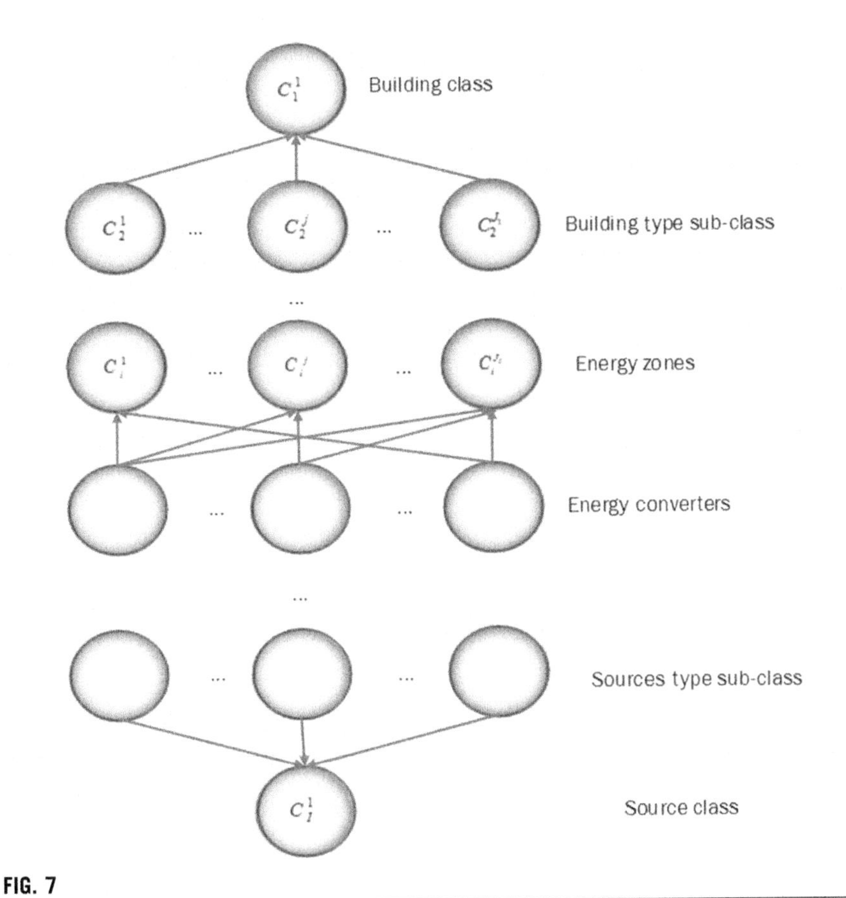

FIG. 7

ESN using semitree structure.

and the dimension of C_i is $J_{i-1} \times J_i$, where J_{i-1} and J_i are the number of classes defined in the predecessor and successor levels, respectively; i is the index of the successor class level; and P_j^i represents the design parameters identified by each class in successor layer. The following algorithm is used to represent the ESN using matrices:

1. The parameters P_j^i are repeated in single row if a child in successor level is connected to multiple parents in one predecessor level.
2. The full presentation of the matrix elements indicates the connections of all the classes in predecessor level with the subclasses in the successor level.
3. The elements replaced by 0 indicate no connection between classes.
4. The elements replaced by 1 indicate that the successor class is related to the predecessor class, but it has no design parameters.
5. No direct connection exists between two nonadjacent classes.

6. If a connection between two nonadjacent classes is required, an additional replicate of the parent class is inserted within the intermediate classes with a connection arrow and element value of 1 indicating no design parameters.

The parameter of the root parent class in level 0 is presented as a 1×1 matrix representing the design parameter of that class. The input to the knowledge base for the ESN is translated to a selection vector with a length equal to the classes count at this level. As the designer navigates through the class levels of the network, a new selection vector is generated. The selection vectors are binary vectors that present routing relations from the top class to the bottom class or object. The length of each vector is K_i. Revisiting Fig. 7, if the selected infrastructure class is *residential apartment*, then the selection vector is given by

$$\begin{aligned} S_0 &= [1] \\ S_1 &= [1 \ \ 0 \ \ 0] \\ S_i &= [0 \ \ 1 \ \ 0 \ \ \ldots \ \ 0 \ \ 1 \ \ 0] \end{aligned} \tag{19}$$

The equation for the possible scenarios can be created by applying the following equation:

$$\begin{aligned} R &= \sum_{i=1}^{I-1} R_i \\ R_i &= S_i C_i \end{aligned} \tag{20}$$

where R represents the summation of the rules in all class levels. If P_j^i is a vector of length K_j^i representing third-dimension knowledge, and the frame structure at each class node has $N_{k_j^i}$ values for each design parameter, then the total number of possible scenarios according to the rule of the level i is given by

$$N_i = \prod_j^J \prod_k^K N_{k_j^i} \tag{21}$$

and the total number of scenarios is given by

$$N = \sum_i^I N_i \tag{22}$$

Eq. (22) shows that increasingly distributing the rules (design parameters) among the class levels decreases the number of generated scenarios. On the other hand, if all the design parameters apply to the lower class level, a large number of scenarios will be generated. In addition, it should be noted that allowing infinite values of any parameter leads to infinite scenarios. For such parameters, the selection of the boundary values in addition to a few intermediate values should be sufficient to generate a reasonable number of scenarios.

13.4.2 RULE-BASED ASSOCIATION WITH ESN

The number of the rules associated with a semantic network commonly grows at an exponential rate with the number of classes, properties, and attributes. The rule base of the ESN could be complicated enough to have countless possible scenarios. The objective during the generation of the rules (scenarios) is to reduce them to the minimum possible number, in order to avoid complexity during the infrastructure energy simulations and evaluations. The best approach is to consider two levels of rule base: a large-scale level and a small-scale level. The large-scale level handles the parent classes that describe the overall infrastructure energy requirements and sources. At this level the rules consider the size of the energy sources, the global energy requirement and classification (thermal, electrical, and gas), and the deployment of the energy sources. On the other hand, the small-scale level considers the distribution of energy sources within energy zones, the relations between energy zones, the energy transformation within the infrastructure, and the connection of energy sources to the loads at each energy zone. It is possible to increase the rule level to reduce the number of rules. A generic scenario is given by

$$R_i = \left(\left(P_1^i \right) \text{AND} \left(P_2^i \right) \ldots \text{AND} \left(P_j^i \right) \ldots \text{AND} \left(P_{J_i}^i \right) \right) \quad (23)$$

and the overall scenario for the infrastructure energy is given by

$$R = \sum_i^I R_i \quad (24)$$

13.4.3 ESN UTILIZATION IN RESIDENTIAL INFRASTRUCTURES

A residential house provides a case study for the developed ESN, as shown in Fig. 8. The system is composed of different energy sources and loads with energy converters. The sources are the gas supply, GG, WT, and PV. The GG is a reliable resource that can be used in emergency cases to ensure the stability of the energy supply. The renewable sources are represented by WT and PV. The loads are classified as light, heat, and electrical loads. The energy converters are the furnace, chiller, and AC/DC and DC/AC converters. The house is divided into four energy zones (basement, kitchen, dining area, and bedroom area) and three facility areas (roof, backyard, and windows). The loads of the house are divided into heating, cooling, lighting, and appliances.

A simple ESN has been developed using the ESN GUI (graphical user interface) to include different classes of infrastructures and facilities. A flexible algorithm allows the designer to add, delete, and modify the required classes when modeling the system. The map of the developed ESN is shown in Fig. 9.

13.5 ESN APPLICATIONS IN INFRASTRUCTURES

13.5.1 ENERGY TECHNOLOGY DEPLOYMENT WITH ESNs

The EDSP is an algorithm that is based on the ESN, which can be used for building energy conservation and management. The structure of the ESN was presented in Ref. [18]; it includes energy domain knowledge with rule-based reasoning to support

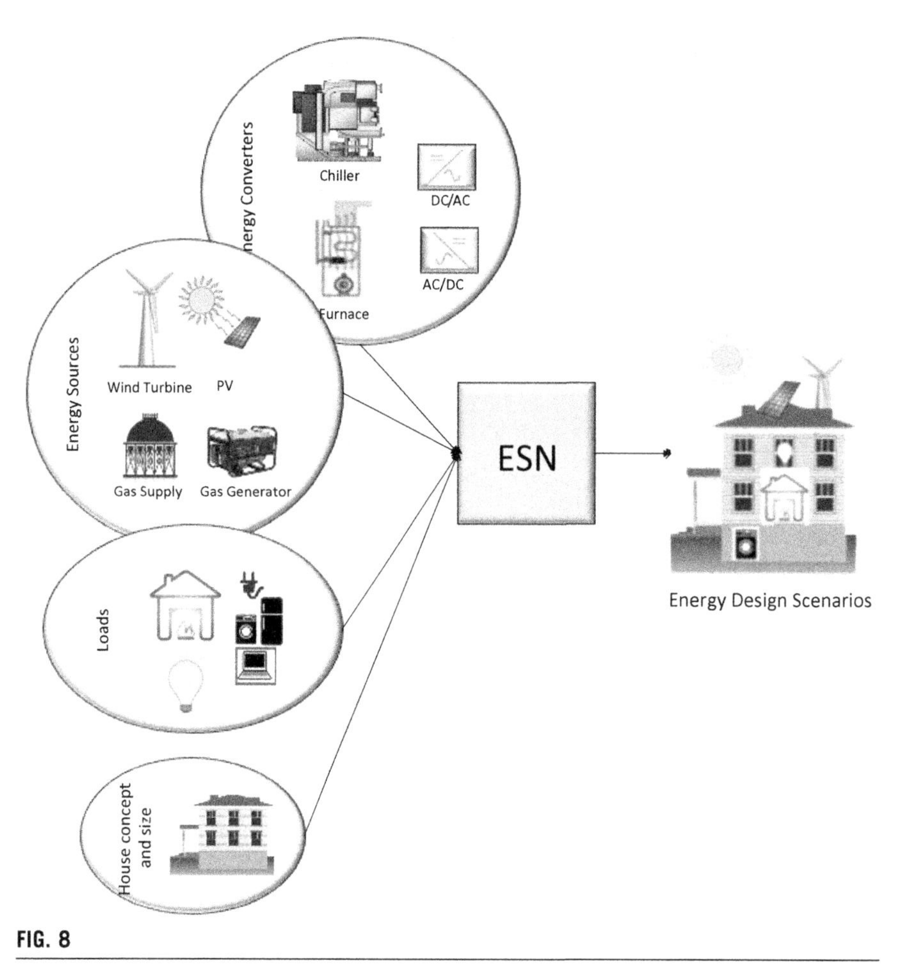

FIG. 8

Case study for ESN.

decisions related to the design and operation of the infrastructure energy system. A dynamic knowledge structure for the ESN was used for energy production chain assessment. The knowledge base of the ESN covers model libraries for the infrastructure that include components and energy technologies. It contains classes of infrastructures types, sizes, thermal zones, energy sources, energy conversion processes, and energy storage. Each class encompasses information about energy components and related parameters. Based on the selection of infrastructure type and size, the ESN's static structure is synthesized as a collection of energy-supply–production paths from generation to loads, including storage. The static ESN structure is unitized by the EDSP to generate detailed scenarios, evaluated using selected KPIs. The following sections discuss the EDSP algorithm.

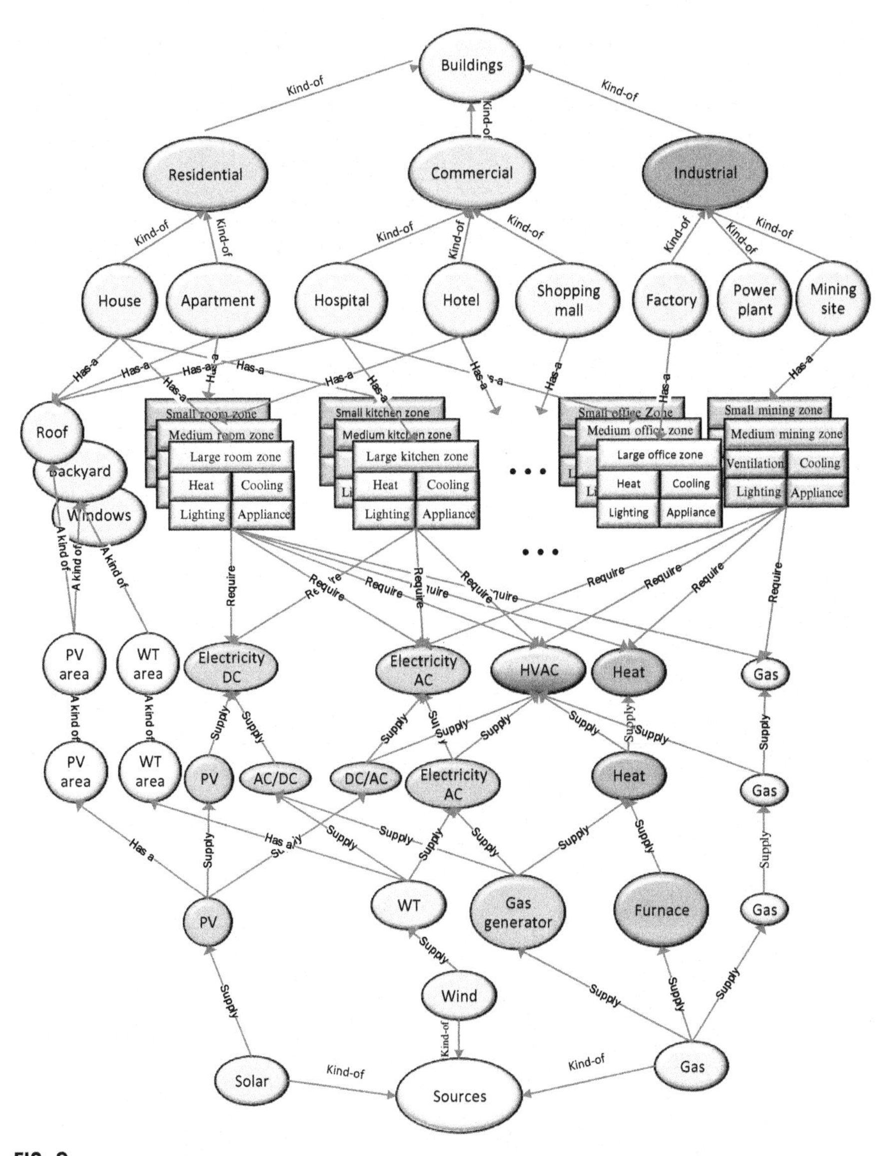

FIG. 9

Energy semantic network (ESN) detailed structure.

13.5.2 EDSP ARCHITECTURE

The EDSP algorithm architecture is shown in Fig. 10. The algorithm starts by receiving inputs about infrastructure information, such as infrastructure type, monthly infrastructure loads (kWh and cost), and infrastructure systems or facilities (available area for energy generation deployment). The input data can be supplied by

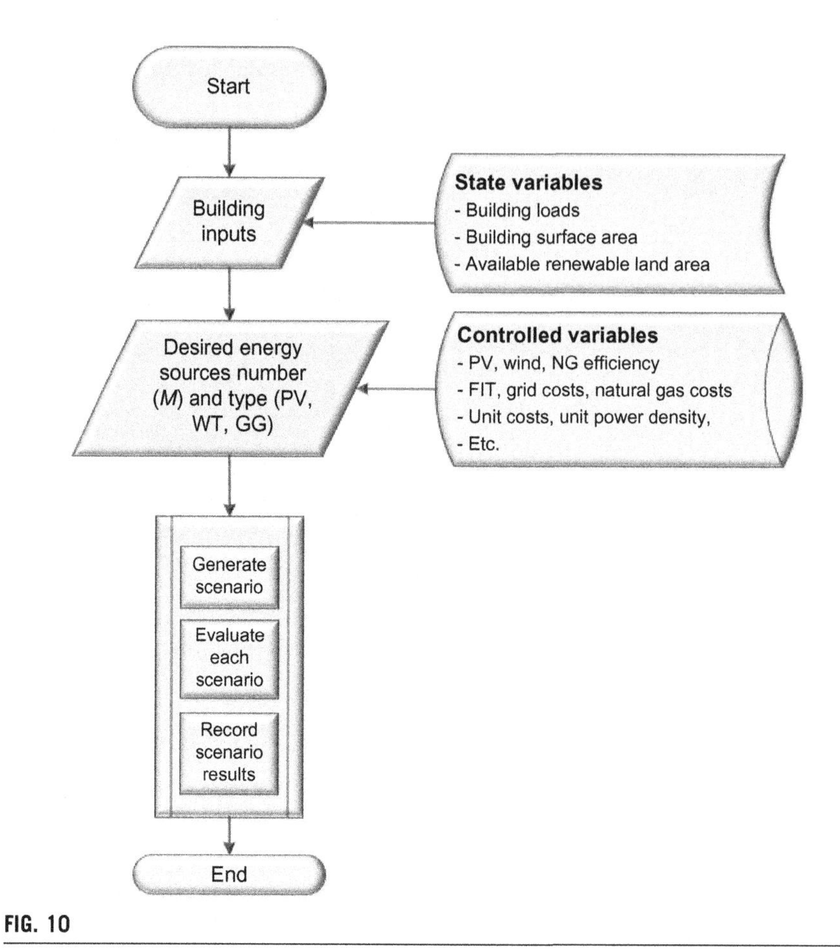

FIG. 10

EDSP algorithm architecture.

the user, as in the case of specific infrastructures with available data, or by a generic model created by the ESN. Then, the user selects the energy sources to be deployed and how they will be applied (FIT or micro-FIT [19]).

Based on the inputs, the EDSP generates a spreadsheet of all possible energy scenarios. For each scenario, rule-based limitations or constraints are applied. These constraints describe the governing regulations and rules for the infrastructure and energy that may affect the design of the relevant energy systems. Only scenarios that satisfy the constraints specified in the rule base are subject to evaluation. The results of the evaluation process are exported to a spreadsheet for analysis and selection. The output data of the EDSP represents KPI values, which are selected for further evaluation. The designer can choose from the different alternative scenarios based on KPI values.

13.5.3 SUBPROCESSES FOR EDSP SCENARIO GENERATION

Energy scenario synthesis is based on three types of DERs: PV, WT, and GG. Fig. 11 shows the flowchart for the EDSP scenario generation subprocess. The structure of the algorithm is based on nested loops for creating energy source deployment scenarios, which are further evaluated using KPI analysis. The algorithm's outer loop starts by assigning a portion of the infrastructure systems and facilities to one energy generation technology. Then the algorithm increments the sizes of the other generation technologies in inner loops, so that the total area used for energy generation is less than the area of infrastructure facility. For each case, the algorithm performs KPI analysis and stores results in a spreadsheet. The outer loop increments the first generation size, which is processed in the inner loops to generate more scenarios.

The number of generated scenarios is given by

$$N = \frac{(L+M)! - L!}{M!} \tag{25}$$

where N is the number of possible scenarios, M is the number of selected energy generators, and L is the given by

$$L = \frac{A}{\Delta a} \tag{26}$$

where A is the facility area of the infrastructure, and Δa is the minimum area increment required to install the generator.

Fig. 12 shows an example of generated scenarios assuming two DERs ($M=2$), infrastructure facility area A in m^2, and minimum increment (Δa) of 1 m^2.

The EDSP creates a spreadsheet with an array of a length equal to the number of scenarios and a row width equal to the number KPIs required for the evaluation process.

13.5.4 ALGORITHM PARAMETERS

This section lists energy technology parameters required for the evaluation and analysis of the scenario synthesis. Table 2 shows the classification parameters used for KPI evaluation and analysis via the proposed algorithm. Notably, these parameters are not subject to change for all generated scenarios, so that all scenarios are evaluated with the same set of KPIs. Some of the listed parameters are subject to change, depending on geographical location, for example, given differences in solar irradiance. In addition, the FIT and electricity grid prices might vary in relation to governmental regulations.

The average power of the three energy technologies: PV, WT, and GG are the parameters subject to change during scenario creation. Each scenario assumes unique settings for the size of the power of the combined PV, WT, and GG.

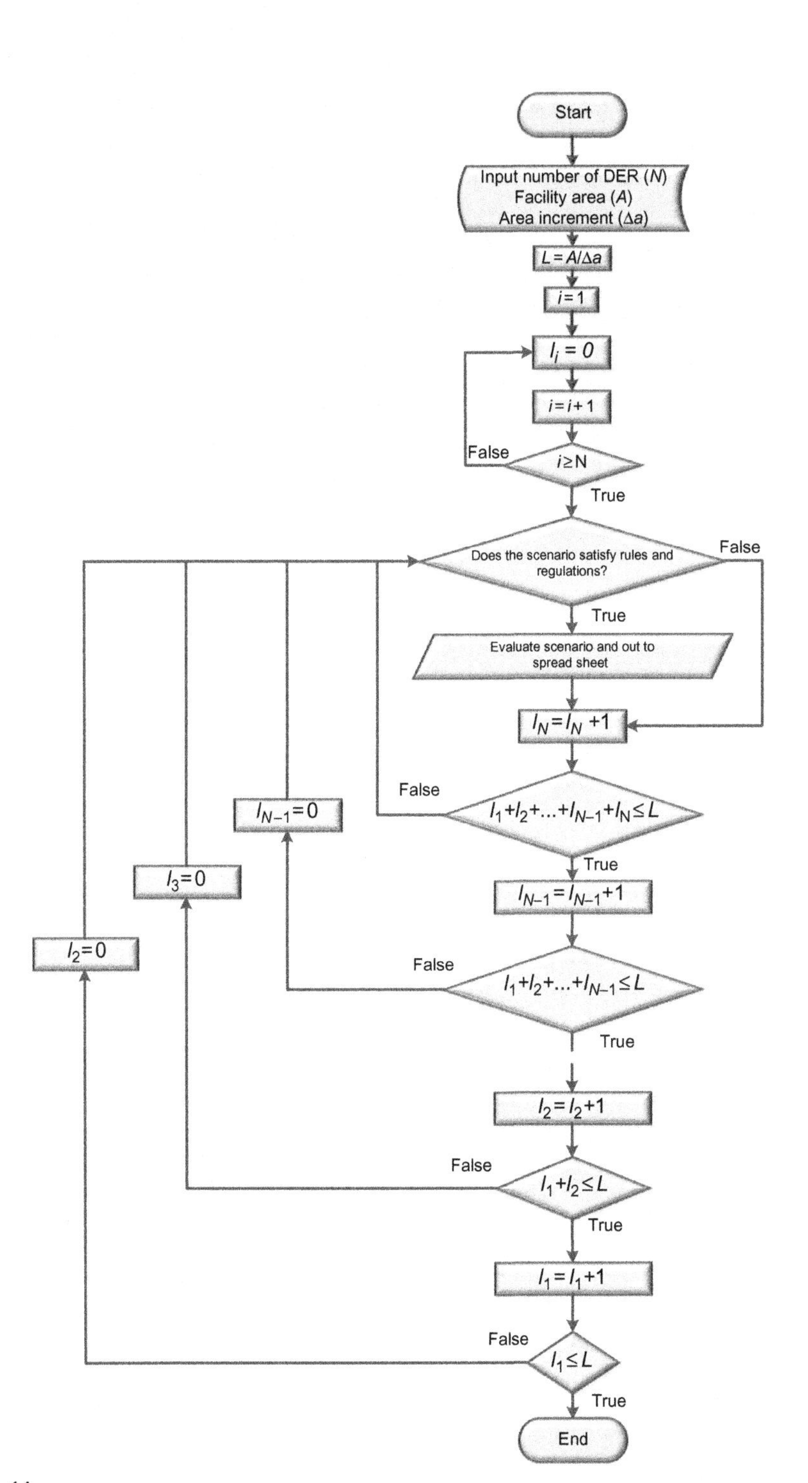

FIG. 11

EDSP scenario generation algorithm.

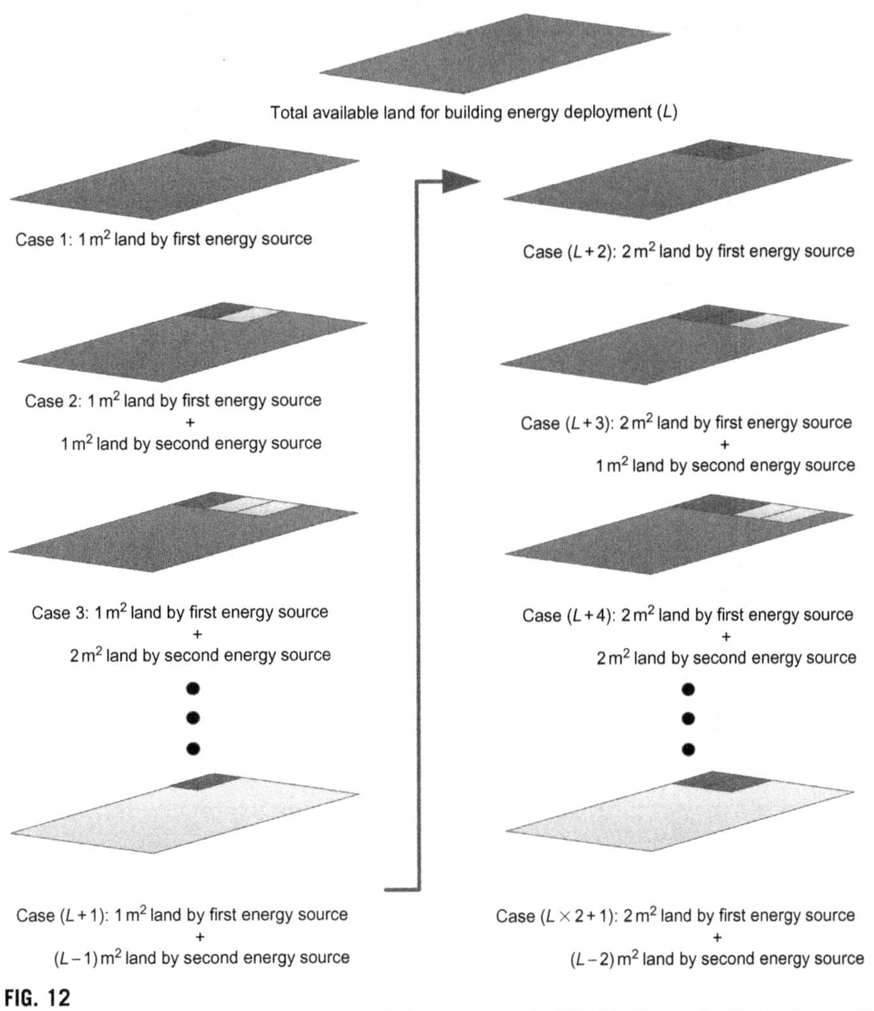

FIG. 12

Scenario example for two DERs.

13.5.5 ESN DEPLOYMENT ASSUMPTIONS

The algorithm is designed to generate possible deployment and interconnection scenarios for generic infrastructures. However, some assumptions have been made in order to achieve feasible evaluation criteria for different scenarios. The assumptions used for the algorithm are as follows:

1. Average unit costs are used to provide a fair estimate of the overall costs.
2. Despite the fact that the size-to-power ratio for each energy generator is different, the area assigned to the unit power for each generator is the same. This is because the technology that requires more operational space, such as PV, requires a

Table 2 Classification of State Variables

Classification	Technology/Other	Parameters
Electrical energy	Solar	- Average unit power - Average unit cost - Capacity factor - Inverter efficiency
Electrical energy	Wind turbine	- Average unit power - Average unit cost - Capacity factor - Mechanical efficiency - Inverter efficiency
Electrical energy	Natural gas generator	- Average unit power - Average unit cost - Mechanical efficiency - Inverter efficiency - Fuel consumption rate
Electrical energy	Grid	- Grid costs of electricity - FIT costs (per source) - Micro-FIT costs (per source) - FIT power ratings - Micro-FIT ratings
Gas energy	Grid	- Grid costs (volumetric costs rates)
Potential energies	Location based	- Average local solar irradiation - Average local wind speed

smaller service and safety zone, as compared to a GG, which requires less operational space and more service and safety area.

3. Solar and wind history data are valid for the algorithm evaluation, assuming they are white Gaussian random variables.
4. All generation technologies generate their maximum output power, according to their ratings and the relevant wind and irradiance profiles.
5. Efficiency factors are used for various technologies (inverter, transformers, etc.).
6. Wind and irradiance directions are perpendicular to wind turbine and PV cells, respectively.

13.5.6 RULES ASSOCIATION WITH THE ESN

The number of scenarios commonly grows at an exponential rate with increases in the number of problem dimensions (facility size, number of energy generation technologies, and number of loads). Hence, computational difficulties should be considered during the evaluation and analysis of different scenarios, even with the powerful computational processors available today. In order to mitigate this problem, rule-based restrictions reduce number of created scenarios. This, in turn, reduces the computational complexity and response time. The rule base represents local laws and regulations applied to the power system and infrastructure. Accordingly, ESN options restrict users to selecting FIT or micro-FIT. The FIT program ranges from 10 to 500 kW, and the micro-FIT ranges from 0 to 10 kW. The user selects one deployment option (FIT or micro-FIT), and, consequently, the algorithm applies restrictions encoded in the rule base to create possible scenarios.

In addition, infrastructure types significantly affect the type of energy generation technology to be used. For example, residential infrastructures, such as townhomes, apartment blocks, and homes, cannot apply the WTs used for farms. Thus, the algorithm adjusts the incremental step size to match infrastructure type. With these rules in place, effective overall scenario creation is achieved using a minimum number of created scenarios. This improves the execution time of the proposed algorithm.

13.5.7 KPI MODELING WITH ESN

Solar PV generation (kWh)

The size of the solar PV generator is given by

$$P_{sr} = l_s \times \Delta a \times P_{savg}/1000 \text{ kW} \tag{27}$$

where P_{sr} is the rated power of the PV, l_s is the normalized size of the PV with respect to incremental area Δa, and P_{savg} is the average power (W) generated by 1 m^2 of available PV panels in the market. The energy annually generated by PV is given by

$$E_s = \sum_{k=1}^{8760} l_s \times \Delta a \times \eta_{savg} \times r_s(k)/1000 \text{ kWh} \tag{28}$$

where η_{savg} is the average efficiency of available PV panels in the market, $r_s(k)$ is the irradiance profile (W/m^2) for the hour k, and the summation is conducted for 1 year.

Wind generation (kWh)

The WT output power is based on the wind power trend shown in Fig. 13 [20–22]. The size of the wind generation is given by

$$P_{wr} = l_w \times \Delta a \times P_{wavg}/1000 \text{ kW} \tag{29}$$

where P_{wr} is the rated power of the wind generation in kW, l_w is the normalized size of the wind generation with respect to incremental area Δa, and P_{wavg} is the average power (W) generated by a WT occupying one square meter. According to the trend shown in Fig. 13, the output power of the WT is given by

$$P_w(k) = \begin{cases} 0 & v_w(k) \leq 3.5 \\ \dfrac{P_{wr}}{11.5}(v_w(k) - 3.5) & 3.5 \leq v_w(k) \leq 14 \\ P_{wr} & v_w(k) \geq 14 \end{cases} \tag{30}$$

The annual energy generated by the WT is given by

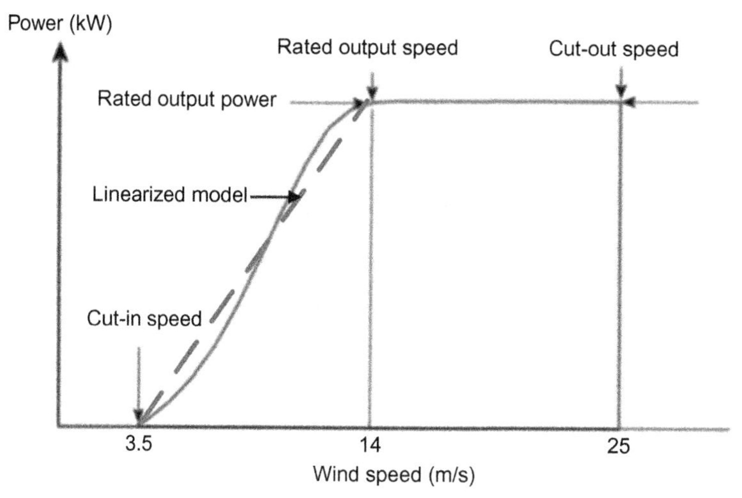

FIG. 13

Typical wind turbine power output with wind speed.

$$E_{\mathrm{w}} = \sum_{k=1}^{8760} P_{\mathrm{w}}(k) \ \mathrm{kWh} \tag{31}$$

where $v_{\mathrm{w}}(k)$ is the wind speed profile (m/s).

Natural gas generation (kWh)

The GG is assumed to be working at its rating capacity during the year. The model also assumes that the surplus energy generated by all generators is exported to the utility grid using the FIT or micro-FIT program. The rated power of the GG is given by

$$P_{\mathrm{gr}} = l_{\mathrm{g}} \times \Delta a \times P_{\mathrm{gavg}}/1000 \ \mathrm{kW} \tag{32}$$

Hence, the energy generated annually by the GG is given by

$$E_{\mathrm{G}} = \sum_{k=1}^{8760} P_{\mathrm{gr}} \ \mathrm{kWh} \tag{33}$$

where P_{gr} is the rated power of the GG, l_{g} is the normalized size of the wind generator with respect to incremental area Δa, and P_{gavg} is the average gas power in watts that can be generated by the GG occupying 1 m^2.

Initial costs

The initial costs of the solar, wind, and gas generation, in dollars, are given by

$$C_{\mathrm{sI}} = P_{\mathrm{sr}} \times C_{\mathrm{sIavg}} \ \$ \tag{34}$$

$$C_{\mathrm{wI}} = P_{\mathrm{wr}} \times C_{\mathrm{wIavg}} \ \$ \tag{35}$$

$$C_{gI} = P_{gr} \times C_{gIavg} \ \$ \quad (36)$$

where C_{sI}, C_{wI}, and C_{gI} are the initial (capital) costs of the power sources of a selected size, and C_{wIavg} and C_{gIavg} are the average initial costs of generation per kW for solar, wind, and gas, respectively.

Operating cost

The operational cost is given by the following equations:

$$\mathrm{fu}_y = \sum_{k=1}^{8760} \mathrm{fu}(k) \ \mathrm{m}^3 \quad (37)$$

$$C_{op} = \mathrm{fu}_y \times C_g \ \$ \quad (38)$$

where fu_y is the yearly fuel consumption (m^3), $\mathrm{fu}(k)$ is the fuel consumed hourly (m^3/h), C_g is the gas price ($\$/\mathrm{m}^3$), and C_{op} is the operational cost ($).

Income

The yearly income, in dollars, is calculated using the amount of energy generated annually from renewables and the electricity price from utility grid. It is given by

$$\text{Yearly income} = (E_s + E_w) \times 0.291 \ \$ \quad (39)$$

Return on investment

The return on investment (ROI) is calculated as follows:

$$\mathrm{ROI} = (C_{sI} + C_{wI} + C_{gI}) / \text{yearly income} \quad (40)$$

Capacity factor

The capacity factor represents the contribution of renewables to the total electricity bill, and it is given by

$$\mathrm{CF} = ((E_s + E_w)/\mathrm{EE}_y) \times 100\% \quad (41)$$

where CF is the capacity factor (%) and EE_y is the yearly consumption of electric energy.

13.6 ESN APPLICATIONS ON RESIDENTIAL HOMES

The following case study illustrates the implementation of the proposed algorithm. The case study focuses on a domestic home in Toronto, Ontario, Canada. The monthly electric bill for the house was available to the authors and is given in Table 4. The algorithm is suitable for any type of infrastructure; however, this house was selected for the case study because of the authors' easy access to the required data for this infrastructure type.

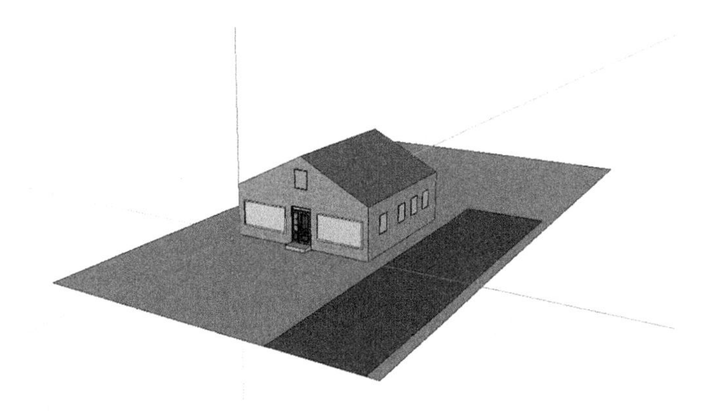

FIG. 14

Case-study home.

Fig. 14 shows the infrastructure selected for the case study. The house is oriented north-to-south, meaning that the front door directly faces north and the back door directly faces south. The available area for renewable energy sources in this scenario is the backyard, which is ~340 m^2. For simplicity, the basement is neglected, on the assumption that it is a single room that includes the necessary HVAC equipment.

The solar irradiance and wind profile history for the Toronto area, in 2013, are used for the evaluation process of the PV and WT, respectively [23–25].

13.6.1 INFRASTRUCTURE DERs

The ESN for the selected DERs is as shown in Fig. 15. The main objective of the scenario evaluation and analysis is to determine the best deployment of DERs in order to provide an effective configuration, according to the desired KPIs for the infrastructure.

As shown in Fig. 15, the infrastructure represents aggregate loads with an available facility area suitable for deploying DERs. The primary step is to consider the three energy generation technologies: solar, wind, and natural gas with unknown sizes. The algorithm's main objective is to generate all possible scenarios that comply with the government and regional regulations presented in the rule base, and then to evaluate those scenarios according to the KPIs indicated in the previous section. The resultant evaluation and analysis provide a clear view of the DER-deployment alternatives within the selected infrastructure.

The appliances are located in the main and upper floors, marked as either M_n, L_n, or A_n, where M stands for motors, L stands for lights, A stands for appliances, and n stands for the index number of each. Table 3 lists the loads along with their descriptions. Originally, the DER was intended to directly supply electricity to appliances within the infrastructure. However, this idea was abandoned because, in Toronto, the

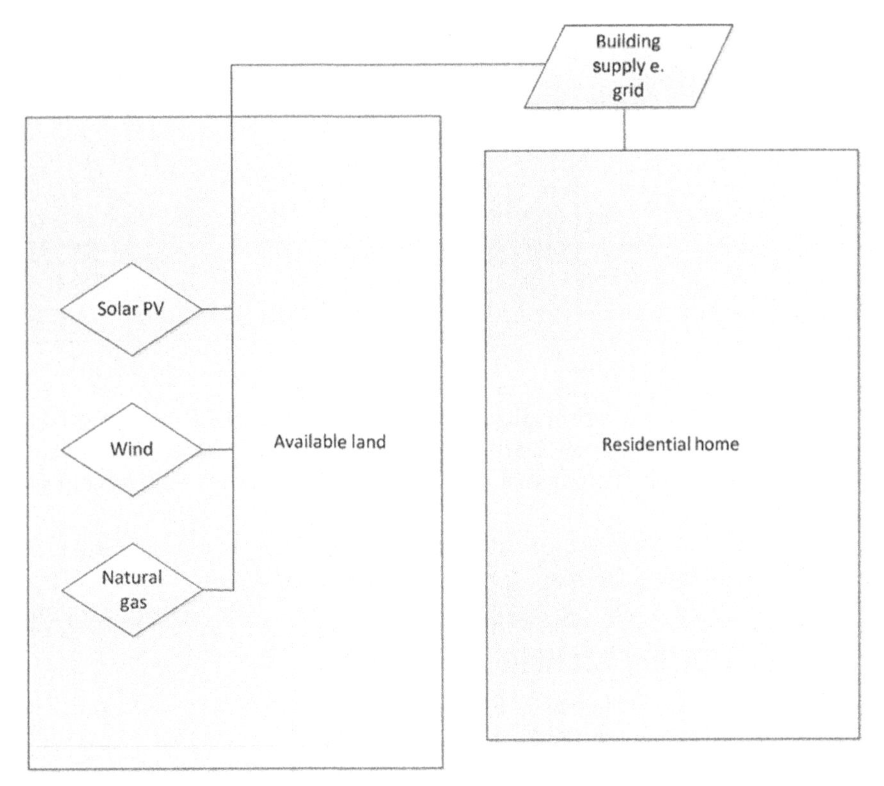

FIG. 15

Loads and DERs.

Table 3 Classification of Loads

Item Number	Description	Item Number	Description	Item Number	Description
L_1	Dining room light	M_1	Motor 1, kitchen	A_8	Television box
L_2	Kitchen light	M_2	Motor 2, bathroom	A_9	Light
L_3	Hallway light	A_1	Refrigerator	A_{10}	Computer
L_4	Bedroom light	A_2	Stove	A_{11}	Clock
L_5	Living room light	A_3	Oven	A_{12}	Television
L_6	Bedroom light	A_4	Microwave	A_{13}	Television box
L_7	Bathroom light	A_5	Coffee maker	A_{14}	Computer
L_8	Hallway light	A_6	Toaster		
L_9	Bathroom light	A_7	Television		

Table 4 Case Study Data

Month	Consumption (kWh)	Electric Bill ($)	Month	Consumption (kWh)	Electric Bill ($)
Jan.	1709	251	Jul.	979	160
Feb.	1724	252	Aug.	912	152
Mar.	1554	230	Sep.	939	146
Apr.	1344	201	Oct.	471	81
May	807	132	Nov.	655	107
Jun.	658	114	Dec.	1157	177

FIT program is available, which allows for a quicker ROI for DERs, as opposed to directly supplying energy to appliances. This is not the case in other scenarios where no FIT program exists. The program code can be easily modified to accommodate these scenarios.

The energy consumption of the infrastructure was calculated based on typical electricity prices during the case study, as shown in Table 4.

13.6.2 EVALUATION RESULTS

The algorithm was executed based on inputs indicated in the previous sections. The results were provided as a spreadsheet, as shown in Table 5. According to case study inputs, 57,971 different possible scenarios can be implemented with different combinations of PV and a natural gas generator for the available area. Table 5 shows the scenarios that have lowest possible ROI.

The results show that increased utilization of natural gas increases the ROI. This is based on the assumption that the natural gas generator operates at 100% of its capacity (which can be easily adjusted within the program logic). The synthesized scenarios are those that generate enough power to satisfy power demands.

Table 6 shows the scenarios with lowest initial cost. The results show that the less-expensive solutions are the ones strictly utilizing PVs. This is due to the lower unit costs associated with solar, as compared with natural gas costs. These scenarios show low ROI, however, due to the negligence of inverter costs, which can easily be incorporated into the programs logic.

Table 7 shows the results of scenarios with a 100% capacity factor. These results can be further extended by using a 100% capacity factor, along with the fastest ROI or lowest initial cost, by extracting such scenarios. The algorithm produces the results on a generalized basis, allowing potential users to flexibly select desired scenarios according to their own particular order.

The results of the case study show two different scenarios. The first scenario has lowest possible initial cost. Table 6 shows the results of the lowest possible initial costs deployed to the case study. Each row represents a different case being applied to the infrastructure. The ROI, capacity factor, annual expect income, and annual

Table 5 Results Ordered to Show Lowest Possible ROI

Solar Land Area (m^2)	Natural Gas Land Area (m^2)	Solar Power Output (kW)	Natural Gas Power Output (kW)	Total Power Output (kW)	Estimated Solar Energy (kWh)	Estimated Natural Gas (kWh)	Estimated Total Renewable Energy Production (kWh)	Estimated Cost of Operation for Natural Gas	Estimated Annual Income R.E.S.	Estimate Initial Cost	ROI	Capacity Factor Total R.E.S. (%)
1	339	0	4630	4630	204	40,561,155	40,561,359	607	2,433,122	1,333,061	1	314,210
1	338	0	4617	4617	204	40,441,505	40,441,709	605	2,425,945	1,329,130	1	313,283
1	337	0	4603	4603	204	40,321,856	40,322,060	603	2,418,767	1,325,200	1	312,356

Table 6 Result Ordered to Show Lowest Possible Initial Cost

Solar Land Area (m^2)	Natural Gas Land Area (m^2)	Solar Power Output (kW)	Natural Gas Power Output (kW)	Total Power Output (kW)	Estimated Solar Energy (kWh)	Estimated Natural Gas (kWh)	Estimated Total Renewable Energy Production (kWh)	Estimated Cost of Operation for Natural Gas	Estimated Annual Income R.E.S.	Estimate Initial Cost	ROI	Capacity Factor Total R.E.S. (%)
1	0	0.17	0	0.17	204	0	204	0	59.36	608.48	10.25	1.58
2	0	0.34	0	0.34	408	0	408	0	118.73	1216.96	10.25	3.16
3	0	0.51	0	0.51	612	0	612	0	178.09	1825.44	10.25	4.74

Table 7 Results Ordered to Show 100% Capacity Factor

Solar Land Area (m^2)	Natural Gas Land Area (m^2)	Solar Power Output (kW)	Natural Gas Power Output (kW)	Total Power Output (kW)	Estimated Solar Energy (kWh)	Estimated Natural Gas (kWh)	Estimated Total Renewable Energy Production (kWh)	Estimated Cost of Operation for Natural Gas	Estimated Annual Income R.E.S.	Estimate Initial Cost	ROI	Capacity Factor Total R.E.S. (%)
62	0	11	0	11	12,648	0	12,648	0	3681	37,725.82	10.25	98.0
63	0	11	0	11	12,852	0	12,852	0	3740	38,334.3	10.25	99.6
64	0	11	0	11	13,056	0	13,056	0	3799	38,942.78	10.25	101.1

energy production rate are provided for further analysis, if required. For example, if budget is ~$1800 for a given infrastructure, scenario 3 in Table 6 is the best option. It should be noted that the ROI for a strictly solar system is ~10 years. This helps verify the results of the simulation calculation and proposed program. Figs. 16–18 show the trends for the total output power, ROI, and initial cost of the available scenarios. The scenarios with low indices are those that indicate less deployment of energy generation technologies in the facility area. Increases in the indices mean that the corresponding energy generation technologies are increasingly deployed in the facility area. Based on the typical hourly load trend [26–29] and the electricity consumption data for the case study, shown in Table 4, the required power is 6 kW. Accordingly, Fig. 16 shows scenarios 30–47, and these scenarios can be considered satisfactory, with output power ranging from 5 to 15 kW. Fig. 17 shows that scenarios 45–47 have better ROI, but their initial costs could range from $10,000 to $20,000, as shown in Fig. 18. Figs. 16–18 help the designer identify the scenarios that best fit the specified requirements.

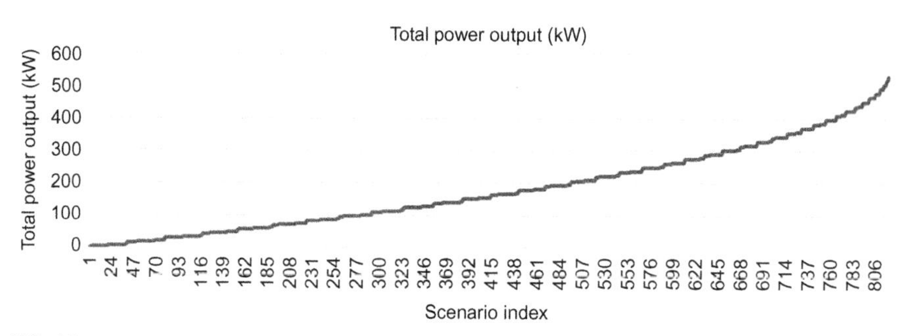

FIG. 16

Total output power for different deployment scenarios.

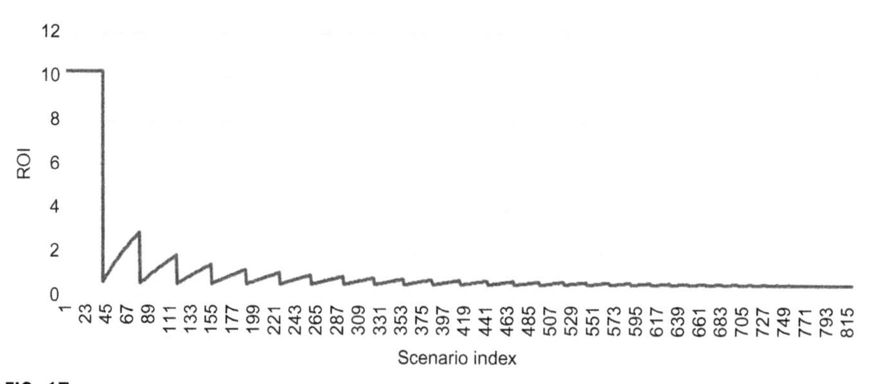

FIG. 17

ROI for different deployment scenarios.

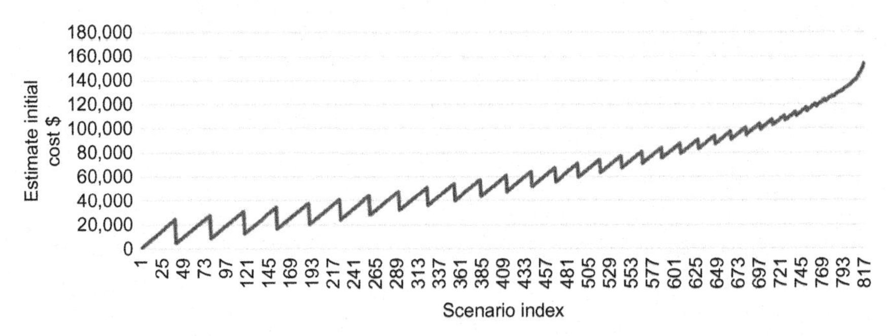

FIG. 18

Estimated initial cost for different deployment scenarios.

13.6.3 RESULT ANALYSIS

This chapter introduces an algorithm for generating possible scenarios for DER deployment. Infrastructure developers and manufacturers of energy resources can use this algorithm to analyze and select the DER deployment scenarios that most suit their needs. Developers face the technical challenges of choosing which DERs to implement within their infrastructures. The program therefore allows the developers to run and evaluate possible scenarios in order to select the most effective DER-deployment scenario to be implemented in the infrastructure, based on the best ROI and other KPIs.

In addition, the proposed program helps manufacturers of DERs to market their products to potential customers, based on accurate performance evaluations. Given any infrastructure specifications, the proposed algorithm can generate and evaluate the best set of DERs for a project. Finding the most effective solution can also help potential consumers adopt the appropriate DERs, and the algorithm can compare all relevant energy technologies (and their combinations), thereby helping decision-makers deploy the most suitable DERs.

13.7 CONCLUSION

This chapter presented an overview of challenges and opportunities of smart energy grids and their implementation using adaptive modeling of energy grid infrastructures using energy semantic networks (ESN). Detailed knowledgebase implementation of ESN has been presented and discussed using selected case studies. Energy conservation scenarios are synthesized for buildings where deployment of different renewable energy technologies are assessed in view of building load profiles and generation capacity. Simulation practices are discussed and used to evaluate smart energy grid infrastructures in view of key performance indicators (KPIs), which are defined and linked to ESN infrastructures. Based on the discussed case studies, the use of ESN demonstrated effective ways to design and operate smart energy grid infrastructures and can ensure highest performance with the use of optimization techniques based on local and global KPIs of smart energy grids.

ABBREVIATIONS

DER	distributed energy resource
EDSP	energy deployment scenarios program
ESN	energy semantic network
FIT	feed-in-tariff
GG	gas generator
IEA	International Energy Agency
KPI	key performance indicator
NG	natural gas electric power
PV	photovoltaic solar power
RES	renewable energy sources
ROI	return on investment
WT	water transfer

SYMBOLS

N	number of scenarios
M	number of selected energy generators
A	facility area
Δa	minimum increment of area to install generator
P_{sr}, P_{wr}, P_{gr}	rated power for solar, wind, and gas generators, respectively (kW)
l_s, l_w, l_g	normalized size with respect to incremental area Δa for solar, wind, and gas generators, respectively
P_{savg}, P_{wavg}, P_{gavg}	average power generated by 1 m^2 of solar, wind, and gas generators, respectively (W)
r_{s}.	solar irradiance profile (W/m^2)
v_w	wind speed (m/s)
P_s, P_w, P_g	power generated by solar, wind, and gas generators, respectively (kW)
E_s, E_w, E_g	energy produced annually by solar, wind, and gas generators, respectively (kWh)
C_{sI}, C_{wI}, C_{gI}	initial (capital) cost of solar, wind, and gas generators, respectively ($)
C_{sIavg}, C_{wIavg}, C_{gIavg}	initial (capital) cost of solar, wind, and gas generators, respectively ($)
fu_y.	yearly fuel consumption (m^3)
C_{op}	operational cost ($)
fu	fuel consumed hourly (m^3/h)
k	hourly sample index ranges from 1 to 8760 (number of hours per year)

REFERENCES

[1] Gabbar HA, Sayed HE, Yamashita Y, Gruver W, Kamel M. Engineering design aspects of hybrid energy supply unit (HENSU). In: 2nd international workshop on computational intelligence & applications, Okayama, Japan, 15 December 2006; 2006. P12:1–12.

[2] Kurohane K, Uehara A, Senjyu T, Yona A, Urasaki N, Funabashi T, et al. Control strategy for a distributed DC power system with renewable energy. Renew Energy 2011;36 (1):42–9.

[3] Wang C, Jia Q-Q, Li X-B, Dou C-X. Fault location using synchronized sequence measurements. Int J Electr Power Energy Syst 2008;30(2):134–9.
[4] Firouzjah KG, Sheikholeslami A. A current independent method based on synchronized voltage measurement for fault location on transmission lines. Simul Model Pract Theory 2009;17(4):692–707.
[5] Orecchini F, Santiangeli A. Beyond smart energy grids—the need of intelligent energy networks for a higher global efficiency through energy vectors integration. Int J Hydrog Energy 2011;36(8):26–33.
[6] Hammons TJ. Integrating renewable energy sources into European grids. Electr Power Energy Syst 2008;30:462–75.
[7] Battaglini A, Lillistam J, Hass A, Patt A. Development of SuperSmart energy grids for a more efficient utilization electricity from renewable sources. J Clean Prod 2009;17:911–8.
[8] Pouresmaeil E, Gomis-Bellmunt O, Montesinos-Miracle D, Bergas-Jane J. Multilevel converters control for renewable energy integration to the power grid. Energy 2011;36:950–63.
[9] Wade NS, Taylor PC, Lang PD, Jones PR. Evaluating the benefits of an electrical energy storage system in a future smart energy grid. Energy Policy 2010;38:7180–8.
[10] Divya KC, Ostergaard J. Battery storage technology for power systems—an overview. Elsevier. Electr Power Syst Res 2009;79:511–20.
[11] Clayton JM, Young FS. Estimating lightning performance of transmission lines. IEEE Trans PAS 1964;83:1102–10.
[12] Bouquegneau C, Dubois M, Trekat J. Probabilistic analysis of lightning performance of high-voltage transmission lines. Electr Power Syst Res 1986;102(1–2):5–18.
[13] Transition to sustainable infrastructures: strategies and opportunities to 2050. Paris: International Energy Agency; 2013.
[14] Schultz CC, Pe CC, Scott G. Energy conservation in existing infrastructures. Eng Syst 2011;28(10):34–40.
[15] Smeds J, Wall M. Enhanced energy conservation in houses through high performance design. Energy Infrastruct 2007;39:273–8.
[16] Hussain S, Gabbar HA, Musharavati F, Pokharel S. Key performance indicators (KPIs) for evaluation of energy conservation in infrastructures. In: Smart energy grid engineering (SEGE), 2013 IEEE international conference, Oshawa, August; 2013.
[17] Kim DLA. Optimizing cost effective energy conservation measures for infrastructure envelope. Energy Eng 2010;107(3):70–80.
[18] Difsa K, Bennstamb M, Trygga L, Nordenstamb L. Energy conservation measures in infrastructures heated by district heating—a local energy system perspective. Energy 2010;35(8):3194–203.
[19] Pitts A, Saleh JB. Potential for energy saving in infrastructure transition spaces. Energy Infrastruct 2007;39:815–22.
[20] Mousa M, Akash B. Some prospects of energy saving in infrastructures. Energy Convers Manag 2001;42:1307–15.
[21] Chedwal R, Mathur J, Das Agarwal G, Dhaka S. Energy saving potential through Energy Conservation Infrastructure Code and advance energy efficiency measures in hotel infrastructures of Jaipur city, India. Energy Infrastruct 2015;92:282–95.
[22] Busch JF, Pont PD, Chirarattananon S. Energy-efficient lighting in Thai commercial infrastructures. Energy 1993;13(2):197–210.

[23] Saberbari E, Saboori H. Net-zero energy infrastructure implementation through a grid-connected home energy management system. In: The 19th electrical power distribution conference (EPDC2014); 2014.
[24] Gabbar HA. Engineering design of green hybrid energy production and supply chains. Environ Model Softw 2009;24:423–35.
[25] Gabbar HA, Eldessouky AS. Energy semantic network for building energy management. Intell Ind Syst 2015;1:213–31. http://dx.doi.org/10.1007/s40903-015-0023-8 .
[26] Marszal AJ, Heiselberg P, Bourrelle JS, Musall E, Voss K, Sartori I, et al. Zero energy infrastructure—a review of definitions and calculation methodologies. Energy Infrastruct 2011;43:971–9.
[27] Hernandez P, Kenny P. From net energy to zero energy infrastructures: defining life cycle zero energy infrastructures. Energy Infrastruct 2010;42:815–21.
[28] Mishra A, Irwin D, Shenoy P, Kurose J, Zhu T. GreenCharge: managing renewable energy in smart infrastructures. IEEE J Sel Areas Commun 2013;31(7):1281–93.
[29] Runge J, Gabbar H. Solar windows control system for an apartment infrastructure in Toronto with battery storage. In: International conference on power engineering and renewable energy (ICPERE), 2014, Bali, December; 2014.

CHAPTER

Advanced optimization methods of micro energy grids

14

A.M. Othman*,†, H.A. Gabbar*

*University of Ontario Institute of Technology (UOIT), Oshawa, ON, Canada**

Zagazig University, Zagazig, Egypt†

14.1 INTRODUCTION

Recently, research on renewable energy has demonstrated significant progress and development of advanced optimization methods of micro energy grids. The main purpose of a micro energy grid (MEG) is to supply specific local loads in certain remote geographical areas or zones with economical and efficient operation using distributed energy resources (DERs), which include generation (such as PV solar cells, wind energy, fuel cells, batteries, and diesel engines) as well as energy storage. MEGs can have islanding operation away from the power utility grid, or the can be in other forms, including grid-connected mode, which has a significant impact and effect. The integration of DERs with transmissions—distribution lines, as well as MEGs—depends on the power electronics and control system technologies.

Lately, modern heuristic optimization techniques have received increasing attention from the MEG application areas. Various controllers—especially adaptive controllers—can provide the most effective control strategies for MEGs based on generation and load profiles. However, effective control strategies can only be achieved via effective optimization algorithms. There is recent research that is based on heuristic optimization methods for achieving the global minimum with convergence rate.

14.2 PARTICLE SWARM OPTIMIZATION

Particle swarm optimization (PSO) is considered an effective technique and is based on a stochastic optimization process that has no explicit knowledge about the gradient of the optimized system. PSO is mainly used for simulating social behavior. PSO applies optimization by keeping a population of proposed solutions (called particles) and later, moving those particles inside the space of the search with respect to a simple formula. The motion of the particles is directed by the best positions in the space of the search, which are originally modified as better positions found by the particles.

Smart Energy Grid Engineering.

14.2.1 CONCEPT OF PSO

PSO utilizes various identical features with evolutionary computational methods. The system starts with a population of stochastic solutions and looks for optimality by updating generations. In PSO, the potential solutions are the particles, which move in the problem space by succeeding the current optimum particles. Each particle maintains track of its coordinates in the space-solution, which is accompanied by the best record it has incurred. The fitness is also memorized; this value is called P_{best}. The best value that is followed by the particle swarm optimizer is estimated by any particle in the neighborhood space of the particle. This position is referred to as L_{best}. When particles are processed by the complete population as their topological neighborhood space, the best value is a global best and is called g_{best}. The PSO idea depends on changing the speed of acceleration for each particle, directing it to its P_{best} and L_{best} locations. The accelerating process is controlled by a stochastic parameter with separate random values being determined for accelerating particles to their P_{best} and L_{best} positions. Many publications are concerned with choosing well-thought-out PSO parameters. The method for tuning PSO parameters is called meta-optimization because another optimization method may be used in an overlapping way to tune the PSO parameters.

PSO technique

PSO depends basically on a randomized velocity corresponding to a potential solution, which called a particle, and it is harmonized with individuals. All the particles in PSO move in the multidimensional problem space, dimension size D, with a speed that is dynamically adapted with respect to the flying experience of particles.

The position of the ith particle is described as $X_i = [x_{i1}, x_{i2}, \ldots, x_{iD}]$, where x_{id} is inside the range $[l_d, u_d]$, d belongs to the range $[1,D]$. l_d is the lower bound and u_d is the upper bound for dth dimension, respectively. The best previous position, which calculates the best fitness value of the ith particle, is saved and represented as $P_i = [P_{i1}, P_{i2}, \ldots, P_{iD}]$, which can also be termed P_{best}.

The concept of the best particle out of all particles in the population is described by the variable g. The position of P_g can also be termed as g_{best}. The speed of the *ith* particle is described as $V_i = [v_{i1}, v_{i2}, \ldots, v_{iD}]$ and is clamped to a maximum speed $V_{\max} = [v_{\max 1}, v_{\max 2}, \ldots, v_{\max D}]$ that is designed by the optimizer. At each time step, the PSO concept regulates the velocity and directs the location of each particle toward its P_{best} and g_{best} locations according to the following equations:

$$v_{id}^{n+1} = wv_{id}^{n} + c_1 r_1^n \left(P_{id}^n - x_{id}^n\right) + c_2 r_2^n \left(P_{gd}^n - x_{gd}^n\right) \tag{1}$$

$$x_{id}^{n+1} = x_{id}^{n} + v_{id}^{n+1} \tag{2}$$

$$i = 1,2,\ldots,m, \quad d = 1,2,\ldots,D$$

where c_1 and c_2 are the positive parameters, described cognitive and social parameters, respectively; m is the size of the swarm, D is the number of members in a particle, r_1 and r_2 are random numbers, uniformly distributed in $[0,1]$, n is the pointer of iterations (generations), and w is the inertia weight, which achieves a balance between global and local investigation, thus leading to a lower iteration to calculate an appropriate optimal solution.

It is specified by the following equation:

$$w = w_{max} - \frac{w_{max} - w_{min}}{iter_{max}} \times iter \quad (3)$$

where w_{max} is the initial weight, w_{min} is the final weight, iter is the current iteration number, and $iter_{max}$ is the maximum iteration number for generations.

The PSO algorithm can be summarized as follows:

1. Initialization for all the particles in the space of the search space stochastically.
2. Assignment (stochastically) of the initial velocities for all the particles.
3. Evaluation of the fitness of all the particles based on previously mentioned objective function.
4. Calculation of the updated velocities for all the particles.
5. Movement of all the particles.
6. Repeat steps 3–5 until an assigned stopping criterion is achieved.

PSO: Controlling velocities

There are many controlling factors that can manage and control the velocity of the particles during the optimization process, which in turn help the PSO to achieve the global minimum of the problem in convergent rate.

- Magnitude of the speed can be increased as a controlling factor.
- Performance can be improper if V_{max} is set the wrong way.
- Several techniques were proposed for controlling the growth of velocities:
  - An effectively updated inertia weight factor.
  - Effectively updated acceleration coefficients.
  - Re-initialization of stagnated particles and swarm size.
  - Robust settings for c_1 and c_2.

14.3 INTEGRATED GA WITH ANFIS

The genetic algorithm (GA) system mainly depends on the initial population, natural selection (NS), mating, and mutations. In the case of continuous fitness function characteristics, it is better to apply the continuous GA because of its accuracy and it requires less memory size. Fig. 1 shows an overview of a continuous GA flowchart.

The fitness trends will control the initial population selection. It is sorted according to the fitness value. Individuals with the highest probability will be chosen. If P_1 and P_2 are chosen according to crossover, the offspring based on crossover are:

$$P'_1 = r \cdot P_1 + (1 - r) \cdot P_2 \quad (4)$$

$$P'_2 = (1 - r) \cdot P_1 + r \cdot P_2 \quad (5)$$

where r is a random number between 0 and 1.

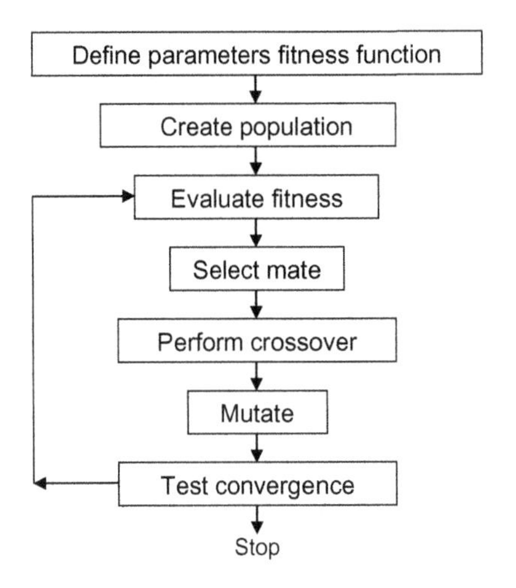

FIG. 1

Flowchart for continuous structure of GA.

Mutation can be considered as a new solution within the generation to allow random effect in various individuals. When the variable P_k of a parent P is decided for mutation, the result will be:

$$P'_k = \begin{cases} P_k + (\text{UB} - P_k) f(\text{gen}) & \text{if a random digit is } 0 \\ P_k - (\text{LB} + P_k) f(\text{gen}) & \text{if a random digit is } 1 \end{cases} \tag{6}$$

where UB and LB are the upper and lower bounds of the individual P_k, respectively, gen is the current generation. $f(\text{gen})$ will give an output value in the range [0,1]. Where the probability of $f(\text{gen})$ increases, at gen increases. The following function is applied to confirm uniform searching in the solution space:

$$f(\text{gen}) = \left(r \cdot \left(1 - \frac{\text{gen}}{\text{gen max}} \right) \right)^b \tag{7}$$

where the r is a stochastic number belongs to the range [0,1], genmax is generation's maximum, and b is called nonuniformity shape.

14.3.1 ADAPTIVE NEURO-FUZZY INFERENCE SYSTEMS

Adaptive neuro-fuzzy inference systems (ANFIS) apply rules that optimize the fuzzy system parameters of a Sugeno system. The Sugeno fuzzy system proposes a systematic method to produce fuzzy rules for certain input/output patterns. The ANFIS training systems use a gradient descent algorithm to optimize the antecedent parameters, and a least-square algorithm to get the consequent parameters. In the first part, a least-square algorithm is applied to adapt consequent parameters, and then the antecedent parameters are modified by backpropagating the errors.

ANFIS result is expressed as:

$$y_m = \bar{w}_1 f_{1m} + \cdots + \bar{w}_n f_{nm} \tag{8}$$

where m is the output number and n is the number of the input membership functions. f_{nm} is expressed as follows:

$$f_{nm} = p_{nm} x_1 + q_{nm} x_2 + r_{nm} \tag{9}$$

So the ANFIS result will be modified to:

$$y_m = \left[\bar{w}_1 x_1 \quad \bar{w}_1 x_2 \quad \bar{w}_1 \quad \cdots \quad \bar{w}_n x_1 \quad \bar{w}_n x_1 \quad \bar{w}_n \right] \begin{bmatrix} p_{1m} \\ q_{1m} \\ r_{1m} \\ \vdots \\ p_{nm} \\ q_{nm} \\ r_{nm} \end{bmatrix} = X W_m \tag{10}$$

Pseudo inverse will be applied to estimate the vector W_m, ie,

$$W_m = \left(X^T X\right)^{-1} X^T y_m \tag{11}$$

The output of the antecedent will be expressed as:

$$a(T+1) = a(T) - \frac{\eta_a}{k} \frac{\partial E}{\partial a} \tag{12}$$

where a is the parameter to be updated, T is the Epoch number, η_a is the learning rate, k is the number of input patterns, and E is the sum of square error.

The antecedent parameters will be modified by derivatives that are estimated by the chain rule as:

$$\frac{\partial E}{\partial a_{iL}} = \frac{\partial E}{\partial y} \cdot \frac{\partial y}{\partial w_i} \cdot \frac{\partial w_i}{\partial \mu_{iL}} \cdot \frac{\partial \mu_{iL}}{\partial a_{iL}} \tag{13}$$

where i is the membership function number, L and y are the sum of ANFIS outputs, and S is the inputs number.

Fig. 2 depicts a novel integrated GA with an ANFIS controller that is applied to self-regulate the proportional-integral-derivative (PID) tri-loop stage for an FACTS device. This can lead to reliable and efficient performance, as it realizes both the criteria of power quality and enhancement of the voltage, and current and power factor profiles. The adaptive controller is shown in structural layers. Artificial neural network (ANN) will cooperate with fuzzy logic to construct the ANFIS system. The GA system will be considered as a cooperative system for the resultant ANFIS to save pretraining patterns to ANFIS to act as the training patterns.

The GA will be able to reach optimal solutions and will realize the desired performance of the system under any operating conditions. An ANFIS merges artificial neural networks with fuzzy logic algorithms, so it has the comprehensive options of both of them. Fuzzy logic (FL) is an effective technique control process, while ANN has distinguished characteristics in classifications and recognitions. ANN are linked with the FL to update the parameters of the fuzzy systems to achieve the best

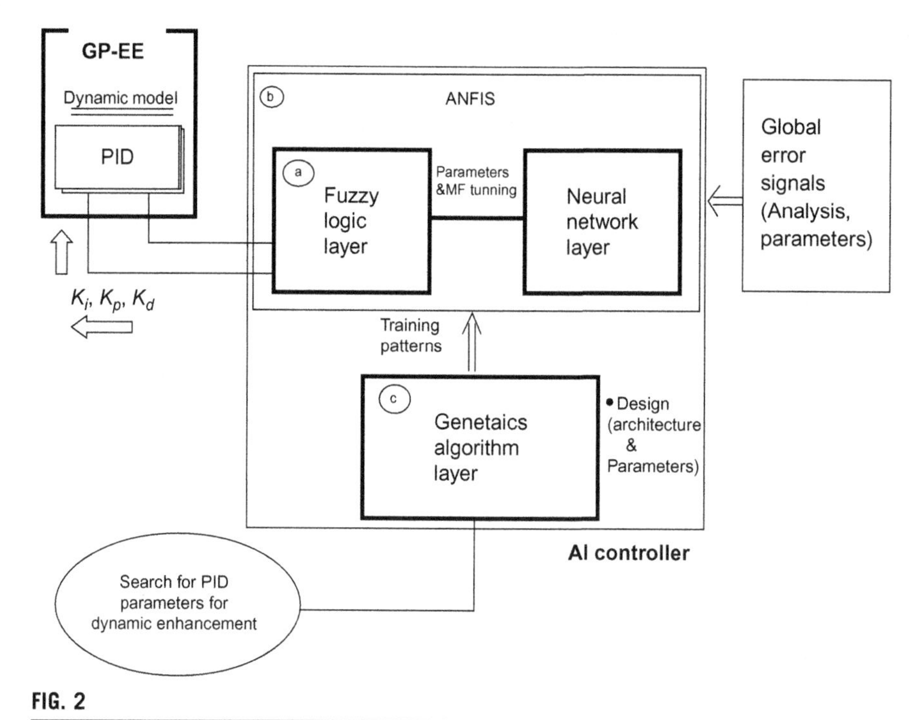

FIG. 2

Structural layers of the adaptive controller.

parameters tuning leading to the requested controlling process. Therefore, the positive behavior of both ANN and FL are realized by merging them, and catching their global benefits.

14.4 ENHANCED BACTERIAL FORAGING OPTIMIZATION

The enhanced bacterial foraging optimization (EBFO) algorithm is developed through the integration of both PSO and bacterial foraging optimization (BFO) to gain the merits of both, while avoiding their demerits. PSO depends on particles that try to converge with keeping to not stop at a local minimum, although sometimes the natural of the problem leads to trap in that local minima. BFO simulates the foraging behavior of bacteria inside of the human paunch. The rule of NS controls the bacteria's behavior, according to "the survival of the fittest." The bacteria that forages well is better than the others, so it survives and the others die. The technique of BFO depends on four processes: *chemotaxis, swarming, reproduction*, and *removal-dispersal*. There are drawbacks for both PSO and BFO. PSO suffers from early convergence before the final settled one, while BFO suffers from fixing the step size based on uncontrolled particle velocities, and also from the random movements of individuals. Enhanced BFO has the ability to handle drawbacks, and to enhance the global optimum. The procedure of EBFO is explained later in a flowchart in Fig. 3.

EBFO starts the searching process with bacteria that has an initial particle velocity, and that is randomly dispersed in the solution space. Using the cell-to-cell swarming option, the fitness function is determined to realize the best positions. Then, there will be updates in the individuals' positions by a chemotaxis option using a velocity factor. In the reproduction stage, the bacterial with the worst health are removed, while the others reproduce by splitting into two. Finally, the removal-dispersal stage involves the removal of the bad bacteria with a very low probability, and best replacements are initialized to achieve the global optimum position.

We will summarize the phases of the implemented EBFO technique:

Phase 1: Definitions for the optimization controlling parameters are as follows:
x: the position of a bacterium
$Ft(x)$: the fitness of a bacterium
N: the total number of bacteria in the population
U: the number of optimized variables
Each bacterium (i) has a position (x) that depends on three components: chemotactic (α), reproductive (β), and removal-dispersal (γ), with consideration of swarm length (λ).
Phase 2: Generation of an initial population for each individual as random initial position $x(i)$, and an initial velocity $v(i)$, which is a random value between $(-1,1)$.
Phase 3: (3a) Calculate the current fitness Ft^i_{current} for each bacterium: $Ft^i_{\text{current}} = Ft(i,\alpha,\beta,\gamma)$, where at the start of the searching there is not much enhancement in the bacterium positions, so the local best fitness is the current value, $Ft^i_{\text{local}} = Ft^i_{\text{current}}$.
(3b) Determine the initial global best fitness in the population $Ft_{\text{global}} = \min\left(Ft^i_{\text{local}}\right)$, that has best global position x^i_{best}.
Phase 4: Start to apply cell-to-cell attractant and repellent actions to the fitness function as follows:

$$Ft^i_{\text{local}} = Ft^i + \sum_{i=1}^{N}\left[-A_1 \times \left(\mathrm{e}^{-T_1 \sum_{n=1}^{U} \left(x_n - x^i_n\right)^2}\right)\right] + \sum_{i=1}^{N}\left[A_2 \times \left(\mathrm{e}^{-T_2 \sum_{n=1}^{U} \left(x_n - x^i_n\right)^2}\right)\right] \tag{14}$$

where A_1 is the quantity rate of attractant actions, T_1 is the diffusion rate of attractant actions, A_2 is the quantity rate of repellent actions, and T_2 is the diffusion rate of repellent actions.
Phase 5: (5a) Update the fitness function by the process of chemotactic:

$$x(i,\alpha+1,\beta,\gamma) = x(i,\alpha,\beta,\gamma) + C(i) \cdot \frac{v^i_\beta}{\sqrt{\left(v^i_\beta\right)^T \cdot v^i_\beta}} \tag{15}$$

$$Ft^i_{\text{global}} = \min\left[Ft^i(x(i,\alpha+1,\beta,\gamma)\right] \tag{16}$$

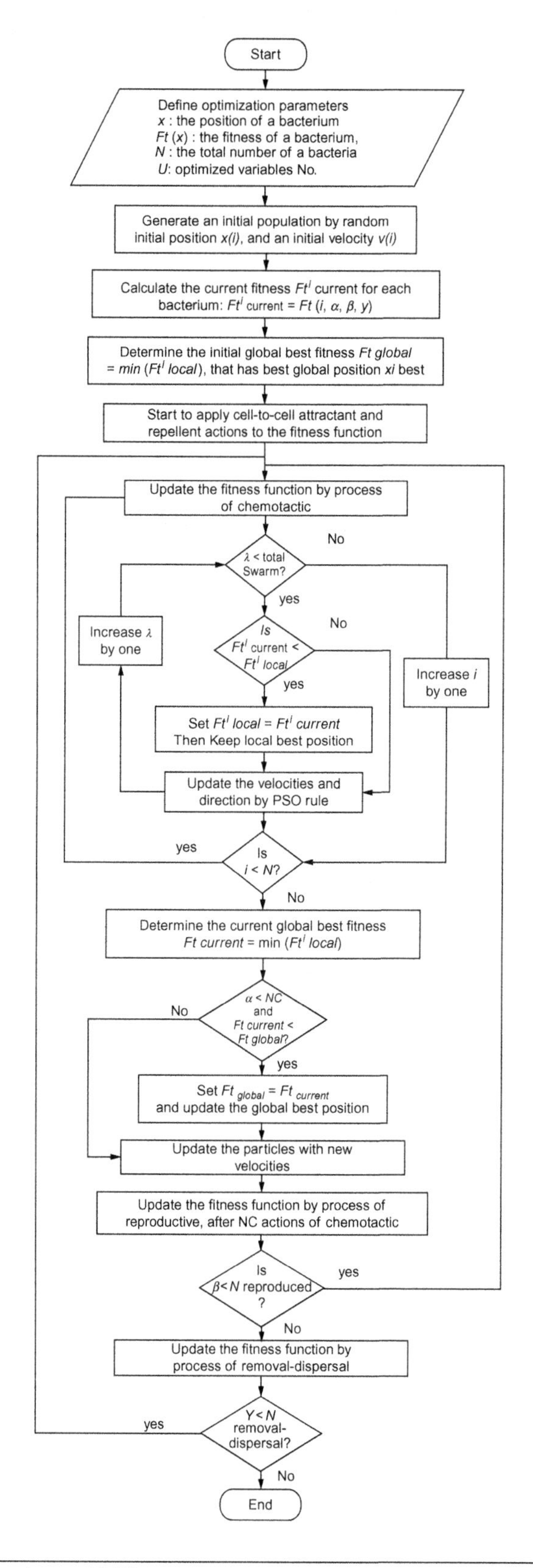

FIG. 3

Flowchart of an enhanced BFO algorithm.

(5b) Update the velocities and the directions based on PSO rules:

$$v^i_{\beta+1} = w \cdot v^i_\beta + c_1 \cdot r_1 \left[x^i_{\text{best}} - x^i_{\text{current}}\right] + c_2 \cdot r_2 \left[x^i_{\text{best}} - x^i_{\text{current}}\right] \tag{17}$$

w, c_1, and c_2 are inertia, cognitive and social constants; r_1 and r_2 are random numbers in the interval $[-1,1]$.

Phase 6: Update the fitness function by the reproductive process, after NC actions of chemotactic, to determine the healthy bacteria that force half of the bacteria to be removed, and then each healthy bacterium to be reproduced (by splitting in two) to keep the population in the same dimension:

$$Ft^i_{\text{healthy}} = \sum_{i=1}^{NC+1} Ft^i(\alpha, \beta, \gamma) \quad \forall N_{\text{reproduced}} = N/2 \tag{18}$$

Phase 7: Update the fitness function by the process of removal-dispersal to simulate adaptation to conditions regarding nutrient absorption, or significant temperature increase. Those conditions lead to the removal-dispersal of some bacteria by probability P_{ed} after a number of NR reproductive actions. This will allow the bacteria stuck in the local minimum to escape and reach the global optimum.

14.5 BACKTRACKING SEARCH ALGORITHM

The backtracking search optimization algorithm (BSA) is a heuristic algorithm working on global minimum based on an iterative population. Applying BSAs is superior in many ways. First of all, the structure of a BSA is quite simple; thus, it is easily applied to various optimization problems. The BSA's mutation process uses only one individual from a previous population, and its crossover process is more complex than in others. The crossover process in BSAs is different than in the other heuristic techniques; the complexity here means it is advanced. In a BSA, crossover process defines the final trial population with better-optimized fitness values, while the initial trial population is defined by the mutation process. The BSA's crossover process has multiple steps to allow and control the number and selection of individuals that will mutate in the next trial. Also, the BSA's crossover process has the option to make individuals of the trial population overflow through defined search space limits. The BSA remembers the population of a randomly selected generation for use in calculating the search-direction matrix. Also, the BSA is a dual-population algorithm that uses both current and historical populations. Then, the BSA's technique has a parameter that controls the trends and directions of searching, and can be adjusted to be large value to work on global search or small value to work on local search. Finally, the BSA's boundary control mechanism is a very efficient tool for realizing the population diversity.

The procedure of the BSA can be summarized in five main processes: initialization, selection-I, mutation, crossover, and selection-II, as described below, and as per Fig. 4.

A. Initialization

The population P of BSA is initialized according to the following equation:

$$P_{ij} \cong U(\text{Lower}, \text{Upper})_j \quad \text{where } i=1,2,3,\ldots,N \text{ and } j=1,2,3,\ldots,D \tag{19}$$

N and D are population size and problem dimension, respectively, U is the uniform distribution.

B. Selection-I

This step calculates historical population P_{old} to be used for determining the search direction by applying the following formula on pervious iterations:

$$P_{ij\text{old}} \cong U(\text{Lower}, \text{Upper})_j \quad \forall \text{ previous iterations} \tag{7}$$

At the beginning of a new iteration, this technique redefines historical population P_{old} as follows:

$$a<b \text{ then } P_{\text{old}} : P|a,b \sim U(0,1) \tag{20}$$

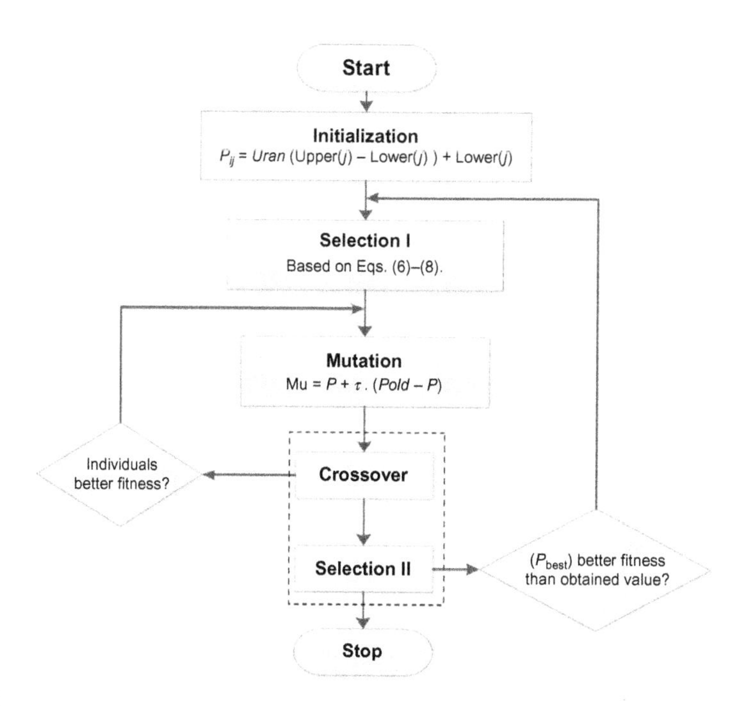

FIG. 4

General structure of BSA.

where a, b are fitness values. Thus, the BSA has a memory that remembers this historical population until it is changed. After determining P_{old}, a change in the order of the individuals in P_{old} occurs by a random shuffling function:

$$P_{old} := \text{permuting}(P_{old}) \tag{21}$$

C. Mutation

The BSA's mutation action presents an initial frame of the trial population mutant:

$$\text{Mu} = P + \tau \cdot (P_{old} - P) \tag{22}$$

where τ is a parameter that controls the trends and directions for searching, and can be adjusted to be a large value to work on global searches, or to be a small value to work on local searches. The BSA uses the historical population to take advantage of the experiences from previous generations.

D. Crossover

Final population T is produced in this step. The initial population is Mu, as is set in the mutation process. Individuals, which have better fitness values during the optimization process, are used to evolve final population individuals. The BSA's crossover process has two steps. The first step determines a binary integer-valued matrix (map) of size $N \times D$ that shows the individuals of T to be processed and updated by P. The BSA's crossover strategy has efficient options that are different from the crossover strategies used in other heuristic algorithms. The first option is based on a mixed rate parameter in the BSA's crossover action to control the number of elements of individuals that will mutate in a trial. The second option allows only one randomly chosen individual to mutate in each trial, giving it the ability to overflow the allowed search space limits as a result of the BSA's mutation strategy.

E. Selection-II

In the BSA's Selection-II stage, a greedy selection is used to update the population with better fitness values than the others. The global minimizer is updated according to P_{best} and the global minimum value is updated to be the fitness value of P_{best}.

14.6 CASE STUDY: DESCRIPTION AND SIMULATION RESULTS

Matlab-Simulink-SimPower platforms are used for the simulations with support of hardware demonstrations that concern with MEG Study System as the following specifications:

- *Utility Grid*: 138 kV, 5 GVA, $X/R = 10$.
- *Microgas Turbine*: $V = 1.6$ kV, $P = 2000$ kW.
- *Wind Turbine* (WT) *Generator*: $V = 1.6$ kV, $P = 1$ MW.

- *PV*: 240 V, 500 kW, Ns = 318, Np = 150,Tx = 293,Sx = 100, Iph = 5, Tc = 20, Sc = 205.
- *Fuel Cell*: 240 V, 2000 kW, number of cells = 220, nominal efficiency, 55%.
- *Battery*: 240 V, rated capacity: 300 Ah, initial state-of-charge: 100%, discharge current: 10, 5 A.
- *Hybrid AC Load 1*: linear load: 0.1 MVA, 0.8 lag pf, nonlinear load: 0.2 MVA, motorized load is an induction motor: 3 phase, 0.3 MVA, 0.85 pf.
- *Hybrid AC Load 2*: linear load: 200 kVA, 0.8 lag pf., nonlinear load: 200 kVA, motorized load is an induction motor: 3 phase, 100 kVA, 0.8 pf.
- *DC Load*: resistive load: 100 kW, motorized load dc series motor: 100 kW.
- *GP-EE(A) [D-FACTS device]*: Cf1 = 50 μF, Rf1 = 0.25 Ω, Lf1 = 3mH.
- *GP-EE(B) [D-FACTS device]*: Cs = 100 μF, Rs = 0.1 Ω, Ls = 10 mH.

The first stage is a planning stage, concerned with an optimal deployment, with respect to capacity sizes and types of DERs and CHP systems within the MEG. The objective is to simultaneously minimize the total net present cost and carbon dioxide emissions. A multiobjective GA is proposed in this stage to handle the planning problem, including the optimization of DER types and capacities. The constraints involve both power and heat demand constraints. Candidate DER technologies are CHPs with various characteristics, boilers, renewable generators (wind and PV), and a main power grid connection. The cost of CHP generators is based on their type and capacity range.

The planning stage
The candidate technologies included are: natural gas turbine (NGT), hydrogen fuel cell (H2FC), natural gas fuel cell (NGFC), WT, and PV. NGT, H2FC, and NGFC are assumed to operate as CHPs (generate heat and power with fixed heat to power ratio). The load profile for a year is divided into four seasons, with each season represented by 1 day, with totals equaling $4 \times 24 = 96$ h. Then, total cost (capital, operational, and maintenance), and CO_2 emission are calculated. This will change based on a daily load profile, Table 1 summarizes the results of optimal planning.
The results of this stage will be fed to the second operational stage to ensure that it works on the optimal patterns of the DERs. The second stage involves operational

Table 1 Optimal Planning Result

Case	NGT (kW)	H2FC (kW)	NGFC (kW)	WT (kW)	PV (kW)	Total Cost ($)	Total CO_2 Emission (kg)
Min cost only	1048	13	0	27	102	1.3679×10^5	8380
Min emission only	391	1300	0	90	6	7.2310×10^5	4306.94
Compromised	1179	751	0	35	406	6.5109×10^5	4876.4

control aspects. GP-EE has been validated for enhancing the performance and power quality aspects of the MEG with DERs.
The operational stage
A novel control adaptation for an error driven, dynamic multi-loop regulator of proposed GP-EE integrated GA with ANFIS system has been applied to improve power quality and energy utilization. The proposed GP-EE schemes with controlled GA with ANFIS are very effective in enhancing power quality, improving power factor, reducing feeder losses and stabilizing the buses voltage, and mitigating the harmonic distortion. The integrated GA and ANFIS system is designed with the parameters shown in Table 2, and then it is applied to get the optimal PID values of GP-EEs, as shown in Table 3.

Table 2 GA and ANFIS System Parameters

GA parameters	
Options. Population type	Double vector
Options. Population size	35
Options. Elite count	Adapted in simulations
Options. Crossover fraction	Adapted in simulations
Options. Migration direction	Forward
Options. Migration interval	35
Options. Migration fraction	0.21
Options. Generations	280
ANFIS parameters	
Number of nodes: 43	
Number of linear parameters: 70	
Number of nonlinear parameters: 31	
Total number of parameters: 83	
Number of training data pairs: 38	
Number of checking data pairs: 29	
Number of fuzzy rules: 22	
Designated epoch number → ANFIS training completed at epoch 325	

Table 3 Values of Optimized PID Controllers Gains

	Optimal Values of PID Controllers Gains		
	K_p	K_i	K_d
GP-EE(A) 1	12	2.4	0.5
GP-EE(A) 2	15	5	0.2
GP-EE(B)	11	1.7	0.2

The digital simulation is carried out with and without the controlled GP-EE in order to show its performance in voltage stabilization, harmonic reduction, and reactive power compensation during normal operating conditions. Also, the enhancement in the power factor with managing the exchange power between MEG and utility grid are all achieved and shown in the following figures. In cases with and without GP-EE, comparisons of harmonics at each bus are made. This includes the dynamic responses of voltage, current, active power, reactive power, apparent power, power factor, and the frequency spectrum for both voltage and current, $(THD)_v$, and $(THD)_i$, at source buses and load buses. Voltage and current harmonic analysis, in terms of the total harmonic distortion (THD), is given (Table 4). It is obvious that the voltage harmonics are significantly reduced, and that the THD of current waveform at each bus is decreased (Figs. 5–10).

Table 4 Percentage THD of Voltage and Current at the Buses

	V_S		V_1		V_L		V_g	
	THD_v	THD_i	THD_v	THD_i	THD_v	THD_i	THD_v	THD_i
Without D-FACTS	0.62	7.25	35.5	21	29.3	36.8	26.7	18.5
With D-FACTS	0.1	4.55	4.82	4.56	4.4	4.92	4.61	4.2

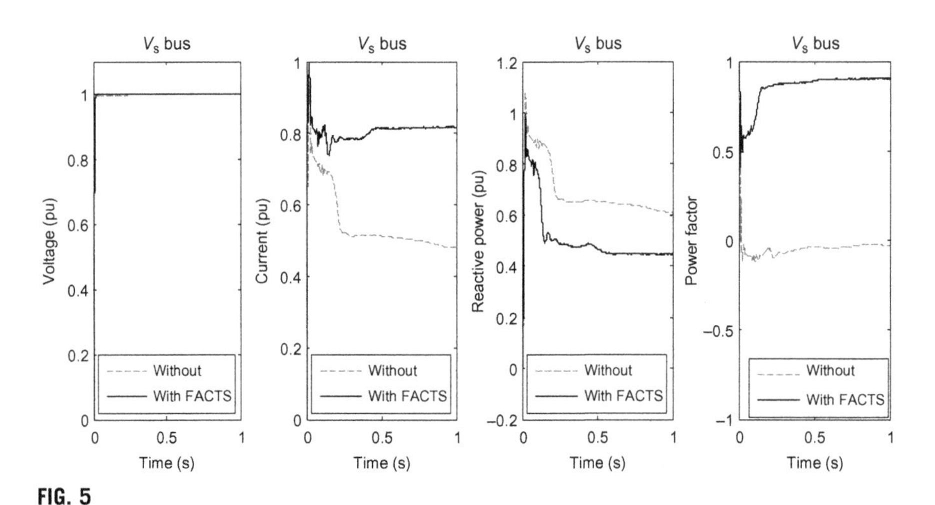

FIG. 5

The voltage, current, reactive power, and power factor at the AC bus V_S, without and with D-FACTS.

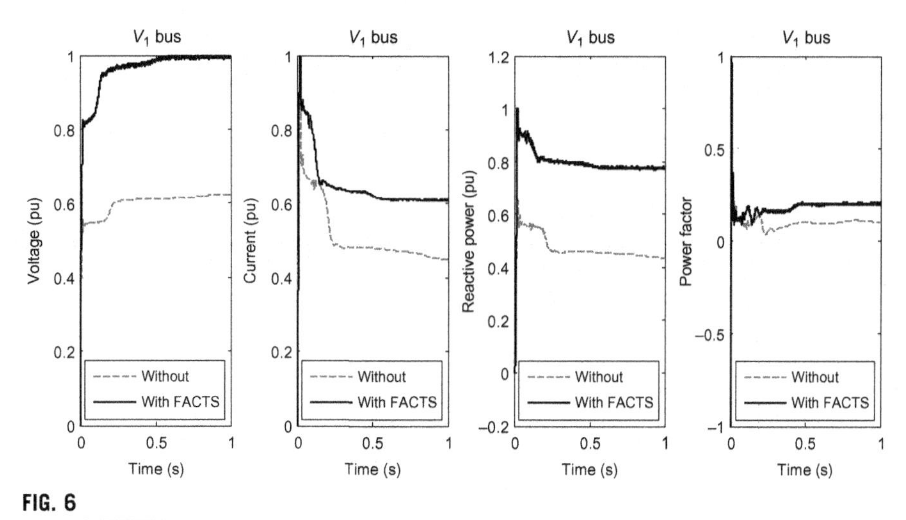

FIG. 6

The voltage, current, reactive power, and power factor at the AC bus V_1, without and with D-FACTS.

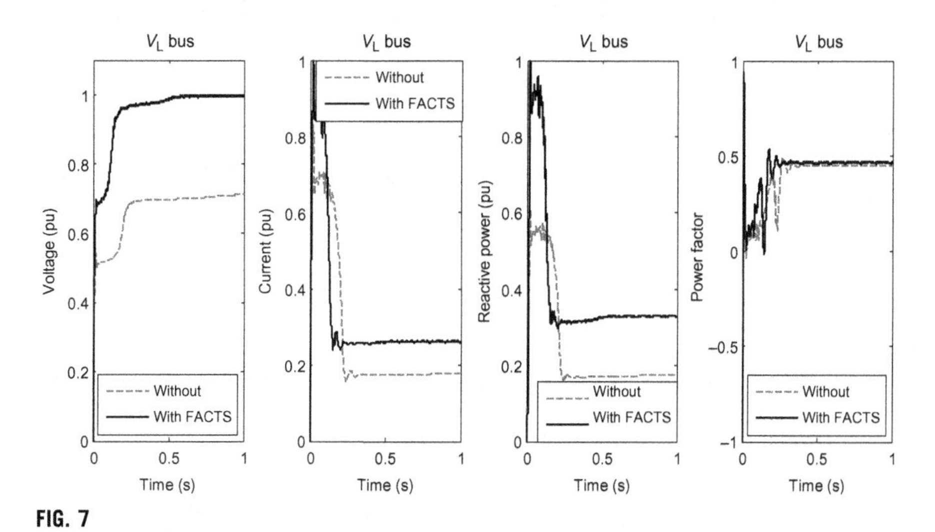

FIG. 7

The voltage, current, reactive power, and power factor at the AC bus V_L.

From all the previous figures, it can be observed that controlled GP-EE mitigates harmonics and can achieve other technical benefits. When comparing the dynamic response results without and with the proposed GP-EE, it is quite apparent that the proposed GP-EE enhances the power quality, improves the power factor, compensates for the reactive power, and stabilizes the buses' voltage.

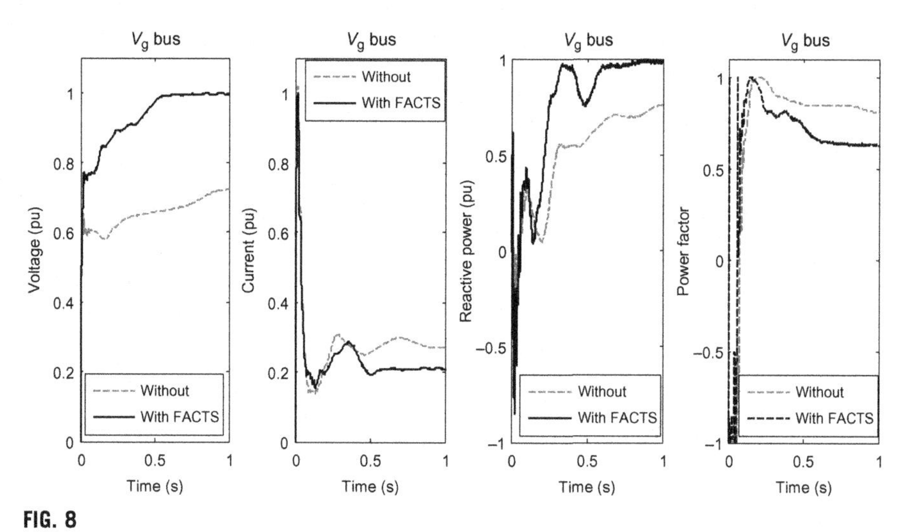

FIG. 8

The voltage, current, reactive power, and power factor at the AC bus V_g.

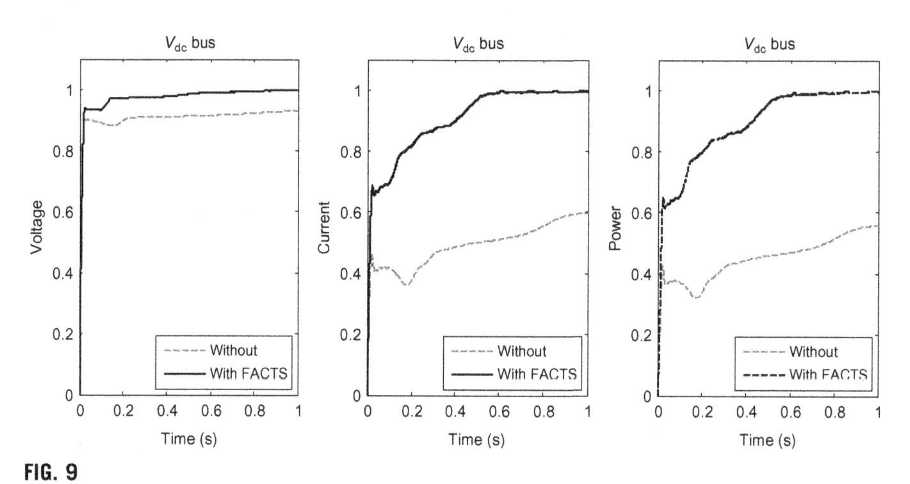

FIG. 9

The voltage, current, and power at the DC bus V_{dc}, without and with D-FACTS.

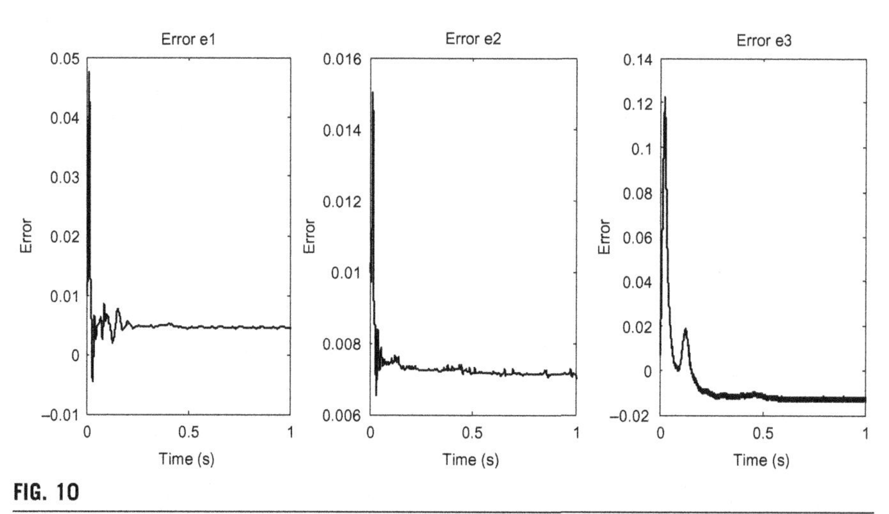

FIG. 10

The global errors e1, e2, and e3 of multiloop controllers of the D-FACTS.

FURTHER READING

[1] Jiayi H, Chuanwen J, Rong X. A review on distributed energy resources and microgrid. Renew Sustain Energy Rev 2008;12:2472–83.
[2] Piagi P, Lasseter RH. Autonomous control of microgrids. In: IEEE power energy society general meeting; 2006. p. 1–6.
[3] Cossent R, Gomez T, Fras P. Towards a future with large penetration of distributed generation. Energy Policy 2009;37(3):1145–55.
[4] Khani D, Sadeghi A, Madadi H. Impacts of distributed generations on power system transient and voltage stability. Electr Power Energy Syst 2012;43:488–500.
[5] Sao CK, Lehn PW. Control and Power Management of Converter Fed Microgrids. IEEE Trans Power Syst 2008;23(3):1088–98.
[6] Baran E, Mahajan R. DC distribution for industrial systems: opportunities and challenges. IEEE Trans Ind Appl 2003;39(6):1596–601.
[7] Liu X, Wang P, Loh PC. A hybrid AC/DC micro-grid and its coordination control. IEEE Trans Smart Grid 2011;2(2):278–86.
[8] Hou H, Zhou J, Zhang Y, He X. A brief analysis on differences of risk assessment between smart grid and traditional power grid. In: 4th international symposium on knowledge acquisition and modeling (KAM); 2011. p. 188–91.
[9] Sharaf A, El-Gammal A. A self adjusting hybrid micro grid PV-FC-diesel scheme for village/island electricity utilization. Int J Power Eng Green Technol 2010;1(2):101–23.
[10] Sharaf A, Wang W, Altas I. Novel STATCOM controller for reactive power compensation in distribution networks with dispersed renewable wind energy. In: IEEE Canadian conference on electrical and computer engineering CCECE'07; 2007. p. 3648–54.
[11] Coello C, Lechuga M. MOPSO: a proposal for multiple objective particle swarm optimizations. In: IEEE proceedings world congress on computational intelligence; 2003. p. 1051–6.
[12] Shi Y, Eberhart R. Empirical study of particle swarm optimization. In: Proceedings of congress on evolutionary computation, vol. 3; 1999.
[13] Berizzi A, Innorta M, Marannino P. Multi-objective optimization techniques applied to modern power systems, In: IEEE power engineering society winter meeting, Jan. 28–Feb. 1, 2001; 2001.
[14] Ngatchou P, Zarei A, El-Sharkawi A. Pareto multi objective optimization: intelligent systems application to power systems, In: Proceedings of the 13th international conference, 6–10 Nov. 2005; 2005. p. 84–91.
[15] Shi Y, Eberhart R. Parameter selection in particle swarm optimization. In: Proceedings seventh annual conference on evolutionary programming; 1998. p. 591–601.
[16] Fadaee M, Radzi M. Multi-objective optimization of a stand-alone hybrid renewable energy system by using evolutionary algorithms: a review. Renew Sustain Energy Rev 2012;16(5):3364–9.
[17] Salas V, Olias E, Alonsob M, Chenlo F. Overview of the legislation of DC injection in the network for low voltage small grid-connected PV systems in Spain and others. Renew Sustain Energy Rev 2008;12(2):575–83.
[18] Cossent R, Gomez T. Towards a future with large penetration of distributed generation: is the current regulation of electricity distribution ready? Regulatory recommendations under a European perspective. Energy Policy 2009;37(3):1145–55.
[19] Falvo MC, Lamedica R, Bartoni R, Maranzano G. Energy management in metro-transit systems: an innovative proposal toward an integrated and sustainable urban mobility system including plug-in vehicles. Electr Power Syst Res 2011;81:.

[20] Pankovits P, Pouget J, Robyns B, Delhaye F, Brisset S. Towards railway-smart grid: energy management optimization for hybrid railway power substations. In: IEEE PES innovative smart grid technologies conference Europe (ISGT-Europe); 2014. p. 12–5.
[21] Cornic D. Efficient recovery of braking energy through a reversible dc substation. IEEE Alstom Transport; 2010.
[22] Okui A, Hase S, Shigeeda H, Konishi T, Yoshi T. Application of energy storage system for railway transportation in Japan, In: International power electronics conference (IPEC); 2010. p. 3117–23.
[23] Leung PCM, Lee EWM. Estimation of electrical power consumption in subway station design by intelligent approach. Appl Energy 2013;101:.
[24] Gulin M, Vašak M, Baotić M. Analysis of microgrid power flow optimization with consideration of residual storages state, In: Proceedings of the 2015 European control conference, Austria; 2015. p. 3131–6.
[25] Yiqiao C, Jiahai W. Differential evolution with neighborhood and direction information for numerical optimization. IEEE Trans Cybern 2013;43(6):2202–15.
[26] Xinggao L, Yunqing H, Jianghua F, Kean L. A novel penalty approach for nonlinear dynamic optimization problems with inequality path constraints. IEEE Trans Autom Control 2014;59(10):2863–7.
[27] Azizipanah-Abarghooee R. A new hybrid bacterial foraging and simplified swarm optimization algorithm for practical optimal dynamic load dispatch. Electr Power Energy Syst 2013;49:414–29.
[28] Rajabi A, Fotuhi M, Othman M. Optimal unified power flow controller application to enhance total transfer capability. IET Gener Transm Distrib 2015;9(4):358–68.
[29] Hong Y, Wei Z, Chengzhi L. Optimal design and techno-economic analysis of a hybrid solar-wind power generation system. J Appl Energy 2009;86(2):163–9.
[30] Anglani N, Muliere G. Analyzing the impact of renewable energy technologies by means of optimal energy planning, In: 9th international conference on Environment and Electrical Eng. (EEEIC), Prague; 2010. p. 1–5.
[31] Patnaik SS, Panda AK. Particle swarm optimization and bacterial foraging optimization techniques for optimal current harmonic mitigation by employing active power filter. J Appl Comput Intell Soft Comput 2012;2012:1–10.
[32] Acharya DP, Panda G, Lakshmi YVS. Effects of finite register length on fast ICA, bacterial foraging optimization based ICA and constrained genetic algorithm based ICA algorithm. J Digital Signal Process 2011;20(3):964–75.
[33] Biswas A, Das S, Abraham A, Dasgupta S. Stability analysis of the reproduction operator in bacterial foraging optimization. J Theor Comput Sci 2010;411:2127–39.

CHAPTER

15

Risk-based lifecycle assessment of hybrid transportation infrastructures as integrated with smart energy grids

N. Ayoub*,†, H.A. Gabbar‡

The University of Missouri, Columbia, MO, United States Helwan University, Cairo, Egypt†*
University of Ontario Institute of Technology (UOIT), Oshawa, ON, Canada‡

15.1 INTRODUCTION

Using electric vehicles (EVs) (eg, battery electric vehicles (BEVs) and plug-in hybrid electric vehicles (PHEVs)), especially in urban driving, are used as an alternative to reduce emissions coming from conventional vehicles [1,2]. The EVs are supposed to be integrated with the traditional transportation system and form what is called a hybrid transportation system (HTS). Yet hosting these cars into the current system and connecting them to a smart grid (SG), both for charge and supply, are associated with a large number of risks. These risks may be associated with establishing transportation facilities and smart grid infrastructure [3]. Hence, in order to realize the maximum potential of such system integration and to reduce the risks associated with it, the environmental and costing lifecycle of the resulted system needs to be thoroughly analyzed, and risks should be identified [4,5]. In this chapter, we present a risk-based lifecycle assessment (RBLCA) for the HTS and its infrastructure integration requirements. In [4], we presented a framework for combining environmental and risk assessment to control and manage risks more effectively and to lead to better operational efficiency. In this chapter, we also focus on both costs and emissions of the lifecycle of HTS and its infrastructure requirements and their associated risks in order to control and manage risks more effectively and achieve better operational efficiency. In the following sections, we present the different terms of RBLCA and a theoretical background of RBLCA frameworks, followed by a literature review and research objectives.

Smart Energy Grid Engineering.

15.1.1 RISKS, LIFECYCLE ASSESSMENT, AND RBLCA DEFINITIONS

Risks can cause adverse effects and consequences in terms of losses that can be economical or environmental [6]. Risks can be caused by natural disasters, human faults, management failure, operational failure, and design errors [3,4]. Other risks are associated with the use of EVs as part of the HTS. These risks are usually due to grid capacity limitations, infrastructure restrictions, and risks created from transportation system electrification [1,3,7]. Other forms of risks related to EVs' operations are the lack of awareness and technical knowledge of how to deal with SGs, eg, when to charge and supply, and how it works [3]. In the literature, lifecycle environmental and costing assessment has been recognized as a useful tool for identifying the impacts of products and making comparisons between products based on their performance [8–10]. Therefore, the ideal case would be to address the identified adverse effects based on the lifecycle stages of the investigated system. In other words, operational, environmental, and economic risks analysis must be incorporated into the lifecycle analysis of the system development project. In this perspective, RBLCA can be defined as "the assessment process that evaluates the environmental and economic impacts and manages their risks along the HTS lifecycle."

15.1.2 LIFECYCLE ASSESSMENT AND RBLCA LITERATURE

The environmental and economic assessment is very popular among researchers for electricity production of energy systems [4,11–14] and to a lesser extent for EVs [9,10,15,16]; however, there is a lack of literature in the studies that consider SG and its integration with EVs from environmental and cost viewpoints. One reason for this trend may be the assumption of ignoring the infrastructure impacts due to its low lifecycle costs and emissions compared to the long running life of the vehicles. Yet building new infrastructure for the new vehicle type, eg, EVs, within a short period may lead to high economic risk and, therefore, should be considered. The emissions mitigation measures of commercial delivery trucks, as vehicle-to-grid (V2G) service providers, were evaluated [17]. They evaluated the battery degradation level and calculated its equivalent emissions, and were proven to have a significant emissions savings effect. The lifecycle emissions savings of V2G technology for use in auxiliary services in five regions were studied by Noori et al. [18]. The authors analyzed the risks of the inherent uncertainties of the system. Egede et al. proposed a framework that contemplated inducing features of the ecological assessment of EVs [16]. The cost and emissions savings, and consequently market penetration, varied from low to significant, depending on the electricity mix in the region. Perujo et al. concluded that with appropriate regulation, EVs can contribute to the overall CO_2 reduction in the transportation sector. This can be applied by intelligent integration of EVs into the SG as decentralized and flexible energy storage systems. Otherwise, EVs could heavily affect the daily requested electric power [19]. Despite such similar studies, to the best of the authors' knowledge, there is no research work devoted to investigating the environmental and cost lifecycle analysis of HTS and its infrastructure.

While an RBLCA framework has not been given due attention by some researchers, it is envisaged as an appropriate approach for providing better and more practical analysis to energy systems [20]. One reason for this anomaly could be the pragmatic nature of the generic approach implied in RBLCA. For example, in addition to the site-specific parameters, the RBLCA application involves the identification of potential exposure routes and probability of vulnerability to provide an effective decision support [21]. Over the years, RBLCA methods have been recently applied on various disciplines such as costing analysis [14,22], maintenance strategies [23,24], energy systems [6,25], decision making [21,26,27], and chemical process design [28,29]. Albeit, risk-based studies that consider lifecycle assessment (LCA) and environmental effects are still lacking in the public literature. For example, Khandoker et al. [20] developed a robust risk-based environmental assessment tool for industrial processes. This tool was used to evaluate different process options at the early design or retrofit stage. The main limitation of applying this methodology is that it does not explicitly consider the role of stakeholders in the decision-making process. Another RBLCA methodology relies on informing the affected public if certain hazards are expected to spread widely beyond certain levels of exposure risk [30]. The authors suggested that a third party should be involved in risk assessment either for the analysis or the independent review of the analysis.

Using IDEF0 as a base for detailed processes design, risk assessment, and LCA was pioneered by several researchers [27–29,31,32] to enable detailed and transparent definitions of the whole design process. The local risks of processes and environmental impacts throughout the products lifecycle have been considered by Yasonari and his colleagues via systematic procedure models for risk-based decision making of small- and medium-sized enterprises (SMEs) [27]. Yasonari et al. have integrated environmental LCA and plant-specific risk assessment using a hierarchical IDEF0 activity modeling approach. The process generation, evaluation, and decision-making activity models were defined and visualized with their associated information flows and data that should be collected by onsite engineers. A case study was conducted on designing a metal cleaning process, reducing chemical risks due to the use of a cleansing agent. This approach is analogous to our proposed framework in relying on providing detailed information for the users of the framework based on predefined objectives and available data resources. Another IDEF0 activity model approach on an LCA-based process design and related activities such as multiobjective decision making has been presented [31]. The whole design process is described hierarchically, starting from the design chemical process as a top activity, and LCA-related activities as subactivities. The economic performance of the designed processes is considered an activity concurrent to LCA, enabling multiobjective evaluation and optimization of the process. A case study is presented on the design of chemical recycling processes of beverage polyethylene terephthalate (PET) bottles with a focus on the multiobjective process assessment. The integration of environmental, health, and safety (EHS) risks using IDEF0 activity modeling in industrial process design is also presented in the literature [28,29]. The different phases of early process design, ie, process chemistry and conceptual design, with process evaluation indicators are identified in the activity model. The model uses EHS aspects as assessment criteria, together with conventional economic and technical indicators. Similar to the framework presented here, the viewpoint of the activity model is that of the design framework user, ie, a design-project manager who leads the team.

To the best knowledge of the authors of the present study, the cited works are the only works that consider an RBLCA approach.

15.1.3 PROBLEM STATEMENT AND RESEARCH OBJECTIVES

The EVs are thought to be a clean method of transportation that can help mitigate greenhouse gases (GHGs) via utilizing the electricity lost in the off-peak hours. They can also work as an energy reserve that can work as a load-shifting subsystem in an SG environment. Hence, EVs are demand-and-supply agents; however, this two-way interactive information and communication technology will require a tremendous advanced infrastructure requirement, such as quick charging stations, advanced two-way metering infrastructure, demand response, etc. [33]. Therefore, there is a need to realize how such system integration will affect the environment as a result of changes made to the current infrastructure. In addition, the economic and environmental risks associated with HTS need to be envisioned to be dealt with. These risks may include a power failure due to the risk of a blackout that may occur, if a large number of EVs are charged during peak hours, or universal deficiencies for batteries' raw materials due to limited supplies or increasing demand constitute risks on system economy. These are just examples of what may happen if things go wrong. Hence, a comprehensive research on the required infrastructure to realize the HTS was carried out. To study the economic and environmental impacts of HTS and deal with its risks discussed earlier, an RBLCA framework is presented to evaluate the HTS infrastructure. The framework is generic and hence can be applied to large projects system or the lifecycle of one of its components from the viewpoint of the project manager and, thus, the information flows are based on the manager's objectives and available data resources. This framework was previously applied to energy supply systems presented in [4].

In Section 15.2 of this chapter, the HTS infrastructure requirements are presented. The RBLCA framework and procedures followed to apply its modules are explained in detail in Section 15.3. The application of the RBLCA on HTS is presented in Section 15.4, followed by conclusions in Section 15.5.

15.2 HTS INFRASTRUCTURE

The penetration of EVs requires an evolutionary change and restructuring in the existing infrastructure and forecasting models to support the newly required functionalities. To fully integrate and deploy EVs, a supply chain infrastructure, modern telecommunications, and sensing technology are required [34]. There is rarity in research that considers infrastructure forecasting for HTS. The energy impact of public charging infrastructure on EVs was examined by Dong and Lin. They concluded that providing a widespread public charging service can reduce PHEVs' gasoline consumption by 30% or more than having home charging infrastructure only. Hence having this infrastructure at home helps increase the energy cost savings for EVs' drivers [35]. For Portugal specifically, the impact of energy supply infrastructures and energy supply chains on CO_2 emissions and energy use, regarding conventional and electrical vehicles was estimated by Lucas and his colleagues [36]. They concluded that the EVs' energy supply infrastructures are more carbon- and energy-intensive per megajoule (MJ) of supplied fuel than conventional ones. Those contradict with the findings of San Román et al. of using EVs as energy storage agents through recharging during off-peak hours. This can make EVs as load-shaving agents in the peak hours using the V2G concept [37]. This contradiction shows the need for a reliable assessment methodology before application. There are several research works that consider the economic environmental impacts of transportation systems and EVs' infrastructures, as in Refs. [10,38–43]. Conversely, it is less extensive to find research work that considers the economic analysis, as well as engineering-side analysis. For example, Schroeder and Traber have estimated the return on investment (ROI) of the fast-charging station and found that the current market is vague for activating a large scale of fast-charging infrastructure due to the high-risk factor of market penetration of EVs [44]. Before presenting the RBLCA framework for HTS, we are presenting the infrastructure requirements in the following sections.

15.2.1 INFRASTRUCTURE ENGINEERING MODEL

Enforcing EVs in the transportation system as well as using them as load-shifting tools require substantial infrastructures related to SGs and transportation systems that should be identified, eg, smart meters and fueling stations. To develop the engineering model for the required infrastructure, we need to estimate each infrastructure layer in the four infrastructure layers shown in Fig. 1. To do this, the expected number of vehicles with their related fueling infrastructure have to be identified. In Fig. 1, we have detailed the required infrastructure for EVs in four layers: natural gas (NG) supply chain infrastructure, fueling infrastructure, charging and discharging infrastructure, and SG infrastructure. It is well known that PHEVs share the characteristics of both conventional EVs having an electric motor and the internal combustion engine vehicles that can run using NG, gasoline, and electricity. Therefore, we have considered natural gas vehicles (NGVs) as well as PHVEs to connect to a gas grid as a consumer and to a power grid as a consumer and supplier. Hence, the HTS will require the following infrastructure components:

1. Supply chain infrastructure: pipelines, trucks, tanks, etc.
2. Fueling infrastructure: fueling stations for all fuel types.
3. Charging and discharging infrastructure: electricity plug-in devices at homes, parks, and fueling stations.
4. SG infrastructure: smart meters.

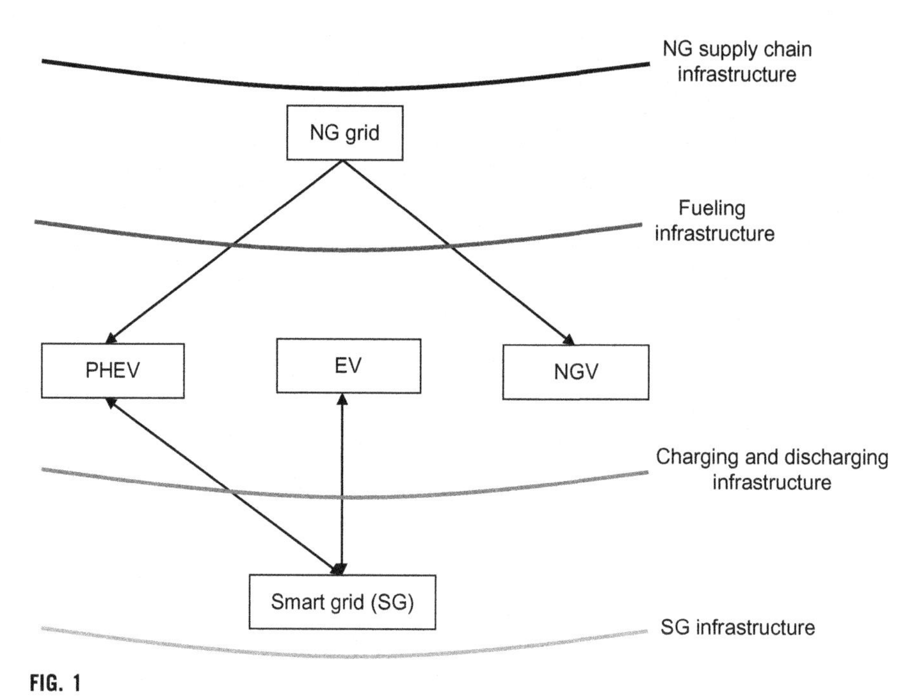

FIG. 1

Schematic representation for the infrastructure engineering model.

15.2.2 FUELING INFRASTRUCTURE

The size of the individual fueling stations is a function of the number of vehicles served, fuel economy, travel distance, and the fueling pattern [45]. According to Ref. [46], the sustainable growth of alternative vehicles during market transition until maturity requires 10–20% of fueling stations to be alternative fueling stations. Due to the problems facing the fueling station business, it is difficult to attain this high percentage that may impose a potential risk of applying the HTS. One way to overcome this issue is applying, by the government, new rules for promoting fueling infrastructure or setting a subsidy scheme. For example, Ontario government helps electric car owners to install charging stations at their homes and businesses to encourage sustainable transportation and fight climate change. Starting Jan. 2013, homeowners and businesses who have received a provincial EV rebate and have installed or are planning to install an EV charging station will be eligible for rebates of up to 1000 USD or 50% of the total purchase and installation cost, whichever is lower [47]. Since infrastructure development is expensive, there is a need to direct investments toward the establishment of fueling facilities in areas resulting in maximum impact [48]. A ratio of 1000 vehicles per station for all alternative fuels and all fueling stations from each type is reasonable capacity [1].

Applying HTSs will lead to an increase in the number of EVs running in the streets. This requires large numbers of charging and discharging stations to be distributed around the city or the region that will apply the system. Those stations will be essential to be installed in both private residences and in the public areas. In advanced countries, charging stations' infrastructure for EVs is not high, as they can be built at every place having access to electricity, including homes, parking lots, gas stations, super markets, etc. [1].

15.3 METHODS

15.3.1 THE PROPOSED RBLCA FRAMEWORK

In this section, the process of applying the three modules of RBLCA is explained in detail.

Module 1: Establish the lifecycle stages model

Understanding and defining the various stages of the studied systems' lifecycle is a critical task that should be done in a comprehensive manner. The following are the phases of applying the first module of the framework. The main application constraints, as shown in Fig. 2, are green credits that may be achieved by applying such a system, the availability of materials required for manufacturing the batteries of the number of the EVs to be integrated into the conventional transportation system, and the stakeholders' beliefs and power, which are vital in policy making decisions. There are also some managerial constraints, such as the expertise required to form the LCA team and the availability of resources that may limit the quality of the data used.

Phase 1: The upper management responsible for the hybrid transportation project realization should decide who is to be involved in performing the RBLCA, ie, the team. The key stakeholders, affecting and affected by the project realization, or those will be at risk if the new hybrid system is operated, have to be identified and represented within the team. Such stakeholders include public representatives, EVs' producers, fleet managers, etc. In addition, the local authorities who are responsible for policy making and urban planning need to be represented in the RBLCA team.

Phase 2: Define the current transportation system specification problems and how EVs will be integrated. Specifically, describe why it is important to use EVs in the city, how they will be used, in which sectors they are supposed to penetrate more, the required infrastructure, and an account of their requirements.

Phase 3: Identify the EVs' alternatives, their power conservation potential, and charging and discharging prerequisites that should be studied by the RBLCA team. The team has to understand all future plans and regulations to set the LCA study targets and identify the environmental and economic impact categories that will be considered in the study. Those impacts are identified based on stakeholders' concerns about possible risks.

Phase 4: Identify the system boundaries of the project and identify the processes that will be included in the study. The quality of the data has to be considered in identifying the system boundaries. The processes of high potential to create risks

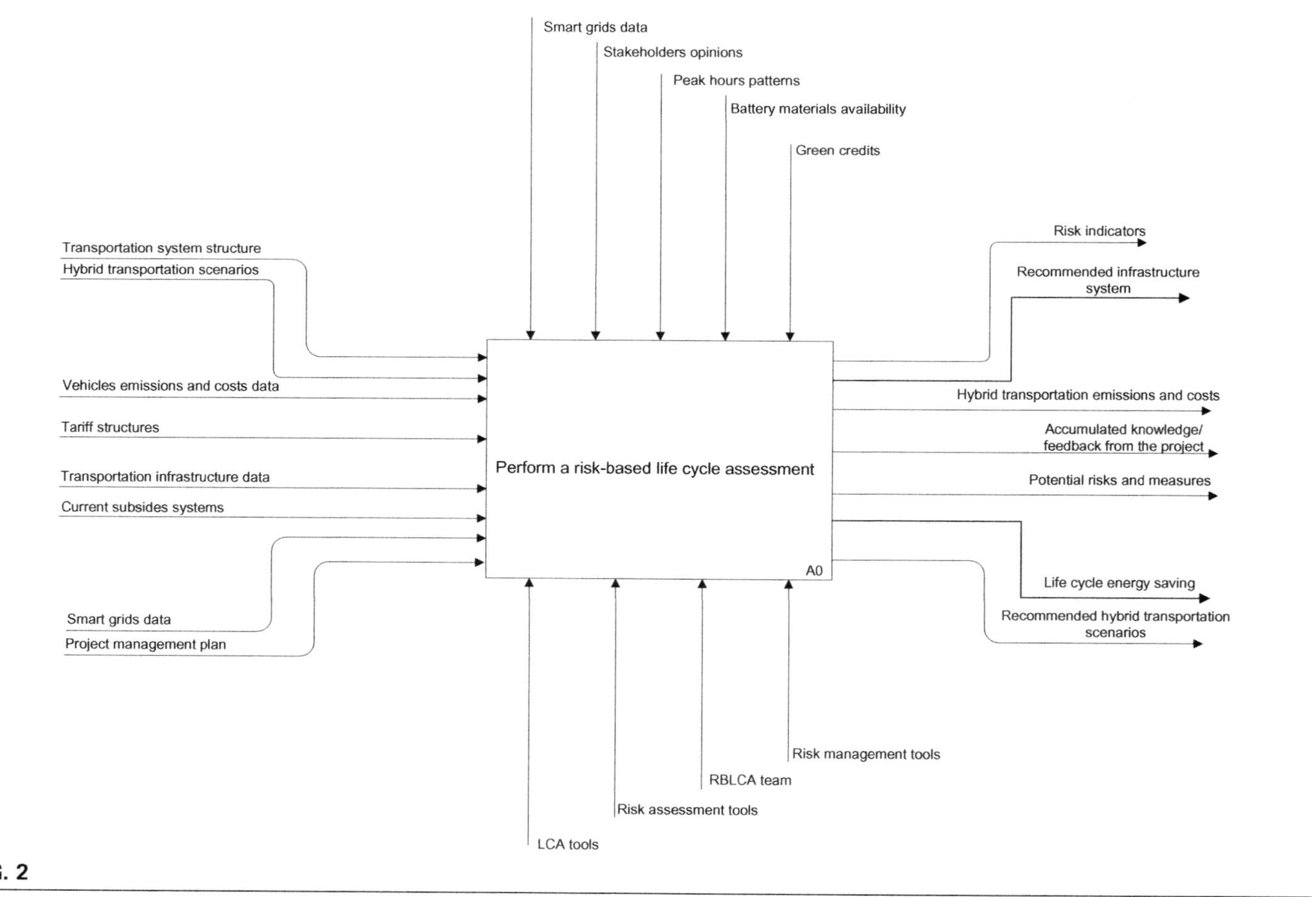

FIG. 2

Risk-based lifecycle assessment in the framework.

along the HTS lifecycle should be included in the system boundaries. The risks identification should consider the viewpoint of stakeholders and local government, in addition to project management.

Phase 5: Perform the data collection process and LCA studies. Prepare the cost and environmental assessment data related to the risk issues raised for the risk assessment. The LCA results can then be analyzed in order to identify potential risks that were not raised by the stakeholders.

The limitations of available information about the newly introduced HTS application and operation processes are typical characteristics of the system design phase. In this phase, the system requirements of vehicles, numbers of batteries, and their materials and infrastructure requirements are identified. The materials and power requirements of the hybrid system are primarily derived from the plans and specifications set in the design phase [49]. Hence, as explained in phases one through four of the framework, the HTS lifecycle study starts from the design process.

Module 2: Identifying risk indicators and performing risk assessment

Risks of constructing the HTS infrastructure, replacing the traditional vehicles, running the EVs and the interaction protocol with SGs should be identified before making any irreversible process, as they may strongly affect the decisions taken by accepting or rejecting the proposed scenario. The risk assessment of HTS is the process made to identify the potential application and operation risks, the chances of risks' occurrences, and the extent of their effects on the environment and the market. Therefore, a set of risk indicators for the different concerns has to be identified by the LCA team, which can be compared to the risk management results. The risk assessment study and indicators depend on the types of risks identified by the stakeholders. They can range from very simple, personal judgments by individuals to results from comprehensive assessments by expert teams using a broad set of tools and information, including historical loss data.

The key to risk assessment is choosing the right approach to provide the needed information without overworking the problem. Phases in this module are constrained by the current subsidies system, SG data, governmental regulation, and risk data output from module 1. The phases of module 2 are applied as follows.

Phase 1: Classify the types of risks based on each stage and operation. List the risks for each stage, identified by the team, and put them in question form. For example, in the design stage, a question to answer may be: what types of EVs-SG interaction protocol are there that may lead to risk reduction? Or, what will be the results in terms of social and economic risks if we apply a specific hybrid transportation scenario?

Phase 2: Determine the information necessary to answer the questions, identify the required resources, the level of details required, who will be responsible for gathering and analyzing the information, and how much it will cost.

Phase 3: Based on the information level and their associated costs, the team should identify the risk analysis tools suitable for each stage. For a low level of details, simple risk assessment tools such as a risk assessment matrix can be used, and for higher levels of details, a more sophisticated tool should be used.

Phase 4: Generate risk-based evidence and set the risk indicators for each stage of the lifecycle.

Module 3: Manage risks on the HTS lifecycle stages

Depending on the type of risk, the aim of risk management is either eliminating or decreasing risks by taking specific measures. Here, care must be taken to optimize the management process cost so it does not exceed the benefits gained. The risk management measures taken should be acceptable by all stakeholders and should not lead to other risks; however, in case of social and environmental risks, it is acceptable, upon the agreement of stakeholders, to have slightly higher costs for managing risks. Therefore, the team has to assess the likely risk management options, identify the methods and means for their management, and set the limits for the risk indicators. These options can accept/reject the risk or find different ways to reduce the risks through the stages of the HTS lifecycle.

Fig. 3 shows that the input information to this module is the identified risks and risk indicators that result from applying the second module. The result of this module represents the supply scenarios, potential risks, measures to eliminate or reduce them, and the recommended environmental-/economic-friendly scenario. It is important to note that the first and second modules should be done in parallel so there is a provision for modifying goals with the steps of the analysis process.

In the next section, a detailed explanation of applying the RBLCA framework is given in the following case study section. In this section, the RBLCA framework is applied to study the lifecycle of a HTS infrastructure and assess the associated risk.

15.3.2 ACTIVITY MODELING IN THE RBLCA FRAMEWORK

In this section, the RBLCA framework can be seen as a general business model, which describes how to execute the RBLCA analysis [29]. The IDEF0 activity models are used to systematically describe practical implementation activities considering inputs, design constraints, mechanisms, and desired outputs at different stages of the framework application [27,28]. Figs. 2 and 3 show the activity models representation in IDEF0. Fig. 2 displays the top level of details required to apply the framework. The input information, the process control measures, the application mechanisms, and tools used to apply the framework and the expected results are presented. Fig. 3 shows the IDEF0 activity model for the three application modules of the RBLCA as A1, A2, and A3. It is important to note that this framework is applicable for large projects thought to produce environmental or economic risks/benefits to their surroundings.

The activity model of the RBLCA has three subactivities as shown in Fig 3. Each subactivity is further divided into more detailed activities as explained here.

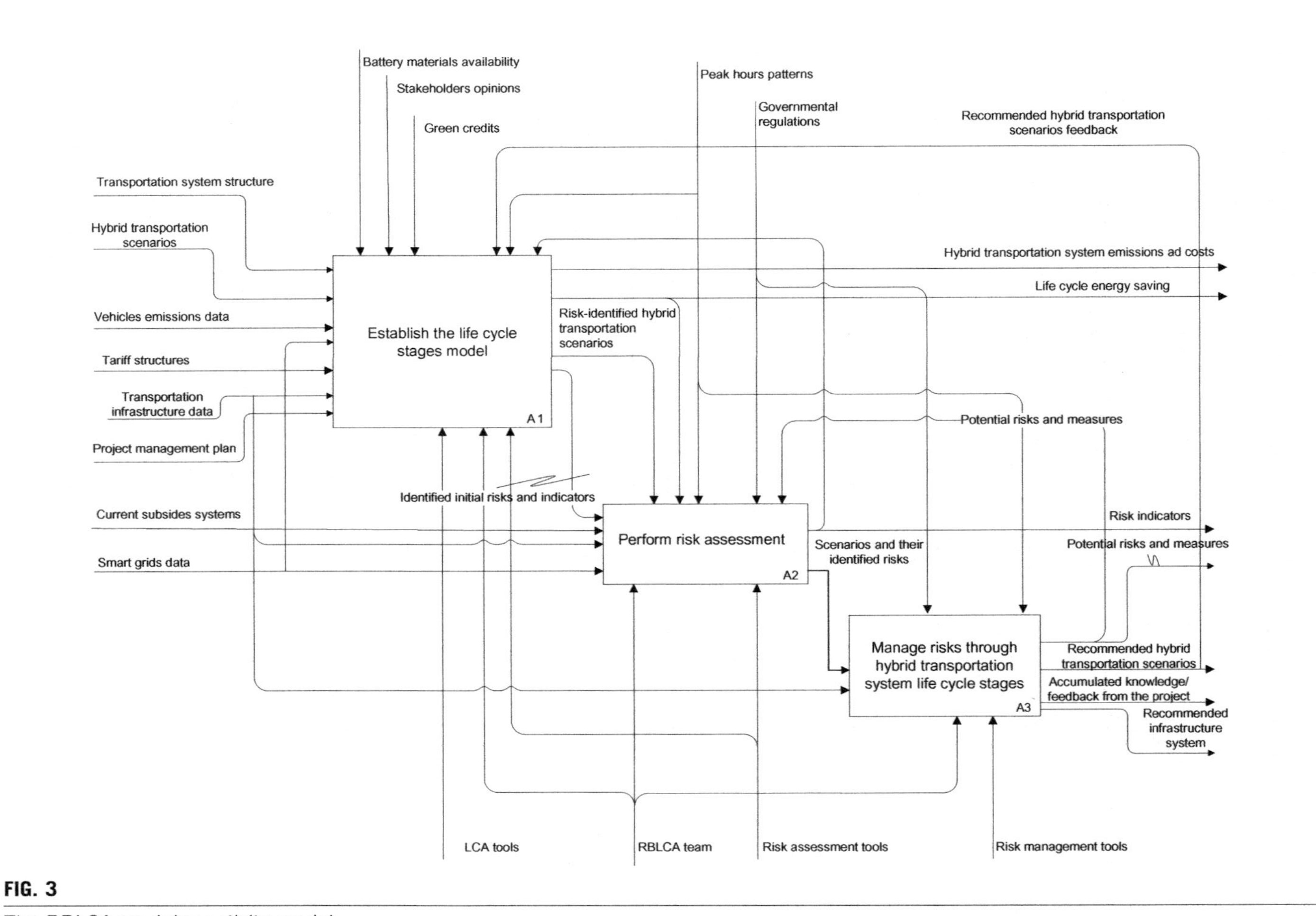

FIG. 3

The RBLCA modules activity model.

The top Activity A0: Perform RBLCA

The objective of this activity, Fig. 2, is to obtain system design alternatives with minimum risks. It is assumed that the HTS design alternatives, EVs share in transportation system, infrastructure, and services are identified before performing this activity. Internal controls, such as time and level of details, and external controls, such as regulations, electricity tariff structure, and materials availability have to be identified. In addition, the tools and methods used in performing the activity, through its subactivities, are identified [31]. The RBLCA is an iterative process in nature as the risk management actions should be reassessed. Hence, feedback information including design modifications or new design alternatives is used to update the RBLCA processes [27].

Activity A1: Establish the lifecycle stages

The aim of this activity (Figs. 3 and 4) is to study the lifecycle steps of the HTS scenarios and identify initial risks that may arise in those steps. The available EVs' technologies and their supporting infrastructure models and technologies are used to define the suitable scenarios and strategies for the system under study. Using those EVs, infrastructure options, project management data, and other inputs to an LCA study is performed, and best scenarios are identified with their LCA emissions and energy-saving potential. The related project management data includes several qualitative and quantitative records, such as process flow charts, process control plans, quality control plans, human resource schedules, activity duration estimates, and enterprise factors [50]. In the subactivity A13, the risk management plan is developed and initial risk identification methods are used to identify initial risks and document their related data. Identifying the initial risks through the lifecycle of the system require using several risk assessment tools and techniques. The process flow approach followed in performing LCA studies makes it easier for the RBLCA team to demonstrate how various elements of the system under study interrelate and identify the causality mechanism.

Activity A2: Perform risk assessment

In this activity (Figs. 3 and 5), the identified risks are assessed qualitatively and quantitatively, and characterized, categorized, and documented. The qualitative and quantitative risk analyses are performed to identify the priorities of tackling risks [51]. The subactivity A21 is set to establish priorities, plan risk responses, and lay the foundation for performing quantitative risk analysis, if required, in a timely and cost-effective way. If the subactivity A22 is not required, the identified risks are entered directly into the "categorize risks" activity. The risks are then categorized to identify how risks will be preserved, which resources will be required, the intended outcomes of the risk management process identified, and who will be in charge of managing risks. Several risk assessment tools are used to complete this activity, including checklist analysis based on accumulated knowledge of similar projects and other sources of information gathering [52], and cause-and-effect diagrams, or Ishikawa diagrams, which are used to find the possible causes of risks [50]. Other techniques can also be used, such as SWOT analysis (strengths, weaknesses, opportunities, and threats), probabilistic analysis, and

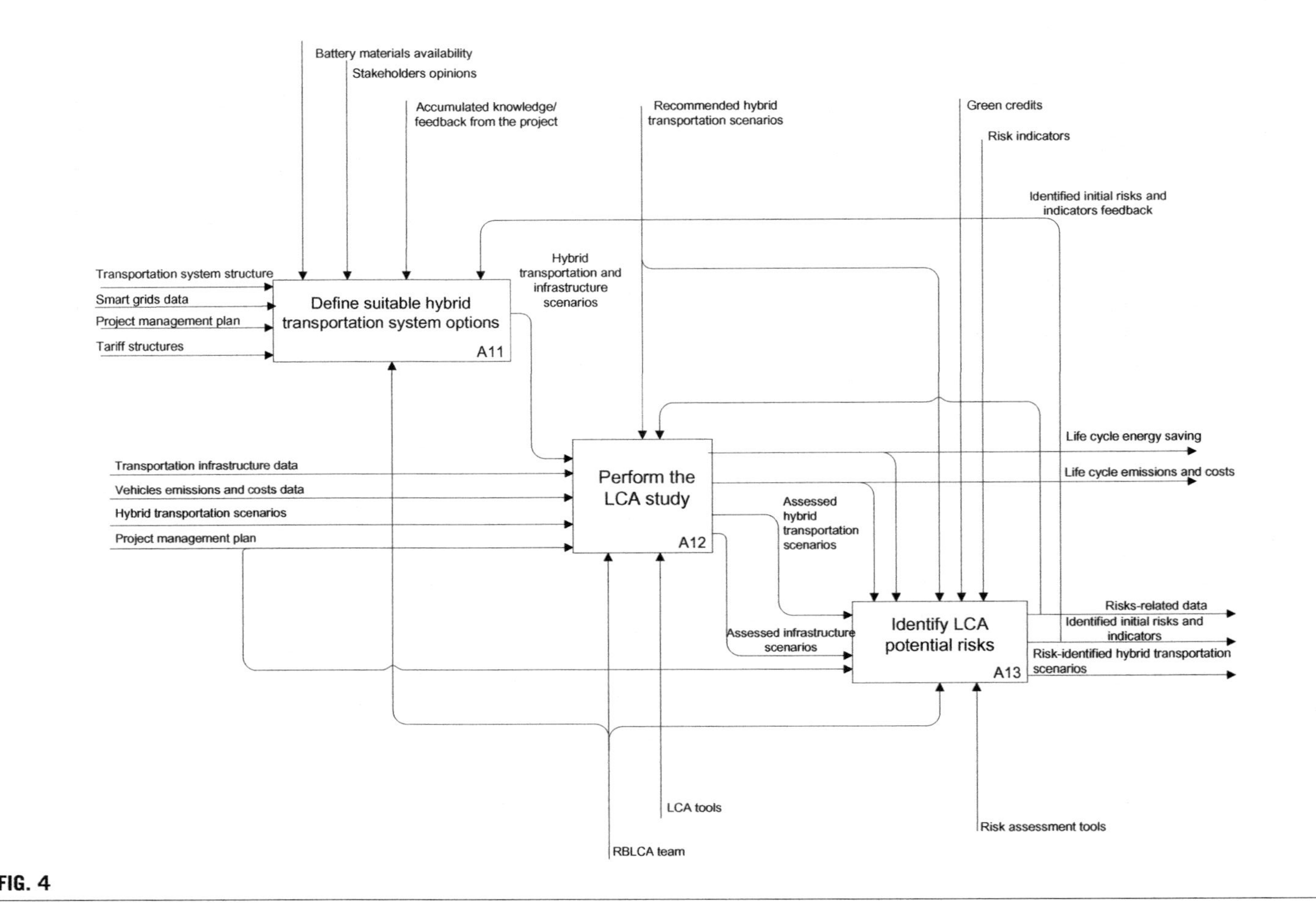

FIG. 4

The detailed activities to establish the lifecycle stages module.

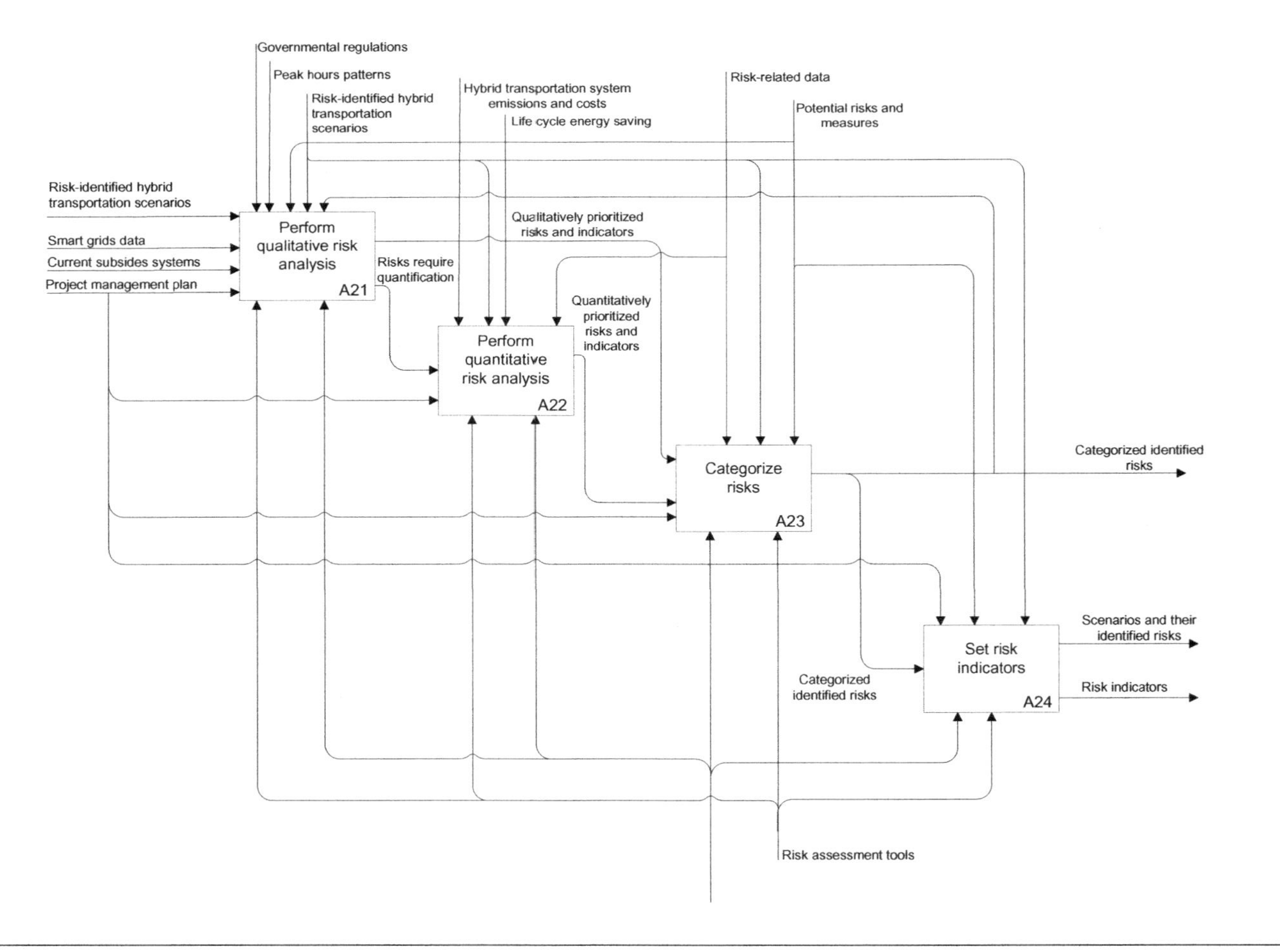

FIG. 5

The detailed activities of the perform risk assessment module.

expert judgment [53]. In the presented case study, the lifecycle emissions and lifecycle energy are used as controls for subactivity A22 and expert opinions to answer the questions raised during risk identification as described in Section 15.5.3 as a risk assessment tool. The two outputs of this activity are fed back to the subactivity A13 (identify LCA potential risks) to ensure the iterative nature of the LCA analysis. The identified risks output from subactivity A24 are also fed back to its subactivity A21, as the qualitative and quantitative risk analyses should be performed regularly throughout the lifecycle [50]. The subactivities A21 and A22 should be repeated, if needed, until the overall potential of risks from system construction and operation has been satisfactorily decreased.

Activity A3: Manage risks through HTS lifecycle stages

This activity (Figs. 3 and 6) aims to manage the categorized risks for each process in the system lifecycle identified in the subactivity A2. The subactivity A31 is devoted toward risk response planning and development of suitable, effective, and economic risk management actions using the risk categories and risk indicators and after getting the agreement of the RBLCA team members. The subactivity A32 is the heart of this risk management activity and uses the risk response plan resulting from subactivity A31 to footpath and control risks. The effectiveness assessment of risk management measures along the system lifecycle is also performed in this subactivity, and possible risk management updates are documented. The initial risk management plan is updated through performing the subactivity A33 (update risk management plan) based on the results of risk control activity (A32). The updated risk management plan, potential risks and measures, risk document updates, economic/environmentally friendly design(s), recommended scenarios, and accumulated knowledge/feedback from the project application are the outputs of this activity.

For interested readers, more explanations about how to apply the subactivities A11–A33 can be found in our previous work [4].

15.4 CASE STUDY OF HTS' RBLCA IN ONTARIO

This case study considers the LCA of the infrastructure needed to realize HTSs. The current transportation system structure, SG data, hybrid transportation scenarios, transportation infrastructure data, and vehicles emissions data are used as inputs to Activity A0. In the first step, the system's functional unit should be defined and all emissions and energy used along its lifetime are normalized to it.

15.4.1 ESTABLISHMENT OF LIFECYCLE STAGES

The project management plan and demand information are used as input to this activity (A1) to define the suitable infrastructure and operational structure (Activity A11) and the economical/environmental performance of the selected HTS (Activity A12), and to identify initial risks and their related data (Activity A13) that will work as an

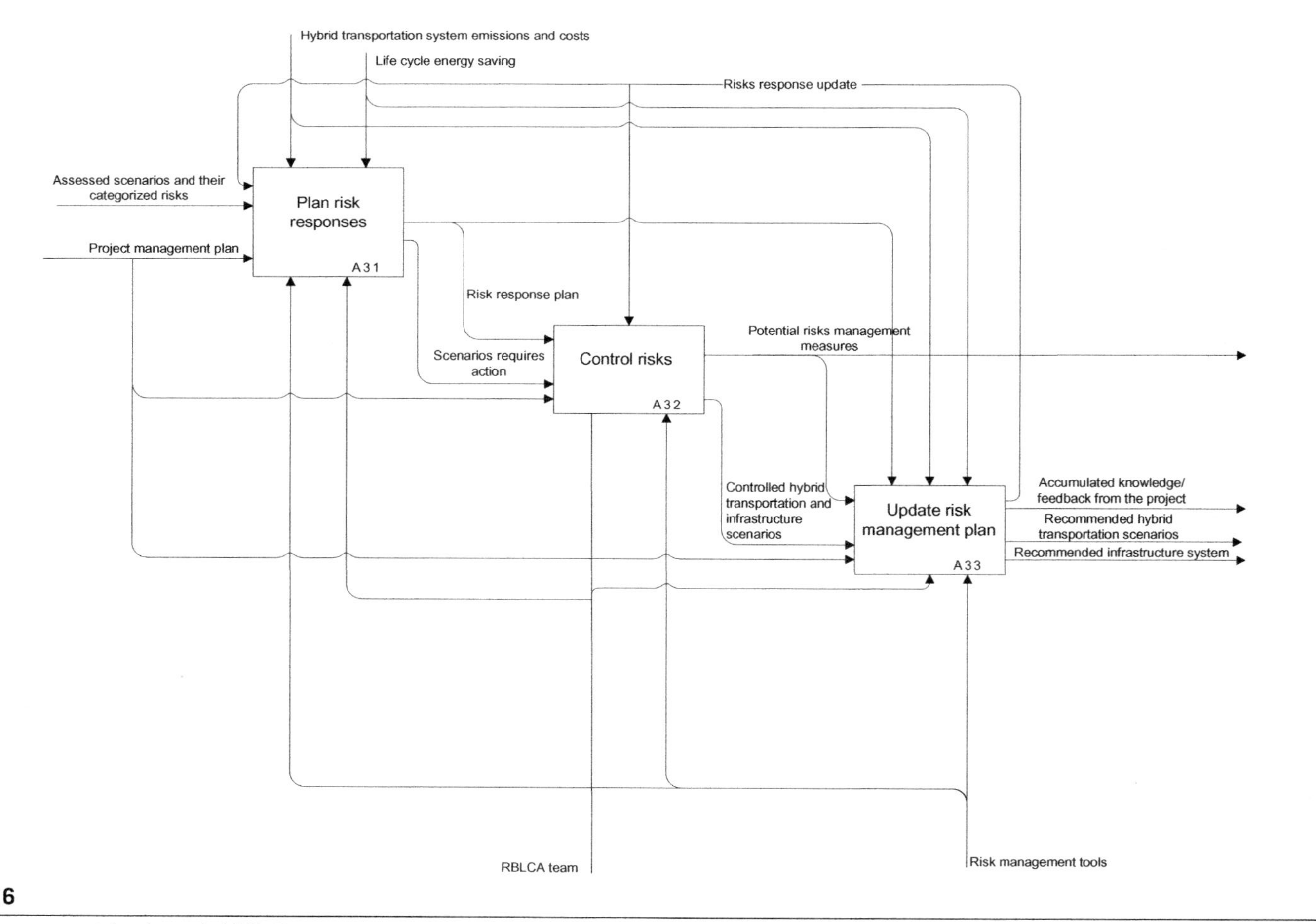

FIG. 6

The manage risks through HTS lifecycle stages module activities.

input to the Activity A2. Hence, three scenarios, as explained below, are used, based on the EVs' penetration potential [1]. The Activity A11 can also be modified based on the identified potential risk feedback from Activity A13. The HTS presented here is a mix of conventional and EVs, if they are available. The EVs are a two-way (charging and discharging) system that should charge from the SG if electricity production is excessive and then supply electricity to the SG during peak hours.

The main aim of the RBLCA study in this case study is to identify and assess the economic/environmental impacts and risks of applying HTS operations through its lifecycle. Fig. 7 shows simplified lifecycle stages for the HTS. The infrastructure and manufacturing stages contain the establishment of all infrastructure used to realize the HTS and the vehicles and EVs' batteries manufacturing. The EVs' batteries manufacturing is considered separately due their economic and environmental risks. The preliminary environmental measures considered in this study are CO_2, as a measure for climate change potential, and particulate matters contents as a measure for human health. The HTS operation stage is mainly governed by the number of vehicles, average distances, and their connection to the SG. This stage imposes the risk of SG blackout if a

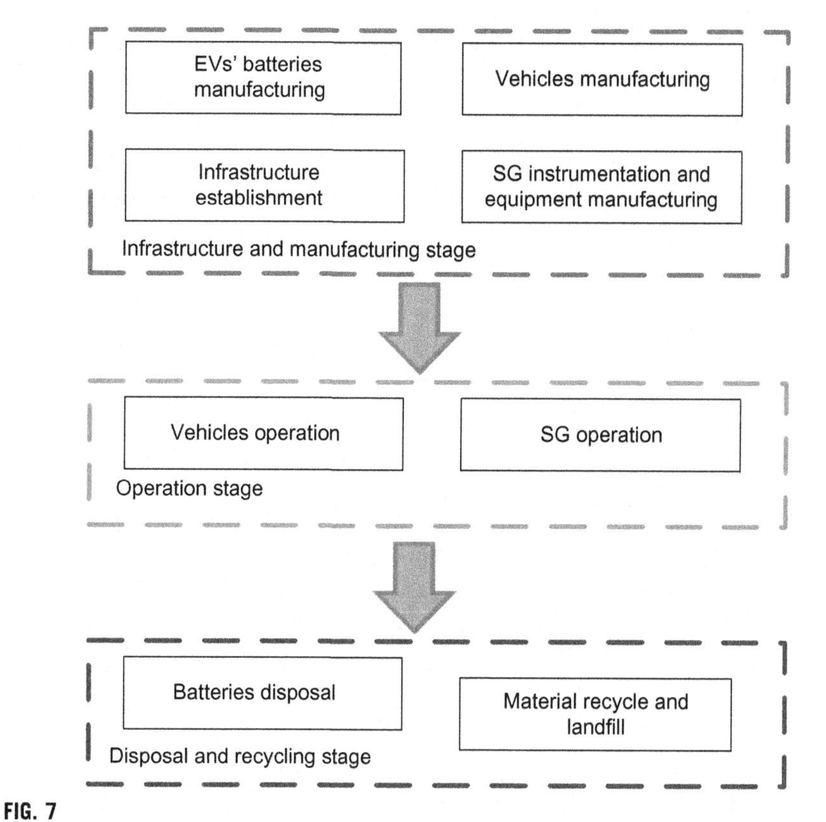

FIG. 7

Simplified lifecycle stages for the HTS.

large number of EVs are charged during peak hours. This stage contains batteries disposal, which contains hazardous materials and end-of-life vehicles' recycling and landfilling. In this chapter, three HTS scenarios for the city of Ontario are considered. In the following sections, the current transportation system in Ontario is presented, followed by the future of integrating the EVs into the current system. Then, the hybrid transportation scenarios are presented. In Section 15.5, some of the study results are presented, focusing on risk assessment and the management module application.

15.4.2 ONTARIO TRANSPORTATION SYSTEM

The number of vehicles in Ontario in 2009 was 7.4 million with an average of 16,000 km of annual distance traveled [54]. Table 1 shows that more than 99% of Canadian vehicles use fossil fuels with gasoline dominating the country's use. Using those country average ratios, the numbers of vehicles in Ontario were calculated by fuel type, as seen in Table 2.

15.4.3 HYBRID TRANSPORTATION FUTURE IN ONTARIO

According to the Ontario Ministry of Transportation, 5% of vehicles will be fueled by electricity in 2020 [47]. Also, Hajimiragha and his colleagues have estimated the market share of PHEVs to be 6% by 2025 [55]. The population number of Ontario in

Table 1 The Estimated Percentages of Vehicle Types Used in Canada by Fuel Type

	All Vehicle Types (%)	Vehicles Up To 4.5 t (%)	Trucks 4.5–14.9 t	Trucks 15 t and Over
Gasoline vehicles	93.94	96.91	26.39	2.50
Diesel vehicles	5.80	2.85	72.23	97.50
Other vehicles	0.26	0.24	1.38	–

Calculated from Office Energy Efficiency. Canadian vehicle survey: 2009 summary report. Canada: Natural Resources Canada's Office of Energy Efficiency; 2009.

Table 2 The Estimated Number of Vehicle Types in Ontario by Fuel Type Based on Country Ratios

	All Vehicle Types	Vehicles Up To 4.5 t	Trucks 4.5–14.9 t	Trucks 15 t and Over
Number of vehicles in Ontario	7,362,689	7,166,834	90,352	105,503
Gasoline vehicles	6,916,857	6,945,444	23,840	2632
Diesel vehicles	426,908	204,459	65,264	102,870
Other vehicles	18,922	16,930	1247	0

2030 will be 16,676,066 following the reference scenario of the Ontario Ministry of Finance [56]. The number of vehicles running on roads will be around 9.4 million based on current calculations of 0.56 vehicles per capita [54]. This large number of vehicles requires significant efforts to be exerted to enforce more environmentally friendly vehicles to the current system as the fossil and green fuels demands of Ontario, in certain year, can be estimated as a function of the number of vehicles running on the road; however, the potential HTS, in any time ahead in the future, is determined based on forecasting the penetration level of new vehicles and the average car lifetime [45]. Following the assumption of 40% penetration level for new vehicles types, Ontario will have a fleet of 3,758,000 vehicles that uses fuels other than gasoline and diesel, eg, EVs, NGV, etc. This will require enormous changes in the fueling infrastructure.

15.4.4 HYBRID TRANSPORTATION SCENARIOS (ACTIVITY A1)

The government of Ontario has several initiatives to enforce green transportation, including subsidies for EV owners. In Jan. 2013, the government announced that the owners of EVs are eligible for up to 1000 USD support to install EV charging units at homes or businesses. Changing the transportation vehicle mix can reduce emissions dramatically [47]. In this work, three scenarios for transportation mix in the year 2030 were considered, as explained in the case study. The reference scenario assumes 12% of vehicles will be EVs by the year 2030 as estimated in Ontario's long-term energy plan [57].

As seen in Tables 1 and 2, the majority of vehicles (97%) are within the light vehicles (up to 4.5 t) category. Therefore, the three scenarios presented here consider the vehicles based on their fuel type only, as shown in Table 3. The table shows the share of each vehicle type in the transportation mix of Ontario in the year 2030.

The first scenario represents the current state of transportation in Ontario. In the second scenario, the data from forecasting the future potential of using EVs, PHEVs, and compressed natural gas (CNG) vehicles are considered [1]. The third scenario considers the high penetration ratio of alternative vehicles in the HTS. The penetration ratio is assumed to be 37% in 2030, and the number of vehicles in Ontario is assumed to reach 9.4 million.

Table 3 Hybrid Transportation Scenarios Based on Fuel Type for Ontario

Vehicle Type	Scenario 1 (%)	Scenario 2 (%)	Scenario 3 (%)
Gasoline and diesel vehicles	81.5	72	63
Compressed natural gas	5.5	8.5	11
EVs	6	9	12
PHEVs	6	9	12
Hydrogen vehicles	1	1.5	2

15.5 RESULTS

15.5.1 VEHICLE OPERATION COSTS AND EMISSIONS

Scenarios lifecycle operation emissions (Activity A1)

The CO_2 emissions reduction using four types of vehicles are shown in Table 4. The benefit factor of hydrogen and CNG vehicles are far below those of gasoline and EVs. The benefit factor is the average vehicle emissions (CO_2/km) used to calculate

Table 4 The CO_2 Emissions Reduction and Costs

Scenarios	Fuel Type			
	Gasoline	**CNG**	**Hydrogen**	**EVs**
Scenario 1				
Number of vehicles	7,656,882	516,722	93,949	1,127,394
Average vehicle emission (kg CO_2/L (kg))	2.392	0.113	0.096	12.8 (kg CO_2/100 km)
Liters/100 km (kg/km for gas)	8.2	5.6	4.665	
Average traveling distance (km)	16,000	16,000	16,000	16,000
Benefit factor	37.409	1.207	0.851	24.412
Vehicles emissions (Mt)	24.030	0.052	0.007	2.309
Costs of CO_2 (M.USD)	286.433	0.624	0.080	27.522
Scenario 2				
Number of vehicles	6,764,362	798,571	140,924	1,691,091
CO_2 emissions/L (kg)	2.392	0.113	0.096	12.8 (kg CO_2/100 km)
Liters/100 km (kg/km for gas)	8.2	5.6	4.665	
Average traveling distance (km)	16,000	16,000	16,000	16,000
Benefit factor	37.409	1.207	0.851	24.412
Vehicles emissions (Mt)	21.229	0.081	0.010	3.463353385
Costs of CO_2 (M.USD)	253.045	0.964	0.120	41.2839
Scenario 3				
Number of vehicles	5,918,817	1,033,444	187,899	2,254,787
CO_2 emissions/L (kg)	2.392	0.113	0.096	12.8 (kg CO_2/100 km)
Liters/100 km (kg/km for gas)	8.2	5.6	4.665	
Average traveling distance (km)	16,000	16,000	16,000	16,000
Benefit factor	37.409	1.207	0.851	24.412
Vehicles emissions (Mt)	18.575	0.105	0.0134	4.618
Costs of CO_2 (M.USD)	221.415	1.247	0.1598	55.044

the CO_2 costs [1]. The average traveling distance in Ontario (16,000 km/year) and the costs of 1 ton of CO_2 is 9.13 Euros (11.92 USD). The CO_2 emissions can be reduced from 26.4 M. tons in the first scenario to 23.3 M. tons in the third scenario, with a reduction rate of 12%. This trend appears because of treating EVs as consumers of electricity only, rather than producers. Charging those vehicles at off-peak hours can be considered as energy storage of lost energy, and supplying it at peak hours should reduce their emission factor by the amount of electricity they supply to the power grid. The cost benefits of applying the second and third scenarios are 19.25 Million USD (M.USD), 36.79 M.USD more than the reference scenario.

Batteries lifecycle emissions (Activities A1, A2)

The average LCA emissions are presented in Table 5. This data is calculated based on data presented in [58]. The emissions from the first three battery types are included in the first three types, eg, NiMH, PbA, and NiCd. Process-specific air, water, and some solid waste emissions are also recorded. It is clear that these emissions values are controlled by heavy metals, both to air and water, which require thorough risk assessment and management panning; however, lack of data availability makes it difficult to compare the five battery technologies. Table 5 also shows there are toxic emissions that may affect human health, such as human toxicity due to nitrogen oxides emissions to air and terrestrial ecotoxicity due to particulate matters emissions.

Scenarios lifecycle operation costs (Activity A1)

Table 6 shows the cost of the first 5 years for vehicle lifetime of five types. The Ford Focus 2012 and Toyota Prius 2013 represent the conventional, and the other three are models for NG and EVs vehicles. As shown in Table 4, the average cost of ownership of a gasoline-powered car is 0.376 USD/km and for the new vehicles [compressed natural gas vehicles (CNGVs) and EVs], the cost of ownership are at 0.423, 0.425, and 0.478 USD/km, respectively. With tax incentives of 7500 USD for EVs, their ownership costs can be competitive and can be reduced to 0.338 and 0.385.

It is important to note that the cost figures of EVs are difficult to be calculated due to the large uncertainties in their future cost trends; however, as seen in the table, the difference in cost/km is a result of the depreciation costs of the new vehicles. If the depreciation costs of the new vehicles are reduced to 40%, the cost/km for the three vehicle types will be reduced to costs near the conventional vehicles as 0.379, 0.335, and 0.372, respectively. The lifetime costs calculated in Table 6 showed that the current tax credit programs are not enough to make EVs competitive as transportation, meanwhile they are successful in the case of PHEVs. The results of CO_2 costs shown in the previous section and results shown in Table 5 make CNG vehicles a perfect candidate for tax credit that enforces their penetration in the transportation system. By the year 2030, the money saved by using the CNGVs and EVs for the third scenario is −777.15, 685.46, and −153.55 M.USD, respectively. This result shows the need for more tax credits to support those vehicles types.

Table 5 Lifecycle Assessment Results Data for Five Vehicle Batteries

Battery Type	VOC (g/kg)	CO (g/kg)	NOx (g/kg)	PM (g/kg)	SOx (g/kg)	CH_4 (g/kg)	N_2O (g/kg)	CO_2eq (kg/kg)	Other Water, Air, and Solid Emissions (mg/kg)
NiMH	1.3	4.9	19.0	20.0	73.9	21.2	0.2	15.4	60 g Al, Ni, Co, etc., to air/water/solid, 0.24 kg slag and 30 g toxics—solid
PbA	1.1	2.9	6.6	5.0	7.5	4.6	0.0	3.9	4.8, Pb; 1.2, Pb
NiCd	2.1	3.9	20.1	12.5	81.7	9.3	0.1	8.3	60, Cd; Co, Ni 40, Cd; Co, Ni
Na/S	1.3	4.9	16.0	23.7	29.3	21.4	0.2	14.9	
Li-ion	0.9	3.0	14.5	19.6	19.7	13.7	0.1	12.9	

Table 6 Comparison of Vehicle Ownership Lifecycle Costs on a 5-Year Basis

Cost Items	Vehicle Type				
	Ford Focus 2012	Toyota Prius 2013	Honda Accord NG Prius 2013	Toyota Prius Plug-In 2013	Ford Focus Electric 2012
Depreciation (%)	37.41	47.36	50.31	61.05	62.13
Fuel cost (%)	21.74	12.98	10.90	8.16	4.99
Financing (%)	3.02	4.25	4.39	4.92	5.49
Insurance (%)	22.34	18.37	18.75	16.24	19.66
Maintenance (%)	4.32	4.61	3.77	4.00	2.79
Fees, taxes (%)	4.45	5.87	5.93	0.00	0.00
Repairs (%)	5.48	5.39	4.79	4.71	4.30
Opportunity cost (%)	1.23	1.17	1.16	0.93	0.64
Cost/km (USD)	0.375	0.377	0.423	0.425	0.478
Tax credit (USD)	0.000	0.000	0.000	7500	7500
Cost/km after tax credit (USD)	0.375	0.377	0.423	0.338	0.385
Difference in lifetime cost (DLTC)	0.000	0.000	0.000	−0.087	−0.009

Calculated from YahooAuto. Autos, price a new car. Available from: http://autos.yahoo.com/; 2012.

15.5.2 HTS FUELING INFRASTRUCTURE (ACTIVITY A1)

Several types of fueling station infrastructures are to be installed. For example, there are two types of CNG and electricity stations: fast-fill, used in retail stations and fleets that have central fueling; and time-fill, used when there is a possibility to fill overnight. Fast-fill fueling is similar to using a gasoline pump and takes minutes to refuel. Time-fill fueling is usually done overnight, in about 5–8 h. A fueling station may be installed wherever NG is available [59]. The hydrogen fueling station considered here is an onsite steam methane reformation production unit that produces 300 kg/day. This station converts NG feedstock into hydrogen using a steam methane reformer (SMR). The SMR is integrated with an NG compressor, blower, water pump, and pressure-swing adsorption (PSA) hydrogen purification system [60]. The NG is supplied to the hydrogen station using pipeline.

HTS scenarios fueling infrastructure costs (Activities A1, A2)

The infrastructure required to enforce the HTS includes fueling stations per fuel type, EV charging equipment, and pipelines for NG transportation. If the share of EVs is high, building new power production plants may be required. Table 7 shows the

Table 7 The Numbers of New Fueling Stations in the Three Scenarios

	Share From Total Vehicles Number in Ontario (%)	Number of New Fueling Stations Infrastructure
Scenario 1		
Electricity charging stations	12.00	1128
CNG fueling stations	5.50	520
Hydrogen fueling stations	1.00	94
Total	18.50	1739
Scenario 2		
Electricity charging stations	18.00	1692
CNG fueling stations	8.50	800
Hydrogen fueling stations	1.50	141
Total	28.00	2632
Scenario 3		
Electricity charging stations	24.00	2256
CNG fueling stations	11.00	1040
Hydrogen fueling stations	2.00	188
Total	37.00	3478

estimated number of fueling station infrastructures that should be installed to cover the demands of each fuel type based on the number of vehicles share in Ontario.

According to the US Department of Energy, costs for installing a CNG fueling station can range from 10,000 USD to 2 M.USD, depending on the size and application [61]. This work assumes that the costs of building fast- and time-fill fueling stations are 1000,000 and 10,000 USD, respectively. A 20% of cost is added for land price sharing with the existing stations. The electricity charging station costs about 2000 USD for the overnight fill (120 V, AC) charging unit and about 70,000 USD for fast-recharging station (240 V, DC) costs about 70,000 USD for a two-plug unit [62]. The current costs of building and operating the fueling stations are supposed constant, assuming inflation rates will diminish the cost reduction due to high penetration rates of new fueling stations. The full investment costs are included, as the equipment has no residual value if a station is closed [63]. The total number of fueling stations in Ontario will be about 9400 stations that should include about 2000 fueling stations for alternative fuels if we assume that 20% of the new fueling stations will supply alternative fuels (CNG, electricity, and hydrogen). Those fueling stations should be built from now until 2030. The installation and connection costs of fueling stations for different fuel types are shown in Table 8.

The connection costs of electric charging stations and time-filling gas stations are assumed to be zero, as the electricity infrastructure and home NG are already established in Ontario. The land price sharing for overnight (time) fueling stations is not

Table 8 The Annual Costs of Building and Connecting Fueling Stations for Different Fuel Types

Station Type	Cost of Installation (USD/year)	Cost of Connection (USD/km/year)	Land Price Sharing (USD/year)
CNG fast-fueling station	183,935	51,093	36,787
CNG time-fueling station	1839	–	–
Electric charging fast-fueling station (2 plugs)	12,875	–	2575
Electric charging time-filling unit	368	–	–
Hydrogen	423,052	51,093	84,610

Calculated from Southern California Gas Company. Building a CNG refueling station. Available from: http://www.socalgas.com/innovation/natural-gas-vehicles/business/station-building.shtml 2012 [cited 27.04.13]; Vaughan V, Shauk Z. Valero, Chesapeake look to build natural gas stations. Available from: http://www.mysanantonio.com/business/article/Valero-Chesapeake-look-to-build-natural-gas-3428155.php; 2012 [cited 27.04.13]; Weinert JX, et al. Hydrogen refueling station costs in Shanghai. Int J Hydrog Energy 2007;32(16):4089–100 using interest rate of 7%.

considered, as they are generally built-in, private commodities. The total costs of fueling stations infrastructure for the three scenarios are shown in Table 9. According to our calculations for the three scenarios, the pipeline transportation distance needed to supply fuel stations ranges from 10 to 30 km. We have used an average of 20 km in our infrastructure calculations. The total investment needed for the three scenarios ranges from 1 to 2 billion dollars per year. The yearly investment in the three scenarios shows high capital costs required for enforcing CNG and hydrogen vehicles. On the reverse, the investments in the fast-fueling electricity charging stations are rather low compared to the costs of installing time-filling stations as well as other types of stations.

15.5.3 IDENTIFYING RISK INDICATORS AND PERFORMING RISK ASSESSMENT (ACTIVITY A3)

Identified risks

The risk categories of systems are attributed to environmental impact categories and the lifecycle stage. The impact categories of the LCA studies are identified by the LCA team based on the project management requirements [4]. There is no explicit rule on categories that are to be used for LCA study [64]. The risk-based LCA study makes the category selection process more comprehensive through identifying impacts and their associated threats, assessing the most relevant risks, and managing them in the same platform. As explained in Section 15.3.1, the risk identification process, in this framework, it is completed in two steps (A13, A2). The initial risks identified in Activity A13 help adjust the system boundaries based on the expertise of

Table 9 Fueling Stations Infrastructure Costs for the Three Scenarios

	Cost of Installation (M.USD/year)	Cost of Connection (M.USD/year)	Land Price Sharing (M. USD/year)	Total (M. USD/year)
Scenario 1				
Electric charging fast-fueling stations (2 plugs)	7.26		1.45	8.71
Electric charging time-filling unit	143.09			143.09
CNG fast-fueling station	47.82	278.97	9.56	336.36
CNG time-fueling station	329.82			329.82
Hydrogen fueling station	39.77	100.86	7.95	148.58
Total infrastructure costs	567.76	379.83	18.97	966.55
Scenario 2				
Electric charging fast-fueling station (2 plugs)	10.89		2.18	13.07
Electric charging time-filling unit	214.63			214.63
CNG fast-fueling station	73.57	429.18	14.71	517.47
CNG time-fueling station	507.41			507.41
Hydrogen fueling station	59.65	151.29	11.93	222.87
Total infrastructure costs	866.16	580.47	28.82	1475.45
Scenario 3				
Electric charging fast-fueling station (2 plugs)	14.52		2.90	17.43
Electric charging time-filling unit	286.18			286.18
CNG fast-fueling station	95.65	557.94	19.13	672.71
CNG time-fueling station	659.63			659.63
Hydrogen fueling station	79.53	201.72	15.91	297.16
Total infrastructure costs	1135.51	759.65	37.94	1933.11

the LCA team, and concerns of the project stakeholders via answering the questions raised as explained in phase one of module two in the framework description. As mentioned earlier, the main problem of batteries emissions is heavy metals, both to air and water, which require thorough risk assessment and management planning. In addition, the toxic emissions impose real threats to human health. An example of the lifecycle stages in each stage of HTS and the related questions is listed in Table 10.

Table 10 Example of Identified Risks and Their Related Impact Categories

Types of Risks	Lifecycle Stage	Risks in Question Form
Social		
Water erosion [65]	HTS infrastructure establishment	What are the measures to reduce water depletion impact?
Noise [66]	HTS infrastructure establishment	What is the acceptable noise limit in the hotel area?
Environmental		
Greenhouse effect ($kgCO_2eq$) [66,67]	HTS production and operation	Do the expected emissions meet allowable limits?
Air pollution (kg of PM) [65]	HTS production	
Water pollution	Vehicle batteries production and disposal	Is there any method to reduce heavy metals and toxic materials from batteries? Are they applicable now or new technologies required?
Utilities are unwilling to assume the risk of ownership:	HTS operations stage	How to convince utility owners to be involved in the EVs penetration process?
Health		
Exposure risk [68]	HTS operations stage	Is there any loss of life expectancy through system operation?
Human toxicity [58]	HTS and infrastructure production and operations stage	What are the expected risks? How can they occur? How can we measure them?
Ecotoxicity [58]	HTS production stage	
Technical		
System design and vehicles mix [69]	HTS design	What are the expected problems of running different types of vehicles at the same time?
Operational and policy		
Technological adaption risks	HTS design	What are the technology selection criteria and rules followed in operating the system?

Continued

Table 10 Example of Identified Risks and Their Related Impact Categories—*Cont'd*

Types of Risks	Lifecycle Stage	Risks in Question Form
Barriers to adoption	HTS design	What are those barriers and how can we overcome them?
Lack of uniform regulation	HTS design and operation	What are the suggested regulations that make the HTS run smoothly?
Conformation with local regulation [69]	HTS design and operation	Are there any incentives by government and how to react if they will change or expire?
Economic risks		
Durability	HTS design	Are the new HTS components durable enough to keep running without problems?
Lifetime	HTS design	What is the expected lifetime for each constituent's of the HTS and their infrastructure and what are the replacement plans?
Initial investment in EV infrastructure	HTS planning	Which private or governmental organizations will finance the transition to the HTS

Risk indicators

A group of initial risk indicators was identified based on the questions raised during initial risk identification in Activity A13. An example of these indicators is shown in Table 11. To answer the questions raised in Table 10 and the indicators shown in Table 11, a set of information about the following items has to be studied:

- Complete HTS design options
- Electricity supply tariffs
- Types of regulations available
- Durability measurements have to be done
- LCA for the system to identify the level of emissions
- Applications plans
- Potential investors analysis

Risk assessment

To identify and prioritize the key risks and key risk indicators, qualitative (Activity A21) and quantitative (Activity A22) risk analysis has to be performed. In this work, the probabilistic analysis and experts' judgments are used to perform the qualitative risk assessment, and no quantitative analysis was done. The initial risk indicators are given points from 1 to 5 (1 means low potential, and 5 high potential) and a qualitative analysis is executed by collecting the opinions of team members, associated stakeholders, and experts in the form of scores. The votes for all risk indicators

Table 11 An Example of Risk Indicators

Risk-Influencing Factors	Risk Indicators	Possible Source of Risk	High-Risk Area
Accidents management plan [70]	Number of accidents	HTS vehicles	Social and health
Failure in system equipment	Failure and response rate	Battery or engine durability components failure	Economic
Air and water pollution	Number of city residents that have asthma	Vehicles that use fossil-based fuel	Health and environment
Noise	Number of complaints by residents and neighbors	Traffic jams and fueling stations operation	Social
Purchase warranty insurance	Availability	No or ineffective warranty	Economic
Reliability	Vehicles' lifetime in service	Vehicles or components failure	Economic

are sorted from highest to lowest to identify the key performance indicators of the HTS. Based on the number of votes, a relative importance index (RII) is calculated. The risk indicators that score more than 3.5 of RII are considered key risk indicators and their related risks have to be considered in the quantitative analysis in Activity A22 [65]. The RII can be calculated from the following equation [65].

$$\mathrm{RII} = \frac{\sum^{5} W_i X_i}{\sum_{i=1}^{5} X_i} \quad (1 \leq \mathrm{RII} \leq 5)$$

where X is the number of respondent to a particular impact; I is the risk score order; and W is the estimated risk by the respondent for this impact.

Also, two dispersion factors (the standard deviation and the range) can be calculated for further analysis by the team while making the final decision about the key risks and their indicators. For example, the risks with high standard deviation or dispersion range need to be worked out before it can be considered as a key risk indicator. In this analysis, the air and water pollution (health risks) and climate change (environmental risks) are assumed to have the highest RII and purchase warranty insurance has the lowest. Therefore, health and environmental risks should undergo risk management processes. The risk score matrix, shown in Fig. 8, is built based on the risks' probabilities of occurrence and RII, and used by the team to categorize risks in Activity A23. Based on the severity of risk category and the risk management plan aims, the values of risk indicators are finalized in Activity A24 based on the risk category to either exclude, negotiate, recommend, or deal with.

	Impact likelihood				
	Rare	Unlikely	Possible	Likely	Imminent
Extremely severe	5	10	16	20	25
Very severe	4	8	12	16	20
Severe	3	6	9	12	15
Slightly severe	2	4	6	8	9
Not severe	1	2	3	4	5

Matrix legend

Indicator status	Not considered	Negotiable	Recommended	Imminent
Color index				
RII values	Less than 2	2 to less than 3.5	3.5 to less than 4	4–5

FIG. 8

Risk score matrix based on the values of relative importance index of risk indicators.

15.5.4 RISK MANAGEMENT MODULE

The risk management processes rely mainly on the quantification of each category identified in Activities A23 and A24 for the associated risks. Based on the risk categories identified in Activity A2, the LCA results and the accumulated knowledge from the project or other similar sources, the RBLCA team should plan the response to the risks in Activity A13. In this case study, we will only consider noise risks, high emissions, and the business risk of using a high share of renewable energies. Controlling risks is then carried out in Activity A32 to either remove or reduce them to an acceptable level based on their score in the risk score matrix and codes and analysis results. For example, water and air pollution from batteries of HTS' EVs comes mainly from their production and can be quantified accurately by material LCA analysis. Hence, the risk management decision would be to mitigate this risk through setting an emission threshold so the harmful batteries can be phased out from the market within a specific time. Furthermore, toxicity level detection equipment should be installed and continuously monitored in the fossil-based fuel vehicles. To manage the emissions and business risks, a multiobjective optimization technique can be used to analyze the HTS' environmental and economic performances and identify the optimal RE-fossil mix. The method of managing risks depends mainly on the cost benefit analysis to ensure the risk management action brings more benefit than it costs and is acceptable by all the stakeholders, and doesn't lead to other risks.

Based on the risk management measures and decisions, the maintenance and risk management plans are updated in Activity A33.

15.6 CONCLUSIONS

This chapter introduced a general risk-based LCA framework with explanations of its phases and application methodology. The three modules of the RBLCA framework are also modeled as IDEF0 activity modeling to ensure better understanding of its application. All presented activity models are described and related to the framework application in a case study. The lifecycle of HTS in Ontario is partially analyzed using the proposed framework. Three HTS scenarios are tested and risks indicators of the HTS usage are analyzed. Then, risk indicators are used to prioritize the risks in the risk assessment module. In the risk management module, the identified risks are included, and some management solutions are discussed. The main focus of this chapter is to present a complete description of the framework and its application on an HTS and not, by any means, a comprehensive LCA study. The proposed framework is a general structure that can be extended to other types of systems. The general structure of the RBLCA framework is applicable for all system types with detailed IDEF0 models modified to include relevant information flows, depending on the system investigated.

REFERENCES

[1] Ayoub NM, Gabbar HA. Hybrid transportation model forecasting and optimisation with practical fuel supply chain. Int J Supply Chain Oper Resilience 2015;1(3):243–79.

[2] T-Raissi A, Block DL. Hydrogen: automotive fuel of the future. IEEE Power Energy Mag 2004;2(6):40–5.

[3] Heinen S, Elzinga D, Kim S-K, Ikeda Y. Impact of smart grid technologies on peak load to 2050. Paris: International Energy Agency; 2011.

[4] Ayoub N, et al. Risk based life cycle assessment conceptual framework for energy supply systems in large buildings. J Clean Prod 2015;107:291–309.

[5] Ayoub N. A multilevel decision making strategy for designing and evaluating sustainable bioenergy supply chains. Process system engineering. Yokohama: Tokyo Institute of Technology; 2007. p. 150.

[6] Gabbar HA, et al. Risk-based performance analysis of microgrid topology with distributed energy generation. Int J Electr Power Energy Syst 2012;43(1):1363–75.

[7] Eising JW, van Onna T, Alkemade F. Towards smart grids: identifying the risks that arise from the integration of energy and transport supply chains. Appl Energy 2014;123:448–55.

[8] Basson L, Petrie JG. An integrated approach for the consideration of uncertainty in decision making supported by Life Cycle Assessment. Environ Model Softw Environ Decision Support Syst 2007;22(2):167–76.

[9] Faria R, et al. Impact of the electricity mix and use profile in the life-cycle assessment of electric vehicles. Renew Sust Energy Rev 2013;24:271–87.

[10] Lin C, et al. Life-cycle private costs of hybrid electric vehicles in the current Chinese market. Energy Policy 2013;55:501–10.

[11] Rydh CJ. Environmental assessment of vanadium redox and lead-acid batteries for stationary energy storage. J Power Sources 1999;80(1–2):21–9.

[12] Schleisner L. Life cycle assessment of a wind farm and related externalities. Renew Energy 2000;20(3):279–88.

[13] Stoppato A. Life cycle assessment of photovoltaic electricity generation. Energy 2008;33(2):224–32.

[14] Weiwei P, et al. Risk-based life cycle cost analysis method for transmission systems. In: 2012 3rd IEEE PES international conference and exhibition on Innovative Smart Grid Technologies (ISGT Europe); 2012.

[15] Archsmith J, Kendall A, Rapson D. From cradle to junkyard: assessing the life cycle greenhouse gas benefits of electric vehicles. Res Transp Econ 2015;52:72–90.
[16] Egede P, et al. Life cycle assessment of electric vehicles—a framework to consider influencing factors. Procedia CIRP 2015;29:233–8.
[17] Zhao Y, Tatari O. A hybrid life cycle assessment of the vehicle-to-grid application in light duty commercial fleet. Energy 2015;93(Part 2):1277–86.
[18] Noori M, et al. Light-duty electric vehicles to improve the integrity of the electricity grid through vehicle-to-grid technology: analysis of regional net revenue and emissions savings. Appl Energy 2016;168:146–58.
[19] Perujo A, Ciuffo B. The introduction of electric vehicles in the private fleet: potential impact on the electric supply system and on the environment. A case study for the Province of Milan, Italy. Energy Policy 2010;38(8):4549–61.
[20] Hossain KA, Khan FI, Hawboldt K. E-Green—a robust risk-based environmental assessment tool for process industries. Ind Eng Chem Res 2007;46(25):8787–95.
[21] ABS Consulting. Marine safety: tools for risk-based decision making. Rockville, MD: ABS Consulting, Government Institutes; 2002.
[22] Padgett JE, Dennemann K, Ghosh J. Risk-based seismic life-cycle cost-benefit (LCC-B) analysis for bridge retrofit assessment. Struct Saf 2010;32(3):165–73.
[23] Khan FI, Sadiq R, Haddara MM. Risk-based inspection and maintenance (RBIM): multi-attribute decision-making with aggregative risk analysis. Process Saf Environ Prot 2004;82(6):398–411.
[24] Chiu C-K, Chien W-Y, Noguchi T. Risk-based life-cycle maintenance strategies for corroded reinforced concrete buildings located in the region with high seismic hazard. Struct Infrastruct Eng 2012;8(12):1108–22.
[25] Gabbar HA, Bedard R. Risk-based performance analysis for regional hybrid fuel with compressed natural gas option. Int J Process Syst Eng 2012;2(2):154–77.
[26] He HZ, Kua HW. Lessons for integrated household energy conservation policy from Singapore's southwest Eco-living Program. Energy Policy 2013;55:105–16.
[27] Kikuchi Y, Hirao M. Hierarchical activity model for risk-based decision making. J Ind Ecol 2009;13(6):945–64.
[28] Hirao M, et al. IDEFO activity modeling for integrated process design considering environmental, health and safety (EHS) aspects. In: Bertrand B, Xavier J, editors. Computer aided chemical engineering. New York: Elsevier; 2008. p. 1065–70.
[29] Sugiyama H, et al. Activity modeling for integrating environmental, health and safety (EHS) consideration as a new element in industrial chemical process design. J Chem Eng Jpn 2008;41(9):884–97.
[30] Sheau-Ting L, Mohammed AH, Weng-Wai C. What is the optimum social marketing mix to market energy conservation behaviour: an empirical study. J Environ Manage 2013;131:196–205.
[31] Sugiyama H, et al. A hierarchical activity model of chemical process design based on life cycle assessment. Process Saf Environ Prot 2006;84(1):63–74.
[32] Kim C-H, et al. The complementary use of IDEF and UML modelling approaches. Comput Ind 2003;50(1):35–56.
[33] Inage S-I. Modelling load shifting using electric vehicles in a smart grid environment. Paris: International Energy Agency; 2010.
[34] Gabbar H, Ayoub N. Green infrastructure and forecasting models for integrating natural gas grid within smart grids. In: IEEE International Conference on Smart Grid Engineering (SGE'12), UOIT, Oshawa, ON, 27–29 August; 2012.

[35] Dong J, Lin Z. Within-day recharge of plug-in hybrid electric vehicles: energy impact of public charging infrastructure. Transp Res Part D: Transp Environ 2012;17(5):405–12.
[36] Lucas A, Alexandra Silva C, Costa Neto R. Life cycle analysis of energy supply infrastructure for conventional and electric vehicles. Energy Policy 2012;41:537–47.
[37] San Román TG, et al. Regulatory framework and business models for charging plug-in electric vehicles: infrastructure, agents, and commercial relationships. Energy Policy 2011;39(10):6360–75.
[38] Baptista PC, et al. Energy and environmental impacts of alternative pathways for the Portuguese road transportation sector. Energy Policy 2012;51:802–15.
[39] Liu J. Electric vehicle charging infrastructure assignment and power grid impacts assessment in Beijing. Energy Policy 2012;51:544–57.
[40] Andress D, Nguyen TD, Das S. Reducing GHG emissions in the United States' transportation sector. Energy Sustain Dev 2011;15(2):117–36.
[41] Bradley TH, Frank AA. Design, demonstrations and sustainability impact assessments for plug-in hybrid electric vehicles. Renew Sust Energy Rev 2009;13(1):115–28.
[42] Al-Alawi BM, Bradley TH. Total cost of ownership, payback, and consumer preference modeling of plug-in hybrid electric vehicles. Appl Energy 2013;103:488–506.
[43] Hackney J, de Neufville R. Life cycle model of alternative fuel vehicles: emissions, energy, and cost trade-offs. Transp Res A Policy Pract 2001;35(3):243–66.
[44] Schroeder A, Traber T. The economics of fast charging infrastructure for electric vehicles. Energy Policy 2012;43:136–44.
[45] Damen K, et al. A comparison of electricity and hydrogen production systems with CO_2 capture and storage—Part B: chain analysis of promising CCS options. Prog Energy Combust Sci 2007;33(6):580–609.
[46] Yeh S. An empirical analysis on the adoption of alternative fuel vehicles: the case of natural gas vehicles. Energy Policy 2007;35(11):5865–75.
[47] Ontario Ministry of Transportation. Electric vehicles. Available from:.http://www.mto.gov.on.ca/english/dandv/vehicle/electric/ev-faq.shtml; 2012 [cited 19.04.13].
[48] Shukla A, Pekny J, Venkatasubramanian V. An optimization framework for cost effective design of refueling station infrastructure for alternative fuel vehicles. Comput Chem Eng 2011;35(8):1431–8.
[49] Junnila S, Horvath A. Life-cycle environmental effects of an office building. J Infrastruct Syst 2003;9(4):157–66.
[50] PMI. A guide to the project management body of knowledge. 5th ed. USA: Project Management Institute; 2013.
[51] Sentjens J, Deakin I, Goudappel E. Greenhouse gas masterplan and risk management. Energy Procedia 2011;4:2028–34.
[52] Sollie OK, et al. An early phase risk and uncertainty assessment method for CO_2 geological storage sites. Energy Procedia 2011;4:4132–9.
[53] Kuselman I, et al. Monte Carlo simulation of expert judgments on human errors in chemical analysis—a case study of ICP-MS. Talanta 2014;130:462–9.
[54] Office Energy Efficiency. Canadian vehicle survey: 2009 summary report. Canada: Natural Resources Canada's Office of Energy Efficiency; 2009.
[55] Hajimiragha A, et al. Optimal transition to plug-in hybrid electric vehicles in Ontario, Canada, considering the electricity-grid limitations. IEEE Trans Ind Electron 2010;57(2):690–701.
[56] Ontario Ministry of Finance. Projected population for Ontario under three scenarios, 2011–2036. Available from: http://www.fin.gov.on.ca/en/economy/demographics/projections/table1.html; 2012 [cited 19.04.13].

[57] Ontario Ministry of Energy. Ontario's long-term energy plan—building our clean energy future. Ontario: Ontario Ministry of Energy; 2010.
[58] Sullivan JL, Gaines L, Systems E. A review of battery life-cycle analysis: state of knowledge and critical needs. Argonne, IL: Argonne National Laboratory; 2010.
[59] Southern California Gas Company. Building a CNG refueling station. Available from: http://www.socalgas.com/innovation/natural-gas-vehicles/business/station-building.shtml; 2012 [cited 27.04.13].
[60] Weinert JX, et al. Hydrogen refueling station costs in Shanghai. Int J Hydrog Energy 2007;32(16):4089–100.
[61] Whyatt G. Issues affecting adoption of natural gas fuel in light and heavy. Richland, Washington, DC: Pacific Northwest National Laboratory; 2010.
[62] Vaughan V, Shauk Z. Valero, Chesapeake look to build natural gas stations, Available from: http://www.mysanantonio.com/business/article/Valero-Chesapeake-look-to-build-natural-gas-3428155.php; 2012 [cited 27.04.13].
[63] Frick M, et al. Optimization of the distribution of compressed natural gas (CNG) refueling stations: Swiss case studies. Transp Res Part D: Transp Environ 2007;12(1):10–22.
[64] Stranddorf HK, Hoffmann L, Schmidt A. Impact categories, normalisation and weighting in LCA. In: T.D.E.P. Agency , editor. Denmark: Danish Ministry of the Environment; 2005.
[65] Ijigah EA, et al. An assessment of environmental impacts of building construction projects civil and environmental research. Civ Environ Res 2013;3(1):93–104.
[66] GORD. Global sustainability assessment system (GSAS). Qatar: Gulf Organization for Research and Development Doha; 2010.
[67] Nishioka Y, et al. Integrating risk assessment and life cycle assessment: a case study of insulation. Risk Anal 2002;22(5):1003–17.
[68] Mui KW, Wong LT, Chan WY. Energy impact assessment for the reduction of carbon dioxide and formaldehyde exposure risk in air-conditioned offices. Energy Build 2008;40(8):1412–8.
[69] Lowder T, et al. Continuing developments in PV risk management: strategies, solutions, and implications. Golden, CO: National Renewable Energy Laboratory; 2013.
[70] Chen Y-Y, et al. The adoption of fire safety management for upgrading the fire safety level of existing hotel buildings. Build Environ 2012;51:311–9.

CHAPTER

Data centers for smart energy grids

16

H.A. Gabbar*, A. Zidan*,†, M. Xiaoli‡

University of Ontario Institute of Technology (UOIT), Oshawa, ON, Canada Assiut University, Assiut, Egypt† China Electric Power Research Institute, Beijing, China‡*

16.1 BACKGROUND

Early computer systems were complex to operate and required a special environment. In the last few decades, information technology (IT) started to grow in complexity and organizations. The term "data center" is used to refer to specially designed computer rooms. Thus, data centers have their roots from the huge computer rooms of the early ages. Companies needed fast and reliable Internet to deploy their systems. Many companies started building large Internet data centers to provide commercial clients with a range of solutions for systems deployment and operation. A data center is a centralized repository for the storage, management, and dissemination of data and information. A data center has a computer system, telecommunications, and a storage system. Due to its importance, a data center includes backup power supply, data communication connections, environmental control (eg, air-conditioning, fire suppression), and various security devices.

16.1.1 REQUIREMENTS FOR DATA CENTERS

IT is a key aspect as companies rely on their information systems to run their operations. If an IT system becomes unavailable, company operations may be impaired or stopped completely. Therefore, a reliable infrastructure for IT operations is required to minimize any chance of disruption. In addition, information security is essential, and a data center should have a secure environment to minimize the chances of a security breach. A data center follows high standards to guarantee the integrity and functionality of its computer environment.

The telecommunications infrastructure standard for data centers specifies the minimum requirements for telecommunications infrastructure of data centers and computer rooms [1]. It defines a standard telecommunications infrastructure for data centers, such as: (1) cabling system for data centers using standardized architecture and media; (2) wide range of applications (LAN, WAN, SAN, channels, and

Smart Energy Grid Engineering.

consoles); (3) current and future protocols; (4) specifications for data center telecommunication pathways and spaces; and (5) standard for data center tiers.

Societies are experiencing rapid IT growth, but their data centers are aging. Thus, there is a trend to modernize data centers to benefit from the enhanced performance and energy efficiency of new IT equipment and capabilities. Data center upgrading includes standardization/consolidation, virtualization, automation, and security:

- Standardize/consolidate: to reduce the number of data centers a large organization may have; also, to reduce the number of hardware, software, tools, and processes within a data center, which can be achieved by replacing aging data center equipment with newer equipment that provides increased capacity and performance.
- Virtualize: to reduce capital and operational expenses, and to reduce energy consumption.
- Automating: by automating provisioning, configuration, patching, release management, and compliance tasks to make data centers run more efficiently.
- Securing: to take into account physical security, network security, data, and user security.

16.1.2 CARRIER NEUTRALITY

Carrier neutrality refers to carrier-neutral data centers that allow interconnection among multiple telecommunication carriers and/or colocation providers. This encourages competition and diversity, as a server in a colocation center can have one provider, multiple providers, or only connect back to the headquarters of the company who owns the server. In a noncarrier-neutral environment, a customer has one option for service and few options for connectivity/colocation. Consumers in a data center tied to one specific carrier may have high prices, limited bandwidth, and lack of competition. Therefore, carrier neutrality is a key factor for outsourcing a data center or seeking interconnection services, as it provides the benefit of both cost efficiency and a greater degree of connectivity.

16.1.3 DATA CENTER TIERS

The telecommunications infrastructure standard for data centers defined four levels (tiers) of data centers in quantifiable manner. The four-tier system provides a simple and effective means for identifying different data center site infrastructure design topologies. It is a standardized method used to define uptime of data centers, which is useful for measuring data center performance, investment, and return on investment (ROI). As shown in Table 1, a tier-1 data center is the simplest one and can be used by small businesses or shops. A tier-4 data center is considered as the most robust and least prone to failures. Tier 4 is designed to host mission-critical servers and computer systems, with fully redundant subsystems (eg, cooling, power, network links, and storage) [2,3].

Table 1 Data Center Tiers [2]

Tier Level	Specifications
1	• Single power and cooling distribution path serving the IT equipment • Without redundant capacity components • Provides availability of 99.671%
2	• Single power and cooling distribution path serving the IT equipment • With redundant capacity components • Provides availability of 99.741%
3	• Multiple active power and cooling distribution paths, but only one path actively serving the IT equipment • With redundant capacity components • All IT equipment is dual-powered and fully compatible with the topology of a site's architecture • Concurrently maintainable site infrastructure • Provides availability of 99.982%
4	• Multiple active power and cooling distribution paths serving the IT equipment • With redundant capacity components • All cooling equipment is independently dual-powered, including chillers and heating, ventilating, and air-conditioning (HVAC) systems • Fault-tolerant site infrastructure with electrical power storage and distribution facilities • Provides availability of 99.995%

16.2 DESIGN CONSIDERATIONS

Data centers are commonly run by large companies or government agencies. Data centers are increasingly used to provide fast-growing cloud-solution services for private and business applications. A data center is the brain of a company and the place where the most critical processes are run. A data center can occupy one room of a building, one or more floors, or an entire building.

16.2.1 DESIGN PROGRAMMING

Design or architectural programming is the research and decision-making process that identifies the scope of work to be designed. There are three main elements to design programming for data centers: (1) facility topology design (eg, space planning); (2) engineering infrastructure design (eg, cooling and electrical power systems); and (3) technology infrastructure design (eg, cable plant).

16.2.2 MODELING CRITERIA

Modeling criteria is the formalized application of modeling to support system requirements, design, analysis, verification, and validation activities starting from the conceptual design phase and continuing throughout the development and lifecycle phases. Modeling criteria can be used to develop future-state scenarios for space, power, cooling, and costs in data centers. Hence, an overall plan can be presented with different design parameters, such as number, size, location, topology, IT system layouts, power, cooling, and configurations. The modeling criteria facilitate the efficient use of the existing mechanical and electrical systems, the expansion of the existing data center, and upgrading of power supplies.

16.2.3 DESIGN RECOMMENDATIONS

Design recommendations and/or plans follow the modeling criteria phase. The design recommendations and/or plans provide a vision of the future of a data center and its physical development. They translate the broad framework (eg, policies, goals, and objectives) into desired outcomes for the design of a data center. The detailed principles for development include: (1) the optimal technology infrastructure and (2) the planning criteria (eg, power capacities, overall power requirements, mechanical cooling capacities, kilowatts per cabinet, raised floor space, and the resiliency level for the facility).

16.2.4 CONCEPTUAL DESIGN

Conceptual design is the first phase of design, as it embodies the design recommendations and/or plans. What-if analysis is a key factor for the conceptual design to ensure that all desired outcomes will be met. During the conceptual design phase, the individual requirements have to be compiled and the targets of the project have to be defined. An optimal process is designed based on this information. For example, layouts should be driven by IT performance requirements, lifecycle costs associated with IT demand, energy efficiency, cost efficiency, and availability.

16.2.5 DETAILED DESIGN

Detailed design follows the conceptual design phase. The detailed design phase elaborates each aspect of the data center by complete description through solid modeling, drawings, and specifications. Thus, the detailed design phase includes the detailed architectural, structural, mechanical, and electrical information and specifications of a data center. At this stage, schematics and construction documents, performance specification, details of all technology infrastructure, detailed IT infrastructure design, and IT infrastructure documentation are produced.

Mechanical engineering infrastructure design

Mechanical systems of a data center include heating, ventilating, and air-conditioning (HVAC), humidification/dehumidification equipment, and pressurization. The design process aims to save space and costs while ensuring that objectives are met.

Electrical engineering infrastructure design

Designing electrical systems includes utility service planning, distribution, switching, bypass from power sources, and uninterruptable power sources. These designs follow energy standards and practices while meeting business objectives.

Technology infrastructure design

Design of technology infrastructure includes the telecommunications cabling systems throughout data centers. The cabling systems cover all data center environments, such as voice, modem, facsimile telecommunications services, premises switching equipment, computer and telecommunications management connections, and data communications. In addition, wide/local area, and storage area networks can be used for linking with other signaling systems, such as fire, security, power, HVAC, and energy management system (EMS).

16.2.6 SITE SELECTION

Site selection indicates the practice of a new data center location. Site selection requires evaluating the merits/demerits of potential locations. There are many location factors that should be considered for data center design, such as proximity to power grids, telecommunications infrastructure, networking services, transportation lines, and emergency services.

16.2.7 MODULARITY AND FLEXIBILITY

Modularity is the degree to which a system's components can be separated and recombined. It measures the strength of division of a network into modules, groups, clusters, or communities. Flexibility refers to designs that can adapt when external changes occur. Therefore, modularity and flexibility are key characteristics for a data center to grow and change over time. A modular data center has modules that are preengineered and standardized to be easily configured and moved as needed.

16.3 DATA CENTERS AND THE SMART ENERGY GRID

A smart energy grid (SEG) has modernized the way energy (eg, electricity and heat) is generated, transported, distributed, and consumed by integrating advanced monitoring, communications, and control in the day-to-day operation of the grid. In the SEG, all entities are connected through communications. As shown in Fig. 1, data collected from sensors and meters can be fed to the utility headquarters and stored in data centers. To realize its goals, the SEG needs data centers to digest the big data due to the volume, variety, and velocity of the data generated by a variety of sources [4].

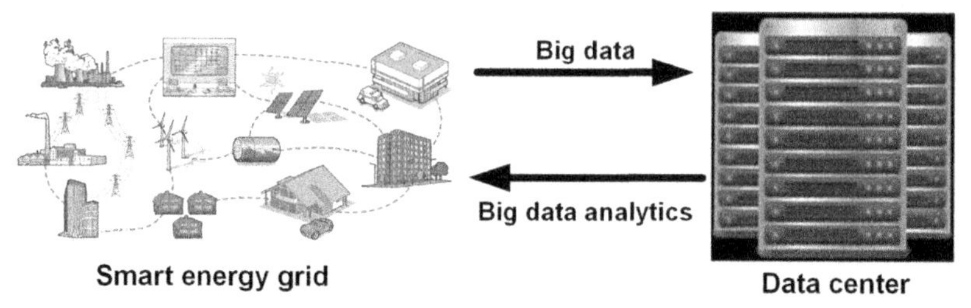

FIG. 1

SEG and data center.

16.3.1 BIG DATA FOR THE SEG ANALYTICS

As shown in Fig. 2, the collected data from smart meters, phasor-measurement units (PMUs), billing, weather sensors, and intelligent electronic devices (IEDs) can be beneficial for many applications, such as predictive analytics, demand-side management, grid awareness, outage detection, asset management, and theft detection. Big data facilitates grid management. For instance, the accessibility to customers' energy usage data expands demand response and improves energy efficiency [5,6]. PMUs have been implemented in SEGs for monitoring, protection, and control. The data from digital relays, PMUs, and IEDs can be used to simulate grid behavior and, hence, to predict upcoming faults or abnormal situations [7]. Therefore, analyzing data from PMUs and IEDs helps to maximize safety and self-healing, ensure service reliability, improve customer service, and prevent outages.

Utilities use predictive analytics based on the SEG data to forecast several parameters that can help operate the grid efficiently, economically, and reliably. Examples are forecasting the energy available from renewable sources, forecasting equipment downtime, accessing power failures, and managing unit commitment. The integration of distributed generating sources, renewable resources, electric vehicles (EVs),

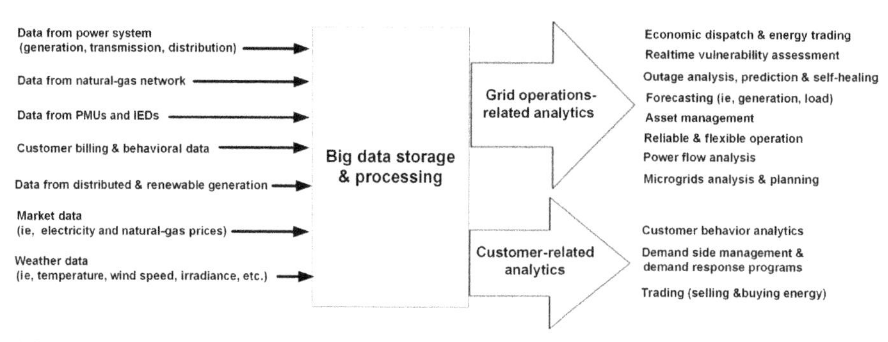

FIG. 2

Big data analytics and SEG.

and variable demand increases the level of uncertainty [8]. Thus, advanced forecasting, load planning, and unit commitment are required to avoid inefficient operation (eg, energy trading or dispatching extra generation). These requirements can be realized through big data as it enhances the grid management and improves the effectiveness and robustness of generation and scheduling.

16.3.2 BIG DATA ANALYTICS MARKET

The Electric Power Research Institute (EPRI) estimated the required investment to realize the vision for a fully functioning smart grid to be around $476 billion [9]. Achieving the benefits from this huge investment depends on analytics platforms to digest unstructured and complex data emerging from different components of the smart grid. Data analytics is one of the fastest growing market segments in North America. The utility data analytics market between 2013 and 2020 is estimated to be $20 billion and it is expected to grow by $4 billion per year until 2020 [10]. Many grid and IT companies (eg, Siemens, ABB, Schneider, IBM, Oracle, and Google) are working to turn big data challenges into opportunities.

Transforming big data into smart intelligence will enable distributed processing by parallelizing the mass of data into portions among thousands of computing machines for realtime and batch processing. For example, Hadoop MapReduce [11] is a popular open-source software framework that supports distributed processing applications as it is capable of delivering the performance for analyzing and managing big data. It is widely used by both government and nongovernment entities. Apache Storm is another free and open-source distributed realtime computation system [12]. Apache Storm can perform analysis of smart metering data, weather forecasts, customer data, and forecasting market pricing and generation. Based on the application, many platforms can be designated for processing extremely large data as distributed processing done over a large number of computing nodes in data centers or in cloud-based data center environments.

16.3.3 ENERGY-EFFICIENT COMMUNICATIONS AND THE SEG

Energy efficiency is using less energy to provide the same service and/or products. With new technologies and information available, governments have great potential to use energy more effectively. Energy efficiency works in many fields, such as communities, industry, transportation, and communication to achieve benefits, such as lowering energy costs, cutting emissions, improving operating performance, and increasing asset values. Energy efficiency has been an integral part to reduce the energy consumption of core and access network equipment. For instance, price-following demand management of SEG can be employed using the communication infrastructure to reduce electricity bills [13].

Generally, the smart grid has three domains in terms of communication coverage and functionality: smart grid home-area network (SG-HAN), smart grid neighborhood-area network (SG-NAN), and smart grid wide-area network (SG-WAN)

[13]. SG-HAN can be a single residential unit with smart appliances, an energy display, power consumption control tool, storage, solar panels, small wind turbines, EVs, and a smart meter. SG-NAN can be a group of houses that are fed by the same transformer. Advanced metering infrastructure (AMI) collects data from SG-NAN and aggregates these data to be sent to SG-WAN, which connects to the utility operator. These communication network domains can be implemented using a variety of communication technologies.

Energy-efficient wireless communications

Wireless communication refers to the transfer of information among two or more points without an electrical conductor. The most common wireless technologies use radio. Wireless communications are employed in many smart grid applications, such as meter data collection, demand management, substation and power line monitoring and protection. Wireless communications can be used in SG-WAN, SG-NAN, and SG-HAN.

(1) Energy-efficient wireless communications for SG-WAN:
Third-generation (3G) and fourth-generation (4G) cellular networks can be used in the SG-WAN. They use the existing infrastructure and provide wide coverage with high data rates. In addition, WiMAX (Worldwide Interoperability for Microwave Access) stands as a strong candidate for SG-WAN communications [13].

(2) Energy-efficient wireless communications for SG-NAN:
SG-NAN is a vital communication domain as it connects utility networks with customer networks. Furthermore, it transmits huge amounts of data and supports a large number of devices [14]. 3G, 4G, and WiMAX can be used in the SG-NAN.

(3) Energy-efficient wireless communications for SG-HAN:
Zigbee is a short-range, low-data rate wireless technology based on the IEEE 802.15.4 standard. Zigbee represents one of the commonly adopted smart home and smart energy standards in SG-HANs. ZigBee supports smart metering, AMI, and provides communication among utilities and household devices. In addition, WiFi represents another candidate for wireless SG-HAN communications.

Energy-efficient wireline communications

Wireline communication refers to the transfer of information over wire-based communication technology. Examples are telephone networks, cable television, Internet access, and fiber-optic communication. Wireline communication technologies can be used to implement SG-HAN and SG-NAN. Powerline communication (PLC) and energy-efficient Ethernet are promising wireline communications. PLC is a communication protocol that uses electrical wiring to simultaneously carry both data and electric power. PLC can be used by utilities for load control and remote metering. PLC can be affected by the disturbances in low-voltage network [15]. Thus, smart grid-driven techniques have not found their way into energy-efficiency approaches in PLC.

Energy-efficient optical networks

Fiber optical networks use signals encoded onto light to transmit information among various nodes of a telecommunication network. They operate from the limited range of a local-area network (LAN) or over a wide-area network (WAN). Also, they can cross metropolitan and regional areas all the way to national, international, and transoceanic distances. Thus, fiber optical networks offer high speed, large bandwidth, and a high degree of reliability. They are widely deployed as the basic physical network infrastructure. Optical communications can be used in SG-WAN or SG-NAN.

16.3.4 SMART GRID-AWARE ENERGY-EFFICIENCY TECHNOLOGIES IN DATA CENTERS

Information and communication infrastructures are significant consumers of electricity. Power consumption of data centers is increasing at remarkable rates. Data centers can benefit from time-of-use (ToU) prices, renewable energy sources, and smart-grid-related concepts to reduce their energy, expenses, and emissions.

Energy-efficient facility management

A data center facility includes the actual facility that houses the servers, uninterruptible power supply (UPS), cooling equipment, and operation room. Most of the power consumed by a data center facility is spent cooling the servers and the UPS. Facility management aims to organize hot aisles and cold aisles appropriately. Hot aisle/cold aisle is a layout design for server racks and other computing equipment in a data center. The goal of a hot aisle/cold aisle configuration is to conserve energy and lower cooling costs by managing air flow. Energy-efficiency of a data center facility is considered independent of smart grid concepts; however, industry encourages the use of eco-modes of UPSs during high electricity prices or to recharge UPSs during overnight hours like EVs.

Energy-efficiency techniques for IT equipment

Energy-efficiency of IT equipment is implemented through energy-efficient server hardware, virtualization, and dynamic workload consolidation [16]. Dynamic frequency and voltage scaling represents the primary energy-efficiency technique for server hardware, as it dynamically adjusts the frequency of the CPU of a server to save power [17]. In addition, virtualization helps running multiple user demands on a single physical host by placing more than one virtual machine on a server. The combination of virtualization and dynamic workload consolidation allow some servers to be left idle and put to sleep (eg, dynamic power management).

Energy-efficiency techniques for multiple data centers

Some data centers are geographically diverse, large-scale Internet data centers, such as the ones owned and operated by Google, Yahoo, Microsoft, Amazon, and eBay. Their energy-efficiency technique is to route user demands toward one or more data centers considering heating/cooling constraints, and to minimize cost or emissions.

(1) Minimizing electricity bills of data centers:
Data center operators can migrate workloads toward data centers that have less expensive electricity according to local ToU tariffs; however, during workload mitigation, users may have delays, especially if the assigned data center is geographically distant and overloaded due to forwarded demands from other data centers. Furthermore, data center operators can buy electricity beforehand (eg, when the price for electricity is low) to compensate for the fluctuations in their workload and electricity prices [18].

(2) Data centers powered by renewable energy:
Data centers can be islanded from the utility grid to be powered from their renewable resources whenever sun penetration or wind power is adequate. For example, the green star network (GSN) testbed of Canada is a network of renewably powered data centers [19]. The GSN testbed aims to explore the feasibility of powering a network of data centers with solar and wind power.

16.3.5 DATA CENTERS FOR HOSTING SEG SERVICES

Interconnected electrical power systems are complex due to their dynamic nature. The data generated by smart grid applications is increasing rapidly. Thus, storage and processing of those data are shifting from utility servers to Internet data centers. A typical example of a smart grid application with a high volume of data is the power system frequency monitoring network (FNET) [20]. FNET collects timestamped measurements from the grid through frequency disturbance recorders and provides functionalities such as realtime event alerts, accurate event location estimation, animated event visualization, and post-event analysis. Measurement data from the FNET are managed in a data center, as seen in Fig. 3 [20]. The FNET data center is a multilayer data management and utilization system. The most powerful component of the data center is the FNET data concentrator, as it has the ability to flag bad data, report missing data, and sound an alarm for abnormal interruptions.

16.4 NETWORK INFRASTRUCTURE

An SEG should satisfy the main functions of system monitoring, supplier-client interactions, energy security, power quality, business intelligence, and data analytics. This requires high-performance data centers for data acquisition, storage, and utilization. The data center is a key section for system monitoring, state analysis, information storage, and other advanced applications.

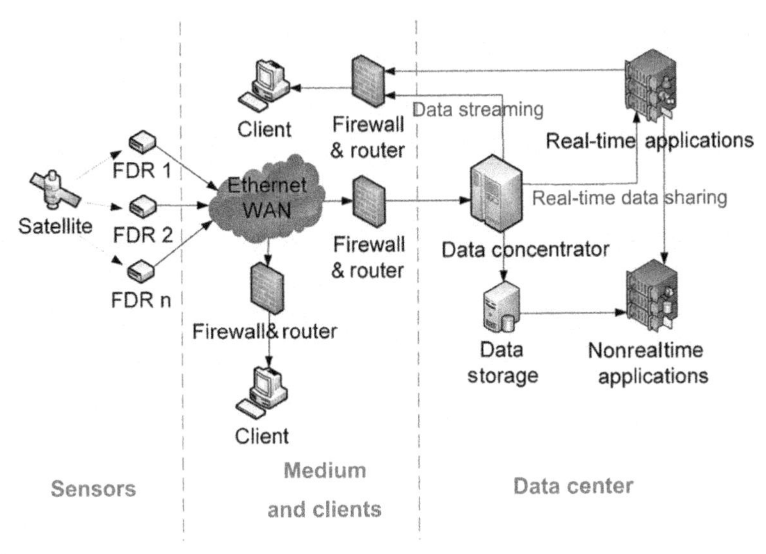

FIG. 3

Building blocks of the FNET system [20].

16.4.1 ARCHITECTURE OF DATA CENTERS

Fig. 4 shows a three-layer hierarchy of a data center that includes a device-layer, communication-layer, and data center-layer. The data center can cover power grid, water, heat, traffic, and other fields, as interested. The components of a data concentrator, data storage, and application servers are supplemented according to the data quantity. Fig. 5 shows another scheme of a data center. The difference between architecture I and architecture II is that advanced applications are divided into realtime applications and nonrealtime applications. Hence, the efficiency of data processing and data concentrator cluster can be improved.

Physical devices

As shown in Figs. 4 and 5, the devices layer includes distributed generations (DGs), loads, and sensors. Sensors collect data/measurements at different timescales to be sent to a data center via a communication network. Each device has the following characteristics: (1) each device can be identified by a specific IP; (2) each device can measure, calculate, save, send and/or receive data, and execute received instructions from its data center; (3) each device has a memory that can save data for a specific period of time; and (4) each device has its space coordinate that enables its location on the geographical map.

Communication network

The communication layer is responsible for transmitting collected data from power energy sources (eg, solar, wind, natural gas turbine, and energy storing devices) and loads to data center. Each communication technology has its advantages and

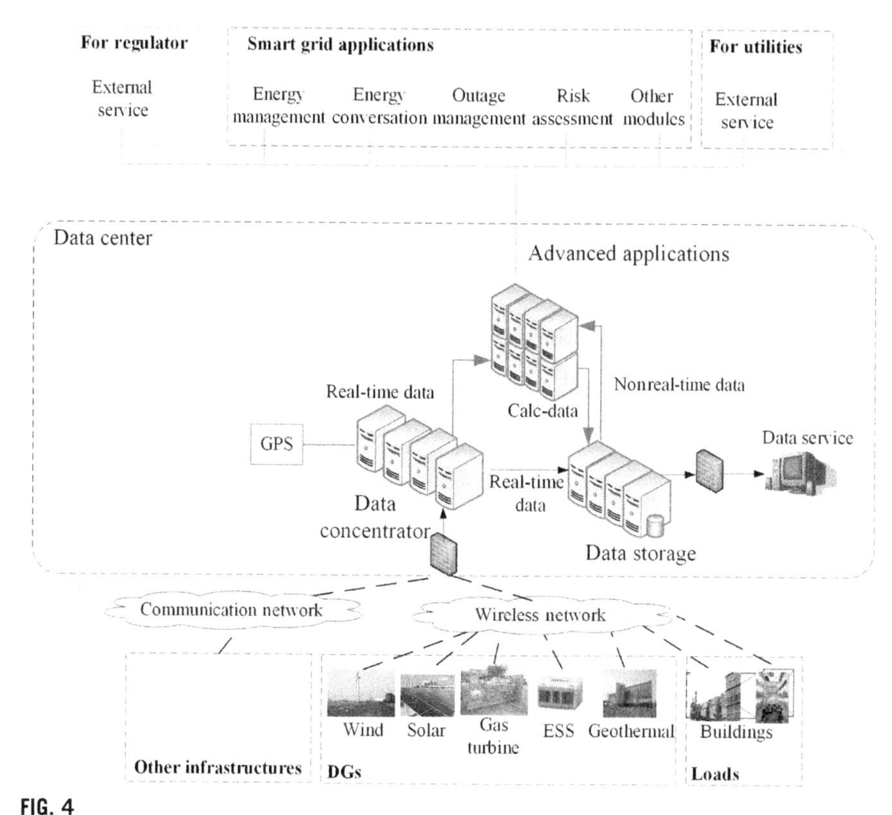

FIG. 4

Architecture I of the data center.

disadvantages, such as optical fiber, Zigbee, and WiFi. A suitable communication technology should meet the demands of data quantity.

Data center

The data center layer includes the data concentrator, database servers, LAN, and information security isolation equipment. Furthermore, a global position system (GPS) is required for data synchronization. The specifications of hardware should meet future expansion and avoid overlapping investments. The function modules of the data center include control and data analysis, EMS, energy conservation, risk and outage management, etc.

(1) Data concentrator:
The data concentrator collects realtime data and sends these data to realtime applications or to data storage (eg, database, reserve historical data) at a regular time. Also, the data concentrator handles the communication protocols of sensors. The realtime database in the read-access memory (RAM) provides

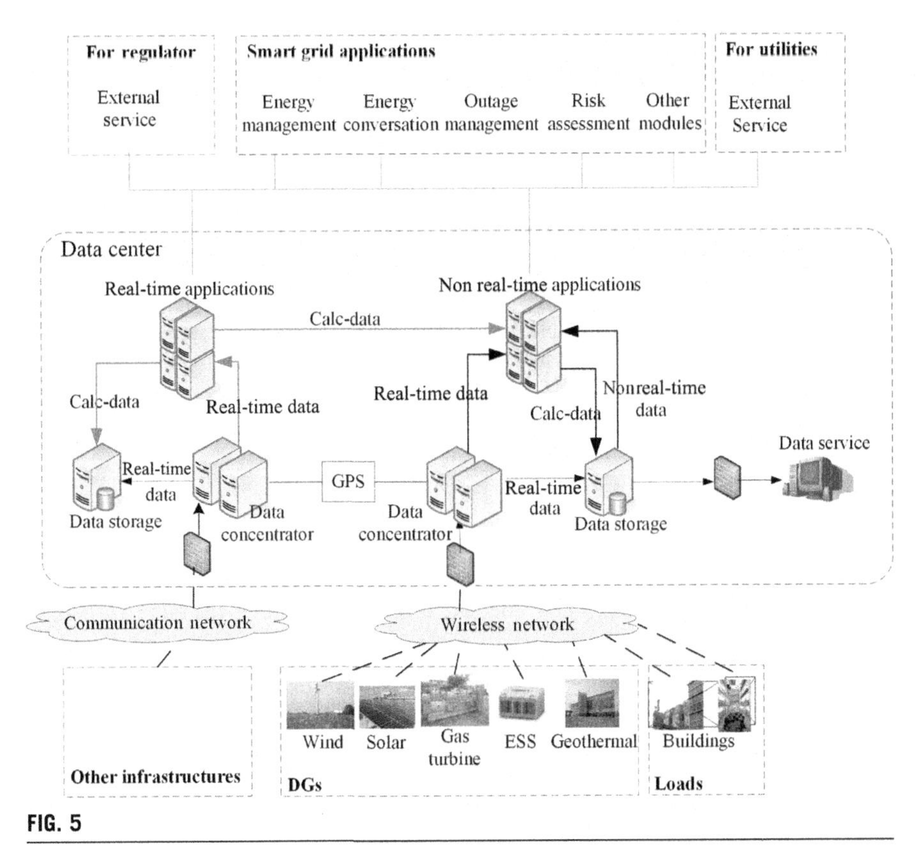

FIG. 5

Architecture II of the data center.

data for realtime application modules. In addition, the calculation/analysis results of realtime applications, the realtime data in RAM of data concentration, and the results of nonrealtime applications are sent to data storage to be saved at different time intervals.

(2) Database:
Most systems use commercial databases, such as Oracle and SQL server; however, these databases may have delays while taking out data from their tables. Thus, some systems use a realtime/historical database, which requires a higher cost. A realtime/historical database is a kind of computer software to conduct automatic acquisition, storage, and display realtime operating data. The use of a realtime database realizes realtime data access, data compression, data graphical display, alarm processing, management reporting, data analysis, data storage, data query, and any other requirement of realtime automatic monitoring systems.

(3) LAN:
LAN provides internal access among all equipment and information security isolation to ensure information safety.

(4) GPS:
GPS provides synchronized timing marks to all data.

(5) Advanced applications:
Advanced applications involve both realtime applications and nonrealtime applications. For example, an EMS and outage management are realtime applications, while energy conservation and risk management are nonrealtime applications. The calculation/analysis results from realtime applications can be used by nonrealtime applications.

(6) Human-machine interface (HMI):
HMI provides interaction interface among operators and the system. For example, a big screen can be used to display the status of any section and/or component in the system, such as DG's status, load level, communication channel status, and equipment profile.

16.4.2 FUNCTIONAL STRUCTURE OF DATA CENTERS

Fig. 6 shows the functional structure of the data center, which includes four sections: modeling, monitoring, analysis, and control. Modeling represents the initialization of the system, such as importing device parameters, building data tables, and networking connection relationships. Monitoring displays realtime states of objects (eg, node, branch, and metering data) in terms of certain key performance indicators. Analysis is a common function for network topology, state estimation, load forecasting, and power flow calculation. In addition, special analysis, such as system operation status identification, risk assessment, reliability, economical, and stability analysis can be included, as they need special algorithms. Making decisions according to states and criteria is also a key function of a data center.

16.4.3 DATA CENTER CAPACITIES

The design stage of a data center capacity includes many variables, such as the housing infrastructure, network feeds, storage, and hardware processing units. The design of data centers is dependent on the balance of two key sets of capacities:

(a) Data center capacities: power, cooling, physical space, weight load, bandwidth, and functional capacities.

(b) Equipment capacities: various devices that could populate the data center in various numbers.

The equipment requirements can be used to determine the size of the center (eg, amount of power, cooling, and cabling). In addition, functional capacity is included, such as physical memory, disk space, how many spindles, database transactions, and any other measures of rack functions.

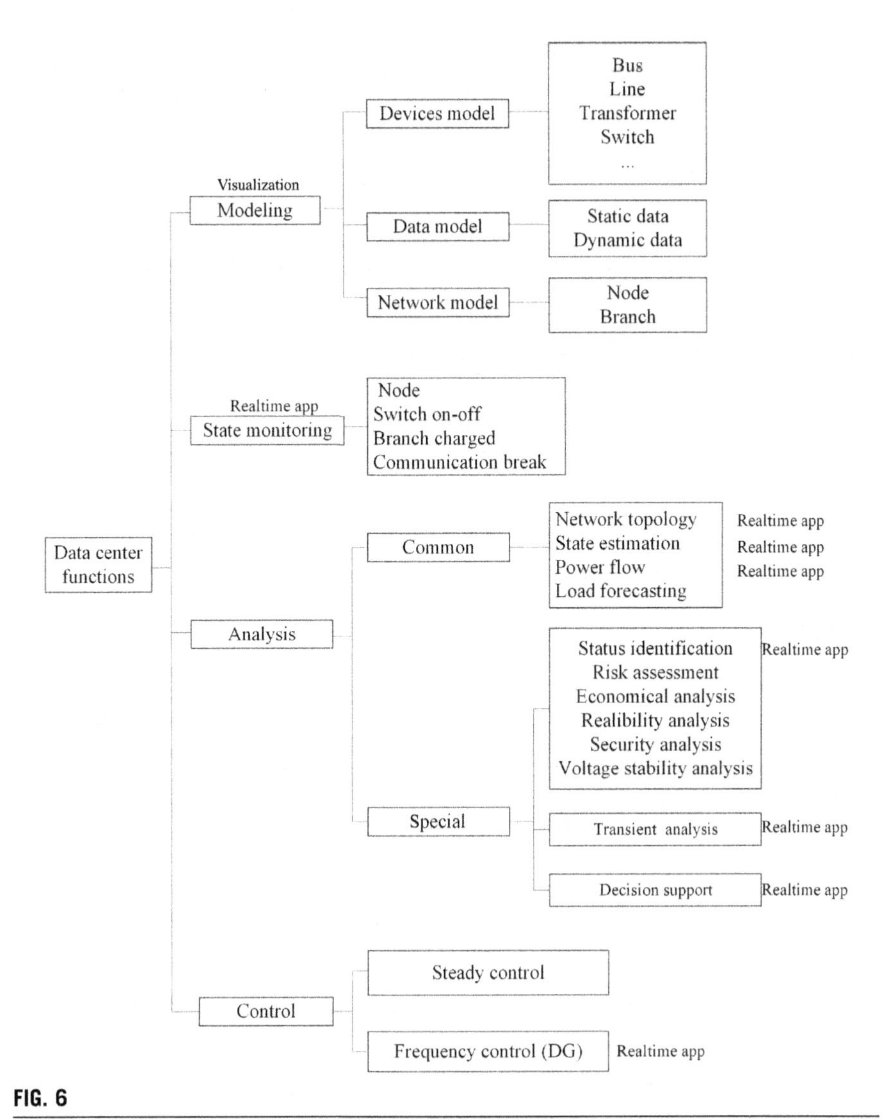

FIG. 6

Functional structure of data center.

16.5 DATA MANAGEMENT AND ANALYTICS FOR UTILITIES

Previously, the data collected by utilities from their customers were limited to a monthly meter reading (eg, one data point a month per customer). Recently, the advent of AMI and the increased sampling frequency have increased the level of data collection. Also, utilities integrate and handle highly granular data from other sources, such as pricing details for demand response and forecasting information for renewable energy.

16.5.1 BIG DATA SOLUTIONS

Big data needs new technologies and new approaches to efficiently process their large quantities within tolerable elapsed times. Traditional relational database technologies are inadequate in terms of response times when applied to large datasets. Thus, big data implementations are leveraging new technologies such as Hadoop as a framework for processing the massive data. Unlike traditional relational database technologies, new big data technologies should be: (1) scalable and easily distributed across multiple servers; (2) open-source technologies with low-cost implementation; (3) robust and reliable; (4) schema-free with easy replication support; and (5) support for structured, semistructured, and unstructured data. Deploying a big data solution ensures program goals are achievable and all components of the solution are scalable to meet current and future requirements.

16.5.2 OPTIONS FOR DATA MANAGEMENT

Utilities have two key issues about big data: where do they put the data and what do they do with it. Therefore, utilities are increasingly creating dedicated data centers. Generally, there are five options for dealing with utility data analytics and, in practice, utilities may use a mix within their data management strategies.

(1) Develop the systems in house:
This option allows each utility to address its challenges and requirements; however, some utilities may lack required development skills and resources to create data analytics systems from scratch. The realistic scenario is that each utility with a large IT capability carries out a certain amount of development and integration of its existing systems. Thus it reduces the need to buy in skills and products from outside.

(2) Buy systems from operational technology vendor:
Operational technology vendors, such as GE Energy and ABB, tackle data integration and provide data analytics systems through their products and technologies. These operational technology vendors are well versed in the issues facing utilities. Hence, they provide products that align with the industrial operational processes and requirements.

(3) Buy systems from IT vendor:
IT vendors, such as IBM, Microsoft, and Oracle, deliver robust design and support applications that are required for full smart grid data analytics; however, they may include unnecessary features that increase costs.

(4) Buy and integrate point products from specialist vendors:
Some components of the smart grid come with their own management systems and dashboards. Examples are eMeter, Itron, and Telvent, as they offer management tools for their products. Some of these products have software tools for smart grid analytics, demand response, customer care and billing, asset management, revenue assurance, and forecasting. These features may be sufficient for the requirements of the utility.

(5) Outsource data analytics to a third-party:
Third-party data analytics service providers take data from utilities and analyze them to provide value-added services to energy customers. Although this option requires data sharing with the third-party, it provides reduced costs and improved analytics.

16.5.3 DATA CENTER MANAGEMENT

Data center management is responsible for overseeing technical and IT issues within the data center. This includes computer and server operations, data entry, data security, data quality, management of services, and applications used for data processing. Data center management integrates into other IT systems for complete data synchronization, including virtual systems, proprietary systems, and automation. Data center management enables a common, realtime monitoring and management platform for all interdependent systems across IT and facility infrastructures.

16.5.4 COORDINATION OF DATA CENTERS UNDER MULTIREGIONAL ELECTRICITY MARKETS

Presently, there is a multielectricity market environment where the price of electricity may exhibit temporal and spatial diversities. In addition, the required cost for powering up and cooling Internet data centers is increasing significantly. To satisfy global user demand, data center providers have their data centers in multiple geographical locations. Each data center hosts a large number of servers with limited computing rate of each server due to its physical limitation. Fig. 7 illustrates the overall architecture of cyber-physical interactions for data centers under the multiregional electricity market environment. The aim is to adjust the on/off status of servers and to develop a dynamic frequency scaling scheme to achieve lower power consumption at each data center location and balance the workloads distributed among locations [21], while at the same time guaranteeing the quality of service experienced by end users.

16.6 DRIVERS AND BARRIERS

16.6.1 DRIVERS

The main driver for adopting data management strategies is the need to handle and analyze the huge amount of data utilities are now faced with. In addition, utilities want to enhance their processes through better analysis of data, such as:

- Business operation efficiency: data analytics will enable utilities to have better asset management and better system planning.

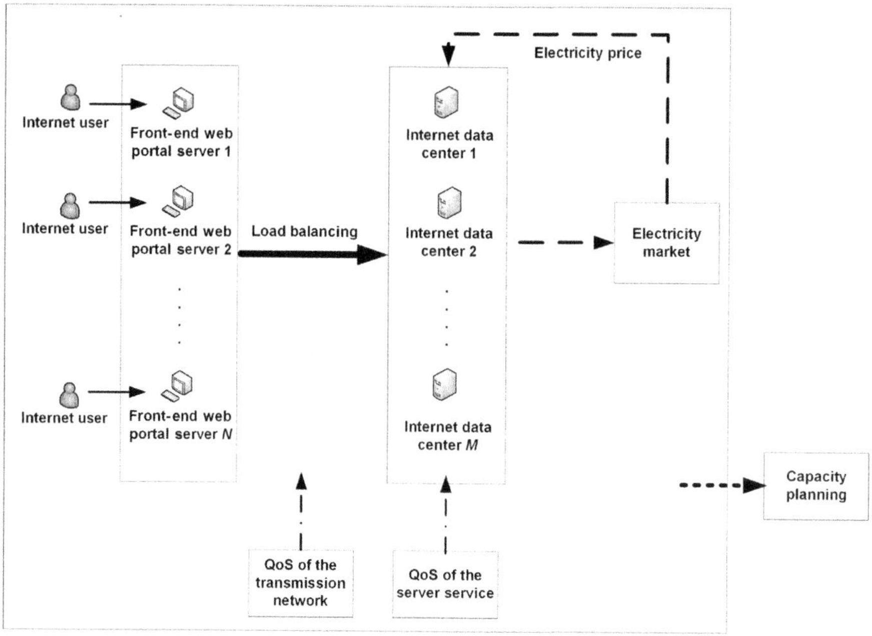

FIG. 7

Overall architecture of cyber-physical interaction for Internet data centers under the multiregional power market environment [21].

- Implementing efficiency measures: realtime data will facilitate ToU rates. Thus, customers can monitor their consumption and save money by shifting their energy use away from times when rates are higher.
- Developing new business models: adding intelligence to the grid operators helps in offering new services, such as energy management, high-speed Internet services, cable TV, and private network links.
- Improving grid resilience and load management: distribution companies will be able to reduce power demand and minimize outages. For example, better data will allow utilities to identify and rectify problems quickly.
- Engaging customers: having more information about customers and their usage patterns is important to deliver better and more tailored service.

16.6.2 BARRIERS

- There are challenges regarding data management around the data itself, such as: (1) how can data be used for decision making; (2) what data will be needed in the future; and (3) what data should be collected and analyzed.
- Budget: availability of cash is a key barrier for many utilities, especially with the lack of clear ROI models.

- Interoperability: as there will be new hardware and software requirements, many utilities are concerned about the capability of technologies to communicate effectively with each other.
- Regulation: the regulatory environment has an effect on data analysis by determining the potential use and value that utilities can derive from data in the first place.
- Data acquisition: because of the large amounts of data flowing in from AMI, data acquisition may represent a barrier for some utilities.
- Human resources: finding talented IT experts is hard at times.
- Market structure: there is likely to be much less incentive for utilities to invest heavily in data management technologies, as it might not be entirely relevant to their business.
- Planning uncertainty: uncertainty may curtail a utility's appetite for investments in data infrastructure.
- Technological complexity: implementing smart grids requires getting a whole new set of technologies.

REFERENCES

[1] Available: http://www.ieee802.org/3/hssg/public/nov06/diminico_01_1106.pdf [accessed 09.03.16].
[2] Uptime Institute Professional Services. Data center site infrastructure tier standard: topology. Available: http://www.gpxglobal.net/wp-content/uploads/2012/08/tierstandardtopology.pdf [accessed 09.03.16].
[3] Arno R, Friedl A, Gross P, Schuerger RJ. Reliability of data centers by tier classification. IEEE Trans Ind Appl 2012;48(2):777–83.
[4] Asad Z, Chaudhry MAR. A two-way street: green big data processing for a greener smart grid. IEEE Syst J 2016;PP(99):1–11.
[5] Chelmis C, Kolte J, Prasanna VK. Big data analytics for demand response: clustering over space and time. In: IEEE international conference on big data, Santa Clara, CA; 2015. p. 2223–32.
[6] Kwac J, Rajagopal R. Demand response targeting using big data analytics, In: IEEE international conference on big data, Silicon Valley, CA; 2013. p. 683–90.
[7] Zidan A, Khairalla M, Abdrabou AM, Khalifa T, Shaban K, Abdrabou A, et al. Fault detection, isolation, and service restoration in distribution systems: state-of-the-art and future trends. IEEE Trans Smart Grid 2016;PP(99):1–16.
[8] Zidan A, El-Saadany EF. A cooperative multiagent framework for self-healing mechanisms in distribution systems. IEEE Trans Smart Grid 2012;3(3):1525–39.
[9] EPRI. Estimating the costs and benefits of the smart grid: a preliminary estimate of the investment requirements and the resultant benefits of a fully functioning smart grid. Palo Alto, CA, USA: Department of Power Delivery & Utilization, Electric Power Research Institute; 2011. Technical Report: 1022519.
[10] Media G. The soft grid 2013–2020: big data & utility analytics for smart grid, GTM Research, Technical Report. 2013.
[11] Apache Hadoop. Available: http://hadoop.apache.org/ [accessed 15.03.16].

[12] Apache Storm. Available: http://storm.apache.org/ [accessed 15.03.16].
[13] Erol-Kantarci M, Mouftah HT. Energy-efficient information and communication infrastructures in the smart grid: a survey on interactions and open issues. IEEE Commun Surv Tutor 2015;17(1):179–97.
[14] Ho Q-D, Le-Ngoc T. Smart grid communications networks: wireless technologies, protocols, issues, and standards. In: Handbook of green information and communication systems. Waltham: Academic Press; 2013. p. 115–46 [Chapter 5].
[15] Hrasnica H, Haidine A, Lehnert R. Broadband powerline communications: network design. Hoboken, NJ: Wiley; 2004.
[16] Çavdar D, Alagoz F. A survey of research on greening data centers. In: IEEE global communications conference, Anaheim, CA; 2012. p. 3237–42.
[17] Liu J, Zhao F, Liu X, He W. Challenges towards elastic power management in internet data centers. In: 29th IEEE international conference on distributed computing systems workshops, Montreal, QC; 2009. p. 65–72.
[18] Rao L, Liu X, Xie L, Pang Z. Hedging against uncertainty: a tale of internet data center operations under smart grid environment. IEEE Trans Smart Grid 2011;2(3):555–63.
[19] Nguyen KK, Cheriet M, Lemay M, Savoie M, Ho B. Powering a data center network via renewable energy: a green testbed. IEEE Internet Comput 2013;17(1):40–9.
[20] Zhang Y, et al. Wide-area frequency monitoring network (FNET) architecture and applications. IEEE Trans Smart Grid 2010;1(2):159–67.
[21] Rao L, Liu X, Ilic MD, Liu J. Distributed coordination of internet data centers under multiregional electricity markets. Proc IEEE 2012;100(1):269–82.

CHAPTER

End-to-end-authentication in smart grid control 17

K.C. Ruland, J. Sassmannshausen
University of Siegen, Siegen, Germany

17.1 INTRODUCTION

The main features of the current energy supply are centralized deployment of energy by powerful power plants, a high-voltage grid through which the energy is distributed to substations, and centralized control of the system. The role of the consumer is passive; it is limited to consuming energy.

The trend toward decentralized energy supply changes both the structure and requirements of power supply systems. There are now a number of smaller and partially inhomogeneous electric generators, such as wind turbines or photovoltaic systems, from where the generated power is fed directly into the low-voltage networks. All these components must be optimally integrated into the power network so that, on the one hand, the power supply is guaranteed, but on the other hand, the power is equally balanced. The control of these systems is more difficult than the control of centralized systems.

A smart grid faces this challenge by automating various processes and enabling communication between its components, for example, consumers, substations, power generators, and energy storage units. Furthermore, fluctuating prices, which are dependent on the current supply and demand on the market, get integrated into the system. This enables intelligent, environmentally friendly, and cost-saving power consumption.

New requirements for the security and reliability of the system result from the decentralized and automated control of networks, which are part of the nation's critical infrastructure.

Fig. 1 shows a schematic illustration of the topology of a smart grid and the flow of energy and data within the system.

As can be seen, there are still formidable power plants and high-voltage distribution networks forming the backbone of the system, but in the base, there are both power generators and consumers. The electric flow on this level can also be bidirectional. For example, storing units can be both charged and discharged, one time-consuming energy, the other time feeding the energy back into the network.

Smart Energy Grid Engineering.

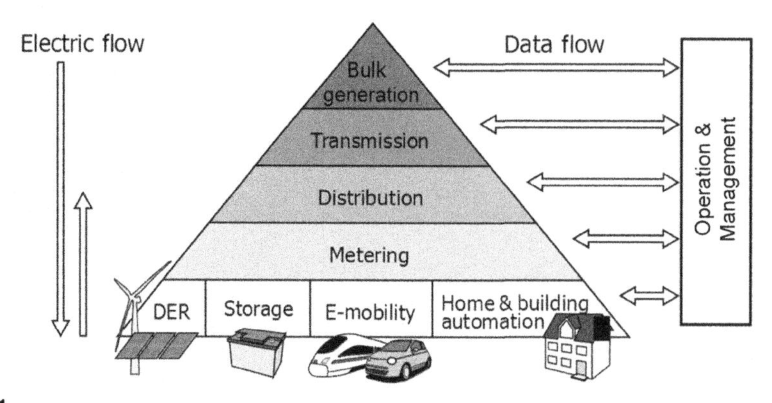

FIG. 1

A simple illustration of a smart grid.

In this contribution, a security solution for the communication between the components of the system is presented. The focus is set on a subarea of smart grids only, namely the automation of substations using mechanisms and protocols described in the standard IEC 61850; however, the field of application of the protocols and the security solution can be extended to cover additional parts of the smart grid. This chapter discusses mechanisms that prevent unauthorized access to resources and make manipulation of data transmission impossible. With the use of digital signatures, the proof of authorship is achieved for control messages that are both sent and received.

Since all messages are logged after execution and reception, they will be accessible later on and can be used to retrace and reconstruct all actions that took place in the system.

17.2 STATE OF THE ART

There are many standards and nonstandard documents covering different aspects of smart grids, each dealing with different areas. This leads to the question of how standards can be classified, so they can be compared regarding their scope and limitations. For this purpose, a model of a smart grid, in which the standards can be classified, is required. Models of smart grids can be used to visualize the scope of existing standards and show gaps between standards where additional standardization is required. The pyramid model used in this chapter is similar to the "smart grid architecture model" described in Ref. [1].

This contribution focuses on two key standards for smart grids: standards IEC 61850 [2] and IEC 62351 [3]. While IEC 61850 deals with the control of substations in electric power networks, the scope of IEC 62351 is security in smart grids; however, the focus is set on security for communication protocols specified in IEC

61850. In addition to these standards, there are more important documents that deal with security in smart grids. In this section, two documents shall be mentioned: "NISTIR 7628 Guidelines for Smart Grid Cyber Security" [4] from the National Institute of Standards and Technology and "Smart Grid Information Security" [5] from European standardization committees. The security solution that will be presented later is based on these standards and documents.

17.2.1 THE STANDARD IEC 61850

The standard IEC 61850 focuses on the automation of substations and introduces rules for the modeling of the system, standardized data models, various communication protocols for different requirements and applications, communication interfaces, and a configuration language for the description of the system's components. The data transfer takes place via ethernet using different protocols for different purposes. For example, a Transmission Control Protocol/Internet Protocol (TCP/IP)-based protocol is used for the communication between clients and server. In this chapter, the focus will be placed on the communication between client and server using the manufacturing message specification (MMS) protocol. Extensions of the communication structures are possible. Fig. 2 again shows the topology of a smart grid. In this illustration, the "operation and management" unit is shown a bit more in detail. The parts covered by IEC 61850 are highlighted to visualize the scope of this standard.

Although the scope of IEC 61850 is the automation of substations, there are now new use cases like the integration of distributed energy resources (DERs) into the electric power system or the communication between intelligent electronic devices

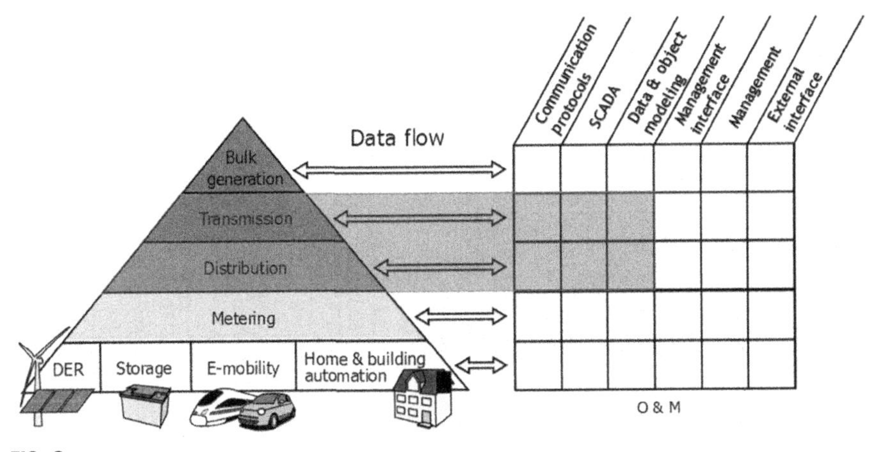

FIG. 2

The visualized scope of the standard IEC 61850. Supervisory control and data acquisition (SCADA) is a system used for monitoring and controlling processes. The areas "communication protocols," "SCADA," and "data and object modeling" are equal to the levels "process," "field," and "station," used in the previously mentioned SGAM.

(IEDs) within buildings that can also be addressed by IEC 61850 [6]. Due to its flexibility, the standard IEC 61850 can be used for wide-area communication beyond the substation.

This flexibility is enabled by standardized data models and services that can be mapped to various protocols. Currently, the standard only defines a mapping to the MMS protocol, but there are other mappings planned, for example, a mapping to web services [7]. With these new use cases beyond substations, there are new requirements for the communication structure, such as scalability or increased privacy. Fig. 3 shows both the initial and the possible future scope of IEC 61850 [8].

(1) *The structure of the system*: the standard introduces the so-called logical nodes, which represent certain functionalities. A logical node can represent, for example, switches in the grid, sensors, communication interfaces, or it can simply contain descriptions of devices. Logical nodes can be summarized to logical devices that represent a real physical device and its functionality.

(2) *The abstract communication service interface (ACSI)*: the ACSI is defined in IEC 61850-7-2 and describes services and functions that can be used to interact with devices. For example, there are services to read or write values, to obtain information about the data model, or transmit files. How the functionality finally is implemented is not specified in the definition of the ACSI. There are various communication protocols to which a mapping of the ACSI can be implemented; the standard IEC 61850-8-1 defines a mapping to MMS. A mapping of the ACSI to web services is also planned, but not published yet [7].

(3) *The MMS protocol*: as mentioned before, the standard IEC 61850 stipulates the use of different protocols for different purposes in substations. There are protocols for the fast transmission of event data, and more complex protocols for the communication between servers and clients. Fig. 4 gives an overview of

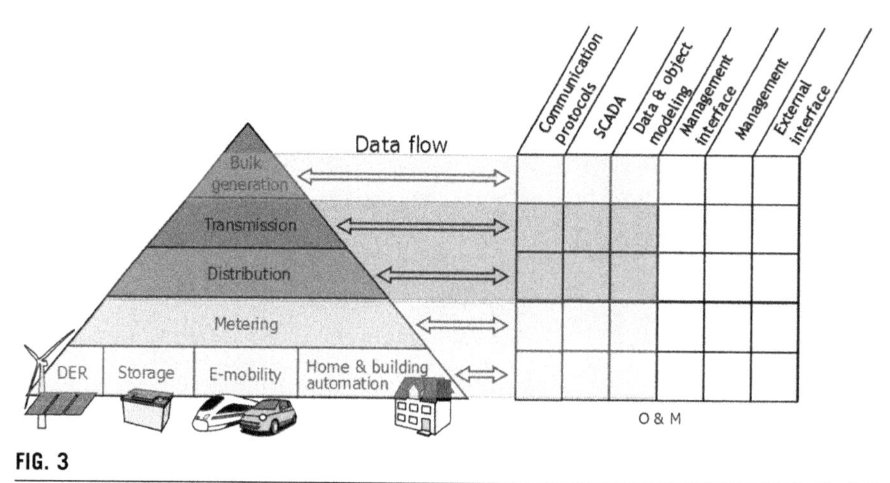

FIG. 3

The areas of a smart grid that could also be covered by IEC 61850.

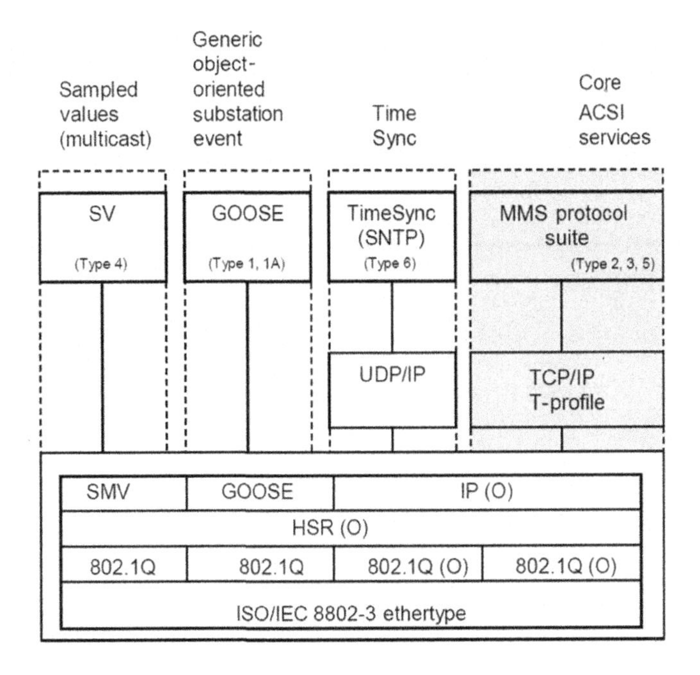

FIG. 4

The MMS protocol among other communication protocols for different purposes. The types of the protocols specify the use of this protocol.

communication protocols specified by the standard IEC 61850. The MMS protocol, which is used for client-server communication and also is the main aspect of this contribution, is highlighted.

The MMS protocol was originally developed by General Motors for the purpose of communication in automated systems in the manufacturing section. MMS is defined in ISO 9506. The MMS protocol is based on the Open System Interconnection protocol stack with association control service element (ACSE) and the Session Control Protocol ISO 8326, which itself is based on TCP/IP [9]. MMS provides a set of functions that allows the client to obtain the data model of the server, read or modify individual values, or even delete entries. In addition, MMS also offers functionality to transfer files. To perform an action, such as changing or receiving a given value, the client sends a request to the server, which processes the request after that, and sends a response back to the client, stating whether the request was processed successfully [10].

The MMS protocol is not only used by IEC 61850 for communication in substations. There are several more areas of application for the MMS protocol in smart grids. The Standard IEC 61400-25 sets the focus on communication and monitoring in wind power plants and also intends the use of the MMS protocol for communication between client and server [11]. Another standard associated with the MMS protocol is IEC 60870-6 describing the Telecontrol Application Service Element 2

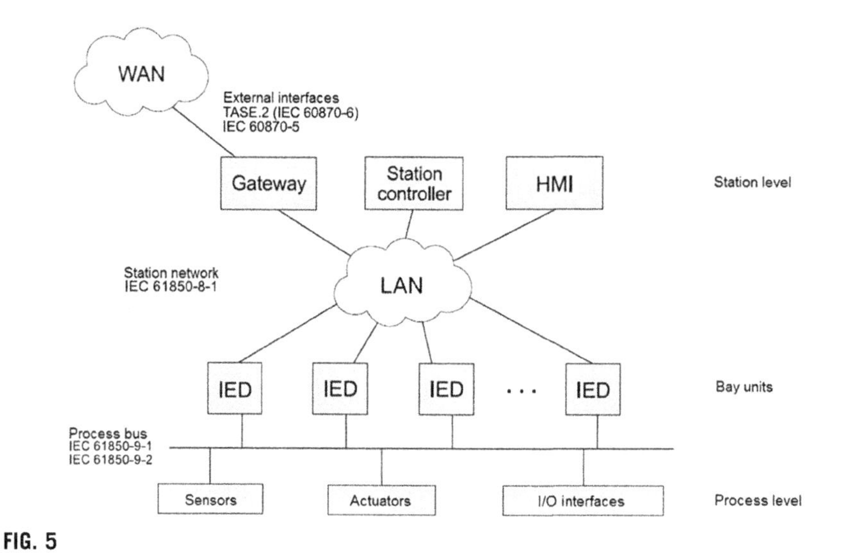

FIG. 5

The network topology of a substation and the standards with their fields of application.

(TASE.2), which is also known as ICCP (inter-control center communication). This standard deals with the communication over wide-area networks among different entities of the smart grid, for example, between the control center and a substation. The wide use of MMS in smart grids requires security measures to protect the communication via MMS. The standard IEC 62351-4 introduces different security measures for the MMS protocols that are going to be presented in part III-A of this contribution.

Fig. 5 shows a typical part of a smart grid, a substation according to IEC 61850.

The figure shows the structure of a substation as it is described in part 5 of IEC 61850. Several networks, each connecting different types of devices, may be seen. At the process bus, which is directly attached to sensors/actuators, low-level protocols like GOOSE/GSSE or streams of raw data are used for data transmission. The protocols used in this area make high demands on the performance since the IEDs are connected directly with sensors and actuators and protection mechanisms are realized on this level. With high performance requirements, for example, low latency and round-trip times, security measures are difficult to realize, since the performance will always be influenced. On the station level, the standard IEC 61850 intends the use of MMS for the communication between the devices. The devices on that level are connected via a local area network, and communication takes place via point-to-point connections. Also, the performance requirements in the station network are not as high as on the process bus. In addition, the transmitted MMS messages are usually significantly longer than GOOSE/GSSE messages, which are kept as short as possible to achieve minimal transmission times. See also IEC 61850-5 [12] and IEC 62351-1 [13] for more details. In Fig. 5, one can also see that there are external

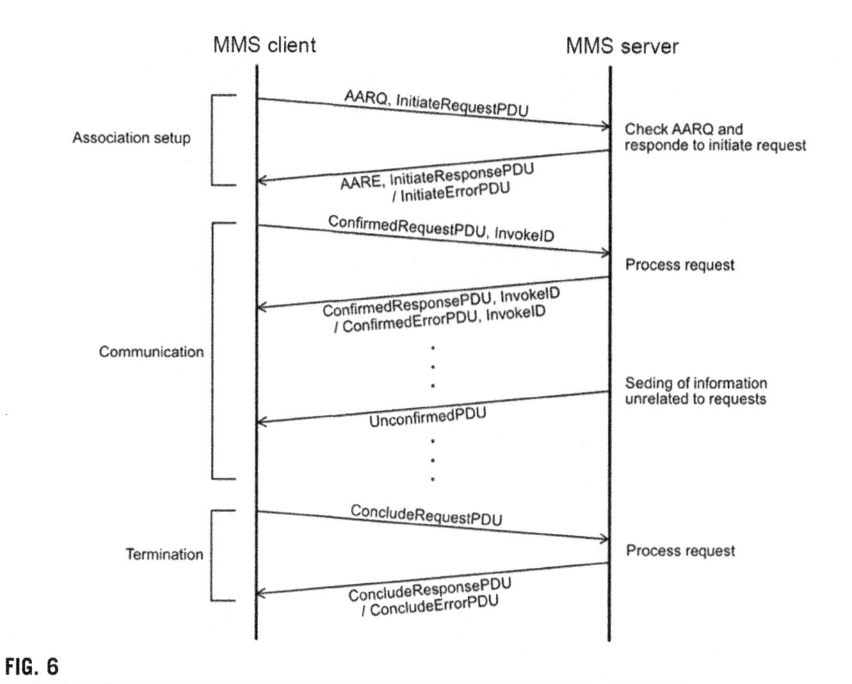

FIG. 6

The communication between client and server and the exchanged PDUs.

connections beyond the substation for wide-area communication. With ICCP/TASE.2, this is also an application field of the MMS protocol. Due to the lower requirements to the performance on the station level and beyond the station, the MMS protocol can be secured with encryption and further measures, providing authenticity of the exchanged data without too many restrictions. Since this contribution sets the main focus on security for the MMS protocol, the communication via MMS shall be explained more in detail. Fig. 6 illustrates the communication between a client and the MMS server.

The association between client and server is set up using ACSE (a detailed description about ACSE can be found in part III-A of this contribution) and the initialization request. Once the association is set up, the client and the server are able to exchange MMS-PDUs (protocol data units). There are two types of services: The so-called confirmed services, consisting of a request to the server, followed by a response of the same, are used by the clients to communicate with the server. Every request contains an "invoke ID," which is unique for an association, and indicates corresponding request/response PDUs. Besides the confirmed services, there are the so-called unconfirmed services, consisting of only one PDU, sent without a response. An example for an unconfirmed service is status information periodically sent to the clients by the server. The structure of the MMS PDUs is defined by using the abstract syntax notation one (ASN.1) syntax and can be looked up in Ref. [14]. According to their definition, MMS-PDUs do not contain any security-relevant information, like names of the entities or message authentication codes (MACs)/signatures. The association between client and server will be released with a conclude request sent to the server by the client. The server can reject the conclude request, for example, if there are still some active operations that have to be finished first. For a more detailed description, see Refs. [9,14].

17.2.2 THE STANDARD IEC 62351

The standard IEC 62351 deals with security aspects in smart grids; there are existing security requirements discussed that have to be met where possible. These are particularly integrity, confidentiality, availability, and nonrepudiation. Furthermore, the standard presents a number of measures, by which these objectives are to be achieved. The recommended security measures vary according to the treated protocol. For example, for time-critical systems, there shall be no time-consuming encryption of the messages, but rather the focus shall be put on data integrity and authenticity instead.

Two profiles are provided: the T-profile, which provides a transport protocol-based security based on Secure Socket Layer (SSL)/Transport Layer Security (TLS); and the A-profile, which provides an application-oriented security for MMS. Fig. 7 shows the difference between the T-profile and A-profile.

In the T-profile, the focus is set on confidentiality using encryption. The T-profile cannot meet security requirements such as nonrepudiation and traceability of transactions. These security mechanisms must be implemented in the A-profile. In general, the communication between the communication partners shall be encrypted

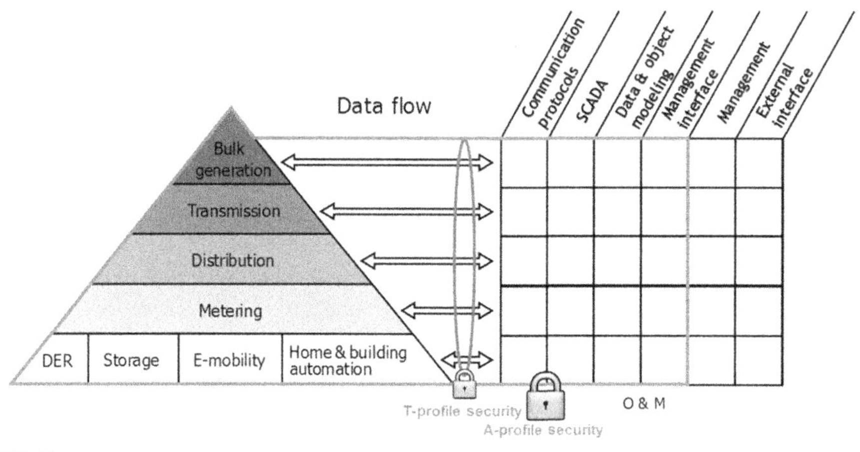

FIG. 7

The T-profile and the A-profile in the smart grid. While security for the T-profile only affects the transmission of the data, the A-profile security is implemented at the application level and affects the whole system.

(using the underlying T-profile), and X.509 certificates shall be used for the identities of the participants. In addition, a system for logging every action and detecting attackers (intrusion detection) is recommended [13]. Although an authentication of the participants takes place when associations are established via ACSE [15], the authenticity of the exchanged MMS protocol elements is not guaranteed. This is precisely the weak point the following security solution treats.

17.2.3 NISTIR 7628 GUIDELINES FOR SMART GRID CYBER SECURITY

Just as the standard IEC 62351, the document "NISTIR 7628 Guidelines for Smart Grid Cyber Security" (which is not a standard) treats with security aspects in smart grids; however, these two documents do not compete. Instead, the focus is set on different areas: While IEC 62351 deals with technical aspects in detail, specifying concrete measures that shall be implemented, the "NISTIR 7628 Guidelines for Smart Grid Cyber Security" are more complete, but leaving out details of the implementation. The document consists of three parts:

(1) *Volume 1: Smart Grid Cyber Security Strategy, Architecture, and High-Level Requirements*: This part contains background information about smart grids and points out the importance of cyber security within the smart grid. It also provides a logical model of the smart grid and discusses multiple high-level security requirements for different parts of the grid.
(2) *Volume 2: Privacy and the Smart Grid*: As the title suggests, this part is about privacy in a smart grid and further about privacy issues that can possibly occur.
(3) *Volume 3: Supportive Analyses and References*: The last part deals with security analysis of smart grids, including describing and classifying various vulnerabilities. Further, a security analysis of the smart grid is presented, and existing security problems are pointed out.

17.2.4 SMART GRID INFORMATION SECURITY (CEN, CENELEC, ETSI)

The document "Smart Grid Information Security" is released by the European Smart Grid Coordination Group and deals (among other aspects) with security in smart grids. The chapter was written under the M/490 Standardization Mandate to European Standardization Organizations with the objective to "enable or facilitate the implementation in Europe of [...] smart grid services and functionalities" [16].

17.3 ADDITIONAL SECURITY REQUIREMENTS

As a summary of the guidelines mentioned before and the standard IEC 62351, five important security requirements can be pointed out:

- Confidentiality
- Integrity

- Availability
- Nonrepudiation of actions
- Traceability of actions

The implementation of the measures described in IEC 62351, in essence authentication at the setup of an association and encryption at the transport level, certainly increases the security level of the system; nevertheless, there is the question of whether these measures alone will suffice to meet all safety requirements and whether the system can still be improved.

In smart grids, it is important that all transactions that took place and status messages that were sent can be traced back and attributed at a later time. For example in case of damage caused by a switching error or incorrect settings, it can be determined when and especially by whom certain actions that led to the error or damage were performed [15,17]. The main precondition for traceability is nonrepudiation of actions that took place. In this section, the focus shall be set on end-to-end-authentication, nonrepudiation and traceability of actions, because there are still security issues regarding these two points

17.3.1 THE WEAK POINT OF IEC 62351

In the previous section, security measures for end-to-end communication protocols described in IEC 62351 were presented: the T-profile, providing confidentiality using encryption on the transport layer; and the A-profile, providing authentication on the application layer. Based on the specifications in IEC 62351, an authentication of the communication partners is only performed at the connection establishment using the ACSE. Part 4 of the standard IEC 62351 describes the authentication of the entities (the client and the server) via ACSE. At the association setup, the client sends an authentication request (AARQ) PDU, which contains an "authentication value" generated by the client to authenticate itself at the server. The server then checks the authentication value received with the AARQ and responds with an authentication response (AARE) PDU. This response PDU contains an authentication value that was generated by the server. The client then checks the authentication value of the AARE. If both entities are authenticated successfully, the association between client and server is established; however, the construction of the authentication value is a weak point of the authentication via ACSE. According to IEC 62351-4, the authentication values for both the AARQ and the AARE PDUs consist of three parts:

- *Signature certificate*: the certificate of the entity that wants to authenticate itself; the signature can be verified with this certificate
- *Time*: the timestamp at the time the authentication value was generated; according to the standard, the timestamp shall as accurate as possible
- *Signature*: according to IEC 62351-4, the signature is calculated only over the value "time" of the authentication value

Before verifying the signature, the recipient of the authentication value must check the certificate using systems like a public key infrastructure or a database of trustworthy certificates. Also, the recipient must check the time value. The standards state that all authentication values older than 10 min shall be discarded. The fact that only the element time is used to generate the signature leads to the problem that an attacker could copy an eavesdropped authentication value to authenticate itself at the server or the client. The standard also recognizes this vulnerability and states that all received authentication values have to be stored. This way duplicates can be detected and discarded (also see [15], Part 5.3). There are, however, still security issues: an attacker can still copy and use an eavesdropped authentication value, for example, an authentication value that was sent by the server to pretend to be the server to other clients in the network. As it can be seen, the authentication only using ACSE is not sufficient, and it is necessary to exchange further security parameters to achieve authentication of the entities of the association.

How these security parameters are exchanged is up to the application: "The method by which these parameters are passed is a local issue" [15]. Also, see Ref. [15], Parts 5.3.3 and 5.3.2 for more details. As a conclusion, it can be said that the authentication at the setup of the association is not sufficient and also not completely secure. After it is performed, the security only depends on the security measures realized on the transport layer (the T-profile).

The standard IEC 62351 intends the use of TLS on the transport layer to secure the communication and achieve confidentiality, integrity, and authenticity of the data exchanged. During the establishment of a TLS connection between two entities, an authentication of both sides is performed and also a common secret is negotiated. From this secret, several keys are derived; there are different keys used for encryption and MAC calculation. The TLS protocol allows the use of MACs to achieve authenticity of the data exchanged. The standard IEC 62351-3 intends the use of MACs to secure the data exchange: "TLS has this capability specified as an option. This standard mandates the use of this capability to aid in countering and detection of man-in-the-middle attacks" [18]. Both the encryption and calculation of MACs are based on symmetric cryptographic algorithms; the authenticity of a particular data packet cannot be proven as it would be possible with digital signatures. Instead, with the use of MACs, the communication participants can only be sure that the data received was sent by the same party the initial key exchange (which included an authentication using certificates) was performed with.

This does not, however, provide end-to-end authentication, since the security measures on the T-profile only affect single connections between single entities. Consider the following setup: two instances, the control center and the substation controller, are permitted to issue commands on the device. They both establish an MMS association and communicate with the MMS server. In this case, the authenticity of the data packets exchanged is ensured as described before; however, the authorship of a logged data packet could not be proven later, because the data packets are not provided by digital signatures by the sender. Fig. 8 illustrates this scenario.

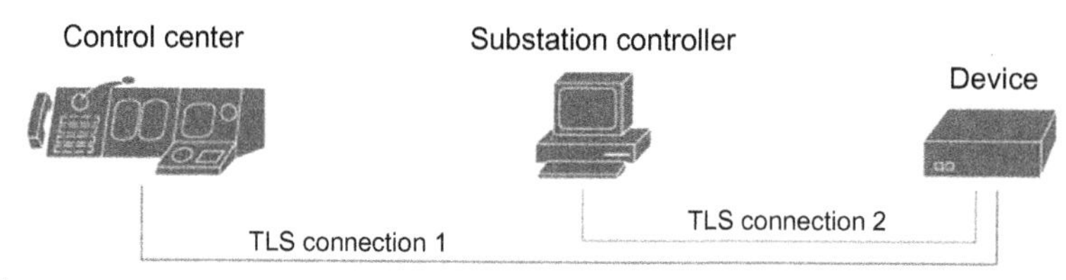

FIG. 8

Only two participants per connection; this system is not very scalable.

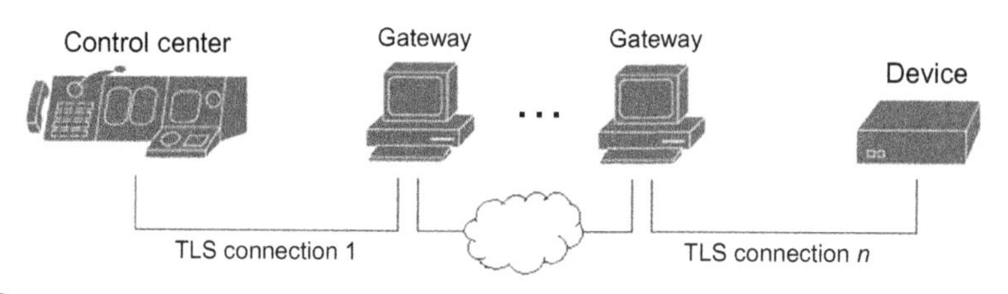

FIG. 9

A more scalable solution: data packets from the control center are forwarded using the connection of the substation controller; the device cannot track back the origin of a data packet.

In a different topology (as is shown in Fig. 9) where the communication takes place across several hops, and data packets are forwarded and reencrypted, the authenticity of origin of a data packet cannot be verified at the end-device and the issued request cannot be traced back later [8].

These security issues lead to the conclusion that the A-profile has to be extended to suffice the initially mentioned security requirements. To verify the authenticity of every data packet, a signature has to be added to every single PDU; however, the ASN.1 definition of the MMS-PDUs does not intend the use of signatures for all data packets. Nevertheless, a signature can be added to every PDU that is sent. A possible way is described in section V as part of an implementation of the security solution.

Another measure for additional security is to restrict actions by server-side access control lists. The "NISTIR 7628 Guidelines for Smart Grid Cyber Security" intends the use of a role-based access control. The proof of identity is based on digital signatures and is performed for each data packet. These measures (basically logging, access control lists, and digital signatures) can be implemented together very well. In the following, the solution will be explained and an implementation of the solution that can be integrated into existing systems will be presented.

17.4 THE SECURITY SOLUTION FOR THE A-PROFILE

The main goal of this security solution is to authenticate the origin of every data packet that is sent and received and further a certain level of nonrepudiation of actions that took place within the system. For this purpose, the data packets are provided with digital signatures by the sender that can later be verified by the recipient using the corresponding certificates; however, only appending digital signatures to every data packet is not sufficient to meet this security requirements. All data packets must be provided with timestamps to detect and prevent replay attacks. It is also mandatory to provide data packets with the distinguished name of the addressee to prevent possible attackers from forwarding eavesdropped data packets to different instances. Also, the name of the sender must be stored together with the signature. Following all extensions, which must be added to the existing system, are listed:

- Every participant of the system must be provided with an identity, which can also be verified using certificates. This may already be given, since the authentication of ACSE and TLS makes use of digital signatures and certificates, too.
- All outgoing data packets must be provided with additional information about the sender and addressee of the particular data packet and also a timestamp and a digital signature delivering proof of origin of the transmitted data.
- Not only the client but also the server has to sign the packets after providing them with the additional information.
- The signature and additional information of all received data packets must be verified. Besides verifying the signature, it must also check whether the data was delivered to the correct destination.
- Before the execution of actions, it is necessary to check whether the client has the required permission using access control lists.
- Both server and client have to log not only the signed data packets received but also the packets sent.
- The log files including the signatures must be accessible by authorized instances.

The extensions are integrated into the system such that the functionalities of both the server and the client do not need to be changed. The functionality of the security solution is implemented as a separate security layer, which is transparent from the view of the server or client. This security layer will be integrated into the MMS protocol stack as an "intermediate layer" (sublayer principle). Figs. 10 and 11 show the principle of the interlayer.

17.4.1 THE NAMES OF THE ENTITIES

In many cases, it is necessary to identify an entity of the system with a distinguished name that is globally unique. This can be, for example, to identify the addressee of a data packet or the author of a data packet so the right certificate for the verification can be selected. This security solution uses certificates that are specified in the standard X.509 [19] (X.509-certificates). Every certificate stores a distinguished name of the issuer of the certificate (ie, the certification authority that issued the specific certificate) in an element called *issuer*. Furthermore, the distinguished name of the certificate's owner is stored within the certificate in an element called *subject*. The structure and construction of the distinguished names is specified by the standard

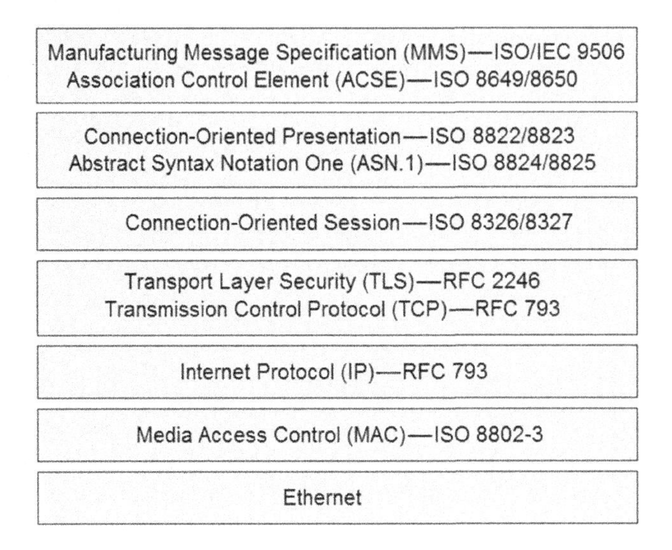

FIG. 10

The nonmodified MMS stack (with encryption at the transport level).

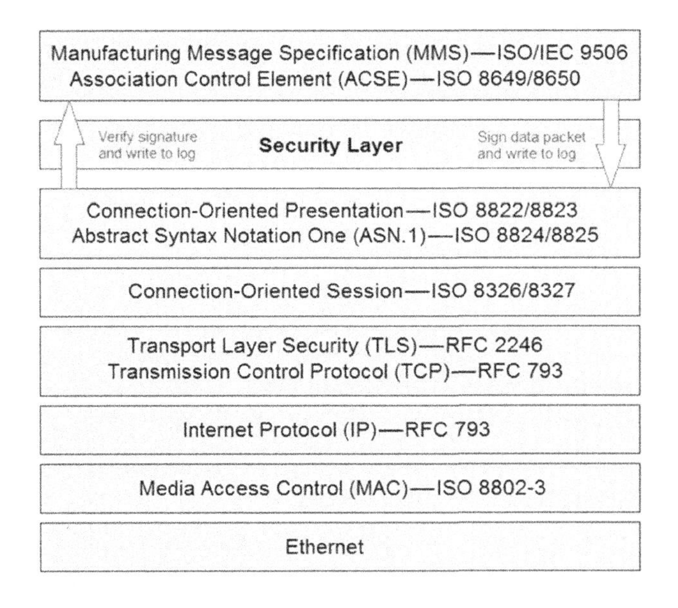

FIG. 11

The modified MMS stack with the added security layer.

X.501 [20]. Under the assumption that every certification authority has its globally unique name and does not issue two certificates with the same element *subject* to two different entities, the concatenation of the elements *subject* and *issuer* is globally unique. This security solution uses this concatenation to generate distinguished names for the systems participants.

17.4.2 SIGNING OF THE DATA PACKETS

The intermediate layer forwards outgoing packets to the lower layers after providing them with a digital signature. For this application, a signature algorithm based on elliptic curve cryptography, like ECDSA will be used. Before the data packet is signed, a timestamp has to be added to prevent replay attacks or to get the correct time of the logged events later. Also, further information about the addressee of the particular data packet has to be added before signing to prevent forwarding of eavesdropped data packets to other participants of the system. Finally, the name of the sender has to be stored in the data packets to enable the recipient of the data packet to select the correct certificate for the validation of the signature. Both the names for addressee and sender shall be the distinguished names of the respective certificates, since this way it is straightforward to find the certificate for the verification of the signature. Also, the names are unique within the system.

17.4.3 THE VERIFICATION OF SIGNATURES

The signature of received data packets can be verified using the corresponding public key retrieved from the appropriate certificate. The certificate can be found using the distinguished name of the issuer that is stored as a data packet. As the number of authorized clients and their identities may be known in advance, the certificate management should not be a problem; the certificates and public keys can be placed in a secure way on the server.

17.4.4 CHECKING THE ADDITIONAL INFORMATION

The checking of the additional information, in essence timestamp and addressee, is comparatively easy. The element containing the addressee must match the recipient's own distinguished name. Also, the timestamp must be checked and recorded. All packets with an older timestamp than the last timestamp received (in the same association) will be discarded to counter replay attacks. Due to the fact that the communication bases on TCP, the right order of the packets can be ensured. When working with timestamps, it is necessary that the clocks of all entities are synchronized and cannot be manipulated. The synchronization of the clocks is out of scope of this chapter.

17.4.5 ACCESS CONTROL LISTS

In addition to verifying the signature, server-side access control lists can be used to classify clients not only as trustworthy/nontrustworthy, but to realize further authorization levels. For example, some clients can be granted full access, whereas other clients are only permitted to read certain values. This security solution also intends the use of access control lists. As described by the "NISTIR 7628 Guidelines for Smart Grid Cyber Security," the implementation uses a role-based access control system in which a role is an accumulation of several rights that are granted to the holder of the particular role. Instances can hold one or more roles.

17.4.6 LOGGING OF EVENTS

The system for the logging is relatively simple: both the incoming as well as the outgoing data packets need to be written to a log file. Due to the use of digital signatures, the authorship for the data packet can be traced back without ambiguity, and since a timestamp was added before signing the packet, the moment the packet was sent can be determined precisely. The servers response should also be logged, ensuring it can later be determined if the corresponding request was processed successfully, that is, whether the request had any effect.

Along with the logging system, an "intrusion detection system" that monitors all failed connection attempts caused by untrusted clients should be implemented; however, an intrusion detection system is out of scope of this contribution.

In the implementation of the system, it has to be ensured that the log files are stored in a protected memory where the files are safe from unauthorized access or being overwritten.

17.4.7 NONREPUDIATION

One of the security requirements was nonrepudiation of actions that took place. The standard ISO 13888-3 deals with nonrepudiation using asymmetric techniques, that is, digital signatures and certificates. This standard introduces so-called tokens that deliver proof of authorship or delivery for messages and are detached from the message itself. Although the approach with the tokens is a bit different, the tokens introduced by ISO 13888-3 and the extended PDUs from this solution are very similar, since the same information is stored: the sender's distinguished name, the addressee's distinguished name, a timestamp, and the digital signature calculated using a specific algorithm and the hash value of the data that shall be signed. According to ISO 13888-3, these elements are sufficient to deliver proof of origin of the data in a scenario where there is no trustworthy third party. Also, see Ref. [21] for more information.

As described up to this point, this security solution is capable of realizing nonrepudiation of origin for all data packets that are transmitted during the communication. The "NISTIR 7628 Guidelines for Smart Grid Cyber Security" distinguishes between several types of nonrepudiation: "Nonrepudiation protects individuals against later claims by an author of not having authored a particular document, a sender of not having transmitted a message, a receiver of not having received a message, or a signatory of not having signed a document" [4]. For the communication via MMS, it may be of interest to receive a proof that a particular message was received (and processed) by the communication partner. The standard ISO 13888-3 calls this nonrepudiation service "nonrepudiation of delivery" and also introduces tokens for this purpose. Basically, the proof of receipt of a particular message consists of a hash value of the received message that is signed together with additional information. Like for nonrepudiation of origin, this additional information consists of the senders'/addressees' distinguished names and a timestamp. As it can be seen, the information needed for this service already is available in this solution.

However, it is more difficult to realize nonrepudiation for delivery for all messages that are sent. This is because the MMS protocol does not intend a response message for every PDU sent. Due to the fact that this security solution only intends to extend the MMS protocol without changing the whole communication flow, it is restricted to realize nonrepudiation of delivery only for the messages the clients send to the server. To achieve nonrepudiation of delivery, the response messages are provided with a hash value (calculated with a collision-resistant hash function) of the corresponding request. This way, the client receives the response message, together with the proof, that the server received the previous request message.

17.5 AN IMPLEMENTATION USING XML SIGNATURES

As mentioned before, the ASN.1 definitions of the data packets do not intend the use of signatures for all data packets. For this reason, the key point in the implementation of the solution is the question of how the data packets can be provided with signatures. Since the data packets are defined using ASN.1. There are several possible encodings of those. Since the security solution uses XML signatures to provide data packets with signatures, the XML encoding rules (XERs) have to be used to encode the data packets. The XML-encoded data packets can be extended with digital signatures and additional information due to the flexibility of the XML data format. Although XML-encoded data packets are longer in comparison to other types of encoding, the process of inserting and verifying XML signatures is comparatively easy, as the software for these purposes already exists and is easy to use. In addition, XML supports the integration of X.509 certificates into the signature block, which allows an exchange of certificates between client and server.

17.5.1 DIFFERENT TYPES OF XML SIGNATURES

There are several types of signatures that can be used to sign XML documents. XML signatures are classified into three types: “enveloped,” “enveloping,” and “detached.” In a detached signature, the signed data is stored externally and the signature only contains a reference to it. For this reason, this type is not suitable for signing data packets. In an enveloping signature, the data is part of the signature, while an enveloped signature is appended to the data that has to be signed [22,23]. Due to the simple structure of the enveloped signatures, these will be used to sign the data packets.

17.5.2 ADDITIONAL INFORMATION EMBEDDED INTO THE SIGNATURE

XML signatures are very flexible and also allow embedding additional information into the signature block. In the implementation of the security solution, a block with additional information about the used key will be stored within the signature block. This block contains a concatenation of the elements *subject* and *issuer* of the X.509-certificate associated with the key used for the signature. With this information, the receiver of the message can easily find the certificate that has to be used in order to verify the signature. Note that both the subject and issuer elements have to be used to identify a certificate, since the subject element itself can be ambiguous.

17.5.3 HOW THE SYSTEM WORKS

To sign a request/response, the respective data packet is read by a parser, a timestamp, and any information about associated packages, issuer, and addressee are added and the document is signed. Finally, the data packet is logged and sent. The verification of a message works similarly: after parsing the message, the signature will be verified and the timestamp will be checked. Besides the verification of a message, its hash value is also calculated and stored, because it may be included in the associated response packet (but only on the server side). The implementation used for the following example calculates the hash value using the SHA-256 algorithm. Prior to further processing, the data packet will be logged and the signature block, timestamp, and any further information removed, since the security layer should be transparent from the view of the other layers.

17.5.4 AN EXAMPLE

The following example demonstrates how a data packet is encoded and provided with a signature. Due to its simple structure and its comparatively short length, the Initiate-Request-PDU that is sent at the establishment of an association will be shown. As mentioned before, ASN.1 is used to describe the data sets that are sent. The following listing only shows the parts of the ASN.1 definition that are required to encode an initiate request. Some parts marked with "..." are omitted for reasons of clarity. For the full document, see Ref. [14].

```
MMSpdu ::= CHOICE
          {

          confirmed-RequestPDU      [0]     IMPLICIT Confirmed-RequestPDU,
          confirmed-ResponsePDU     [1]     IMPLICIT Confirmed-ResponsePDU,
          confirmed-ErrorPDU        [2]     IMPLICIT Confirmed-ErrorPDU,
          unconfirmed-PDU           [3]     IMPLICIT Unconfirmed-PDU,
          rejectPDU                 [4]     IMPLICIT RejectPDU,
          cancel-RequestPDU         [5]     IMPLICIT Cancel-RequestPDU,
          cancel-ResponsePDU        [6]     IMPLICIT Cancel-ResponsePDU,
          cancel-ErrorPDU           [7]     IMPLICIT Cancel-ErrorPDU,
          initiate-RequestPDU       [8]     IMPLICIT Initiate-RequestPDU,
          initiate-ResponsePDU      [9]     IMPLICIT Initiate-ResponsePDU,
```

```
        initiate-ErrorPDU          [10]    IMPLICIT Initiate-ErrorPDU,
        conclude-RequestPDU        [11]    IMPLICIT Conclude-RequestPDU,
        conclude-ResponsePDU       [12]    IMPLICIT Conclude-ResponsePDU,
        conclude-ErrorPDU          [13]    IMPLICIT Conclude-ErrorPDU
```

```
Initiate-RequestPDU ::= SEQUENCE
        {
        localDetailCalling                 [0] IMPLICIT Integer32 OPTIONAL,
        proposedMaxServOutstandingCalling [1] IMPLICIT Integer16,
        proposedMaxServOutstandingCalled [2] IMPLICIT Integer16,
        proposedDataStructureNestingLevel [3] IMPLICIT Integer8 OPTIONAL,
        mmsInitRequestDetail               [4] IMPLICIT InitRequestDetail
        }
InitRequestDetail ::= SEQUENCE
        {
        proposedVersionNumber   [0] IMPLICIT Integer16,
        proposedParameterCBB    [1] IMPLICIT ParameterSupportOptions,
        servicesSupportedCalling [2] IMPLICIT ServiceSupportOptions
        }
ParameterSupportOptions ::= BIT STRING
        { ...
        }
ServiceSupportOptions ::= BIT STRING
        { ...
        }
```

The application creates a data structure representing an initiate request PDU and initializes its values. After this, an encoding function will be invoked in order to encode the previously created data structure with encoding rules specified by the application. For this solution, the XERs are used. The function creates an array of bytes containing the encoded data. In this case, the data is simply an XML document describing the initiate request PDU. The following listing shows this XML document as it is passed to the next layer, the previously described security layer. Some parts marked with "..." are omitted for reasons of clarity.

```
<MmsPdu>
    <initiateRequestPdu>
        <localDetailCalling>65000</localDetailCalling>
        <proposedMaxServOutstandingCalling>5</proposedMaxServ
          OutstandingCalling>
        <proposedMaxServOutstandingCalled>5</proposedMaxServ
          OutstandingCalled>
```

```
        <proposedDataStructureNestingLevel>10</proposedDataStructure
         NestingLevel>
        <mmsInitRequestDetail> <proposedVersionNumber>1</proposed
         VersionNumber>
            <proposedParameterCBB>
                11110001000
            </proposedParameterCBB>
            <servicesSupportedCalling> 111011100001110000000000000
            000000000010000001000000000000000...
            </servicesSupportedCalling>
        </mmsInitRequestDetail>
    </initiateRequestPdu>
</MmsPdu>
```

Now, a timestamp, the name of the addressee, and the signature have to be added by the security layer. For this purpose, the byte array (which contains the XML document) is parsed by an XML parser returning a document object model, which will be expanded by adding three nodes, one containing the current timestamp, the next the name of the addressee, and the last one the signature. The signature is added last, since the timestamp and the name of the addressee have to be a part of the signed document. The next listing shows the final document as it will be logged and sent to the server.

```
<MmsPdu>
    <initiateRequestPdu>
        <localDetailCalling>65000</localDetailCalling>
        <proposedMaxServOutstandingCalling>5</proposedMaxServ
         OutstandingCalling>
        <proposedMaxServOutstandingCalled>5</proposedMaxServ
         OutstandingCalled>
        <proposedDataStructureNestingLevel>10</proposedDataStructure
         NestingLevel>
        <mmsInitRequestDetail> <proposedVersionNumber>1</proposed
         VersionNumber>
            <proposedParameterCBB>
                11110001000
            </proposedParameterCBB>
            <servicesSupportedCalling> 111011100001110000000000000
            000000000010000001000000000000000...
            </servicesSupportedCalling>
        </mmsInitRequestDetail>
    </initiateRequestPdu>
<timestamp>1307232715962046O7</timestamp>
<addressee>The addressee's distinguishedname</addressee>
```

```
<Signature xmlns="http://www.w3.org/2000/09/xmldsig#">
<SignedInfo>
<CanonicalizationMethod Algorithm="http://www.w3.org/TR/2001/REC-xml-c14n-
20010315"/>
<SignatureMethod Algorithm="http://www.w3.org/2001/04/xmldsig-more#ecdsa-
sha256"/>
</SignedInfo>
<SignatureValue>8L2BZWqElXJAQr5AWUznbxnhMCFe0fGPxPv4q6P0Ooz0P
UYS4buHioaKCvhTqQaR 9UBs6RDY9Hq4csvM4NCTnQ==</SignatureValue>
<KeyInfo>
<KeyName>IssuerName,SubjectName</KeyName>
</KeyInfo>
</Signature>
</MmsPdu>
```

As it can be seen, this is the original MMS-PDU as it can be seen in the previous listing extended with the additional information. There are two separate nodes containing the timestamp and the name of the addressee. It would also be possible to store the name of the sender in an additional node; however, in this example, implementation of the name of the sender is stored in the Element *keyValue* of the signature. The signature was created using ECDSA with a 256-bit key using SHA-256 as hash algorithm. Note that the elements "IssuerName" and "SubjectName" contained by the KeyName element refer to the corresponding elements of the X.509 certificate associated with the signing key.

The NISTIR 7628 guidelines recommend using keys and curves with a security level not fewer than 112 bits for the use until the year 2029 and not fewer than 128 bits beyond the year 2030 [4]. The guidelines also name certain curves and hash algorithms that should be used.

The reception of a data packet works the other way around: the data will be parsed by an XML parser, the signature will be verified, the data packet will be logged, and both signature and timestamp, as well as any information about associated data packet's will be removed before the data packet finally will be passed to the upper layer, where it is decoded using a suitable function. In this implementation, the decoding function is an XML decoder that reads the XML data back into a data structure. The data structure contains exactly the data that was initially encoded and sent by the communication partner.

17.5.5 ADVANTAGES OF XML SIGNATURES

The XML data format and XML signatures are very flexible; it is possible to change signing algorithms or signature lengths without changing the architecture of the system. It is also possible to embed additional information about certificates that have to be used to verify the signature. Besides the signature itself, additional elements like timestamps and hash values of associated data packets can be included. An implementation using XML has the additional advantage in that the solution with XML signatures can easily be transferred to other systems that use XML-based communication protocols. An example of this is the planned extension IEC 61850-8-2 of the standard IEC 61850. There, besides the mapping of the ACSI services to MMS (IEC 61850-8-1) [10], a mapping to web services using the XML-based Extensible Messaging and Presence Protocol (XMPP) is planned as part of IEC 61850-8-2 [7]. The standard IEC 61400-25, which deals with the communication and control in wind farms, also defines the use of XML-based web services.

17.6 PERFORMANCE OF THE SYSTEM

The last part of this chapter covers the performance of the system, specially the performance of the extended version compared to the version without the security solution. There are two aspects that will be looked at: on one hand, the latency caused by the additional cryptographic functionality, and on the other hand, the influence on the performance of the system caused by the different encoding of the data.

17.6.1 THE SETUP

For the first test, we used an MMS server that runs on a common desktop computer. The MMS server manages some simulated devices, modeled as a data structure, as described by IEC 61850. This data structure can be accessed by the clients. The server also simulates changing sensor values that can be obtained by the clients. The MMS server is connected via 100 M bit ethernet with the device that runs the client. In this case, the client runs on a common PC as well. The client realizes a supervisory control and data acquisition (SCADA) system that periodically obtains the current sensor values from the server, causing a constant data flow in the network. On the server side, the following values are measured:

- The time taken for decoding, encoding, signing, and verifying of the MMS-PDUs
- The time taken for processing the request

On the client side, the following values are measured:

- The time taken for decoding, encoding, signing, and verifying of the MMS-PDUs
- The time taken for the whole request to the server

The time taken for the processing of the request on the server side will be always the same, since it is not affected by the extension. The latency caused by the network can be obtained by subtracting all other time values from the time the request took in total.

17.6.2 THE SYSTEM WITHOUT EXTENSION

First, the client and the server without extensions were started. The following table shows the results for the measured values:

Encoding/decoding (server side)	<1 ms
Processing of request (server)	1 ms
Encoding/decoding (client side)	<1 ms
Total time	22 ms

The values of the times for signing/verifying are omitted, since there is no signing and verifying of messages in the original version. As it can be seen, the processing and encoding/decoding is very fast, and almost the whole latency is caused by the network. Fig. 12 shows an MMS-PDU that was captured on the server side using the tool "Wireshark."

As it can be seen, there are two MMS-PDUs that belong together: the request PDU and the corresponding response PDU. Since the frame was captured on the server side, the times of the two frames are nearly the same. The figure shows the request PDU; it contains the name of the variable that contains the sensor value to be obtained. Request PDU and response PDU together are 227 bytes long and, as it can be seen, data this size can be transmitted very fast over ethernet.

```
Length  No.   Time            Protocol   Info
  136   9336  330.3660820(MMS            confirmed-RequestPDU
   91   9337  330.3713760(MMS            confirmed-ResponsePDU

+ Frame 9336: 136 bytes on wire (1088 bits), 136 bytes captur
+ Ethernet II, Src: LiteonTe_51:5f:2a (74:de:2b:51:5f:2a), Ds
+ Internet Protocol Version 4, Src: 192.168.178.37 (192.168.1
+ Transmission Control Protocol, Src Port: 56200 (56200), Dst
+ TPKT, Version: 3, Length: 82
+ ISO 8073/X.224 COTP Connection-Oriented Transport Protocol
+ ISO 8327-1 OSI Session Protocol
+ ISO 8327-1 OSI Session Protocol
+ ISO 8823 OSI Presentation Protocol
- MMS
  - confirmed-RequestPDU
      invokeID: 198
    - confirmedServiceRequest: read (4)
      - read
        - variableAccessSpecificatn: listOfVariable (0)
          - listOfVariable: 1 item
            - listOfVariable item
              - variableSpecification: name (0)
                - name: domain-specific (1)
                  - domain-specific
                      domainId: testmodelSENSORS
                      itemId: TTMP1$MX$TmpSv$instMag$f
```

FIG. 12

A captured MMS-PDU.

17.6.3 THE EXTENDED SYSTEM

The extension was implemented using the Xerces-C++ XML Parser for the work with XML documents, such as adding nodes or parsing documents, the library OpenSSL was used for cryptographic operations, and the "Apache XML Security" library was used for the work with XML signatures. There are several things that may have a negative influence on the performance: On the one hand, two "heavy" libraries, the XML parser and the OpenSSL library; and on the other hand, a significantly increased PDU length caused by the XML encoding and the additional information embedded within every PDU sent. The following table shows the results for the extended version of the system.

Encoding/decoding (server side)	1 ms
Signing/verifying (server side)	5 ms
Processing of request (server)	1 ms
Encoding/decoding (client side)	1 ms
Signing/verifying (client side)	8 ms
Total time	65 ms

As can be seen, the total time increases with the extended version. The encoding and decoding takes a bit more time compared to the original version, but considering the fact that instead of a dedicated DER-parser (distinguished encoding rules) now a "heavy" XML parser is used, the time difference is surprisingly small. The signing and verifying of the signatures and additional information are fast, too. The calculation of a signature is a bit faster than the verifying of the signature, which may be caused by the fact that along with the verifying of the signature, the access control lists, the field containing the addressee's name, and the timestamp are checked, too. The highest increase of time can be seen at the transmission time, that is, the time needed to transfer the data over the network. Fig. 13 shows a capture of the network traffic that was caused by the extended version of the system.

This time, the captured PDUs are not marked as MMS-PDUs, since the encoding and the structure have changed, and the tools do not recognize the data as MMS-PDUs anymore; however, in the raw data, the XML-encoded data can be found, as well as the corresponding XML signatures. It can also be seen that the data size significantly increases. While in the original version, the data size of request/response PDU was around 200 bytes in total, the data for the same request/response is now about 2 KB. This seems to be the cause of the significantly increased transmission time that was measured before.

The test scenario presented here was very simple. There are several more tests needed to allow more reliable statements about the system's performance in realistic scenarios. Although the increased data size does not affect the transmission time in the same ratio, the network can be congested much faster when the extended system is used. There are solutions for the compression of XML data for scenarios with

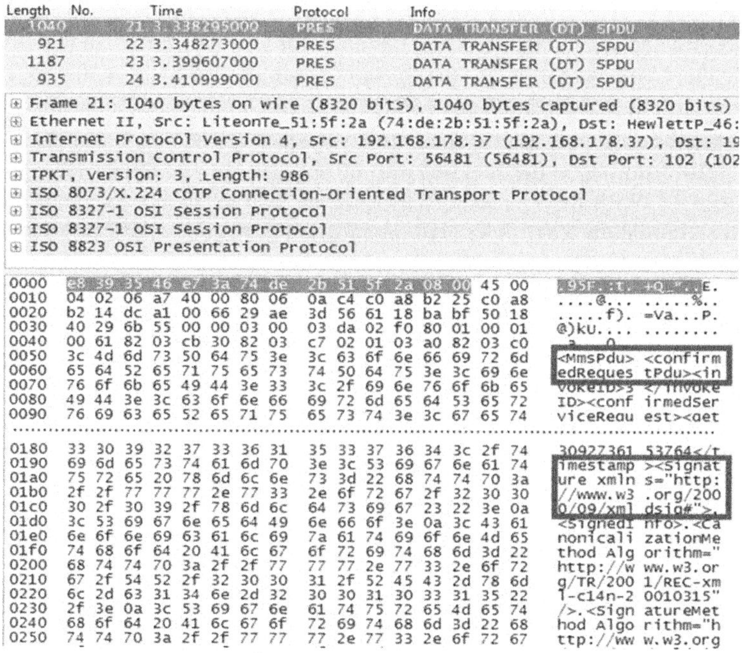

```
Length  No.   Time          Protocol  Info
1040    21    3.338295000   PRES      DATA TRANSFER (DT) SPDU
921     22    3.348273000   PRES      DATA TRANSFER (DT) SPDU
1187    23    3.399607000   PRES      DATA TRANSFER (DT) SPDU
935     24    3.410999000   PRES      DATA TRANSFER (DT) SPDU

⊞ Frame 21: 1040 bytes on wire (8320 bits), 1040 bytes captured (8320 bits)
⊞ Ethernet II, Src: LiteonTe_51:5f:2a (74:de:2b:51:5f:2a), Dst: HewlettP_46:
⊞ Internet Protocol Version 4, Src: 192.168.178.37 (192.168.178.37), Dst: 19
⊞ Transmission Control Protocol, Src Port: 56481 (56481), Dst Port: 102 (102
⊞ TPKT, Version: 3, Length: 986
⊞ ISO 8073/X.224 COTP Connection-Oriented Transport Protocol
⊞ ISO 8327-1 OSI Session Protocol
⊞ ISO 8327-1 OSI Session Protocol
⊞ ISO 8823 OSI Presentation Protocol

0000  e8 39 35 46 e7 3a 74 de  2b 51 5f 2a 08 00 45 00   .95F.:t. +Q_*..E.
0010  04 02 06 a7 40 00 80 06  0a c4 c0 a8 b2 25 c0 a8   ....@... .....%..
0020  b2 14 dc a1 00 66 29 ae  3d 56 61 18 ba bf 50 18   .....f). =Va...P.
0030  40 29 6b 55 00 00 03 00  03 da 02 f0 80 01 00 01   @)kU.... ........
0040  00 61 82 03 cb 30 82 03  c7 02 01 03 a0 82 03 c0   .a...0.. ........
0050  3c 4d 6d 73 50 64 75 3e  3c 63 6f 6e 66 69 72 6d   <MmsPdu> <confirm
0060  65 64 52 65 71 75 65 73  74 50 64 75 3e 3c 69 6e   edReques tPdu><in
0070  76 6f 6b 65 49 44 3e 33  3c 2f 69 6e 76 6f 6b 65   vokeID>3 </invoke
0080  49 44 3e 3c 63 6f 6e 66  69 72 6d 65 64 53 65 72   ID><conf irmedSer
0090  76 69 63 65 52 65 71 75  65 73 74 3e 3c 67 65 74   viceRequ est><get
......................................................................
0180  33 30 39 32 37 33 36 31  35 33 37 36 34 3c 2f 74   30927361 53764</t
0190  69 6d 65 73 74 61 6d 70  3e 3c 53 69 67 6e 61 74   imestamp ><Signat
01a0  75 72 65 20 78 6d 6c 6e  73 3d 22 68 74 74 70 3a   ure xmln s="http:
01b0  2f 2f 77 77 77 2e 77 33  2e 6f 72 67 2f 32 30 30   //www.w3 .org/200
01c0  30 2f 30 39 2f 78 6d 6c  64 73 69 67 23 22 3e 0a   0/09/xml dsig#">.
01d0  3c 53 69 67 6e 65 64 49  6e 66 6f 3e 0a 3c 43 61   <SignedI nfo>.<Ca
01e0  6e 6f 6e 69 63 61 6c 69  7a 61 74 69 6f 6e 4d 65   nonicali zationMe
01f0  74 68 6f 64 20 41 6c 67  6f 72 69 74 68 6d 3d 22   thod Alg orithm="
0200  68 74 74 70 3a 2f 2f 77  77 77 2e 77 33 2e 6f 72   http://w ww.w3.or
0210  67 2f 54 52 2f 32 30 30  31 2f 52 45 43 2d 78 6d   g/TR/200 1/REC-xm
0220  6c 2d 63 31 34 6e 2d 32  30 30 31 30 33 31 35 22   l-c14n-2 0010315"
0230  2f 3e 0a 3c 53 69 67 6e  61 74 75 72 65 4d 65 74   />.<Sign atureMet
0240  68 6f 64 20 41 6c 67 6f  72 69 74 68 6d 3d 22 68   hod Algo rithm="h
0250  74 74 70 3a 2f 2f 77 77  77 2e 77 33 2e 6f 72 67   ttp://ww w.w3.org
```

FIG. 13

A captured extended MMS-PDU.

limited bandwidth, like WAP Binary XML (WBXML). It would also be possible to integrate WBXML into the implementation of the system to investigate how far the data transmitted can be compressed using XML data.

17.7 CONCLUSION

Although the initial scope of the standard IEC 61850 is the automation of substations, there are now new use cases in a smart grid that can also be addressed by IEC 61850 due to the flexibility of this standard. The new use cases create new requirements to the systems and security in general that cannot be met with the existing security measures. These new security requirements, arising as a summary of the standard IEC 62351, the NISTIR 7628 guidelines, and other documents dealing with smart grid security, can only be met when the existing systems and protocols are improved. Important security requirements, such as nonrepudiation and authenticity of every message sent between the participants, can be met with the security solution that was presented within this chapter. With the use of certificates for the identity of the communication partners and per message authentication using digital signatures and additional information like timestamps, the authorship of the messages cannot be disputed, and end-to-end-authentication is ensured for every data packet that is sent.

Using a logging system, transactions that took place within the system can be traced and reconstructed at a later date. XML signatures are a good choice for that purpose, because the implementation of the system is simple, and the solution can be transferred to other applications using XML-based communication protocols. Due to the implementation as an intermediate layer, the integration in existing systems is straightforward, since the functionality of both server and client does not need to be changed. With the increasing influence of web technologies in smart grids, a security solution for XML-based communication protocols will be of great importance for further development of smart grids.

REFERENCES

[1] CEN, CENELEC, ETSI Smart Grid Coordination Group. Smart grid reference architecture; 2012.
[2] IEC 61850: Communication networks and systems in substations, TC 57.
[3] IEC 62351: Power systems management and associated information exchange—data and communications security, TC 57.
[4] The Smart Grid Interoperability Panel, Cyber Security Working Group. NISTIR 7628 guidelines for smart grid cyber security. Gaithersburg, MD: U.S. Department of Commerce, National Institute of Standards and Technologies; 2010. August.
[5] CEN CENELEC ETSI Smart Grid Coordination Group. Smart grid information security; Nov 2012.
[6] Office of the National Coordinator for Smart Grid Interoperability. NIST framework and roadmap for smart grid interoperability standards, release 1.0. NIST Special Publication 1108; Jan 2010. Available at, http://www.nist.gov/public.affairs/releases/upload/smartgrid interoperabilityfinal.pdf.
[7] Englert H. Neue Kommunikationskonzepte für den Netzbetrieb-aktuelle Entwicklungen in der IEC 61850, Smart Grids Forum, Hannover Messe; 2014. Available online at, https://www.vde.com/de/smart-grid/forum/beitraege/Documents/201404-09-neue-kommunikationskonzepte-englert.pdf [accessed 03.07.15].
[8] Fries S, Hof H-J, Dufaure T, Seewald MG. Security for the smart grid—enhancing IEC 62351 to improve security in energy automation control. Int J Adv Security 2010;3 (3 & 4):169–83.
[9] Systems Integration Specialists Company Inc. Overview and introduction to the manufacturing message specification (MMS); 1995. Available at, http://www.sisconet.com/downloads/mmsovrlg.pdf.
[10] IEC 61850-8-1: Communication networks and systems in substations part 8-1: specific communication service mapping (SCSM)—mappings to MMS (ISO 9506-1 and ISO 9506-2) and to ISO/IEC 8802-3.
[11] IEC 61400-25: Communications for monitoring and control of wind power plants, TC 88.
[12] IEC 61850-5: Communication networks and systems in substations part 5: communication requirements for functions and device models, Dec. 2005.
[13] IEC 62351-1: Power systems management and associated information exchange—data and communications security part 1: communication network and system security—introduction to security issues.

[14] Systems Integration Specialists Company Inc. SISCO MMS syntax; 1994. Available at, http://www.sisconet.com/downloads/mmsabstractsyntax.txt.
[15] IEC 62351-4: Power systems management and associated information exchange—data and communications security—part 4: profiles including MMS.
[16] Smart Grid Mandate M/490 EN. Standardization mandate to European Standardisation Organisations (ESOs) to support European smart grid deployment. European Commission Directorate-General for Energy; 2011. Mar. 1.
[17] Verband der Elektrotechnik, Elektronik und Informationstechnik. VDE-Positionspapier Smart Grid Security Energieinformationsnetze und systeme; 2014. Available at, https://www.vde.com/de/InfoCenter/Studien-Reports/Seiten/Positionspapiere.aspx.
[18] IEC 62351-3: Power systems management and associated information exchange—data and communications security—part 4: profiles including TCP/IP.
[19] International Telecommunication Union ITU. Information technology open systems interconnection—the directory: public-key and attribute certificate frameworks. Series x: data networks, open system communications and security directory, International Telecommunication Union; 2012. ITU-T Recommendation ITU-T X.509.
[20] International Telecommunication Union ITU. Information technology—open systems interconnection—the directory: models. Series x: data networks, open system communications and security directory, International Telecommunication Union; 2012. ITU-T Recommendation ITU-T X.501.
[21] ISO/IEC 13888-3 IT security techniques non-repudiation—part 3: mechanisms using asymmetric techniques.
[22] Dournaee B. XML security. RSA Press series, Osborne, KS: McGraw-Hill; 2002.
[23] XML signature syntax and processing. 2nd ed. June 2008. W3C Recommendation. Available at http://www.w3.org/TR/xmldsig-core/.

FURTHER READING

[1] IEC 61850-1: Communication networks and systems in substations introduction and overview.
[2] RFC 2246—The TLS protocol version 1.0, 1999. Available at http://tools.ietf.org/html/rfc2246.

CHAPTER

SCADA and smart energy grid control automation

18

K. Sayed*[,†], H.A. Gabbar*

*University of Ontario Institute of Technology (UOIT), Oshawa, ON, Canada**

Sohag University, Sohag, Egypt[†]

18.1 INTRODUCTION

In this chapter, supervisory control and data acquisition (SCADA) systems for a smart power grid are explained, with discussion about the efficacy and challenges in the integration process and the automation systems. The smart grid SCADA system integrates the existing renewable energy sources (RES) system with digital information processing and advanced telemetry systems. It is clear that the increased integration and automation of the electric microgrid and utility grids present new development aspects of energy management. Automation systems are computer-based control systems used to control and monitor sensitive processes within various industries and critical infrastructures such as electric grid, energy grids, natural gas pipelines, water pipelines, and wastewater industries. The SCADA control systems might perform additional automation functions such as switching circuit breakers (CB), adjusting protection relays settings, operating switches, and adjusting flow valves to regulate the fuel flow [1–4]. Typically, control systems collect site measurement and operational data from the electric substation, and then process, display, and analyze this information. The remote control commands to local or remote outstations are issued from the master station control center.

Using control systems in power system applications has become increasingly attractive during the past few decades. There are two major types of control systems: SCADA systems and distributed control systems (DCS). Typically, DCS systems are used within one generating station or over a specified geographic zone. SCADA systems usually are used for electricity transmission extended geographically over long distances or radial distribution. Utilities, however, might use a DCS in traditional power generation stations and an SCADA system in transmission and distribution substations, although SCADA also is applicable in large-scale renewable energy systems, such as wind and solar farms. SCADA systems for renewable energy are computer-aided control systems, sometimes called renewable energy management systems (EMS).

Smart Energy Grid Engineering.

EMS have become more advanced recently because of the intelligence and high capabilities of the applications software. The requirements for data acquisition electronic devices and the associated communications systems within the control center were extended to the limits that computer and communication technologies could offer at the time. Specially designed systems and devices had to be developed to fulfill the power system application requirements taking advantage of advanced information and new communication technologies. Recent trends in control industry deregulation have changed the requirements of the electricity supervisory control center and have uncovered its weakness. In the past, conventional control centers were inflexible, independent, too centralized, and closed by today's standards. The evolving changes in recent power system operational needs demand a distributed control center that is decentralized, flexible, integrated, and opened. Today's control centers are moving in that direction with varying degrees of success [5–10].

As the backbone of large-scale renewable power SCADA systems should have all of the design elements to accommodate the multifaceted nature of distribution automation and the distribution management system (DMS) applications. A smart grid SCADA system's main function is to assist distributed generation, switching procedure, alarming, telemetry, event logging, measurement recording, and remote control of outstation equipment.

A modern SCADA system should support the engineering plan by providing entrance to power system data without affecting any operational workstation. SCADA systems are known for their effective support for exporting power system network data, and we will review briefly the SCADA technologies utilized in today's control centers to enable them to be more distributed. With the rising of the Internet age, the trend in information and communication technologies is moving toward microgrid and grid computing and Web services, or microgrid services, and so a microgrid service-based future control center will be needed.

Renewable energy systems have gained more popularity over the years because of the incessant failure and general unreliability of the power grids and microgrids. Renewable energy forms a major source of energy in distributed generation systems; the energy generated can be integrated into the existing power grid or it can be used for domestic microgrid consumption. Even though renewable sources are in abundance and inexhaustible, their occurrence at a quantity enough for power generation at all times is not guaranteed because of variations in climatic conditions, thereby jeopardizing the chances of relying on them as the only source of energy. This prompted research and development in the areas of power generation and storage of energy in order to improve the efficiency of such systems. Such research has led to a drastic reduction in the cost of these systems, which convert renewable energy into electrical energy.

The increasing size of photovoltaic (PV) power plants all over the world has made their operation and maintenance tasks much more complex than they were a few years ago [10]. Many of these PV plants are equipped with advanced SCADA systems in order to collect the necessary information to assess their performance, such as meteorological data, information from the PV farm field, and PV inverters [11–14].

The great amount of data provided by those SCADA systems, however, requires the development of new procedures capable of handling all this data and providing accurate information about the performance, failures, and long-term trends. Little information is available today about experiences in automatic failures detection and performance evaluation of large-scale solar PV plants.

SCADA systems are essentially process control systems (PCS) that gather, monitor, and analyze real-time environmental data from a simple residential building or a complex large-scale PV or wind farm power plant. PCSs are designed for microgrid automation or power distribution systems based on a predetermined set of data and conditions, such as generated/consumed energy or power grid management. Some systems consist of one or more types of data acquisition devices such as remote terminal units (RTUs) and/or programmable logic controllers (PLC) that are connected to any number of field devices such as sensors, digital meters, protection relays, and station alarms (batteries, chargers and fire alarm).

SCADA systems are composed of the following equipment:

1. *Outstation hardware*: Remote substation control devices, such as state of charge, current transformer (CT), voltage transformer (VT), fuel valves, and CB that can be controlled locally or remotely.
2. *Local substation processors*: Collect data from the field instruments and hardware equipment, including the RTU, PLC and intelligent electronic devices (IEDs), such as digital relays and digital meters. The local processor is responsible for dozens of analog and digital inputs/outputs from IEDs and switchgear equipment.
3. Digital instruments: Usually installed in the field or in a facility and sense conditions such as current, voltage, irradiance, temperature, pressure, wind speed, and flow rate.
4. Communications devices: Could be either short-range or long-range communications. Short-range communications are installed between local RTUs, instruments, and operating equipment. These are relatively short distance cables or wireless connections that carry digital and analog signals using electrical properties, such as voltage and current, or other settled industrial communications protocols. Long-range communications are installed between local processors RTU/PLC and host servers. This communication typically uses methods such as leased telephone lines, microwave, satellite, frame relay network, and cellular packet data.
5. Host computers/servers: Host computers, such as data acquisition computer server, engineering/operation workstations. As the central point of monitoring and control, they will be in the control room or master station. The operation workstation is where an engineer or operator can supervise the process, as well as receive system alarms, review data, and exercise remote control.

Fig. 1 displays a high-level overview of SCADA architecture, where the remote stations might be an electric substation, the SCADA network on one segment, with another organization network on differing network segments. With the progress in

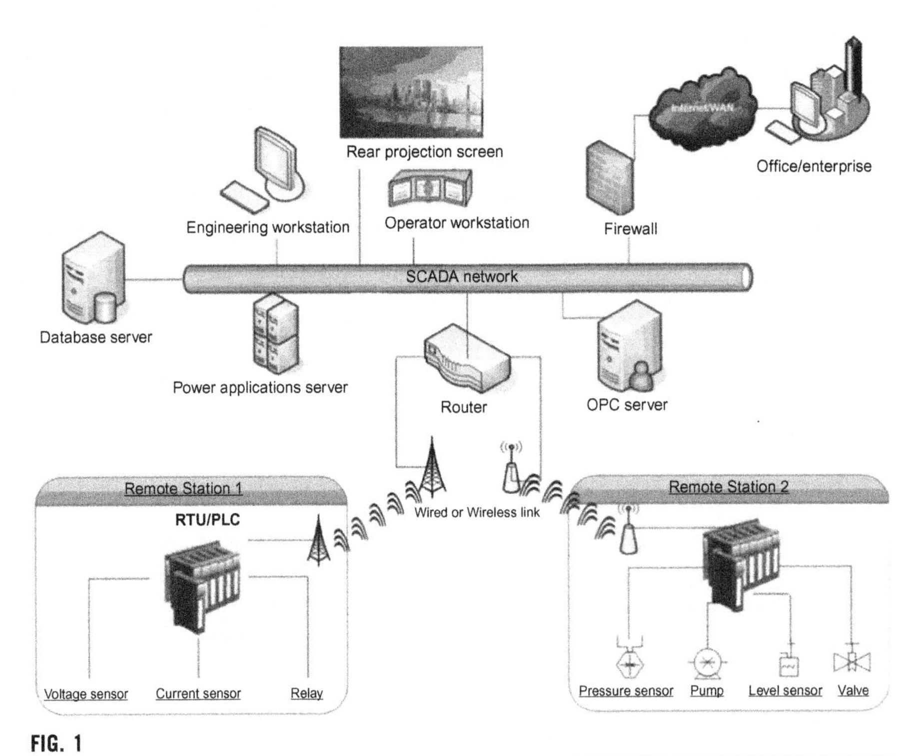

FIG. 1

SCADA system network.

digital computing, the integration of digital IEDs plays a substantial role in industrial manufacturing, wherein a factory uses PLCs/RTUs to control the devices, and develops large, complicated systems in which intelligent systems are part of the manufacturing plant control systems.

Most often, an SCADA system will monitor and make slight changes to function optimally. SCADA systems are considered closed-loop control systems and operate with comparatively little human interference. One of the key advantages of SCADA is its ability to supervise a whole system in a real-time environment. This is simplified by data acquisitions, including meter reading and checking statuses of sensors, that are communicated at regular, small-time intervals. An SCADA system as an industrial automation system acquires data from instruments and sensors located at remote sites and transmits data to master station central site for either controlling or monitoring purpose. The collected data from sensors and instruments usually is viewed on one or more SCADA host computers at the central site. Based on the data received from the remote substations, automated or operator-driven supervisory commands can be pushed to remote substation control devices, usually called outstation or field devices.

Large-scale and customer-premise grid-connected solar PV power or wind farm must be equipped with fully functional automatic SCADA system. Communication media are the key to data transmission to ensure the efficiency of the specific

SCADA system, so intensive research and testing of the communication technology are necessary. The SCADA technology of a solar PV power plant is still in the development stage, and the communication solution also is self-contained and immature. Because of the single method, lack of network management, and low level of integration, current communication solutions of SCADA systems are difficult to adjust to accommodate increasing business needs, and so research is focusing on a new solution [15–20].

18.2 SMART GRID CONCEPT

The smart grid framework is composed of and concerned with distributed intelligence, including data decentralization, renewable distributed generation and energy storage, and distribution system automation. Also of concern are customer partnership and interaction, microgrids, and high-demand devices. The smart grid is, by definition, about real-time data monitoring and active microgrid management via rapid two-way digital communication through the implementation of technological solutions to the power delivery infrastructure. Integration exists between microgrids and within the electric utility, renewable power generating devices, consumer loads devices, and third-party entities, either as consumers, vendors, or regulatory organizations. Smart grid comprises an intelligent monitoring system that observes the flow of electrical energy throughout the power network and incorporates the use of cables or transmission lines to manage power fluctuations, losses, and cogeneration integration from solar, fuel cell (FC) and the wind.

Generally, most effective smart grids can monitor/control residential home devices that are noncritical during peak power consumption times to reduce power demand, and return their function during nonpeak hours. Proposals for optimization include smart microgrids, smart power grid, and intelligent grid. In addition to normalizing electric demand, the ability to manage power consumption peaks can support in avoiding brown-outs and black-outs when power demand exceeds supply, and allow for maintaining critical loads and devices under such conditions. Fig. 2 displays a high-level communication flow between different components in a smart grid.

The smart grid initiative has made significant progress toward the modernization and growth of the electric utility infrastructure and aims to integrate it into today's advanced communication era, both in function and in architecture. That evolution comes with a number of organizational, socioeconomic, technical, and cybersecurity challenges. The expansion and depth of those challenges are significant, and a number of regulatory organizations have taken up the initiative to bring their own standards and requirements into alignment with these new challenges. The initiative also has offered many opportunities to explore new areas to take advantage of data communication among distributed and remote electric networks and their devices.

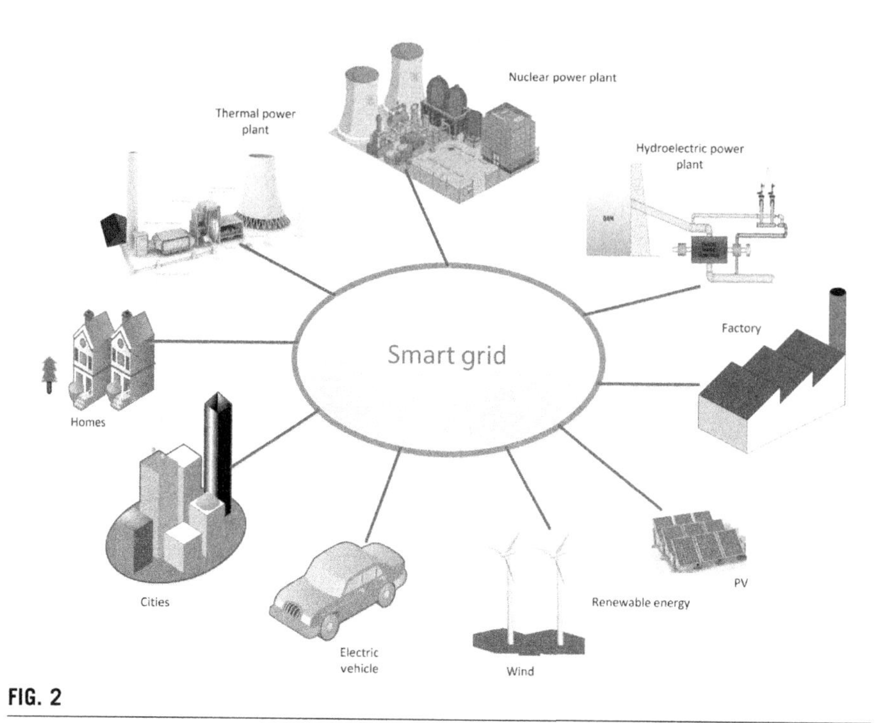

FIG. 2

Smart grid architecture.

18.3 SMART GRID/SCADA INTEGRATION

Integrating SCADA into the smart grid can be accomplished by electrical, communications and data networks, and allows for distributed and central aggregation of information and control over the entire electrical utility network as depicted in Fig. 3. SCADA empowers electricity consumers by interconnecting EMS to authorize customers to manage their own demand of energy and control costs. It allows the grid to be self-healing by responding automatically to power quality issues, power outages, and power system faults. SCADA monitors and optimizes the grid assets, while minimizing operations and maintenance costs. The smart grid, intelligence and control need to exist along the entire power supply chain. This includes electricity generation and transmission from beginning to delivery end-points at the customer's side, and includes both fixed and mobile devices in the smart grid architecture.

Digital communications in a smart grid take place over a diversity of technologies, protocols, and devices that comprise wireless and wired networks, and radio data communications, fiber optics networks, power line carriers, distribution line carriers, and satellite. SCADA software supports the dynamic grid management that encompasses monitoring the line segments and other control points in the electric network. To be fully efficient, every segment of the power line and piece of

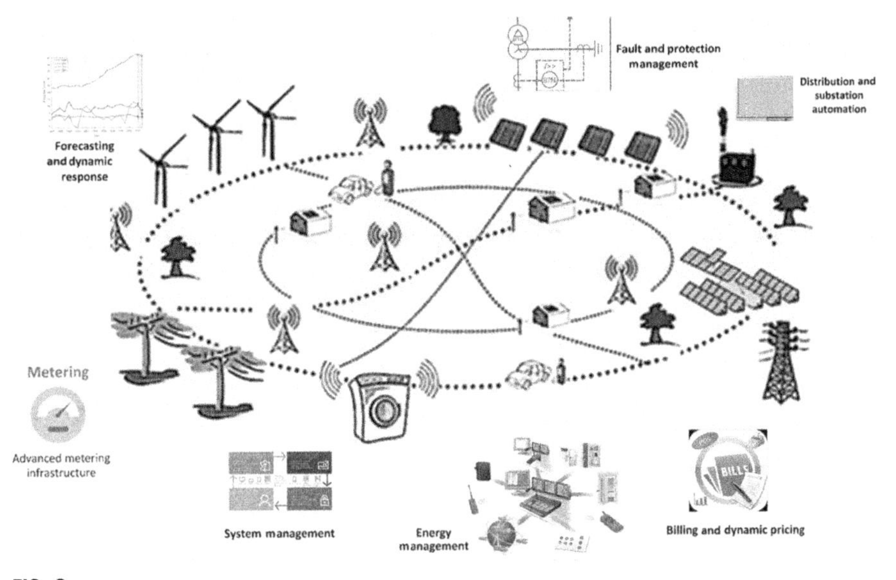

FIG. 3

SCADA/smart grid integration.

equipment in the distribution system should be monitored, in addition to allowing customers to observe and control their own loads and their energy consumption. Then a considerable amount of data has to be analyzed, organized, and utilized for both online and offline decision-making software that can be installed on power applications servers.

Decentralized SCADA software is important because of the significance of SCADA devices, data acquisition, and computation ability, which prevent a centralized data acquisition solution. The IEDs will develop the organization, collection, and data analysis necessary for decision making and switching management, data routing, and other required control actions that might be necessary based on the status of operation. This functionality subsists either as part of the devices' firmware or via RTU configuration functions and settings within each device.

18.4 SCADA APPLICATIONS IN POWER SYSTEM

A SCADA system is widely used in a power system to collect, analyze, and observe the power system data effectively. As the power system deals with power generation, transmission, distribution, and renewable energy sectors, monitoring and control are the main aspects in all these areas. Electric utilities detect current flow and line voltage, to monitor the status of CB, and to take sections of the power grid online or offline. Thus, the SCADA oversight of the power system improves the overall efficiency of the system, by saving costs and time. This can be achieved by optimizing the operation, minimizing loss, and supervising and controlling the generation and transmission systems. The SCADA's function in the power system network provides greater system reliability and stability for integrated grid operation. The power system automation system offers contingency-based fast load shedding, power control and SCADA functionality for the electrical system. These applications might be supplied by different vendors and applications such as:

- Generation/transmission/distribution monitoring and control system
- Generation/transmission/distribution control system
- Generation/transmission/distribution integrated control system
- Generation/transmission/distribution protection and control system
- Power management system
- Switching management system
- Load management system

18.4.1 SCADA FOR POWER GENERATING STATIONS

Advanced control and communication devices are used to deliver an optimum solution for each and every operation with flexible and advanced control structures. This can be done by using of PLC controllers, RTU data acquisition devices and advanced communication links along with SCADA software and hardware in power generating plants and stations. Fig. 4 shows the structure diagram of SCADA in power generation as it supervises several tasks of operation, including protection functions, controlling and monitoring scheduled and unscheduled maintenance procedures. The control functions of SCADA system in power generation include:

- Continuous monitoring of speed and frequency
- Observation of the dynamic status of CB, switches, and protection relays
- Planning of generation operations
- Control of active and reactive powers
- Wind/steam/gas turbine protection
- Monitoring of fuel system and station auxiliary services
- Voltage- and frequency-based load scheduling
- Processing of historical data for parameters related to generation
- Observation of weather stations in case of wind and solar plants.

The SCADA provides an integrated group of control, management, and supervision functions for traditional and renewable power generation stations. These functions include:

- Generator control, including MPPT for solar PV station and yaw control in wind turbines
- CB on/off control
- Integration with protection relays including adaptive protection for microgrids
- Synchronization function between generators
- Transformer and tap-changer control according to the network conditions.

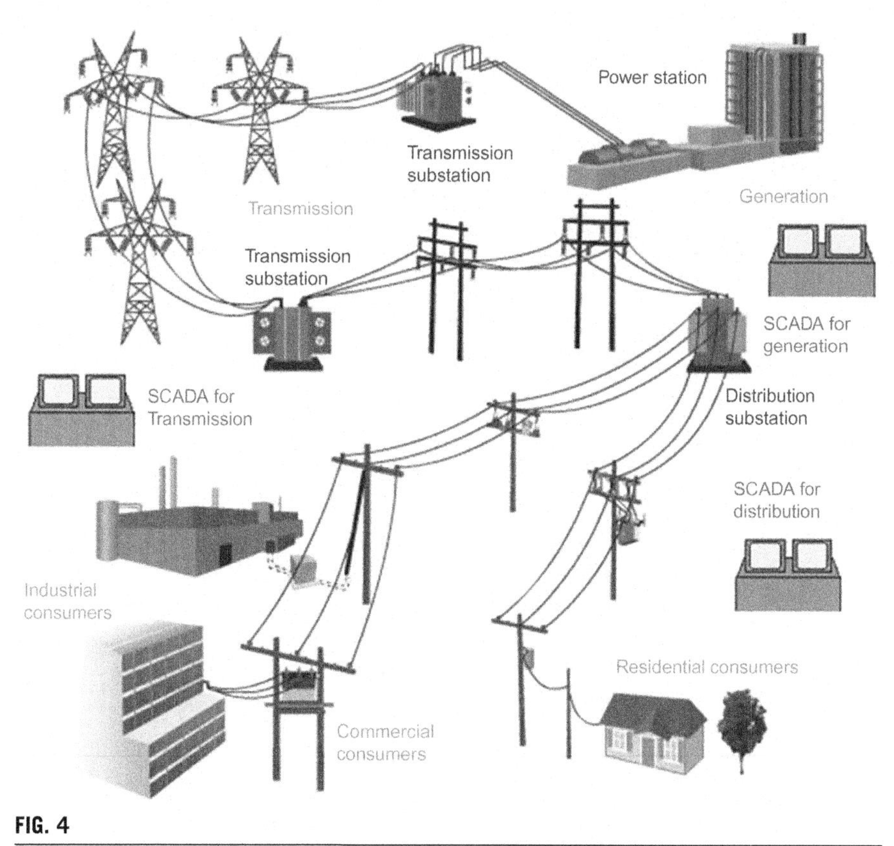

FIG. 4

SCADA for electrical power industry.

18.4.2 SCADA FOR POWER DISTRIBUTION SYSTEMS

EMS deal with electric energy from transmission/distribution substations to the different load types with the use of high/medium voltage cables and transmission lines. Most of the electrical power distribution utilities depend on manual labor to carry out distribution network tasks, such as connection or interruption power to loads, fault removing, service restoration, and taking the voltage and current readings by observation every hour. SCSADA implementation in the power network not only decreases the manual work operation but also eases smooth automatic operations with minimized load disturbance. Fig. 5 shows the schematic diagram of SCADA in a power distribution system where it acquires whole data from various substation components and from pole-mounted distribution transformers at remote locations and processes the corresponding data and status information.

Data acquisition devices such as RTU or PLC in electrical substations continuously monitor the substation CB and transformers and transfer this data to a centralized SCADA system main computer. When a power outage occurs, the SCADA

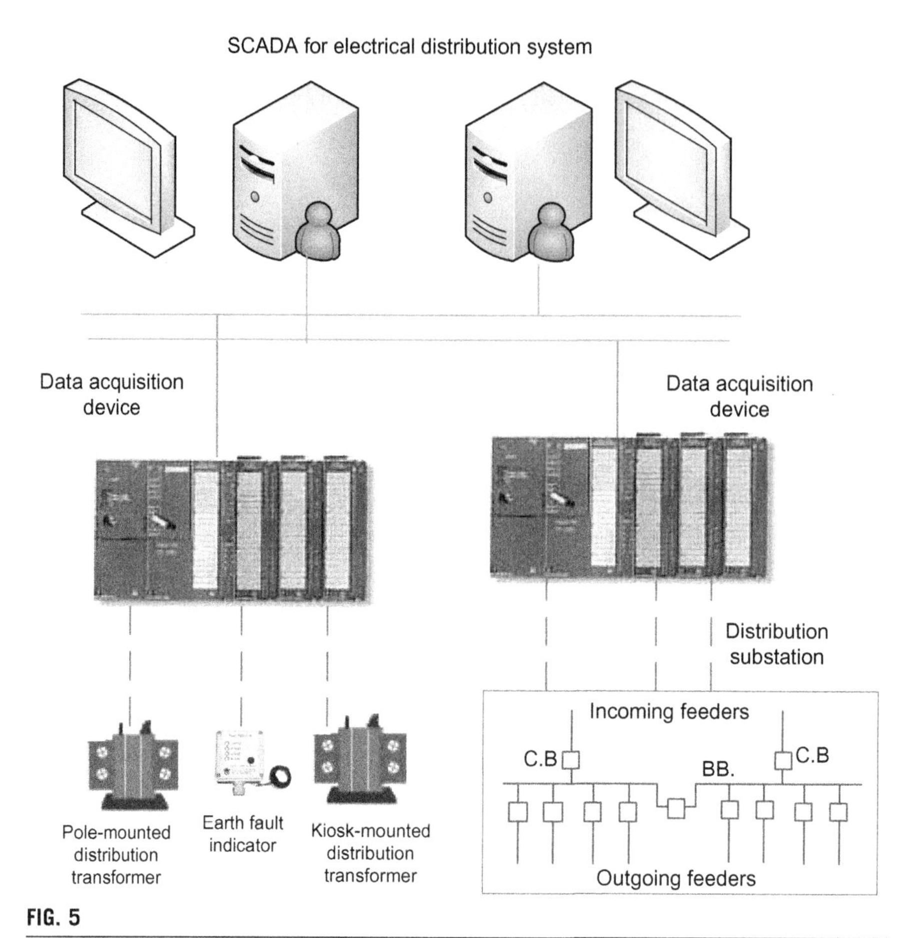

FIG. 5

SCADA for a power distribution system.

system detects the fault type and location, without waiting for customers' calls. SCADA gives an alarm when a control point exceeds or violates the limits. The events and alarms are passed to the operators to identify and analyze, depending on the network operation condition. The SCADA in substations automatically controls the CB, transformers, bus couplers, and earthing switches for exceeding parameter limits; therefore, continuous inspection of network status and parameters are performed regularly without a line worker. Some of the SCADA functions in power distribution system are given as follows:

- Improving power quality by improving power factor and harmonics contents
- Limiting peak load demand
- Real-time monitoring and controlling of power network
- Trending and alarming to enable fixing outage problems

- Historical data and archiving
- Quick response to customer service calls
- Motor control and integrating with SCADA
- Power control algorithms
- Load shedding and load restoration.

18.5 SCADA IN SOLAR PV PLANTS

PV power generation system can be divided into standalone PV systems and grid-connected PV systems. Grid-connected PV power plants consist of a PV array, converter, EMS, and other several parts. A typical distributed network of PV power plants is shown in Fig. 6. An SCADA system can be employed to be a subsystem of EMS in PV power plants. Its core part is an RTU. An SCADA system could manage the PV system using data streams in the range of several thousand measures per second. SCADA software configured to meet solar PV application requirements provides flexibility in controlling and monitoring the different PV plant components, including inverters, trackers, CB substations, and meters.

In the case of a small-scale solar PV system, it is important to assess how much energy the system can produce according to a specific location, orientation, and plant power conversion efficiency. An efficient monitoring system is important to account for the amount of energy produced by a PV system in real time, and to guarantee the forecast conversion efficiency will be accurate throughout the PV panel's service life. A digital meter also is used to measure voltage, current, frequency, active/reactive power, as well as the energy produced. Data loggers supply data into the SCADA database, which can store it for use later. Data loggers record solar radiation data, which defines the number of sunny days in a specific location to be used for prediction and statistics; solar PV system voltage, current, and battery charging to estimate total power production and system efficiency; and solar thermal collection, which provides relevant information for estimating capital costs. The digital meters have serial port communication interfaces such as RS485/RS232 or Ethernet that allow PC to access the data. Most of the electricity supply providers in the world have developed guidelines and standard criteria for interconnecting renewable energy resources into their electric distribution networks. During minimum loading conditions, solar PV installations efficiently reduce the customer load and might inject energy back to the utility grid using a net energy metering package. In IEEE P1547.6, the IEEE recommended a group of standard guidelines to guide PV system users in design and installation of PV systems that can be connected to the utility grids.

Industrial SCADA software should meet its PV application requirements. This SCADA software should provide flexibility in controlling and monitoring the various PV plant components and measurements, including maximum power point trackers, inverters, substations, CB, and meters. For monitoring purposes, the system

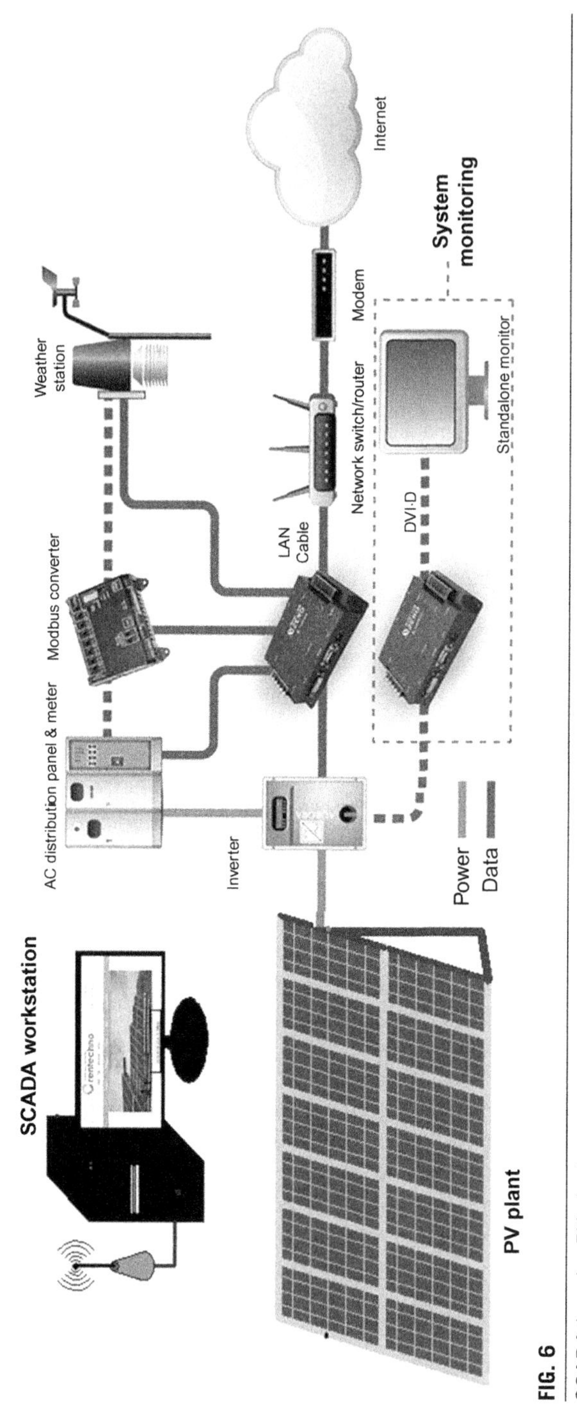

FIG. 6
SCADA in solar PV plants.

records any event and generates alarms so that the engineering staff can order switching action or change the process of plant operation.

An SCADA system monitors PV farm performance by comparing it with the PV plant design datasheet, which includes PV modules' voltage, current at maximum power, open circuit voltage, short circuit current, DC-AC inverter, and charge regulators specifications provided by the manufacturer. PV manufacturers also give electric parameters such as the number of strings and string lengths. The PV model is continuously supported by local weather data and calculates real-time energy production with the full capacity rating. A comparison is made automatically between the calculated and the real production figures supplied by the station data logger. This comparison gives an accurate designation of the station's operational performance and checks the operation status frequently at a predefined time interval. Recent monitoring and performance evaluation of solar PV stations have become highly important because of the high maintenance and operation cost and also because of a reduction of energy produced from aging and degradation during the life cycle of PV plant components. The use of an SCADA system can be essential to ensure high efficiency, less downtime, and fault diagnosis of a solar PV power station during its entire lifecycle.

From a design point of view, it is important to understand how data acquisition in the PV system works, starting from the PV module output. In large solar PV installations, multiple solar PV modules are combined together in parallel to multiply the string output currents to higher levels using DC combiner. The string combiner boxes have consolidated transducer units that measure the analog values of DC voltage and current and then calculate the power. The data acquisition device (e.g., RTU) makes the collected data available through a serial communication port or wireless communication for interfacing the SCADA system via Modbus. In some cases, the RTUs installed at the site substation can be connected to multiple string combiner boxes on the multi-drop serial communication port, RS485. While on the AC side, DC/AC inverter offers RS485 ports to allow a proper connection for the SCADA interface. The communication software drivers collect time-stamped data from control boxes and RTUs for computer processing, violation alarming, report generation, archiving storage, and displaying in real time. All the DC and AC side measurements, status, substation alarms and failures are acquired continuously. The SCADA system also interacts with digital protective relays, digital power meters, environmental monitoring station, low tension (LT), and high tension, control panels, inverters, DC switches, charge regulators, transformers, and, in general, any devices installed in the PV plant. In order to make PV applications effective, scalable, and potentially sustainable, other design aspects of the SCADA platform can be taken into account. These aspects include a dynamic configuration, redundancy for data protection, standalone and client-server configurations, historical and real-time values, graphical trends, oscillography analysis, and advanced management of system alarms. Regarding standards compliance, communication protocols such as IEC 61850 and DNP3 are key elements to if SCADA is required to communicate with several PV plant devices. The application software such as the graphical user interfaces (GUI), report generator, switching management, and scheduler, will enable easy access of all data. Another option is Web-access functionality, which provides all kinds of capabilities and can access a remote site.

18.6 IMPROVING WIND-FARM OPERATION USING SCADA

The SCADA application for wind farm operation can improve the actual power conversion efficiency. Modern wind farms contain a variety of measurement devices, SCADA components, and communication systems. For reliable, safe, and automated control, individual wind turbines use microprocessors with closed-loop control algorithms to regulate internal control, which include pitch, yaw, generator, and supervisory controllers. The safety system is hardwired independently of the main controller for fail safe, i.e., for emergency shutdown. The historical data about the wind farm response because of wind speed variation and wind gusts can be gathered in the wind energy assessment study. This study includes wind resource measurement, data analysis, site assessment, and wind performance evaluation, due diligence and forecasting [21]. The SCADA dataset is composed of four stages: wind speed, energy conversion system, effect of temperature, and vibration issues [21,22]. The SCADA data usually are analyzed for fault identification to aid the operation experts in a fault isolation and service restoration. The monitoring of the wind farm enables modeling the power curve of a specific wind turbine. The graphical analysis is a method for extracting and processing information on a daily basis from the meteorological mast database and from the SCADA systems database. The graphical method also is an appropriate and meaningful way to highlight the most regular power production and nacelle wind speed variations along the wind farm. It is important to monitor the farm performance in order to detect possible malfunctioning of individual turbines and quantify the influence of variations on the electrical output power production.

Reducing operation and maintenance costs can be achieved by improving wind turbine performance. Lessening these costs can reduce the payback period greatly, help level the cost of electricity, provide a motive for investors, and support spreading this type of clean energy source [22]. Maintenance costs can be reduced by using SCADA technology to control and monitor onshore and offshore wind turbines. Wind turbines usually work in stiff, remote environments, and, for this reason, need maintenance on a regular schedule. The outage maintenance or unscheduled maintenance because sudden failures can be costly because of lost production time and in addition to maintenance expenses. As wind turbines age, parts can fail and energy production degrades, thereby increasing maintenance costs as a percentage of power production. The field data is analyzed to enable preventive maintenance rather than rely on time-interval-based maintenance. The monitoring of wind turbines discovers failures before they reach a serious state, increases equipment lifetime, keeps the electrical and mechanical equipment working at rated capacity, enables better planning and logistics, and can reduce the need for regular maintenance.

Condition monitoring systems for wind farms have focused on detecting failures in the main generator, bearings, and gearbox, which are the highest cost components of a wind turbine [21,22]. These are standalone systems that require the installation of sensors and hardware. An SCADA-based condition monitoring system uses collected data to prevent catastrophic failures by an early alarm of primary faults. Different approaches are available to use SCADA data to predict the health of operational wind turbines [23–29]. Wind turbine reliability remains an important focus for wind turbine owners, operators, and manufacturers [30], and condition monitoring can improve reliability and reduce downtime [31]. The use of SCADA data for wind turbine condition monitoring has a number of advantages, in particular, the data already are collected, so no additional hardware is required.

Yaw controller rotates the wind turbine into or out of the essential to mitigate fatigueloads, maintains optimum electrical energy production, unwinds power and control cables between nacelle and base or generator controller which adjusts torque to maximize electrical power output and maintain rotor speed below the rated one. It is also actively dampen drive train torsional vibrations by applying small ripple torque closer to drive train natural frequency and phase angle.

Wind farm supervisory controller is necessary to operate wind turbine autonomously from one operational condition to another at the start-up, power production, and shut-down conditions. The primary functions of the proposed control and management system are: (1) Supervision and control the interconnection of the wind turbine power plant to the utility grid, (2) Control the performance of the generator and power converters output, (3) Optimizing the energy conversion efficiency of the wind turbine, (4) Providing system performance measurements for operator evaluation, (5) Achieving safe shutdown under normal and fault conditions, (6) Operation and monitoring of auxiliary equipments such as cooling of gearbox, fans and pumps, heaters for cold weather applications and lubrication pumps.

The SCADA server might be located on site or in a control room, depending on communication access to the wind turbines, which means one SCADA might monitor multiple wind turbines regardless of their location. An IP connection could link single turbines or small clusters of turbines so they are continuously monitored and controlled by an SCADA server in the control room, Fig. 7. The OLE for process control (OPC) server is a software program that converts the PLC communication protocol into the OPC protocol. The client software is the application program that needs to connect to the hardware, such as a power plant human machine interface (HMI). To get data, the OPC client uses the OPC server and send control commands to the hardware.

SCADA software enables monitoring wind energy systems, and allows controlling individual turbines, as well as the entire wind farm. The collected data help in blade design, aeroacoustic measurements, and ways to obtain optimum performance. SCADA functions are divided into three main functions that cover all areas of an integrated SCADA solution:

- Real-time monitoring and control of wind farms
- Handling events and alarms for the wind farm subsystems
- SCADA servers that collect and process the system data.

The wind farm is heavily instrumented and provides a lot of data about aspects such as wind speed, turbine output power, blade angle, stall level, and yaw. The wind turbines can be accessed via a suitable communication method such as wireless

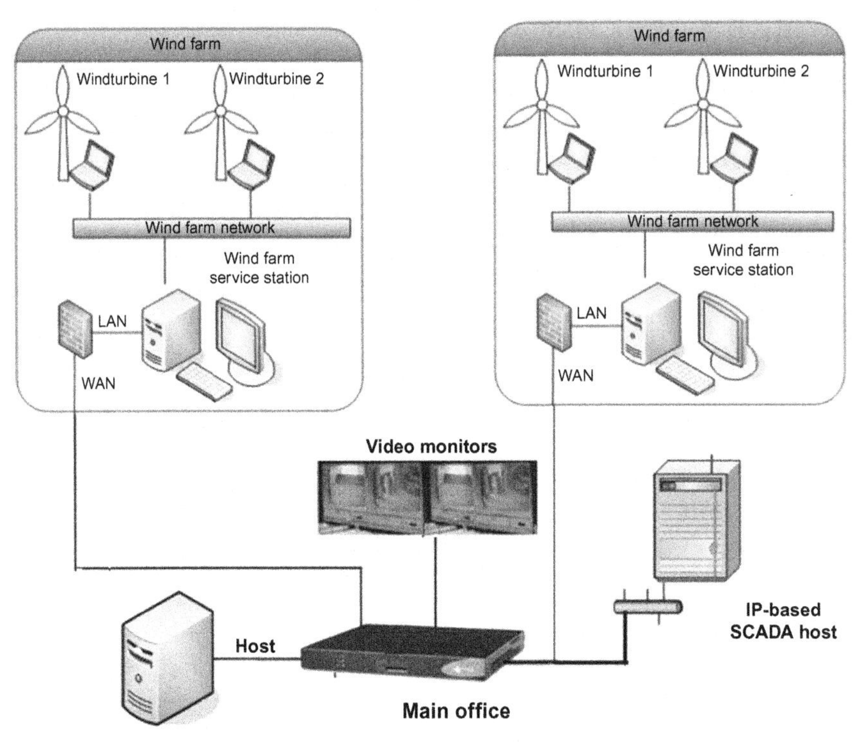

FIG. 7

Schematic diagram of SCADA system in wind farm.

communication, telephone modem, or even transport control protocol/Internet protocol (TCP/IP). The SCADA software should provide a complete control and monitoring of each wind turbine and the entire wind farm. A graphical overview of the wind farm shows the status of each wind turbine and related measurements such as voltage, current, wind speed, and production data. To determine the corresponding alarms, the software should be supported by the turbine specifications, including all relevant parameters of the used wind turbine, including the rated wind speed, temperatures, pitch angle, generator and rotor speed, rated electrical parameters, yaw control system.

18.7 FUEL CELLS CONTROL AND MONITORING

Fuel cell (FC) is a DC power generator that converts the fuel chemical energy (hydrogen, natural gas, methane, methanol) and an oxidant (air or oxygen) directly into electrical energy. While several FC technologies are available, the most popular and practical technology for a small to medium-sized standby power supply is the proton exchange membrane (PEM) FC that produces electricity through an electrochemical reaction using hydrogen and oxygen. This electrochemical reaction happens without any combustion process. An FC operates electrochemically through the use of an electrolyte, like a generator that operates as long as the fuel is supplied.

FCs are designed to provide stable electrical power while operating over a wide range of power and environmental conditions. Advantages of FCs include high efficiency, simple, quiet and clean, low maintenance and noise with few moving parts, no harmful emissions, and economical cost for stationary standby power generation [32–34]. FC systems are load-following because fuel consumption depends on the load. FC power systems are designed for a wide range of customers, including hospitals, hotels, universities, utilities, and water treatment facilities. The applications of next generation high temperature FC products, such as diesel-fueled marine ship FCs, combined-cycle FC power plants, and next generation solid oxide FCs (SOFC). The PEM FC's higher efficiency, environmental friendliness, and modularity have made this system one of the most attractive candidates for both transportation application and stationary standby power generation.

The criteria for selecting an FC standby system for a specific application include the output power requirements, frequency and duration of power failures, response time to the load, environmental restrictions, and service requirements. FCs can be used as the sole backup/emergency power solution in many critical applications, and they also can provide an added degree of protection for a site using resilience solutions. FCs can offer rack-mounting options within an equipment shelter as well as reinforced, environmentally friendly, outdoor cabinets for flexibility to meet electrical network demand. The ability to refuel FCs allows the system to run continuously as long as required during extended outages. In fact, FCs can be the selected backup power source at sites with comparatively low power loads and power failures that can last for hours or even days.

18.7.1 INTEGRATING AN FC INTO AN ELECTRICAL NETWORK

Another advantage of an FC that makes it attractive for use in standby power environments is that an FC produces DC power. It is similar to a standby rectifier power source, Fig. 8, because the power provided from the FC can directly feed the site's DC power bus. In a power outage situation, the FC automatically turns on, providing DC power formerly provided by the utility through rectifiers. This means FCs can work for long reserve times as a standby power supply in customer critical applications. FC systems are proposed to work in parallel with the traditional DC power system components. FCs can be integrated easily into an existing power network or can be designed into a new standalone network location. They can also be part of a clean power hybrid system also composed of solar and/or wind power. A variety of FCs are capable of serving loads in a variety of critical geographical locations because of hot and cold weather design features.

The SCADA supervisory controller for FC systems is necessary to operate the fuel cell and monitor all the auxiliary systems such as electrolyzer and cooling and auxiliary systems. The typical applications of FC SCADA system are:

- Cell voltage monitoring of fuel cell stacks
- Cell voltage monitoring of battery packs
- Cell voltage monitoring of electrolyzers or other electrochemical multi-cells

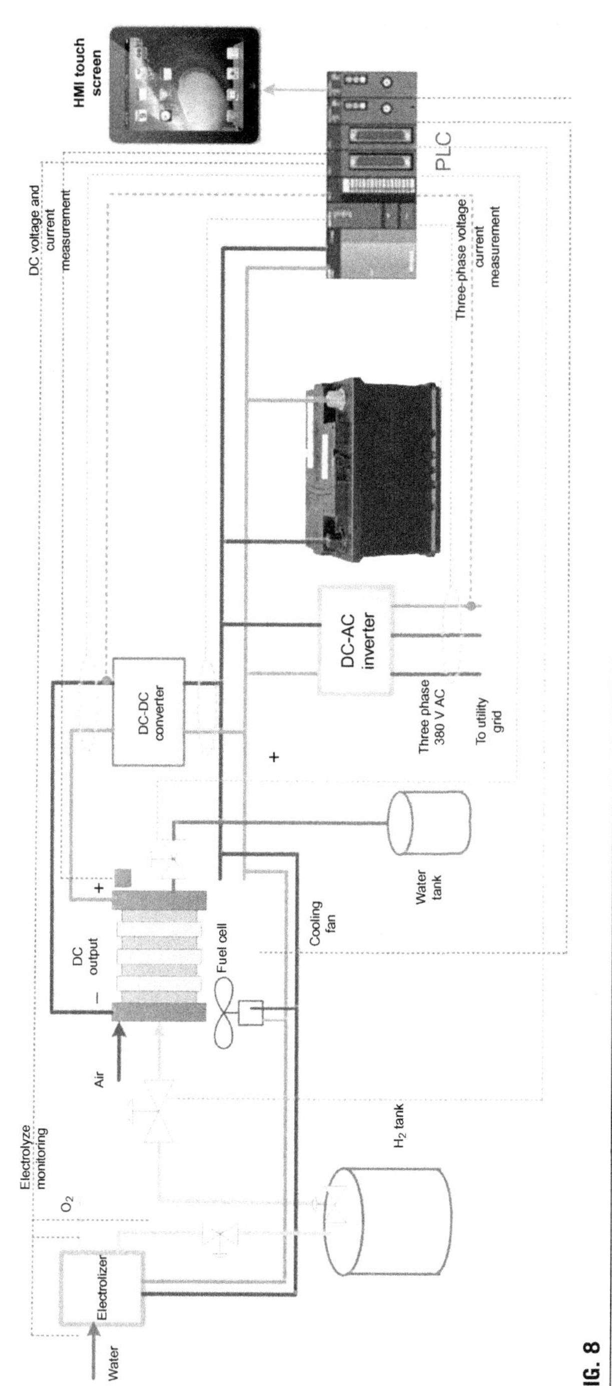

FIG. 8

Fuel cells to provide backup to their communications equipment.

- Temperature monitoring of up to four locations
- Digital input monitoring for coolant flow loss or other alarm conditions
- Protection functions using solid state relays and control output for auxiliary relay
- Operating limits and monitoring of temperature, voltage and current measurements
- Automatic shutdown of relay and contact outputs for cell over-voltage, cell under-voltage, and cell over-temperature conditions
- Monitoring indicators of power, alarm, low-supply voltage, and station alarms

18.7.2 BENEFITS OF THE FC SCADA SYSTEM

SCADA systems can provide solutions for the monitoring and control of their FC systems. The solution includes HMI/SCADA application; real-time automation software; and the alarm and events management system. An FC power system requires a user-friendly, robust HMI, secure to allow monitoring and remote-site control of the FC power stations as standalone/grid-connected systems. The SCADA system also can provide the FC power system with alarm management and real-time functionality. The dispatchers at the control room can acknowledge or take action when they alerted by an event or alarm. The Internet-enabled HMI control promotes ease of use and remote control/monitoring of that kind of power system. Another benefit is the integration into existing microgrid systems. SCADA delivers the facilitated Internet-based remote access HMI. The SCADA delivers also an event/alarm management that is required for remote monitoring/control of a standby FC power station working in either standalone or hybrid modes.

Because of the specific nature of FC systems, measurement, supervision, and control devices are substantial. It is interesting to find a system enable the real-time monitoring and control the performance of a PEM FC system. The control process occurs according to the instructions recommended by the FC stack manufacturers. The automation of FC station is based on an SCADA system that acquires data and monitors the input/output measurements and control signals [32,33]. The SCADA system stores data in an organized process database. The SCADA platform consist of three main elements: an SCADA server, the FC system controller, and memory storage device. The SCADA server provides access to the system's electrical parameters and configures interface through serial communication. The SCADA user can access the data for analyzing, reporting, or making engineering tasks for system expansion. In the beginning, the main screen of the SCADA system enables users to select the role in which they wish to work (operator, engineer, or administrator). The software tool has a detailed description of the FC plant and rated input/output values, rated operating parameters, proper control interface, and station operation. The SCADA software traces the dynamic changes in the site and sends data to the master station control room, which supervises the whole system.

The FC current and voltage transducers are being connected to analog input terminals of RTU, and thus monitoring voltage and current, as shown in Fig. 9. If an FC is run above its rated current for long periods of time, it can contribute to equipment

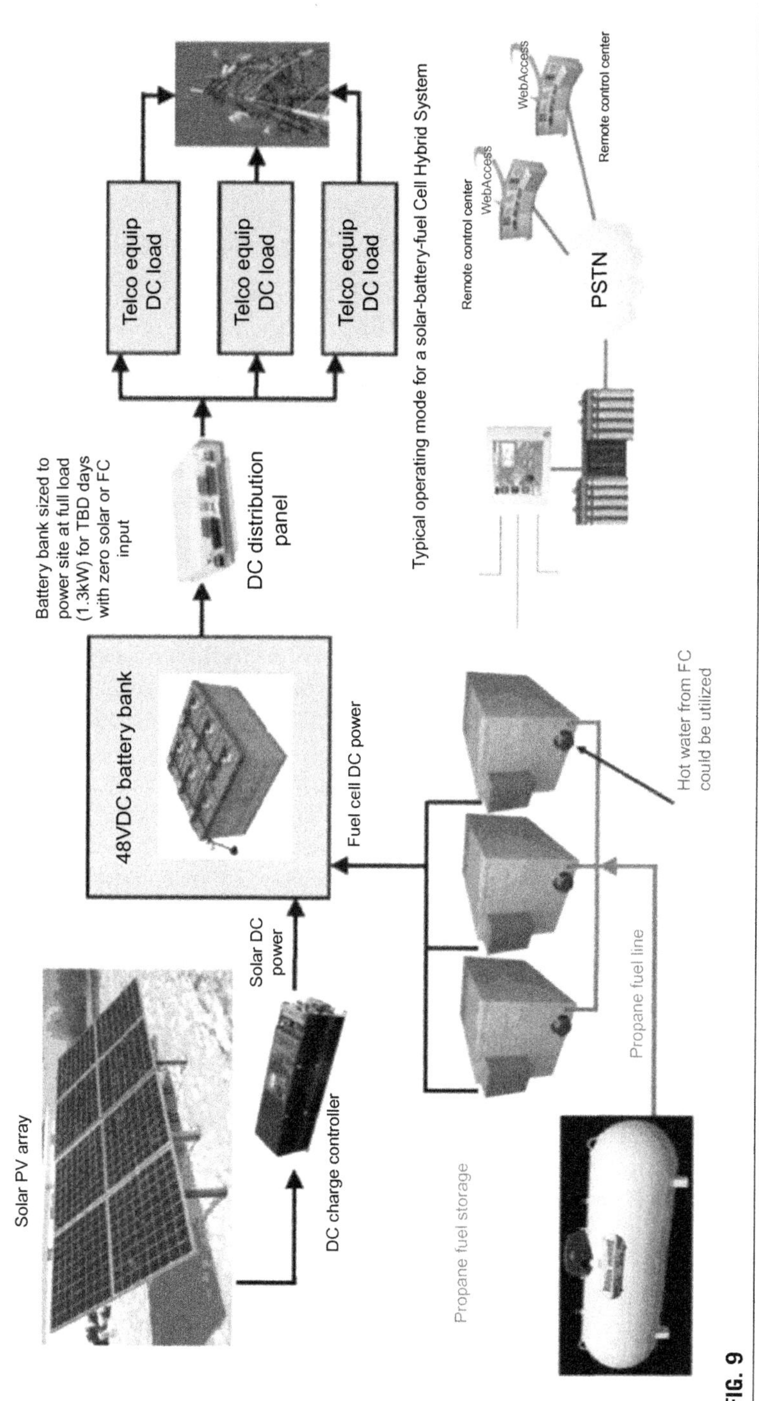

FIG. 9

Schematic diagram of SCADA for standalone fuel cell hybrid system.

degradation, so monitoring the fuel voltage guarantees that increases in FC stack current doesn't drive the cells into reversal. Cell reversal results in heat, and can result in a cell-to-cell short circuit and an electrical arcing. A blocking diode is placed in series to prevent reversal currents, and a protective relay is added to protect the FC system from external faults. The SCADA is responsible for receiving information from all the subsystems and enable them to work simultaneously. An external power source is required to supply all the devices in the balance of plant, including the FC. This external power source is necessary because the balance of plant devices should be powered even before the FC system begins working, and because of the extra time required for the FC system to reach steady-state operation and deliver rated electrical power [32].

18.7.3 CONTROL AND MONITORING SYSTEMS

Several attempts have been made to design FC systems to significantly improve their performance, reliability, efficiency and durability. In some types of PEM FCs, the main reason of a short lifecycle and degradation are thermal stress and water, corrosion, reduction of fuel and oxidant, and chemical reactions of the FC components. Thermal management is essential at cold start-ups, while bad water management can result in dehydration or flooding [33]. For these reasons, the operating procedure of FC system should be designed to ensure high efficiency, good performance, durability and safety. These procedures are established by the supervisory control system. The SCADA must observe the measured values of different variables and keep them within the admissible range to avoid sudden FC damage and irreversible performance degradation. If the FC system gets in a critical alarm condition, it is required to carry out a forced shutdown. The robustness and resilience nature of the network is essential, regardless of the type of communications network that is used for SCADA systems. Recent natural disasters have demonstrated the importance of communication systems to microgrid and utility grids. To have a resilient power network, the communications equipment also must provide resilience operation and reliability. FCs increase power reliability of communication networks by providing backup/standby power to many sites throughout the distribution networks.

18.8 USING SCADA IN HYBRID POWER SYSTEMS

Energy demand—either in standalone or grid-connected modes of power systems—is increasing steadily. The challenge is to meet that increasing demand for power, while acknowledging the public interest in global environmental concerns, such as the effect of greenhouse gases and global warming, and the reduction in fossil fuel resources. These problems can be solved by recent research and development of alternative RES. Introducing RES, such as solar PV, FC, and wind energy has an excellent potential of contribution to power generation. These sources are clean and abundantly available in nature, and offer many advantages over conventional sources of power generation. Combining two or more RES forms a hybrid power

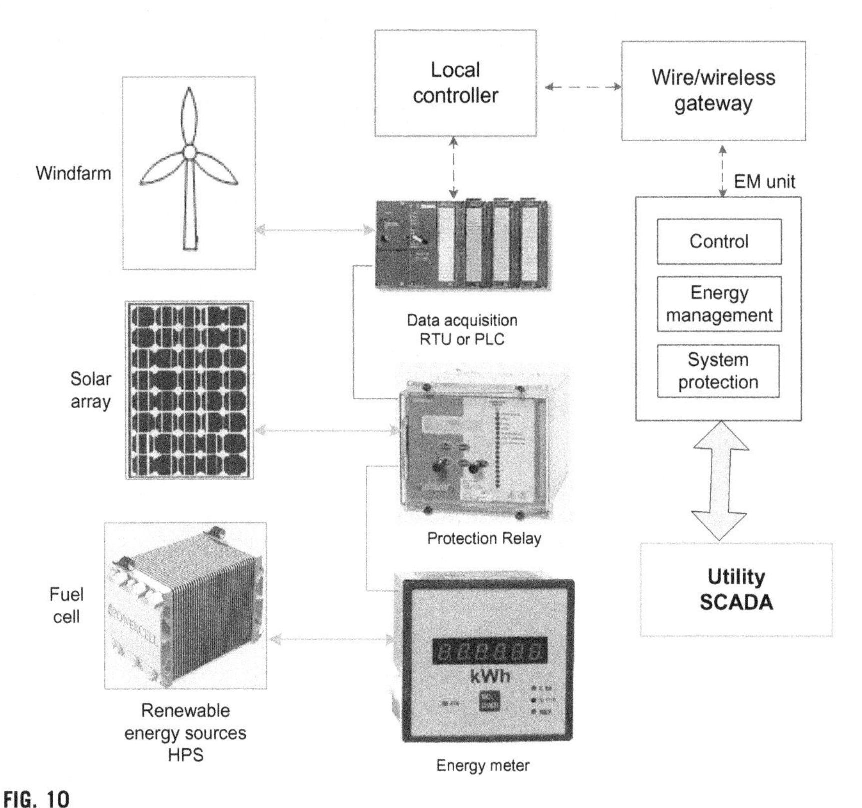

FIG. 10

Schematic diagram of renewable energy-based hybrid power system.

system (HPS). A schematic diagram of a hybrid renewable power system is shown in Fig. 10. The reliability of an HPS mainly relies on the dynamic behavior of the RES, so it is important to analyze these dynamic characteristics in real time for long time periods. A major challenge lies in the development of a real-time control scheme for the HPS. To test the HPS controllers, a controller must be able to interface with a hardware simulator and process the inputs and outputs in real time. Today, the improvements in communications systems have prompted the implementation of the HPS using PLCs and RTU control schemes by a centralized supervisory control platform, which is commonly known as an SCADA system. All the modules should be modeled in the PLC environment. PLC and SCADA system communicate with each other over a dedicated protocol that can be transmitted over a serial port or Ethernet.

The control structure of HPS is shown in Fig. 10. The monitoring of PV cells and FCs is done in PLC and is interfaced to SCADA/EMS so that the operator can observe the parameters easily and control the parameters according to the changes in system requirements. The data of HPS is collected through PLC and connected to the control room through a communication protocol. The detailed connection of RES, such as

HPS, is shown in Fig. 10. The PLC is interfaced to the energy management control unit to get control actions according to the respective microgrid loads and environmental conditions. The entire system is connected to SCADA for supervision and control. The control room contains various I/O consoles, such as engineering and operator consoles. The engineering console is responsible for adding new points or new IED devices to the system.

18.9 SCADA SYSTEM ELEMENTS

The real-time data collected by the SCADA system is then passed to the planning engineers for consideration in development studies of the distribution system. Because of growth of the electricity distribution industry, utilities make annual investments to improve efficiency of the electric distribution system to satisfy the growing load requirements. Real-time data enable engineers to effectively plan the annual capital expenses needed to face the needs of the growing electric network. Electrical power quality issues include reduction of harmonic content to an acceptable total harmonic distortion (%THD) that is defined by international standards. This monitored information is used as a key performance guide of the electric power network.

In general, of the main elements of an SCADA system are:

1. Host equipment or master station hardware
2. Communication infrastructure (communication networks and equipments)
3. Field devices or outstation hardware to support control operations and telemetry requirements of a DMS platform

The components of the SCADA automation system in the master station can be divided into four major areas:

1. SCADA and database servers
2. DMS/EMS applications and server(s)
3. Trouble management/switching management applications servers
4. Front end processors (FEP) or communication servers

The above components of SCADA system usually are intended to perform four functions:

1. Data acquisition
2. Networked data communications
3. Data display and presentation
4. Control functions

18.9.1 HOST COMPUTER SYSTEM

The host computer network usually is known as a master station or simply control room, or SCADA control center. The central host computer could be a standalone single computer or a network of computer servers. This computer network provides process control remotely or locally for all SCADA system devices, supporting requested supervisory control and a remote method of acquiring measurement data, alarms and events for monitoring power supply processes. The SCADA host platforms also provide functions for dynamic graphical displays, alarming, logging, variables trending, and historical data storage.

The essential elements of SCADA platform are:

1. Host servers (redundant server network with backup/failover capability)
2. Communication FEP
3. Full functionality graphics user interfaces
4. Relational database servers for archiving power system historical events. The data/Web servers are enabled for access to system points in real time (measurements values, CB status and events)

The major elements and system components of the typical master station or control center are illustrated in Fig. 11.

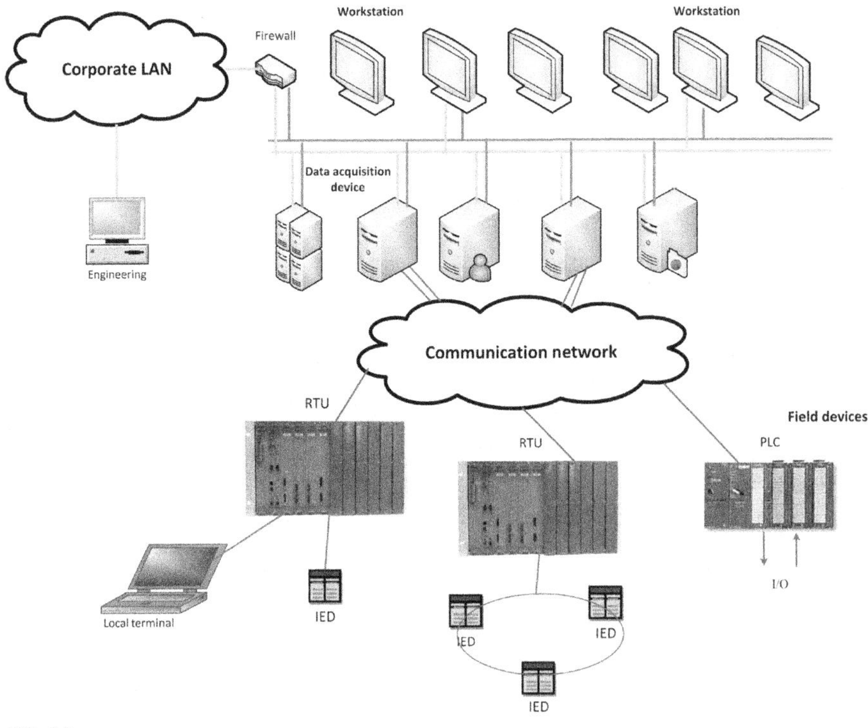

FIG. 11

Master station system architecture.

SCADA servers

As SCADA systems have proven their vital value in operation during stormy weather conditions, fault isolation, service restoration, and daily operations, the reliability of SCADA has become a requirement for highly reliable, available, and efficient-performance systems. Redundant server network hardware operating in a "live" backup/failover mode is required to withstand the high availability design criteria. High-performance servers with abundant physical memory are also required, in addition to, redundant array of independent disks hard disk storage systems.

Communication servers or FEP

The current state of the host to field communications devices still depends heavily on serial communications interfaces. This requirement is satisfied by using the FEP, which can be organized in several forms based on data bus architecture and operating systems. The location of the FEP in relation to the SCADA server can vary based on system requirements. Some configurations locate the FEP on the LAN with the SCADA server, but in other cases, existing communications hubs might dictate that the FEP reside at the communication hub. The incorporation of the wide area network (WAN) into the architecture requires a more robust FEP application to sustain reliable communications as compared to LAN configuration. In general, the FEP will include three functional devices:

1. A network/CPU board
2. Serial cards
3. Possibly a time code receiver

The FEP's functionality should include the ability to download configuration and scan tables. The FEP also should support the ability to report limit violations for those analog values that have changed by a user-defined amount. FEP network and SCADA servers should have the ability to deal with worst-case conditions (i.e., all points changing outside of the dead-band limits), which typically happens during severe system disturbance conditions.

18.9.2 FULL GRAPHICS USER INTERFACE

The recent trend in the GUI is a full graphics (FG) user interface. Although character graphics consoles still are used by many utilities, SCADA vendors are aggressively moving their platforms toward an FG UI. FG displays prove the ability to display power system dynamic microgrids and networks, renewable power plants, along with the electric distribution substations in a geographical or semi-geographical perspective. The advantage of using an FG interface becomes evident in particular for electrical energy control centers, as SCADA is developed beyond the substation control room where feeder diagrams become critical to distribution operations.

18.9.3 RELATIONAL DATABASES, DATA SERVERS

A relational database is defined simply as a collection of data points organized as a group of formally described tables where data can be entered or reassembled in many different ways without having to reorganize the database tables. Power system quantities such as CB status, bus coupler status, digital alarms and feeder loading (MW, MWH, MVAR, and three-phase ampere loading), and bus volts support valuable information to the energy systems planning engineer.

The availability of event data loggers is important in fault analysis. Using relational databases, data servers, and Web servers by the operation and engineering functions provide access to power network information and data while preventing the SCADA server from non-operating personnel.

18.9.4 HOST TO OUTSTATION HARDWARE COMMUNICATIONS

Serial communications to outstation devices can be implemented through several media: copper wire (RS485/RS232), fiber, radio, leased line, and even satellite. Leased telephone circuits, fiber, and satellites, however, have a comparatively high cost, and new radio technologies offer an attractive communications solution. One of such technologies is the multiple address radio system (MAS). The MAS usually operates in the range of 900 MHz and is omnidirectional, enabling radio coverage in an area with radius up to 25 miles, depending on terrain. A single MAS master radio can collect data from many remote sites to a concentrator. Communication protocol and bandwidth can limit the number of RTU that can be connected by a master radio. The protocol limit is simply the address range supported by the protocol. Bandwidth limitations can be substituted by the use of effective protocols, or by slowing down the scan rate to include more remote units. Combining spread-spectrum and point-to-point radio with MAS offers a chance to address-specific communication issues. MAS radio currently is preferred to packet radio; MAS radio communications tend to be more suitable for smaller timeout values on communication responses, scan time, and controls.

18.9.5 FIELD DEVICES

Distribution automation (DA) or DMS outstation devices are multifeatured installations with an extended range of control, operations, planning, switching, and system performance issues for the utility personnel. Each device provides specific SCADA functionality, supports system control operations, includes fault detection, collects planning data, and records power quality information. These devices usually are found in the electrical power substation and at specific locations along the transmission line. The multifeatured capability of the DA device increases its ability to be used in the electric power system. The functionality and operations capabilities supplement each other with respect to the control and operation of the electrical microgrid and grid.

18.9.6 MODERN RTU

Modern RTU is modular architecture with advanced capabilities to support data processing functions. The modular RTU design configurations make it available for installations ranging from a small point count site, such as a pole-mounted distribution transformer, to a very large number points in a large power substation. Modern RTU modules include expandable analog input/output, digital input/output points, accumulated input units, and communication cards with power supply. RTU installation requirements are met by assembling the required number of RTU cards or rack-mounted modules to accommodate the analog, control, digital, and communication links for the site to be supervisory controlled. The packaging of the reasonable point count RTU is chosen for the distribution line requirement. The SCADA for substation automation has the option of installing one RTU in one cabinet with connections to the substation IEDs or installing a master RTU and number of slave RTUs at different locations within the substation and connected to the master RTU with fiber-optic communications. The distributed RTU modules are integrated to a data concentrating unit, which, in turn, communicates with the host SCADA computer system.

The modern RTU can accept AC inputs from a variety of measurement (digital meter) and protection devices (digital relays) through a serial port. The RTU can take the measurements as hardwired from CT, potential or VT, station service transformers, and transducers. Direct AC inputs with the processing capability of the modern RTU support fault current detections and harmonic content measurements. The fault location algorithm can be embedded in the RTU firmware, giving the RTU the ability to report the direction, magnitude, and duration of fault current with time tagging of the fault event to 1-millisecond resolution. Monitoring and reporting of harmonic content in the distribution feeder also could be included in the RTU. The digital signal processing capability of the RTU supports the necessary calculations to report THD for each of the three-phase voltage and current measurement at the automated distribution feeder or substation site. The RTU configuration software can include the logic capability to support the development of control algorithms to meet specific operating requirements. Automatic transfer control schemes have been built using automated switches and RTUs with the high processing capability. This high capability provides another benefit to the design engineer when developing the quality of service and addressing critical load concerns. The logic capability in the RTU has been used to create an algorithm to control switched capacitors for operation on the distribution feeder. The capacitors are switched on at zero voltage crossing and switched off at zero current crossing. This algorithm can be specified to switch the capacitors for various system specifications, such as voltage, power factor, reactive load, and time. The remote control capability of the RTU allows the distribution operator to take control of the capacitors to meet inductive load needs. These new logic and input capabilities have allowed the modern RTU to become a dynamic device with increased processing capabilities, uses, and applications.

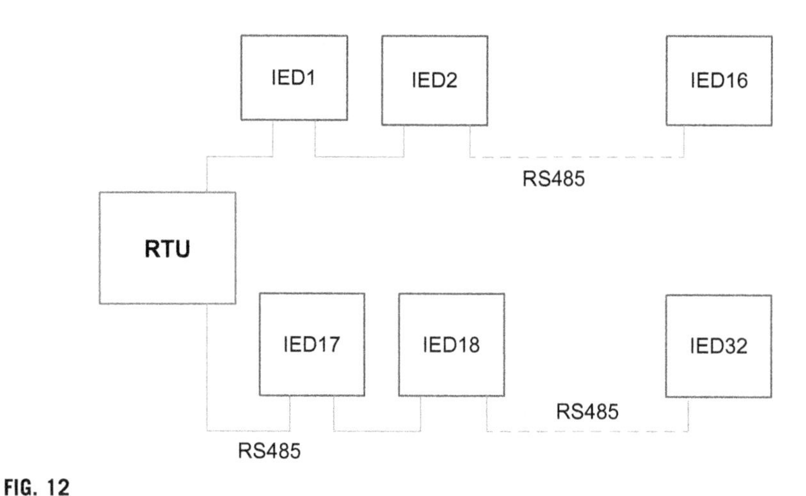

FIG. 12

Connection of IEDs to the RTU using serial port RS485.

18.9.7 PLCS AND IEDs

Programmable logic controllers (PLCs) and IEDs are the main site components of the distribution automation system, which meet specific operating and data gathering requirements (see Fig. 12). The IEDs in substation include digital protection relays and digital meters. While there is some overlap in capability with the modern RTU, the PLCs can be integrated with the RTUs in the substation to assist in its remote operation. The typical PLC can support serial communications to an SCADA server. The RTU has the capability to communicate via an RS485 interface with the PLC. IEDs include digital meters, digital relays, and switchgear on specific substation equipment, such as breakers, switches, regulators, and load tap changer on power transformers. The IEDs have the ability to communicate with an SCADA server through serial or Ethernet ports, the IEDs typically report to the modern RTU via an RS-485 interface or via status output contact points. RS-485 is preferable to RS-232 because it supports long distance, high speed, and several IEDs can be connected in series. Recently, IEDs become attractive in the automation process because of improved communication capabilities.

18.9.8 RECENT TRENDS IN SCADA COMMUNICATIONS

As in many traditional data communication networks many types of communication methods are supported from PLC/RTU to central/distributed SCADA systems via local and wide area networks. A local area network is included within a small geographical area, such as an industrial building or a campus and might include a few buildings in close proximity. A wide area network, however, integrates many LANs spread widely across different cities at least 100 km (67 miles) apart [35]. These types of WANs include the following:

- Analog point-to-point and multipoint modem networks
- Frame relay type point-to-point and multipoint networks
- Wireless radio/satellite communication networks
- Fiber-optic-based communication networks

An SCADA network currently might be built around many of the above possible collection of networks and transmission protocols. Communication solutions for SCADA platform, include the physical wiring, network and communication protocols of the local and wide area networks such as Ethernet and frame relay. TCP/IP suite is a different, but similar open standard used by all).

The RTU/PLC protocols are emerging as practical standards in SCADA systems such as Modbus and distributed network protocol (DNP). Fieldbus, including technologies such as Modbus, CANbus, Profibus, LonWorks, and many other technologies, is mainly used to solve data communication between intelligent sensors, digital meters, controllers, digital protection relays, actuators and other outstation devices. It currently is the most popular used and comparatively functional industrial communications technology.

The point-to-point Modbus protocol has become a practical standard for RTU and PLC communications. During communication on a Modbus network, the protocol specifies how each controller will know the device address, recognize a message addressed to it and determine the required action to be taken and evolve any attached data to it. DNP, a member limited protocol used in some power systems, has various versions, with the latest being version 3.0. The DNP association has rules, which tend to limit the utilization of the protocol, and main SCADA software suppliers have been slow in implementing this protocol.

Industrial Ethernet is technically compatible with IEEE 802.3 standards and is widely used in industrial SCADA and control systems. The industrial Ethernet, however, should meet the requirements of the industrial field in real-time, material selection, reliability, product strength, and environmental applicability. Its main technical advantages are wide range of applications, the high speed of communication, security of control network, and support of a variety of physical media and topologies.

Telecommunications

In the world of communication technology two types of networks exist: circuit switched and packet switched. The circuit switched network establishes direct connection between two or more stations by means of switches, normally done with telephone dial-up modem networks. There is, however, a general shift toward a packet style operation where the data is handled in packets prefixed with some addressing. In a packet-switched network, data is routed through best possible route in a complex meshed public, private, or local area network. Packet switched networks are more cost-effective because a dedicated network is not requested from start to finish.

Wireless networks

Wireless networks come in many flavors and styles. The mechanisms of large-scale remote industries can be controlled more effectively and safely using wireless SCADA, which is the most economical and time-saving technology. SCADA is used in the power industry with the perfect human machine interface. It has solved many issues related to supervision, data acquisition, controlling, and monitoring. It has divergent applications such distribution management, energy management, power plant management, water treatment, and oil and gas distribution system. SCADA also has enabled grid monitoring so the power can be bought and shared on a national basis. The application of SCADA is beneficial to the power systems sector, as well. These wireless networks include

- Satellite networks
- Licensed VHF and UHF point-to-point and multipoint radio
- Spread spectrum license free (900 MHz, 2.4, 5.8, and 24 GHz)
- Point-to-point microwave

Within a narrow range of buildings or a campus, wireless data can be moved from node to node with privately owned spread spectrum radio networks. Broader ranges require some form of public network. The most common method is dial up.

Power line carrier communication

Power line carrier (PLC) communication technology transmits data through the power line. The PLC technology is making use of high-voltage power lines (10 kV or above, and low-voltage power lines 220 or 380 V) for the development and promotion of remote meter reading. The use of PLC is expanding into the distribution lines for load control and even into households to control lighting, alarm systems and air conditioning and heating. The main application, however, is protective relaying for transmission lines. A channel is used in line protection so that both ends of a circuit are cleared at high speed for all types of faults, including end zone faults. A PLC channel also can be used to provide remote tripping functions for power transformer protection, shunt reactor protection, and remote breaker failure relaying.

The main advantage of PLC is eliminating the difficulty of installing additional dedicated communication cables. The disadvantages of PLC are obvious, too. The power line is a very bad channel for communication, mainly because of interference and signal attenuation. Interference comes from power electronic devices, low-voltage load, switch operation, and broadcast signals. In such a noisy environment, it is difficult to ensure data quality. Signal attenuation is brought about by the complicated structure of the power grid, which gives signals multiple transmission paths. The power line communication environment, therefore, is too severe to ensure reliability and stability. The solar PV power station has many power electronic devices, such as inverter, static var compensator, and static var generator. These devices arouse harmonic interference into the AC power line, so PLC should not be used on the AC power line. The DC power line between PV convergence box and inverter, however, has less interference and a single transmission path, and can implement PLC technology.

18.10 CONCLUSIONS

In this chapter, the smart grid initiative was explored, and integration of SCADA systems into the smart grid was described, including an overview of the problem domain as a whole. The evolution of the smart grid initiative to improve the electric utility power infrastructure has brought with it a number of opportunities for improving efficiencies and performance. Along with those benefits, however, come challenges in the effort to assure safety, security, and reliability for microgrids, utilities, and consumers alike. One of the considerations in designing the capabilities of the smart grid is the integration of supervisory control and data acquisition (SCADA) systems to allow a utility to remotely monitor and control network devices as a means of achieving reliability and demand efficiencies for utility sectors. Communication technologies in smart grid with renewable energy sources are explored.

ABBREVIATIONS

CANbus	Controller area network
CB	Circuit breakers
CT	Current transformer
DA	Distribution automation
DAC	Data acquisition computer server
DC/AC	Direct current/alternating current
DCS	Distributed control systems
DMS	Distribution management system
DNP	Distributed network protocol
EMS	Energy management system
FC	Fuel cell
FEP	Front-end processors
FG	Full graphics
Fieldbus	A family of industrial computer network protocols
GUI	Graphical user interface
HMI	Human-machine interface
HPS	Hybrid power system
I/O	Input/output
IEDs	Intelligent electronic devices
LAN	Local area network
LonWorks	Local operating network; a networking platform
MAS	Multiple address radio system
Modbus	A serial communications protocol originally published by Modicon (now Schneider Electric)
MPPT	Maximum power point tracking
OLE	Object linking and embedding

OPC	OLE for process control
PCS	Process control systems
PEM	Proton exchange membrane
PLC	Programmable logic controllers
PLC	Power line carrier
Profibus	Process field bus; a standard for fieldbus communication
PV	Solar photovoltaic generation
RAID	Redundant array of independent disks (hard disk storage system)
RES	Renewable energy sources
RTU	Remote terminal unit
SCADA	Supervisory control and data acquisition
SG	Smart grid
SOC	State of charge
SOFC	Solid oxide fuel cells
TCP/IP	Transmission control protocol/Internet protocol
THD	Total harmonic distortion
VT	Voltage transformer
WAN	Wide area network

REFERENCES

[1] Kato D, Horii H, Kawahara T. Next-generation SCADA/EMS designed for large penetration of renewable energy. Hitachi Rev 2014;63(4):151–5.

[2] Collier SE. Ten steps to a smarter grid. IEEE Ind Appl Mag 2010;16(2):62–8.

[3] Schneider Electric. SCADA systems. Kanata, ON: Telemetry and Remote SCADA Solutions; 2012. White paper.

[4] IEEE Std. 999-1992. IEEE recommended practice for master/remote supervisory control and data acquisition (SCADA) systems. New York: Institute of Electrical and Electronics Engineers; 1992.

[5] Wu FF, Moslehi K, Bose A. Power system control centers: past, present, and future. Proc IEEE 2005;93(11):1890–908.

[6] Chen C, Chen Y. Intelligent identification of voltage variation events based on IEEE Std 1159-2009 for SCADA of distributed energy system. IEEE Trans Ind Electron 2014;62(4):2604–11.

[7] Cresta M, Gatta FM, Geri A, Maccioni M, Mantineo A, Paulucci M. Optimal operation of a low-voltage distribution network with renewable distributed generation by NaS battery and demand response strategy: a case study in a trial site. IET Renew Power Gen 2015;9(6):549–56.

[8] Guozhen H, Tao C, Changsong C, Shanxu D. Solutions for SCADA system communication reliability in photovoltaic power plants. In: IEEE 6th international power electronics and motion control conference. IPEMC '09; 2009. p. 2482–5.

[9] Muñoz M, de la Parra Í, García M, Marcos J, Pérez M. A tool for the performance evaluation and failure detection of Amareleja PV plant (ACCIONA) from SCADA. In: 2015 17th European conference on power electronics and applications (EPE'15 ECCE-Europe); 2015. p. 1–9.

[10] Terada H, Onishi T, Tsuchiya T. Proposal of environmental adaptation for the next-generation distribution SCADA system. In: 2012 China international conference on electricity distribution (CICED); 2012. p. 1–4.

[11] Tsagaris A, Triantafyllidis DG. Data monitoring system for supervising the performance assessment of a photovoltaic park. In: 2012 IEEE 13th international symposium on computational intelligence and informatics (CINTI); 2012. p. 385–9.

[12] Chen C-S, Tsai C-T, Lin C-H, Hsieh W-L, Ku T-T. Loading balance of distribution feeders with loop power controllers considering photovoltaic generation. IEEE Trans Power Syst 2011;26(3):1762–8.

[13] Jaganmohan Reddy Y, Ramsesh A, Raju KP, Pavan Kumar YV. A novel approach for modeling and simulation of hybrid power systems using PLCs and SCADA for hardware in the loop test. In: International conference on sustainable energy and intelligent systems (SEISCON 2011); 2011. p. 545–53.

[14] Jennings C. PG&E's cost-effective photovoltaic installations. In: Photovoltaic specialists conference, 1990, conference record of the twenty first IEEE, vol. 2; 1990. p. 914–8.

[15] Yao H, Peng Q, He W, Zhang X. Integrated communication technology for supervisory control and data acquisition system of PV power station. In: 2012 s international conference on intelligent system design and engineering application (ISDEA); 2012. p. 1277–80.

[16] Mohammed OA, Nayeem MA, Kaviani AK. A laboratory based microgrid and distributed generation infrastructure for studying connectivity issues to operational power systems. In: IEEE PES general meeting; 2010. p. 1–6.

[17] Terada H, Onishi T, Tsuchiya T. A monitoring point selection approach for power distribution systems. In: 2013 8th international conference on system of systems engineering (SoSE); 2013. p. 190–5.

[18] Marinelli M, Sossan F, Isleifsson FR, Costanzo GT, Bindner H. Day-ahead scheduling of a photovoltaic plant by the energy management of a storage system. In: 2013 48th international universities' power engineering conference (UPEC); 2013. p. 1–6.

[19] Jurišić B, Holjevac N, Morvaj B. Framework for designing a smart grid testbed. In: 2013 36th international convention on information & communication technology electronics & microelectronics (MIPRO); 2013. p. 1247–52.

[20] Jayasuriya DC, Rankin M, Jones T, de Hoog J, Thomas D, Mareels I. Modeling and validation of an unbalanced LV network using Smart Meter and SCADA inputs. In: 2013 IEEE TENCON spring conference; 2013. p. 386–90.

[21] Castellani F, Garinei A, Terzi L, Astolfi D, Gaudiosi M. Improving windfarm operation practice through numerical modelling and supervisory control and data acquisition data analysis. IET Renew Power Gen 2014;8(4):367–79.

[22] Kusiak A, Li W. The prediction and diagnosis of wind turbine faults. Renew Energy 2011;36:16–23.

[23] Carvalho H, Gaião M, Guedes R. Wind farm power performance test, in the scope of the IEC 61400-12.3. In: EWEC 2010 European wind energy conference and exhibition proceedings, Warsaw, Poland; 2010. p. 4180–90.

[24] Gill S, Stephen B, Galloway S. Wind turbine condition assessment through power curve copula modeling. IEEE Trans Sust Energy 2012;3(1):94–101.

[25] Paiva LT, Veiga Rodrigues C, Palma JMLM. Determining wind turbine power curves based on operating conditions. Wind Energy; 2013.

[26] Kusiak A, Verna A. Monitoring wind farms with performance curves. IEEE Trans Sust Energy 2012;4(1):192–9.

[27] Schlechtingen M, Ferreira SI, Achiche S. Using data-mining approaches for wind turbine power curve monitoring: a comparative study. Sust Energy 2013;99:1–9.
[28] Yang W, Court R, Jiang J. Wind turbine condition monitoring by the approach of SCADA data analysis. Renew Energy 2013;53:365–76.
[29] Wilkinson M, Darnell B, van Delft T, Harman K. Comparison of methods for wind turbine condition monitoring with SCADA data. IET Renew Power Gen 2014;8(4):390–7.
[30] McMillan D, Ault GW. Quantification of condition monitoring benefit for offshore wind turbines. Wind Energy 2007;31:267–85.
[31] García Márquez FP, Tobias AM, Pérez JMP, Papaelias M. Condition monitoring of wind turbines: techniques and methods. Renew Energy 2012;46:169–78.
[32] Segura F, Andújar JM. Step by step development of a real fuel cell system. Design, implementation, control and monitoring. Int J Hydrog Energy 2015;40(15):1–13.
[33] Eker Kahveci E, Taymaz I. Experimental investigation on water and heat management in a PEM fuel cell using response surface methodology. Intl J Hydr Energy 2014;39(20).
[34] Andu JM, Segura F, Duran E, Renter LA. Optimal interface based on power electronics in distributed generation systems for fuel cells. Renew Energy 2011;36(11):2759–70.
[35] Kalapatapu R. SCADA protocols and communication trends. In: The instrumentation, systems and automation society ISA, presented at the ISA, 5–7 October 2004. Texas: Reliant Center Houston; 2004. www.isa.org.

Glossary

A

AC Alternating current: Electric current that reverses its direction many times per second at regular intervals.

Activity modeling Map of the activities that make up a process and showing their interconnections and interactions, inputs and outputs, types of resources assigned, and the nature and extent of constraints and controls.

Adaptive controllers Control method used by a controller that must adapt to a controlled system with parameters that vary, or are initially uncertain.

B

Backtracking search General algorithm for finding all (or some) solutions to some computational problems, notably constraint satisfaction problems, that incrementally build candidates to the solutions, and abandons each partial candidate.

Bacterial foraging optimization algorithm Inspired by the group foraging behavior of bacteria such as *Escherichia coli* and *Myxococcus xanthus*. BFOA is inspired by the chemotaxis behavior of bacteria that will perceive chemical gradients in the environment (such as nutrients) and move toward or away from specific signals.

Batteries life cycle emissions Emissions to environment over the lifetime of batteries.

Battery storage Collection method used to store electrical energy on a large scale within an electrical power grid.

BBC Big bang crunch: Optimization algorithm inspired by one of the theories of the origin of the universe.

Braking energy Energy recovery mechanism that slows a vehicle or object by converting its kinetic energy into a form that can either be used immediately or stored until needed.

C

CA Clustering algorithm: Algorithm used to classify a given data set through a certain number of clusters.

CAN Controller area network (CAN bus): Standard designed to allow devices to communicate with each other in applications without a host computer.

CB Circuit breakers: Electrical switches designed to protect an electrical circuit from damage caused by overcurrent, overload, or short circuit.

Combined heat and power Simultaneous generation of electricity and useful heating and cooling from the combustion of a fuel or a solar heat collector.

Communication Sending and receiving signals especially by means of electrical or electroacoustic devices and electromagnetic waves.

Control loop Controller seeks to maintain the measured process variable at set point in spite of unmeasured disturbances.

Convergence rate In numerical analysis, the speed at which a convergent sequence approaches its limit.

Converter Electrical or electromechanical device for converting electrical energy.

Coordination Organization of the different elements of a complex body or activity so as to enable them to work together effectively.

CT Current transformer: Electric device that produces an alternating current (AC) in its secondary, which is proportional to the AC in its primary. Designed for measurement, known as instrument transformers.

D

DA Distribution automation: Application of intelligent control over electrical power grid functions to the distribution level and beyond. Can be enabled via the smart grid.

DAC Data acquisition computer server: Central host computer used to gather and analyze real-time data.

Data analytics Science of examining raw data with the purpose of drawing conclusions about that information.

Data center Facility used to house computer systems and associated components, such as telecommunications and storage systems.

Database Collection of information that is organized so that it can easily be accessed, managed, and updated.

DCS Distributed control systems: Control system for a process or plant, wherein control elements are distributed throughout the system.

DG Distributed generator: Generator located close to the particular load that it is intended to serve. Characteristics generally include an operating strategy that supports the served load and interconnection to a distribution or subtransmission system.

Diesel generators Combination of a diesel engine with an electric generator (often an alternator) to generate electrical energy.

Distributed energy resources (DER) Smaller power sources that can be aggregated to provide power necessary to meet regular demand.

Distributed energy Generated or stored by a variety of small, grid-connected devices referred to as distributed energy resources (DER) or distributed energy resource systems.

Distributed generation Generally refers to small-scale (typically 1 kW to 50 MW) electric power generators that produce electricity at a site close to customers or that are tied to an electric distribution system.

Disturbance Interruption of a settled and peaceful condition.

DMS Distribution management system: Collection of applications designed to monitor and control the electric distribution system. Improves reliability and quality of service in terms of reducing outages, minimizing outage time, maintaining acceptable.

DMSP Decision made by the system approach planner.

DNP Distributed network protocol: Set of communications protocols used between components in process automation systems. Main use is in utilities, such as electric and water companies.

DNS Distribution network system: Interconnected group of lines and associated equipment for the local delivery of low-voltage electricity between the transmission network and end users.

Droop control Includes a frequency droop controller and a voltage controller.

E

Economical cost Combination of gains and losses of any goods that have a value attached to them by any one individual.

Effective strategies Tend to have certain common characteristics. The importance of these characteristics varies with the situation.

Efficiency State or quality of being efficient.

Efficient operation Operation with high ratio between the input to run a operation and the output gained from the operation.

Electric vehicle Use one or more electric motors or traction motors for propulsion and run by a large rechargeable battery.

Electric vehicles Vehicles that use one or more electric motors or traction motors for propulsion and run by a large rechargeable batteries.

EMS Energy management system: System of computer-aided tools used by operators of electric utility grids to monitor, control, and optimize the performance of the generation and/or transmission system.

Enabling technologies An invention/innovation that can be employed to drive major and rapid change in the capabilities of a system.

Energy economics Broad scientific subject area that includes topics related to supply and use of energy in societies.

Energy management Planning and operation of energy production and energy consumption units. Objectives are resource conservation, climate protection, and cost savings.

Energy semantic network Creates and evaluates possible scenarios of energy system structure of buildings (energy generation, conversion, and conservation measures).

Energy storage Capture of energy produced at one time for use at a later time. A device that stores energy comes in multiple forms, including chemical, gravitational potential, electrical potential, electricity, elevated temperature, latent heat, and kinetic.

Energy Property of objects that can be transferred to other objects or converted into different forms.

ESS Energy storage system: System to capture the energy produced at one time for use at a later time.

Explicit knowledge Knowledge that can be readily articulated, codified, accessed, and verbalized, and easily transmitted to others. Most forms of explicit knowledge can be stored in certain media.

F

FACTS technology Flexible AC transmission systems (FACTS) technology: Used to improve power quality and efficiency.

FC Fuel cell: Device that converts the chemical energy from a fuel into electricity through a chemical reaction.

FEP Front-end processors: Small computer that interfaces to the host computer through a number of networks, such as SNA, or a number of peripheral devices.

FG Full graphics: Type of graphical user interface (GUI).

Fieldbus Name of a family of industrial computer network protocols.

Filter compensators System that improves an undesirable response in a feedback and control system. It has a building block in control technique.

Flywheel Heavy revolving wheel in a machine that is used to increase the machine's momentum and thereby provide greater stability or a reserve of available power during interruptions in the delivery of power to the machine.

Fuel cell Cell producing an electric current directly from a chemical reaction.

Fuel chain Various supply chains corresponding to each fuel that can be utilized for the HTT. The construction of these chains depends on several parameters, such as economic situation, nature of the applications and target consumers, and fuel characteristics.

Fuel supply chain models The model that emphasizes material and information flow between fuel manufacturers and their trading partners.

FW Fireworks algorithm: Optimization algorithm inspired by observing a fireworks explosion.

G

GA Genetic algorithms: Search heuristic method that mimics the process of natural selection. This method is routinely used to generate useful solutions to optimization and search problems.

Gas emission Emission of greenhouse gases, such as CO_2, that trap heat in the atmosphere.

Genetic algorithm Adaptive heuristic search algorithm based on the evolutionary ideas of natural selection and genetics. As such, it represents an intelligent exploitation of a random search used to solve optimization problems.

Global minimum Smallest overall value of a set, function, etc., over its entire range. It is impossible to construct an algorithm that will find a global minimum for an arbitrary function.

Green technology Encompasses an evolving group of methods and materials, from techniques for generating energy to nontoxic cleaning products.

Grid modes Microgrid connects to the distribution system and takes energy from the grid to power its load.

GUI Graphical user interface: Type of interface that allows users to interact with electronic devices through graphical icons and visual indicators.

H

Heuristic optimization Technique designed for solving a problem more quickly when classic methods are too slow, or for finding an approximate solution when classic methods fail to find an exact solution.

HMI Human machine interface: Space where interactions between humans and machines occur to allow effective operation and control of the machine from the human end.

HPS Hybrid power system: System that combines more than one power generating energy source, such as a photovoltaic/wind hybrid system.

HS Harmony search: Evolutionary algorithm inspired by the improvisation process of musicians.

Hybrid fuel Hybrid fuel options are available by new natural gas (NG), hydrogen, and electricity fuel options while keeping current fuel options.

Hybrid local loads Combination of AC and DC loads that concentrate in certain consumption areas.

Hybrid transportation systems A system for moving people or goods consisting of different types of vehicles, e.g., electric vehicles, conventional vehicles, etc.

Hybrid transportation Combines a conventional internal combustion engine propulsion system with an electric propulsion system. They use a diesel-electric and are also known as hybrid diesel-electric transportation.

I

I/O Input/output: Devices used by a human user (or other electronic system) to communicate with a computer or electronic device.

IDEF0 Type Zero Integrated DEFinition methods used to model a wide variety of automated and nonautomated systems.

IEDs Intelligent electronic devices: Microprocessor-based controllers of power system equipment, such as protection relay or digital meter.

Information technology Application of computers and Internet to store, retrieve, transmit, and manipulate data or information in the context of a business or other enterprise.

Infrastructure engineering model Model that coordinates all of the systems for realizing a given engineering project.

Initial risks Potential risks identified in the early stage of the design process of the system.

Interconnected grids Integrated grids at a regional scale or greater that operate at synchronized parameters and are tied together during normal system conditions.

Inverter Electronic device or circuitry that changes direct current (DC) to alternating current (AC).

Islanded connected power Condition in which a distributed generator (DG) continues to power a location even though the electrical grid power from the electric utility is no longer present.

L

LAN Local area network: Computer network that interconnects local computers within a limited area.

Large microgrid Local energy assets, resources, and technologies combined into a system that's designed to satisfy the host's electric requirements, and as complex as integrating variable DERs in a balanced net-zero system.

Life cycle operation costs The total costs of operating a system along its lifetime.

Linearization A linear approximation of a nonlinear system that is valid in a small region around the operating point.

Load profiles Graph of the variation in the electrical load versus time. A load profile will vary according to customer type (typical examples include residential, commercial, and industrial), temperature, and holiday seasons.

LONWORKS Local operating network: Networking platform created to address the requirements of control applications.

M

MAS Multiple address radio system: Preferred communication medium for supervisory control and data acquisition (SCADA) and distribution automation (DA) radio communications. A typical system consists of a master station and several remote stations.

Micro energy grid Small local grid at low voltage (LV) or medium voltage (MV) level that includes loads, a control system, and a set of energy sources, such as distributed generators and energy storage devices.

Microgrid A small-scale power grid that can operate independently or in conjunction with the area's main electrical grid.

Microgrid Discrete energy system consisting of distributed energy sources (including demand management, storage, and generation) and loads capable of operating in parallel with, or independently from, the main power grid.

MILP Mixed integer linear programming: Problem with linear objective function; bounds and linear constraints, but no nonlinear constraints.

MODBUS Serial communications protocol originally published by Modicon (now Schneider Electric).

Modulated power Addition of information to an electronic or carrier signal. A carrier signal is one with a steady waveform with constant amplitude and frequency.

Monitoring A system monitor is a hardware or software component used to monitor resources and performance in a computer system.

MPPT Maximum power point tracking: Control method used to extract maximum possible power from solar panels.

Multienergy systems Energy supply systems, such as electricity, heat, natural gas, and hydrogen.

N

NPV Net present value: Current value of a sum of money, in contrast to some value it will have in the future when it has been invested at compound interest.

O

OPC Open platform communications: Series of standards and specifications for Process Control industrial telecommunication.

Operating costs Expenses that are related to the operation of a business, or a device, component, piece of equipment, or facility. Cost of resources used by an organization just to maintain its existence.

OPF Optimal power flow: Determine the best way (usually with minimum operating cost) to instantaneously operate a power system.

Optimization Maximizing or minimizing a real function by systematically choosing input values from within an allowed set and computing the value of the function.

P

Particle swarm optimization Population-based stochastic optimization technique developed by Dr. Eberhart and Dr. Kennedy in 1995, inspired by social behavior of bird flocking or fish schooling. PSO shares many similarities with evolutionary computation techniques.

PEM Proton exchange membrane: Type of fuel cell that can be made from either pure polymer membranes or from composite membranes in which other materials are embedded in a polymer matrix.

PLC Programmable logic controllers: Digital controller used to automate typically industrial electromechanical processes, such as control of machinery on factory assembly lines.

Plug-in hybrid electric vehicles Vehicles that combine a gasoline or diesel engine with an electric motor and large rechargeable batteries.

Power flow Numerical analysis of the flow of electric power in an interconnected system. A power-flow study usually uses simplified notation such as a one-line diagram and per-unit system, and focuses on various aspects of AC power parameters, such as voltages, voltage angles, real power, and reactive power. It analyzes the power systems in normal steady-state operation.

Power management Feature of some electrical appliances that turn off the power or switch the system to a low-power state when inactive.

Power quality Often defined as the electrical network's or the grid's ability to supply a clean and stable power supply. In other words, power quality ideally creates a perfect power supply that is always available, has a pure noise-free sinusoidal wave shape, and is always within voltage and frequency tolerances.

PQ control Used to keep the output active and reactive power constant. Consists of P and Q controller.

Production chains Sequence of processes involved in the production and distribution of energy.

PROFIBUS Process field bus: Standard for fieldbus communication.

Protection Branch of electrical power engineering that deals with the protection of electrical power systems from faults by isolating faulted parts from the rest of the electrical network.

PSO Particle swarm optimization: Population-based stochastic optimization technique inspired by social behavior of bird flocking or fish schooling.

PV Photovoltaic: Solar power technology that turns sunlight directly into electricity.

Q

QP Quadratic programming: Mathematical optimization problem in which there is a quadratic function of several variables subject to linear constraints on these variables.

R

RAID Redundant array of independent disks: Hard disk storage system.

Railway infrastructures Important to supply chain infrastructure, modern communications channels, and sensors systems to effectively apply the integrated SEG.

Reactive power Quantity of "unused" power that is developed by reactive components, such as inductors or capacitors in an AC circuit or system.

Regional transportation Transportation system that contains different conventional vehicles that operate using diesel and gasoline. Also, contains sustainable, environment friendly vehicles such as hydrogen vehicles, electric trains, electric vehicles, and others.

Reliability Overall consistency of a measure. A measure is said to have a high reliability if it produces similar results under consistent conditions.

Reliability The ability of a system or component to function under stated conditions for a specified period of time.

Renewable energy sources Energy generated from renewable natural resources, such as sunlight, wind, rain, tides, and geothermal heat. Renewable energy technologies include solar power, wind power, hydroelectricity, biomass, and biofuels.

Renewable energy Generally defined as energy that is collected from resources that are naturally replenished on a human timescale, such as sunlight, wind, rain, tides, waves, and geothermal heat.

RES Renewable energy sources: Energy sources that are naturally replenished on a human time scale, such as those derived from solar, wind, geothermal, or hydroelectric action.

Risk assessment Systematic process of evaluating potential risks that might be involved in a projected activity or undertaking.

Risk based life cycle assessment Technique used to assess risks and environmental impacts associated along the life cycle of a product and uses standard measures to manage them.

Risk indicator Measure used in management to indicate how risky an activity is.

Risk management processes The systematic application of management policies, procedures and practices to the tasks of establishing the context, identifying, analyzing, assessing, treating, monitoring, and communicating.

RTUs Remote terminal units: Microprocessor-controlled electronic device that interfaces objects in the physical world to a distributed control system or SCADA system.

S

SCADA Supervisory control and data acquisition: System for remote monitoring and control, gathering of data in real time from remote locations in order to control equipment and conditions.

Scheduling Short-term determination of the optimal output of a number of electricity generation facilities to meet the system load, at the lowest possible cost, subject to transmission and operational constraints.

Small modular reactor Nuclear power plants that are smaller in size (300 MWe or less) than current generation base load plants (1000 MWe or higher). These smaller, compact designs are factory-fabricated reactors.

Smart buildings Control and automation of lighting, heating, ventilation, air conditioning (HVAC), appliances, and security. Modern systems generally consist of switches and sensors connected to a central hub.

Smart energy grid An electrical grid that includes a variety of operational and energy measures, including smart meters, smart appliances, renewable energy resources, and energy efficiency resources.

Smart grid Electrical grid that includes a variety of operational and energy measures, including smart meters, smart appliances, renewable energy resources, and energy efficiency resources.

Smartgrid An electricity supply network empowered by the application of information technology, tools, and techniques, including meters, sensors, real-time communications, software, and remote-controlled equipment, to improve grid reliability and efficiency.

SOC State of charge: Method that gives precise estimation of battery state of charge from direct measurement variables.

SOFC Solid oxide fuel cells: Electrochemical conversion device that produces electricity directly from oxidizing a fuel. SOFC are characterized by their solid oxide electrolyte material.

Solar photovoltaic Converts sunlight directly into electricity. PV gets its name from the process of converting light (photons) to electricity (voltage), which is called the PV effect.

Stochastic process Random process evolving with time. In probability theory, a stochastic process is a time sequence representing the evolution of some system represented by a variable whose change is subject to a random variation.

Stochastic programming A mathematical programming in which some of the data incorporated into the objective or constraints is uncertain. Uncertainty is usually characterized by a probability distribution on the parameters. Although the uncertainty is rigorously defined, in practice it can range in detail from a few scenarios (possible outcomes of the data) to specific and precise joint probability distributions.

Super capacitor High-capacity electrochemical capacitor with capacitance values much higher than other capacitors (but lower voltage limits) that bridge the gap between electrolytic capacitors and rechargeable batteries.

Super conducting Phenomenon of exactly zero electrical resistance and expulsion of magnetic flux fields occurring in certain materials when cooled below a characteristic critical temperature.

Sustainability Endurance of systems and processes. Organizing principle is sustainable development, which includes the four interconnected domains: ecology, economics, politics, and culture.

System architecture Conceptual model that defines the structure, behavior, and more views of a system.

System boundaries Borders that separate the use cases that are internal to a system from the actors that are external to the system.

System gradient Generalization of the usual concept of derivative to functions of several variables.

T

TCP/IP Transmission control protocol/Internet protocol.

THD Total harmonic distortion: Measurement of the harmonic distortion present on a signal waveform and is defined as the ratio of the sum of the powers of all harmonic components to the power of the fundamental frequency of this signal.

Time variant System that is not time invariant (TIV). Generally, its output characteristics depend explicitly upon time.

Transportation system System for moving people or goods consisting of conventional vehicles only.

U

Utility network Interconnected network for delivering electricity to consumers. It carries power to demand centers, and distribution lines that connect individual customers.

V

Vehicle to grid System in which electric vehicles communicate with the power grid to sell demand response services by either returning electricity to the grid or by throttling its charging rate.

Voltage quality Maintaining the near sinusoidal waveform of power distribution bus voltages and currents at rated magnitude and frequency and to determine the fitness of electric power to consumer devices.

VT Voltage transformer: Electric device used to meter and protect high-voltage circuits. Converts high voltage to low voltage.

W

WAN Wide area network: Telecommunications network or computer network that extends over a large geographical distance.

WD Water drop: Swarm-based nature-inspired optimization algorithm.

Wind turbines Devices that convert the wind's kinetic energy into electrical power. The term appears to have been adopted from hydroelectric technology (rotary propeller). The technical description of a wind turbine is aerofoil-powered generator.

Index

Note: Page numbers followed by *f* indicate figures, *t* indicate tables, and *b* indicate boxes.

E

G

H

I

N

O

P

S